AUTOPHAGY - A DOUBLE-EDGED SWORD - CELL SURVIVAL OR DEATH?

Edited by **Yannick Bailly**

Autophagy - A Double-Edged Sword - Cell Survival or Death?
http://dx.doi.org/10.5772/50855
Edited by Yannick Bailly

Contributors

Bassam Janji, Edmund Rucker, Thomas Gawriluk, Amber Hale, Dan Ledbetter, Ricky Harminder Bhogal, Gerardo Hebert Vázquez-Nin, Patricia Silvia Romano, Tassula Proikas-Cezanne, Daniela Bakula, Gary Warnes, Aiguo Wu, Yian Kim Tan, Hao A. Vu, Kah-Leong Lim, Gui-Yin Lim, Rubem F. S. Menna-Barreto, Thabata Duque, Xênia Souto, Valter Andrade-Neto, Vitor Ennes-Vidal, Yannick Bailly, Satoru Noguchi, Anna Cho, Tonghui Ma, Azhar Rasul, Nikolai Viktor Gorbunov, Daotai Nie, Djamilatou Adom, Tanaka, Yuko Hirota, Keiko Fujimoto, Ana Esteves, Sandra Cardoso, Michiko Shintani, Kayo Osawa, Ioannis Nezis, Malgorzata Gajewska, Jeannine Mohrlüder, Dieter Willbold, Oliver Weiergräber, Cindy Miranti, Eric Nollet

Published by InTech

Janeza Trdine 9, 51000 Rijeka, Croatia

Publishing Process Manager Oliver Kurelic
Technical Editor InTech DTP team
Cover InTech Design team

Additional hard copies can be obtained from orders@intechopen.com

Autophagy - A Double-Edged Sword - Cell Survival or Death?, Edited by Yannick Bailly
p. cm.
ISBN 978-953-51-1062-0

For EU product safety concerns:
IN TECH d.o.o., Prolaz Marije Krucifikse Kozulić 3, 51000 Rijeka, Croatia,
info@intechopen.com or visit our website at intechopen.com.

Meet the editor

Yannick Bailly was born in 1956. During his PhD at the University of Strasbourg, and the University of P. & M. Curie in Paris, he deciphered the fish gill nervous system. Afterwards, at the CNRS in Paris, he greatly improved the understanding of synapse elimination in the developing rodent cerebellum. Since 1994, after returning to Strasbourg as a CNRS Research Director, his laboratory has become renowned for its expertise in ultrastructural neuroanatomy. He is also the Scientific Director of the Electron Microscopy Department of the Plateforme d'Imagerie in vitro of the Institut des Neurosciences Cellulaires et Intégratives. Yannick Bailly has made major contributions concerning the synaptic localization of cardinal molecules involved in major neurodegenerative diseases, such as amyloid precursor proteins and presenilins in Alzheimer's disease and prion protein. His group has provided valuable insight into neuronal death mechanisms involved in brain pathologies, in particular in prion diseases, reviewed in a chapter of this book.

Contents

Preface

Autophagy has recently benefited from rapid research progress in the field, and this master regulator of cell homeostasis is currently viewed as a valuable biomedical marker for a number of physiological processes and pathological mechanisms underlying major diseases.

Autophagy is known to exert cytoprotection in different cellular contexts, and autophagy induction generally prolongs life. Nevertheless, autophagy is necessary for tissue removal and can trigger cell death in certain situations. These opposed cytoprotective and cell death initiating roles, as well as tissue and time-dependent regulation of autophagy underscore the complexity of the autophagy pathway, and the importance of elucidating the molecular mechanisms controlling autophagy in cell survival and death. Based on the significant effects of autophagy deficiency on the development and pathogenesis of several disorders in animal models, recent research has yielded amazing results with autophagy-targeted pharmacological treatments of diseases. As recently stated by researchers in this field, the reality of autophagy-targeted therapy is now closer than ever expected or predicted.

This book focuses on autophagy relationships with cell death and disease, highlighting the most challenging aspects of current research, and the latest insights into the molecular mechanisms underlying autophagy.

Recent years have seen a growing interest in the different routes to cell death. Although apoptosis and autophagy have been previously considered as two different cell death pathways, one currently envisions a continuum of cell death mechanisms because it is now recognized that autophagy can induce apoptosis. Indeed, when the autolysosomal pathway is deregulated, autophagy can lead to cell death, either as a precursor of apoptosis in apoptosis-sensitive cells, or as a destructive cell digestion process. Whereas autophagy can selectively degrade survival factors and thereby initiate cell death, autophagy can also activate apoptosis by selectively degrading apoptotic inhibitors. This novel idea that autophagy comes into play in the balance between survival and death has major implications in the design of strategies for counteracting the pathophysiological processes. Further understanding of how autophagy is regulated should promote new therapeutic strategies that can ultimately treat a number of diseases, including myopathies, lysosomal storage diseases, cancers, infectious diseases, diabetes, liver diseases, as well as major neurodegenerative diseases which involve impaired autophagic elimination of misfolded proteins (Alzheimer's, Parkinson's, Huntington's and prion diseases). If autophagy induction is to be considered as a promising therapeutic strategy for neurodegenerative diseases, the dark side of autophagy must be taken into account. For the moment, it remains unclear whether deficits in autophagy provoke neurodegeneration or result from the neurodegenerative status. The data suggest that disrupting autophagy goes hand in hand with neurodegeneration, and a cause and effect relationship may contribute to neuronal damage. Transient, short-termed autophagy is protective, but turns deleterious when autophagy is chronically activated or excessively maintained in neurons. As reviewed in several chapters of the present book, this double-edged nature of autophagy will ultimately be critical for the development of autophagy-targeted therapeutics, not only for neurodegenerative diseases, but also for infectious diseases and cancer, where pathogens and cancer cells hijack the autophagic machinery for their survival and proliferation.

Yannick Bailly
Neuronal Cytology and Cytopathology,
Institute of Cellular and Integrative Neurosciences,
Department of Neurotransmission & Neuroendocrine Secretion,
University of Strasbourg, France

Section 1

New Insights into Mechanisms of Autophagy

Chapter 1

Role of Human WIPIs in Macroautophagy

Tassula Proikas-Cezanne and Daniela Bakula

Additional information is available at the end of the chapter

http://dx.doi.org/10.5772/54601

1. Introduction

Eukaroytic cellular homeostasis is critically secured by autophagy, a catabolic pathway for the degradation of cytoplasmic material in the lysosomal compartment. Macroautophagy, one of the major autophagic pathway, is initiated upon PtdIns(3)P generation by activated PtdIns3KC3 in complex with Beclin 1, p150 and Atg14L. Subsequently, specific PtdIns(3)P-effector proteins permit the formation of double-membrane vesicles, autophagosomes, that sequester the cytoplasmic material. Autophagosomes then communicate and fuse with the lysosomal compartment for final cargo degradation. Members of the human WIPI family function as essential PtdIns(3)P-binding proteins during the initiation of macroautophagy downstream of PtdIns3KC3, and become membrane proteins of generated autophagosomes. Here, we discuss the function of human WIPIs and describe the WIPI puncta-formation analysis for the quantitative assessment of macroautophagy.

Autophagy (auto phagia: *greek*, self eating) is an ancient cellular survival pathway specific to eukaryotic cells. By promoting a constant turn-over of the cytoplasm, the process of autophagy coevoled with the endomembrane system to secure the functionality of organelles. Primitive eukaryotic cells employed the autophagic pathway to survive periods of nutrient shortage and to degrade invading pathogens [1,2]. The survival function of autophagy has been experimentally proven by landmark studies such as the analysis of essential autophagic factors in *C. elegans*, demonstrating that autophagy defines the life-span of eukaryotic organisms [3], and the characterization of mice deficient for essential autophagic factors, demonstrating that autophagy functions to compensate for nutrients and energy during the post-natal starvation period [4].

The survival function of autophagy is based on the three major autophagic pathways, macroautophagy, microautophagy and chaperone-mediated autophagy (CMA) that coexist in eukaryotic cells [5]. In the process of microautophagy, proteins or organells are non-selectively engulfed by the lysosome through lysosomal membrane invagination and vesicle scission [6]. CMA specifically targets cytoplasmic proteins containing the KFERQ-like consensus motif for an Hsp70-assisted transport to the lysosomal compartment and an LAMP2-assisted import into the lysosomal lumen in higher eukaryotic cells [7]. The process of macroautophagy is hallmarked by the formation of autophagosomes, double-membrane vesicles that sequester the cytoplasmic cargo (membranes, proteins, organells) and that communicate with the lysosomal compartment for final degradation. Constitutively active on a low basal level, macroautophagy stochastically clears the cytoplasm and promotes the recycling of its constituents. In addition, upon a variety of cellular insults that lead to organelle damage and protein aggregation, macroautophagy is specifically induced and engaged to counteract toxicity.

The cytoprotective function of the three major forms of autophagy critically prevent the development of a variety of age-related human diseases, including cancer and neurodegeneration. However, autophagic pathways also play a vital role in the manifestation of certain pathologies, thus it is of urgent interest to monitor and understand the differential contribution of autophagic pathways to both human health and disease [5].

2. The process of macroautophagy

Central to the process of macroautophagy is the formation of autophagosomes that sequester and carry the cytoplasmic cargo – membranes, proteins and organelles - to the lysosomal compartment for subsequent degradation and recycling (Figure 1). For decades, the membrane origin of autophagosomes was uncertain [8]. Recently, a variety of independent studies provided evidence that multiple membrane sources should in fact become employed for the formation of autophagosomes [9]. Upon a hierarchical recruitment of autophagy-related (Atg) proteins [10], membrane origins are thought to undergo membrane rearrangements, including the formation of ER-associated omegasome structures [11], from which autophagosomal precursor membranes (phagophores) emerge [12,13]. By communicating with the endosomal compartment, the phagophore membrane is proposed to elongate and close to form the autophagosome that robustly sequesters the cytoplasmic cargo within a double-membraned vesicular structure [13]. Next, autophagosomes mature through communication with the endosomal/lysosomal compartment and the degradation of the sequestered cargo occurs in autolysosomes, fused vesicles of autophagosome and lysosomes [13]. Interestingly, *kiss and run* between autophagosomes and lysosomes has also been demonstrated to contribute to cargo final degradation [14].

The level of autophagosome formation is crucially balanced by the activity of the mTOR complex 1 (mTORC1), which inhibits macroautophagy when positioned at peripheral lysosomes and which releases its inhibitory role when positioned at perinuclear lysosomes [15] (Figure 2). The inhibition of mTORC1 mediates the activation of phosphatidylinositol 3-kinase class III (PtdIns3KC3 or Vps34) that phosphorylates PtdIns to generate PtdIns(3)P, an essential phospholipid for the forming autophagosomal membrane [16].

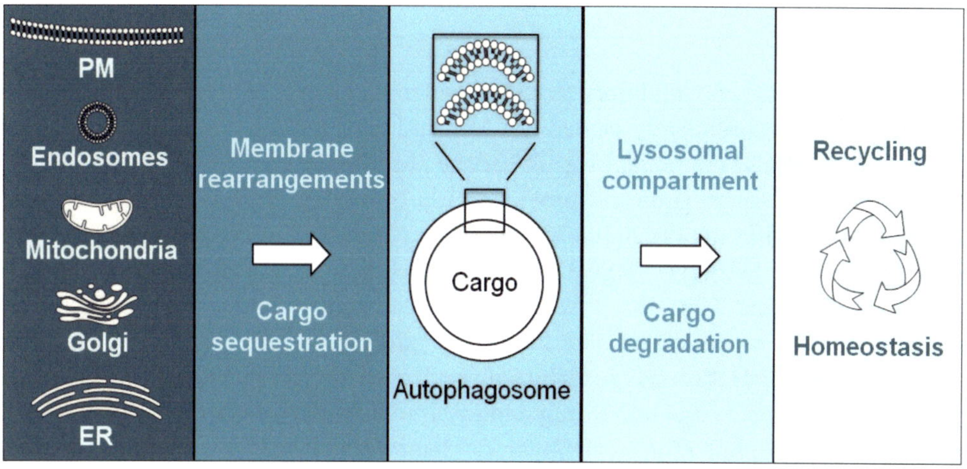

Figure 1. An overview of the process of macroautophagy.

PtdIns3KC3 functions in canonical macroautophagy in complex with Beclin 1 (Atg6 in yeast), p150 (or Vps15) and Atg14L [17], the latter recruiting PtdIns3KC3 to the ER [18] where membrane rearrangements are initialized by the PtdIns(3)P effector proteins DFCP1 [11] and WIPIs [19,20]. Macroautophagy can also be induced by non-canonical entries, e.g. independent of Beclin 1 [20,21,22].

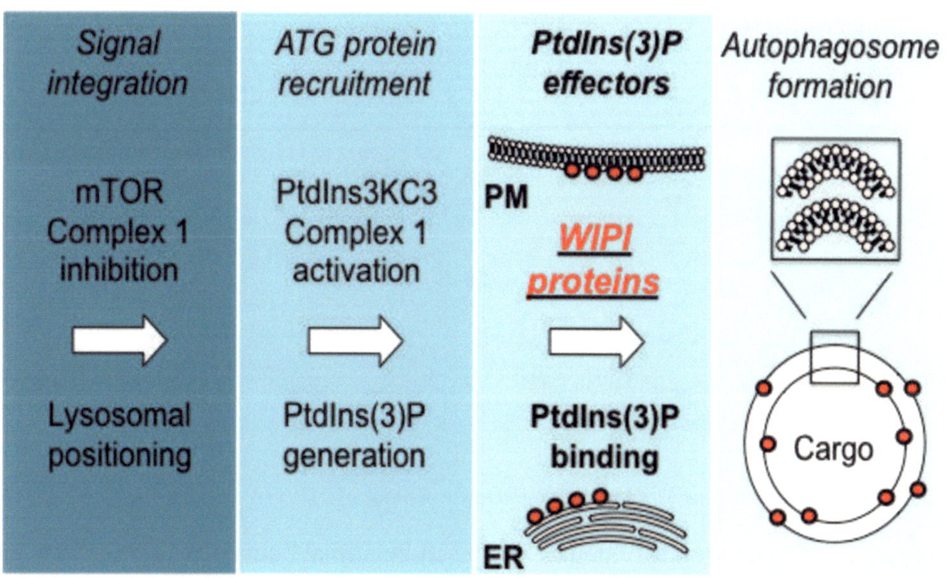

Figure 2. Evolutionarily conserved WIPIs function as essential PtdIns(3)P effectors to regulate macroautophagy.

3. Human WIPIs

By screening for novel p53 inhibitory factors, we identified a partial, uncharacterized cDNA fragment [23] and subsequently cloned the corresponding full-length cDNA from normal human liver and testis [19]. Using BLAST it became apparent that the isolated cDNA should be part of a novel human gene and protein family consisting of four members, which we subsequently cloned from normal human testis and placenta [19]. We proposed to term this novel human family WIPI (WD-repeat protein interacting with phosphoinositides) based on the following findings. First, the primary amino acid sequence suggested that the WIPIs contain seven WD40 repeats [19,24] that should fold into 7-bladed beta propeller proteins with an open Velcro configuration, as shown by structural homology modeling [19]. Second, WIPIs represent novel domains that specifically bind to PtdIns(3)P and PtdIns(3,5)P_2 [19,24,25,26]. Third, a comprehensive bioinformatic analysis demonstrated that the human WIPI family identified belongs to an ancient protein family of 7-bladed beta propellers with two paralogous groups, one group containing human WIPI-1 and WIPI-2, and the other group containing WIPI-3 and WIPI-4 [19,26]. Jeffries and coworkers found that WIPI-1 (WIPI49) should function in mannose-6-phosphate receptor trafficking [24], and our own studies demonstrated that WIPI-1 functions during macroautophagy in human tumor cells [19].

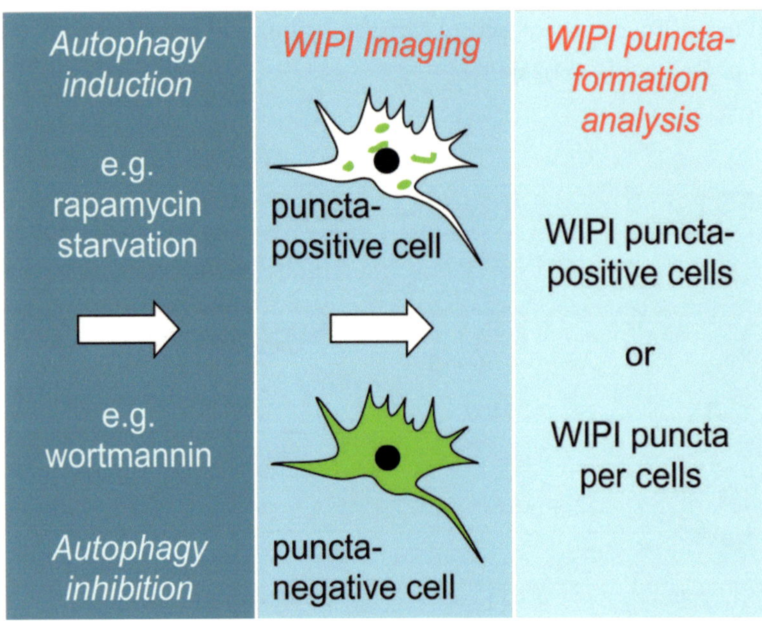

Figure 3. Assessing macroautophagy by WIPI puncta-formation analysis.

All human WIPI genes are ubiquitiously expressed in normal human tissue, but show high levels in skeletal muscle and heart [19]. Moreover, in a variety of human tumor types the abundance of all WIPI genes was shown to be aberrant when compared to matched normal samples from the same patient; WIPI-1 and WIPI-3 seemed to be more abundant, and WIPI-2 and WIPI-4 less abundant in the tumor [19]. In human tumor cell lines the abundance of the four WIPIs also differs [19,26]. However, the contribution of WIPIs in tumor formation is as yet uncharacterized.

During the process of macroautophagy, essential PtdIns(3)P effector functions (Figure 2) have been assigned to members of the human WIPI family [10,19,22,26,27,28], according to the ancestral function of yeast Atg18 [25,29,30,31,32]. Moreover, human WIPIs were also shown to be involved in pathogen defense by promoting the degradation of internalized bacteria in the lysosomal compartment [33,34,35], and to further contribute to Parkin-mediated mitophagy [36].

Upon the initiation of autophagy (Figure 2) WIPI-1 and WIPI-2 specifically bind to generated PtdIns(3)P at phagophore membranes [10,19,26,37]. In addition, WIPI-1 and WIPI-2 also bind, although to a lesser extend, to PtdIns(3,5)P_2 [24,26,37], however with unknown functional consequences. Phospholipid binding is mediated by evolutionarily conserved amino acids positioned in blade 5 and 6 of the beta-propeller structure of human WIPI proteins [19] and yeast homologs [38,39,40]. Further, WIPI-1 and WIPI-2 act as PtdIns(3)P effectors upstream of both the Atg12 and LC3 ubiquitin-like conjugation systems, hence regulate LC3 lipidation [10,22,26,37] which is required for the elongation of the phagophore. Moreover, both WIPI-1 and WIPI-2 become membrane proteins of formed autophagosomes and probably also of autolysosomes [41].

From the further specific localization of WIPI-1 and WIPI-2 upon the induction of macroautophagy, conclusions about the membrane origin of WIPI-positive autophagosomes can be concluded (Figure 2): i) as WIPI-1 specifically accumulates at the ER and at the plasma membrane (PM) upon starvation-induced macroautophagy, both of these membrane systems might contribute to phagophore and autophagosome formation, ii) as WIPI-2 also accumulates at the plasma membrane upon starvation, and in addition to membranes close to the Golgi, a differential engagement of particular membrane systems for autophagosome formation might be mediated by the different WIPIs [41].

4. WIPI-1 puncta-formation analysis

The specific protein localization of WIPI-1 at both phagophores and autophagosomes has been employed for the quantitative assessment of macroautophagy in mammalian cells [37] and extended for usage of automated fluorescent image acquisition and analysis [22,34,42]. Upon the induction of macroautophagy, e.g. by rapamycin administration or starvation (Figure 3), WIPI-1 accumulates at autophagosomal membranes, termed puncta. Upon the inhibition of autophagy, e.g. by wortmannin treatment, WIPI-1 is distributed throughout the cytoplasm. Under nutrient-rich conditions few WIPI-1 puncta-positive cells are observed and this assessment reflects basal macroautophagy. To visualize endogenous WIPI-1, indirect immunofluorescence with specific anti-WIPI-1 antibodies is conducted. Alternatively, overexpressed WIPI-1 fusion proteins, e.g. tagged to GFP, can also be employed to quantify the status of macroautophagy. Both, the number of cells displaying WIPI-1 puncta and the number of WIPI-1 puncta per cell can be used to assess macroautophagy [43,44].

5. Outlook

The notion that human WIPIs function as essential PtdIns(3)P effectors in macroautophagy needs to be addressed in more molecular detail as follows: i) analysing the individual contribution of the WIPIs to phagophore formation, ii) defining the function of WIPIs at autophagosomes and autolysosomes and iii) identification of WIPI interacting proteins and the signaling network regulating the PtdIns(3)P effector function of WIPIs. As WIPIs are aberrantly expressed in human tumors, the role of WIPIs during tumorigenesis, in particular the regulation of gene expression in normal and tumor cells is of further current interest. Moreover, the identification of compounds that permit a direct interference with the specific binding of WIPIs to PtdIns(3)P might become suitable in the future to specifically modulate macroautophagy in anti-tumor therapies.

Abbreviations

ATG, autophagy related; CMA, chaperone-mediated autophagy; PtdIns(3)P, phosphatidylinositol 3-phosphate; PtdIns(3,5)P2, phosphatidylinositol 3,5-bisphosphate; PtdIns3KC3, phosphatidylinositol 3-kinase class III; mTOR, mammalian target of rapamycin; mTORC1, mTOR complex 1; WIPI, WD-repeat protein interacting with phosphoinositides.

Acknowledgements

We thank the German Research Foundation (DFG, SFB 773) for grant support to TP-C, and the Forschungsschwerpunktprogramm Baden-Wuerttemberg (Kapitel 1403 Tit. Gr. 74) for supporting the doctoral thesis of DB.

Author details

Tassula Proikas-Cezanne and Daniela Bakula

*Address all correspondence to: tassula.proikas-cezanne@uni-tuebingen.de

Autophagy Laboratory, Department of Molecular Biology, Interfaculty Institute for Cell Biology, Eberhard Karls University Tuebingen, Germany

References

[1] Levine B (2005) Eating oneself and uninvited guests: autophagy-related pathways in cellular defense. Cell 120: 159-162.

[2] Yang Z, Klionsky DJ (2010) Eaten alive: a history of macroautophagy. Nat Cell Biol 12: 814-822.

[3] Melendez A, Talloczy Z, Seaman M, Eskelinen EL, Hall DH, et al. (2003) Autophagy genes are essential for dauer development and life-span extension in C. elegans. Science 301: 1387-1391.

[4] Kuma A, Hatano M, Matsui M, Yamamoto A, Nakaya H, et al. (2004) The role of autophagy during the early neonatal starvation period. Nature 432: 1032-1036.

[5] Mizushima N, Levine B, Cuervo AM, Klionsky DJ (2008) Autophagy fights disease through cellular self-digestion. Nature 451: 1069-1075.

[6] Mijaljica D, Prescott M, Devenish RJ (2011) Microautophagy in mammalian cells: revisiting a 40-year-old conundrum. Autophagy 7: 673-682.

[7] Kaushik S, Cuervo AM (2012) Chaperone-mediated autophagy: a unique way to enter the lysosome world. Trends Cell Biol 22: 407-417.

[8] Juhasz G, Neufeld TP (2006) Autophagy: a forty-year search for a missing membrane source. PLoS Biol 4: e36.

[9] Mari M, Tooze SA, Reggiori F (2011) The puzzling origin of the autophagosomal membrane. F1000 Biol Rep 3: 25.

[10] Itakura E, Mizushima N (2010) Characterization of autophagosome formation site by a hierarchical analysis of mammalian Atg proteins. Autophagy 6: 764-776.

[11] Axe EL, Walker SA, Manifava M, Chandra P, Roderick HL, et al. (2008) Autophagosome formation from membrane compartments enriched in phosphatidylinositol 3-phosphate and dynamically connected to the endoplasmic reticulum. J Cell Biol 182: 685-701.

[12] Hayashi-Nishino M, Fujita N, Noda T, Yamaguchi A, Yoshimori T, et al. (2009) A subdomain of the endoplasmic reticulum forms a cradle for autophagosome formation. Nat Cell Biol 11: 1433-1437.

[13] Yla-Anttila P, Vihinen H, Jokitalo E, Eskelinen EL (2009) 3D tomography reveals connections between the phagophore and endoplasmic reticulum. Autophagy 5: 1180-1185.

[14] Jahreiss L, Menzies FM, Rubinsztein DC (2008) The itinerary of autophagosomes: from peripheral formation to kiss-and-run fusion with lysosomes. Traffic 9: 574-587.

[15] Korolchuk VI, Saiki S, Lichtenberg M, Siddiqi FH, Roberts EA, et al. (2011) Lysosomal positioning coordinates cellular nutrient responses. Nat Cell Biol 13: 453-460.

[16] Obara K, Ohsumi Y (2011) PtdIns 3-Kinase Orchestrates Autophagosome Formation in Yeast. J Lipids 2011: 498768.

[17] Matsunaga K, Saitoh T, Tabata K, Omori H, Satoh T, et al. (2009) Two Beclin 1-binding proteins, Atg14L and Rubicon, reciprocally regulate autophagy at different stages. Nat Cell Biol 11: 385-396.

[18] Matsunaga K, Morita E, Saitoh T, Akira S, Ktistakis NT, et al. (2010) Autophagy requires endoplasmic reticulum targeting of the PI3-kinase complex via Atg14L. J Cell Biol 190: 511-521.

[19] Proikas-Cezanne T, Waddell S, Gaugel A, Frickey T, Lupas A, et al. (2004) WIPI-1alpha (WIPI49), a member of the novel 7-bladed WIPI protein family, is aberrantly expressed in human cancer and is linked to starvation-induced autophagy. Oncogene 23: 9314-9325.

[20] Codogno P, Mehrpour M, Proikas-Cezanne T (2012) Canonical and non-canonical autophagy: variations on a common theme of self-eating? Nat Rev Mol Cell Biol 13: 7-12.

[21] Gao P, Bauvy C, Souquere S, Tonelli G, Liu L, et al. (2010) The Bcl-2 homology domain 3 mimetic gossypol induces both Beclin 1-dependent and Beclin 1-independent cytoprotective autophagy in cancer cells. J Biol Chem 285: 25570-25581.

[22] Mauthe M, Jacob A, Freiberger S, Hentschel K, Stierhof YD, et al. (2011) Resveratrol-mediated autophagy requires WIPI-1-regulated LC3 lipidation in the absence of induced phagophore formation. Autophagy 7: 1448-1461.

[23] Waddell S, Jenkins JR, Proikas-Cezanne T (2001) A "no-hybrids" screen for functional antagonizers of human p53 transactivator function: dominant negativity in fission yeast. Oncogene 20: 6001-6008.

[24] Jeffries TR, Dove SK, Michell RH, Parker PJ (2004) PtdIns-specific MPR pathway association of a novel WD40 repeat protein, WIPI49. Mol Biol Cell 15: 2652-2663.

[25] Dove SK, Piper RC, McEwen RK, Yu JW, King MC, et al. (2004) Svp1p defines a family of phosphatidylinositol 3,5-bisphosphate effectors. EMBO J 23: 1922-1933.

[26] Polson HE, de Lartigue J, Rigden DJ, Reedijk M, Urbe S, et al. (2010) Mammalian Atg18 (WIPI2) localizes to omegasome-anchored phagophores and positively regulates LC3 lipidation. Autophagy 6.

[27] Lu Q, Yang P, Huang X, Hu W, Guo B, et al. (2011) The WD40 repeat PtdIns(3)P-binding protein EPG-6 regulates progression of omegasomes to autophagosomes. Dev Cell 21: 343-357.

[28] Vergne I, Roberts E, Elmaoued RA, Tosch V, Delgado MA, et al. (2009) Control of autophagy initiation by phosphoinositide 3-phosphatase Jumpy. EMBO J 28: 2244-2258.

[29] Guan J, Stromhaug PE, George MD, Habibzadegah-Tari P, Bevan A, et al. (2001) Cvt18/Gsa12 is required for cytoplasm-to-vacuole transport, pexophagy, and autophagy in Saccharomyces cerevisiae and Pichia pastoris. Mol Biol Cell 12: 3821-3838.

[30] Barth H, Meiling-Wesse K, Epple UD, Thumm M (2001) Autophagy and the cytoplasm to vacuole targeting pathway both require Aut10p. FEBS Lett 508: 23-28.

[31] Stromhaug PE, Reggiori F, Guan J, Wang CW, Klionsky DJ (2004) Atg21 is a phosphoinositide binding protein required for efficient lipidation and localization of Atg8 during uptake of aminopeptidase I by selective autophagy. Mol Biol Cell 15: 3553-3566.

[32] Efe JA, Botelho RJ, Emr SD (2007) Atg18 regulates organelle morphology and Fab1 kinase activity independent of its membrane recruitment by phosphatidylinositol 3,5-bisphosphate. Mol Biol Cell 18: 4232-4244.

[33] Kageyama S, Omori H, Saitoh T, Sone T, Guan JL, et al. (2011) The LC3 recruitment mechanism is separate from Atg9L1-dependent membrane formation in the autophagic response against Salmonella. Mol Biol Cell 22: 2290-2300.

[34] Mauthe M, Yu W, Krut O, Kronke M, Gotz F, et al. (2012) WIPI-1 Positive Autophagosome-Like Vesicles Entrap Pathogenic Staphylococcus aureus for Lysosomal Degradation. Int J Cell Biol 2012: 179207.

[35] Ogawa M, Yoshikawa Y, Kobayashi T, Mimuro H, Fukumatsu M, et al. (2011) A Tecpr1-dependent selective autophagy pathway targets bacterial pathogens. Cell Host Microbe 9: 376-389.

[36] Itakura E, Kishi-Itakura C, Koyama-Honda I, Mizushima N (2012) Structures containing Atg9A and the ULK1 complex independently target depolarized mitochondria at initial stages of Parkin-mediated mitophagy. J Cell Sci 125: 1488-1499.

[37] Proikas-Cezanne T, Ruckerbauer S, Stierhof YD, Berg C, Nordheim A (2007) Human WIPI-1 puncta-formation: a novel assay to assess mammalian autophagy. FEBS Lett 581: 3396-3404.

[38] Krick R, Busse RA, Scacioc A, Stephan M, Janshoff A, et al. (2012) Structural and functional characterization of the two phosphoinositide binding sites of PROPPINs, a beta-propeller protein family. Proc Natl Acad Sci U S A 109: E2042-2049.

[39] Baskaran S, Ragusa MJ, Boura E, Hurley JH (2012) Two-site recognition of phosphatidylinositol 3-phosphate by PROPPINs in autophagy. Mol Cell 47: 339-348.

[40] Watanabe Y, Kobayashi T, Yamamoto H, Hoshida H, Akada R, et al. (2012) Structure-based Analyses Reveal Distinct Binding Sites for Atg2 and Phosphoinositides in Atg18. J Biol Chem 287: 31681-31690.

[41] Proikas-Cezanne T, Robenek H (2011) Freeze-fracture replica immunolabelling reveals human WIPI-1 and WIPI-2 as membrane proteins of autophagosomes. J Cell Mol Med 15: 2007-2010.

[42] Pfisterer SG, Mauthe M, Codogno P, Proikas-Cezanne T (2011) Ca2+/calmodulin-dependent kinase (CaMK) signaling via CaMKI and AMP-activated protein kinase contributes to the regulation of WIPI-1 at the onset of autophagy. Mol Pharmacol 80: 1066-1075.

[43] Proikas-Cezanne T, Pfisterer SG (2009) Assessing mammalian autophagy by WIPI-1/Atg18 puncta formation. Methods Enzymol 452: 247-260.

[44] Klionsky DJ, Abdalla FC, Abeliovich H, Abraham RT, Acevedo-Arozena A, et al. (2012) Guidelines for the use and interpretation of assays for monitoring autophagy. Autophagy 8: 445-544.

Chapter 2

Atg8 Family Proteins — Autophagy and Beyond

Oliver H. Weiergräber, Jeannine Mohrlüder and Dieter Willbold

Additional information is available at the end of the chapter

http://dx.doi.org/10.5772/55647

1. Introduction

In eukaryotic cells, macroautophagy is now recognised as the most important mechanism for degradation of long-lived proteins and complete organelles, thus enabling cells to sustain their function under conditions of stress, such as nutrient deprivation, hypoxia or the presence of intracellular pathogens (for recent reviews, see [1,2]). While the core machinery is conserved in all eukaroytes [3], it is becoming more and more evident that upstream regulation and interfacing with other cellular pathways can differ significantly, depending on the species and cell type investigated.

Proteins of the Atg8 family are essential factors in the execution phase of autophagy. The yeast *Saccharomyces cerevisiae* only possesses a single member (the eponymous Atg8); in higher eukaryotes and a few protists, however, the family has expanded significantly, in exceptional cases including products of as many as 25 genes [4].

For more than ten years, our group has been investigating structure and function of Atg8 family proteins, with special emphasis on the $GABA_A$ receptor-associated protein (GABARAP). In this review, we will first give a concise outline of the biology of these molecules and of important milestones in their investigation, supporting their roles both in the autophagic machinery and in general membrane trafficking events. The remainder of the text shall illustrate the recent progress in our understanding of the structure and function of GABARAP and related proteins. In particular, we will discuss the identity of potential binding partners and the structures of resulting complexes, as assessed by X-ray crystallography, NMR spectroscopy and comparative modelling.

2. Biology of Atg8 family proteins

During the past two decades, more than 30 autophagy-related proteins have been identified in yeast as components of the Atg (autophagy) and Cvt (cytoplasm to vacuole targeting) pathways [5]. Mammalian cells contain counterparts for most of these proteins as well as some additional factors that are specific to higher eukaryotes. Genetic analysis unveiled Atg proteins 1 to 10, 12 to 14, 16 to 18, 29 and 31 to be essential for the formation of canonical autophagosomes [3]. They have been grouped into several functional units, including the Atg1/ULK (unc-51-like kinase) complex, the class III phosphatidylinositol 3-kinase (PI3K) complex, and the Atg12 and Atg8/LC3 conjugation systems [6].

Upon starvation, inhibition of the protein kinase target of rapamycin (TOR) results in activation of the Atg1/ULK complex, which is the most upstream unit in the hierarchy [7], and of the class III PI3K complex. The latter generates phosphatidylinositol 3-phosphate (PI3P) at the site of autophagosome formation, which is termed the pre-autophagosomal structure (PAS) in yeast and probably corresponds to the ER-associated omegasome in mammals. The function of PI3P in autophagy is still incompletely understood; this lipid is known to be important for the recruitment of downstream effector proteins, and its amount and spatial distribution are tightly regulated [8].

Hierarchical analysis of yeast Atg proteins indicates that the two ubiquitin-like conjugation systems act more downstream in autophagosome biogenesis. Atg12 is activated by the E1-like enzyme Atg7 and subsequently transferred to its target Atg5 via the E2-like enzyme Atg10 [9]. The resulting conjugate interacts with Atg16, which mediates generation of a 2:2:2 complex [10]. This assembly is a marker of the PAS and the expanding phagophore but dissociates upon autophagosome closure [11,12]. As outlined below, the Atg12 conjugation system is functionally coupled to the Atg8/LC3 system.

Similar to other ubiquitin-like modifiers, Atg8 and its mammalian orthologues are synthesised as precursor proteins with additional amino acids at their C-termini. These are proteolytically cleaved by cysteine proteases (Atg4 in this case), yielding truncated products (form I) with a conserved terminal glycine residue. Intriguingly, Atg8/LC3 proteins are finally attached to phospholipids rather than polypeptides: after processing by the E1-like Atg7 and the E2-like Atg3, they are covalently linked to phosphatidylethanolamine (PE) [13,14], resulting in protein-phospholipid conjugates (form II) that are supposed to be membrane-associated. This modification is reversible, and delipidation of Atg8/LC3 proteins is again mediated by Atg4 [15,16].

The Atg12-Atg5-Atg16 complex exhibits E3-like activity for Atg8/LC3 proteins by promoting their transfer from Atg3 to PE [17,18]. Since this complex has been found to associate only with the outer surface of the isolation membrane, Atg8/LC3 lipidation is supposed to occur there [12]. Atg14 and Vps30, two components of the class III PI3K complex, were shown to be required for the recruitment of the Atg16 complex (and thus Atg8-PE) to the PAS [7]. The precise mechanism of these regulatory functions, however, remains to be elucidated.

The first Atg8 protein to be identified was mammalian LC3B (initially termed LC3), which to the present day has remained the most extensively studied member of the family. It was reported in 1987 to associate with microtubule-associated proteins (MAPs) 1A and 1B [19] and was first implicated in the modulation of MAP1 binding to microtubules [20]. While the phenomenon of cellular autophagy has been observed as early as 1957 [21], it took more than four decades until the involvement of LC3B in this process was recognised [22].

Yeast Atg8 has been first described in the late 1990s [23]; since its gene was isolated as a suppressor of autophagy defects (hence its original name Aut7), its essential role in the autophagy pathway was immediately evident. This functional assignment was aided by the absence of partially redundant paralogues in yeast. In contrast, mammalian cells possess several family members which, based on amino acid sequence similarities, can be divided into two subgroups [24]. In humans, LC3A (with two variants originating from alternative splicing), LC3B, LC3B2 and LC3C constitute the LC3 subfamily, whereas GABARAP, GABARAPL1/GEC1, GABARAPL2/GATE-16 and GABARAPL3 form the GABARAP subfamily. They are expressed ubiquitously with moderate variations between different tissues. In this context, it is noteworthy that the expression of GABARAPL3 has been demonstrated on the transcriptional level only [25]; the corresponding open reading frame might therefore represent a pseudogene.

As with LC3B, the cellular functions originally ascribed to GABARAP subfamily proteins were not obviously related to autophagy. GATE-16, for instance, was initially found to be involved in intra-Golgi protein transport and was later shown to promote these processes by linking NSF (N-ethylmaleimide sensitive factor) to a SNARE (soluble NSF attachment receptor) protein on Golgi membranes [26,27]. GABARAP was identified in 1999 as an interaction partner of $GABA_A$ receptors [28]. Further investigations revealed that GABARAP is essential for $GABA_A$ receptor trafficking to the plasma membrane [29]. Interaction with integral membrane proteins turned out to be a recurrent theme in GABARAP research, as this protein was found to also associate with the transferrin receptor, the AT1 angiotensin receptor, the transient receptor potential vanilloid channel (TRPV1) and the κ-type opioid receptor [30-33]. Analogous to GABARAP, its closest relative GABARAPL1/GEC1 also interacts with the $GABA_A$ receptor and the κ opioid receptor [34,35]. Association with NSF has been confirmed for GABARAP [36] and GEC1 [35], in addition to GATE-16. Finally, it is interesting to note that all Atg8 proteins investigated thus far appear to show affinity for tubulin [37,38], suggesting physical association with microtubules.

With the rapid evolution of autophagy research in recent years, our knowledge about the cellular functions of Atg8-like proteins has grown dramatically. In particular, it is now well-established that the mammalian orthologues as a group are just as indispensable for the autophagy process as Atg8 is for yeast, and that this function strictly depends on lipid conjugation. Consequently, knockout of Atg3 or overexpression of a dominant-negative Atg4 mutant result in unclosed isolation membranes with altered morphology [39-40]. It is important to realise, however, that the individual members of the family perform both distinct and overlapping functions, and the precise definition of these activities has remained a challenging task.

An addition to their roles as essential components of the autophagic machinery, Atg8 family proteins have also emerged as valuable tools for the investigation of this process. Among the large number of autophagy-related polypeptides identified to date, Atg8 and its homologues are unique in that they remain associated with mature autophagosomes and thus are commonly exploited as *bona fide* markers for this organelle [24].

3. Structural foundations

While the discovery of the first Atg8 family protein (LC3B) dates back to the late 1980s, the three-dimensional structures of these molecules have remained elusive for many years. In particular, extensive searches of sequence databases at that time did not reveal entries with known fold.

The situation changed in 2000, when the crystal structure of bovine GATE-16 was published [41]. In the absence of obvious templates for molecular replacement, structure determination required experimental phase information, which was acquired using multiple-wavelength anomalous dispersion. Availability of this data not only speeded up the subsequent X-ray structure determination of related proteins, but was also seminal in that it revealed several previously unexpected properties which are shared among Atg8-like proteins and have since proven crucial for the biological function of this family.

The GATE-16 structure features a compact ellipsoid fold belonging to the α+β class (Figure 1). Among the most unexpected findings was a striking similarity to the ubiquitin superfamily fold in the C-terminal two-thirds of the polypeptide (coloured dark blue). This portion is also known as the β-grasp fold. Usually, it comprises a four-stranded β-sheet of the mixed type (i.e. the two inner strands are arranged in parallel and are flanked by antiparallel outer segments) and one or two helices shielding the concave face of the sheet. In fact, this similarity to ubiquitin is just another manifestation of the well-established notion that during the evolution of proteins, chain topologies and three-dimensional folds are conserved much more stringently than primary structures. In addition to the β-grasp fold, GATE-16 contains an N-terminal extension with two additional helices, which are attached to the convex face of the β-sheet. This stretch is a hallmark of all Atg8-like proteins distinguishing them from other members of the ubiquitin superfamily.

Based on significant similarities in amino acid sequences, Paz et al. have claimed that the three-dimensional fold found for GATE-16 should be shared by all other family members including GABARAP, LC3 and yeast Atg8. In principle, this assumption has proven entirely valid; current evidence indicates, however, that conformational dynamics might differ among family members. Specifically, the N-terminal helical extension has been found to attain alternative conformations, at least under certain experimental conditions (see below).

Another peculiar observation in the GATE-16 structure was the presence of a region containing partially solvent-accessible hydrophobic residues, which are flanked by basic side chains. Since these apolar residues are located on either side of strand β2, i.e. the exposed edge of the β-sheet, they are now commonly assigned to two hydrophobic patches (hp), with hp1 and hp2

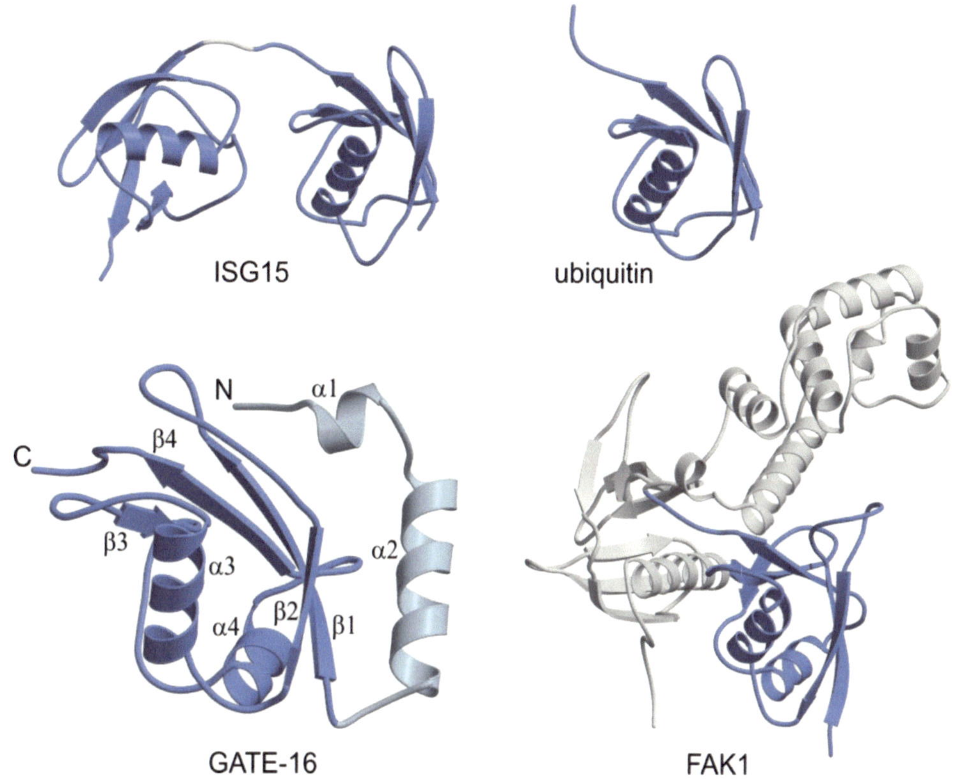

Figure 1. The characteristic fold of Atg8 family proteins, as exemplified by the GATE-16 crystal structure (PDB ID 1EO6 [41]), contains the β-grasp motif (dark blue) which is a hallmark of the ubiquitin (PDB ID 1UBQ [42]) superfamily. While all Atg8 homologues contain a unique N-terminal extension (light blue), different arrangements are found in other ubiquitlin-like proteins, such as in ISG15 (PDB ID 1Z2M [43]), which features a β-grasp tandem, or focal adhesion kinase 1 (FAK1 – only FERM domain is shown, PDB ID 2AEH [44]), which has a more complex domain structure. All proteins are presented as ribbon models, which were generated using MOLSCRIPT [45] and RASTER3D [46], using secondary structure assignments given by DSSP [47].

extending towards helices α2 and α3, respectively. In the crystal structure, these sites are involved in the formation of lattice contacts by interacting with phenylalanine side chains (F_{115} and F_{117}, respectively) at the C-termini of neighbouring molecules. Since the residues constituting this basic/hydrophobic face are highly conserved among Atg8 family proteins, they were proposed to also mediate protein-protein interactions *in vivo*. Last but not least, the two molecules present in the asymmetric unit diverge in the conformations of their C-terminal tails, which are detached from the globular β-grasp fold to a different extent. This observation provided a first hint at the dynamic character of this segment.

The second Atg8 family member to be investigated in structural terms was GABARAP. As the name implies, this protein has been originally identified as a binding partner of a $GABA_A$ receptor subunit [28]. The proposed functional connection to a major pharmacological target attracted great interest in the scientific community, and several groups commenced to work on the three-dimensional structure of GABARAP. Finally, it was published as many as five times independently, by our lab [48] as well as by others, involving either X-ray crystallography [37,49,50] or NMR spectroscopy [51]. As expected, all GABARAP structures exhibit the same overall fold as previously found for GATE-16, with one exception. Coyle et al. described a second crystal form obtained under high-salt conditions, in which the N-terminal stretch including helix $\alpha1$ was rotated away from the GABARAP core, with a proline residue (P_{10}) apparently serving as a hinge. In fact, this segment assumed an extended conformation and formed a lattice contact with the hydrophobic surface patches of a neighbouring molecule [37]. The authors speculated that this interaction might allow for the formation of extended scaffolds supporting the clustering of associated membrane proteins (such as $GABA_A$ receptors) while at the same time physically linking them to the microtubule cytoskeleton via a binding site in helix $\alpha2$. However, experimental evidence endorsing this conclusion is still unavailable.

Owing to the very nature of NMR spectroscopy, which aims to model ensembles of structures satisfying experimental distance restraints, the method is particularly well suited to assess dynamic properties of folded polypeptides. In the case of GABARAP, NMR spectra recorded in our lab revealed line broadening and/or signal splitting for backbone amide groups in several segments, which is indicative of conformational exchange on an intermediate to slow (millisecond to second) time scale [48]. These regions, which comprise the majority of helices $\alpha1$ and $\alpha2$ together with adjacent loops, are closely apposed in the three-dimensional structure and appear to be centred on P_{10}; the hydrophobic surface patches undergo conformational exchange as well. Finally, a recent diffusion-ordered spectroscopy (DOSY) NMR study suggested the presence of temperature-dependent conformational transitions in the GABARAP molecule, with associated changes in diffusion and self-association properties [52]. Although the unfolding of helix $\alpha1$ observed in one of the X-ray structures [37] may have been induced by the crystallization conditions, favouring a peculiar lattice contact, our data confirm that the N-terminal portion of GABARAP exhibits an equilibrium of two or more conformations. In fact, preliminary NMR relaxation dispersion experiments indicate that this conformational exchange comprises at least two processes on different time scales (C. Möller, M. Schwarten, P. Ma, P. Neudecker, unpublished data).

Subsequently, the three-dimensional structures of other members of the Atg8 family have been determined, including GEC1, which is closely related to GABARAP, LC3A isoform 1, LC3B [38], and yeast Atg8 itself [53,54]. The latest addition to this list is LC3C, the crystal structure of which has been determined in complex with an autophagy receptor [55]. While all these structures displayed the expected overall fold, they did also add to the controversy regarding the flexibility of the N-terminal subdomains. For instance, the NMR structure of LC3B did not show any indication of fluctuations around the $\alpha1$-$\alpha2$ hinge [38], which is at variance with our findings for GABARAP. The structure of Atg8 turned out to be particularly interesting in this respect. The protein is more difficult to handle than other family members because of its tendency to aggregate at concentrations required for NMR structure determination. Notably, we found that resonances corresponding to the N-terminal part of Atg8 were broadened or even undetectable, resulting in the absence of distance restraints between helix $\alpha2$ and the ubiquitin-like core of the molecule. Consequently, structure calculations yielded an ensemble of models in which the $\alpha2$ region partially retained its helical conformation, but its position with respect to the β-grasp fold was poorly defined [53]. In an independent approach to the

Atg8 structure, Kumeta et al. noted that Atg8 differed from other family members in that the $\alpha 2$ helix-terminating proline (P_{26}) was replaced by a lysine. A $K_{26}P$ mutation not only reduced the aggregation propensity of the molecule, but also stabilised the structure of the helical subdomain, with a corresponding improvement in spectral quality [54]. In summary, current evidence suggests that conformational polymorphism may be an intrinsic property of the N-terminal subdomain, at least in a subset of Atg8-like proteins. It is important to note, however, that the functional significance of these observations still needs to be established.

The available sequence and structural data for yeast Atg8 and its human orthologues are summarised in Table 1.

Protein name	Uniprot ID	Isoforms	X-ray structures	NMR structures
*Sc*Atg8	P38182	1		2KQ7 [52], 2KWC [53]
GABARAP	O95166	1	1GNU [50], 1KJT [49]	1KOT [48], 1KLV [51]
GABARAPL1	Q9H0R8	1	2R2Q	
GABARAPL2	P60520	1	1EO6 [41]	
GABARAPL3	Q9BY60	?		
MAP1LC3A	Q9H492	2	3ECI	
MAP1LC3B	Q9GZQ8	1		1V49 [38]
MAP1LC3B2	A6NCE7	1		
MAP1LC3C	Q9BXW4	1	3VVW [54]	

Table 1. Overview of yeast (*Sc*) and human Atg8 family members. GABARAPL1 and GABARAPL2 are also known as GEC1 and GATE-16, respectively. Note that the existence of the GABARAPL3 protein in cells has not been established yet. With the exception of LC3C, which has only been investigated in a heterodimeric complex, the PDB entries listed are those featuring the respective protein as the single polypeptide component.

4. A common paradigm of GABARAP-ligand interaction

As a matter of course, investigation of the biological functions of Atg8-like proteins has been and continues to be closely connected to the search for interaction partners in their cellular environment. One of the largest sets of interaction data is available for GABARAP; we shall therefore consider these results in more detail.

Shortly after its discovery, numerous proteins have been reported to bind to GABARAP, including candidates as diverse as NSF [36], tubulin [28], ULK1 [56], transferrin receptor [30], phospholipase C-related inactive protein type 1 (PRIP-1 [57]), glutamate receptor-interacting protein 1 (GRIP1 [58]), gephyrin [59] and DEAD box polypeptide 47 (DDX47 [60]). However, data on the mode of interaction, let alone structures of the respective complexes, were not available. In order to gain insight into the binding specificity of GABARAP, we have screened a phage-displayed random dodecapeptide library with a recombinant glutathione S-transferase (GST)-GABARAP fusion protein [61]. While this approach did not yield a single dominating sequence, several peptides were obtained multiple times, and side chain preferences at certain positions were clearly evident. Specifically, multiple sequence alignment of the phage display-selected peptides revealed a highly conserved tryptophan residue. Besides this tryptophan at sequence position i, aliphatic residues at positions i+1 and i+3, an aromatic residue at position i+2 and a proline at position i+4 or i+5 seemed to support GABARAP binding. The positions on the N-terminal side of the tryptophan were less conserved, but a certain preference for hydrophilic and charged amino acids was obvious. These observations inspired two different experimental strategies, which were directed towards artificial (model) ligands and to native interaction partners, respectively, and their modes of GABARAP binding.

4.1. Model ligands

First of all, the preference for tryptophan and aromatic residues at positions i and i+2, respectively, prompted the use of small-molecule indole derivatives as probes in a quantitative saturation-transfer difference NMR study [62]. We were able to locate two indole binding sites displaying different affinities, which essentially mapped to the conserved hydrophobic patches identified previously on the GABARAP surface.

At the same time, the highest-affinity peptide found in the phage display screen was selected as a prototype ligand and its interaction with GABARAP was investigated in detail [63]. This candidate (termed K1) with the sequence DATYTWEHLAWP bound to GABARAP with a K_D of 390 nM, as determined by surface plasmon resonance (SPR) measurements. Notably, its presence caused extensive changes to the $^{1}H^{15}N$-heteronuclear single quantum coherence (HSQC) spectrum of GABARAP, indicating significant alteration in the chemical environment of numerous amide groups. We have determined the three-dimensional structure of the complex by X-ray crystallography (Figure 2, left). The peptide ligand (drawn as coil) makes close contact with the GABARAP molecule in its entire length, burying approx. 620 Å^2 of solvent-accessible surface. Based on structural properties, the peptide can be divided into three parts: The N- and C-terminal segments (red) assume an extended conformation; residues 1 to 5 roughly align with helix $\alpha3$ of GABARAP, while residues 10 to 12 are apposed to the central β-sheet, with their backbone approximately perpendicular to strand β2. These two stretches are connected by a short 3_{10} helix (residues 6 to 9, grey). Overall, the interaction is dominated by side chain-side chain contacts. It is interesting to note that the 3_{10} helix formed by the backbone of W_6EHL_9 brings the side chains of W_6 and L_9 in close proximity; together they interact with a set of exposed aromatic and aliphatic side chains roughly located between the central β-sheet and helix $\alpha3$ (i.e. hp2). Finally, the C-terminus of the peptide is anchored by the side chain of W_{11}, which contacts residues belonging to hp1, bounded by the β-sheet and helix $\alpha2$ [63]. As expected, the involved GABARAP residues largely coincide with those affected in our HSQC titration and with the preferred binding sites for indole derivatives [62].

4.2. Native binding partners

In a complementary approach, we aimed at defining novel physiological binding partners of GABARAP. Since the phage display screens yielded a diverse yet related set of peptides, a position-specific scoring matrix (PSSM) was determined from the sequence alignment as an accurate representation of the consensus properties and the tolerance for exchanges at individual positions. Database searches using this PSSM revealed several potential matches, including calreticulin (CRT) and the heavy chain of clathrin (CHC). For calreticulin, the significance of the interaction could be demonstrated by pull-down experiments using immobilised protein, which associated with endogenous GABARAP from brain extracts [61]. Moreover, immunofluorescence staining of neuronal cells confirmed a colocalization of both proteins in punctuate structures, possibly corresponding to a vesicular compartment. We therefore set out to investigate this interaction in more detail. First of all, SPR measurements revealed tight binding with a very low off-rate, which prevented regeneration of the chip using standard protocols, and a dissociation constant of 64 nM. In accordance with these findings, the $^{1}H^{15}N$-HSQC spectrum of GABARAP was altered dramatically after addition of calreticulin; due to the large decrease in overall signal strength, however, identification of interacting residues was not feasible by NMR spectroscopy.

To overcome this problem, we resorted to the investigation of smaller calreticulin fragments for their GABARAP binding capabilities [64]. The three-dimensional structure of full-length calreticulin has not yet been determined; based on the structure of is paralogue calnexin [65], it is predicted to consist of N- and C-terminal segments contributing to a globular domain, and an intermediate part, the so-called P domain (proline rich domain), which forms an arm-like structure. Since the putative GABARAP binding motif is located at the proximal end of the proline rich region, we selected both an undecapeptide comprising the core interacting residues (positions 178 to 188) and the complete arm domain (residues 177 to 288) for this study. First of all, our measurements indicated an increase in affinity with peptide length: CRT(178-188), the P domain and full-length calreticulin yielded dissociation constants of 12 μM, 930 nM and 64 nM, respectively. These observations suggest that the undecapeptide does contain the primary interaction sites, but the P domain as well as the globular domain of calreticulin provide additional contacts. Moreover, replacing the tryptophan in the WDFL motif by alanine turned out to dramatically reduce the GABARAP affinity of the peptide. These data were confirmed by NMR experiments, which revealed strong alterations to the GABARAP spectrum after addition of the calreticulin P domain or CRT(178-188), but not of its mutant. Again, hydrophobic pockets hp1 and hp2 on the GABARAP surface were identified as major binding sites for the two calreticulin fragments investigated.

Finally, we determined the three-dimensional structure of the GABARAP-CRT(178-188) complex (Figure 2, right). The peptide ligand assumes an extended conformation in close contact to GABARAP, and as expected, the interaction is dominated by apolar contacts involving hp1 and hp2. The overall orientation of the peptide, however, turned out to differ substantially from the one found previously for the artificial K1 ligand. Specifically, the central portion of the peptide establishes main chain hydrogen bonds to strand β2, and therefore represents an intermolecular extension of the central β-sheet, whereas main chain-main chain

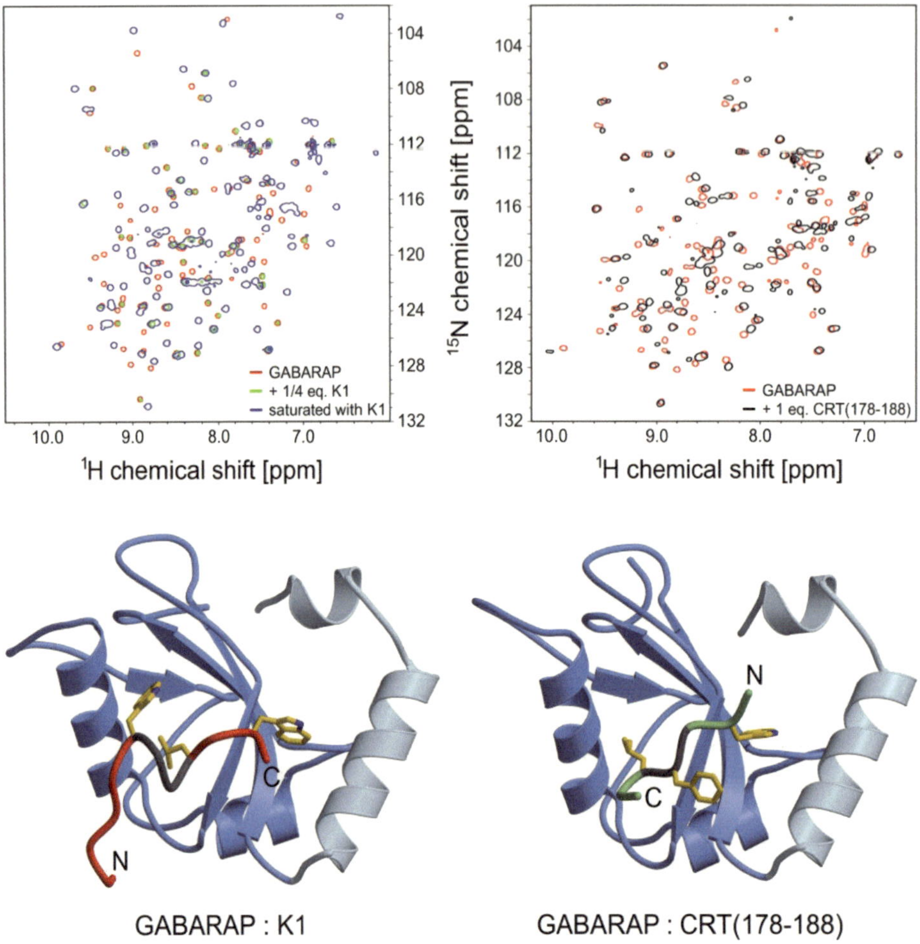

Figure 2. Ligand binding mode of GABARAP. Addition of peptides K1 (left) and CRT(178-188) (right) causes significant alterations in the $^{1}H^{15}N$-HSQC spectrum of ^{15}N-GABARAP (top panels). The dosage of unlabelled peptide in each experiment is given in stoichiometric equivalents (eq.) of the GABARAP amount. Bottom panels illustrate the crystal structures of the two complexes. The GABARAP molecule is drawn as ribbon model with colouring as introduced in Figure 1. Ligand peptides are depicted as coil with secondary structure elements marked in dark grey; hydrophobic side chains contacting the apolar patches of GABARAP are shown in stick mode.

contacts are virtually absent in the K1 complex. Moreover, it is interesting to note that the sequence arrangement of hydrophobic ligand residues interacting with hp1 and hp2 is reversed between the two complexes. Consequently, in CRT(178-188) the tryptophan and leucine residues of the WDFL motif associate with hp1 and hp2, respectively, while in the K1 peptide the N-terminal of the two tryptophans anchors to hp2.

Despite these differences, complex formation with the calreticulin peptide results in conformational changes in the GABARAP molecule that are qualitatively similar to those observed in the K1 complex. Specifically, insertion of apolar side chains into hp1 and hp2 implies a rearrangement of hydrophobic core residues, leading to outward displacement of helices $\alpha2$ and $\alpha3$ to different extents.

http://dx.doi.org/10.5772/55647

Since attempts to cocrystallise GABARAP with full-length calreticulin or its P-domain have been unsuccessful, we have built a homology model which makes use of available data on the soluble portion of calnexin [65] and the calreticulin P-domain [66], in addition to the GABARAP-CRT(178-188) complex structure. The major GABARAP interaction site is located at the N-terminal junction between the globular domain and the arm domain of calreticulin. While the corresponding residues appeared to be disordered in the X-ray structure of calnexin, our data indicate that in calreticulin, at least after binding of GABARAP, this portion protrudes from the base of the P domain, assuming a well-defined conformation [64].

Although our observations suggest a biological significance of the calreticulin-GABARAP complex, its precise function has been difficult to define. This is largely due to the seemingly incompatible subcellular locations of the two molecules. GABARAP is a cytosolic protein which gets associated with the cytosolic leaflet of intracellular membranes during autophagosome generation, while calreticulin is well-known as a soluble chaperone of the ER lumen [67]. In recent years, however, it has become clear that calreticulin is not absolutely restricted to the ER, but has distinct functions in other cellular compartments, such as the cytosol, the nucleus and the plasma membrane. Intriguingly, these calreticulin fractions appear to be derived from the ER pool; export into the cytosol involves a retrotranslocation process that is distinct from the pathway used for proteasomal degradation of misfolded proteins [68]. Based on these considerations, several scenarios involving a GABARAP-calreticulin complex may be envisaged. Inspired by preliminary experimental evidence [69,70], we have speculated that cytosolic calreticulin may cooperate with GABARAP to enhance transport of N-cadherin to sites of cell-cell contact at the plasma membrane. Similarly, integrin α subunits have been demonstrated to contain binding sites for calreticulin [71], and association of $\alpha3\beta1$ integrins with GABA receptors [72] suggests a possible connection with GABARAP. While in both cases the precise function of the complex still needs to be established, the presence of calreticulin may introduce calcium dependence to the respective cellular process.

The heavy chain of clathrin is another potential GABARAP interaction partner identified in our laboratory; it has been found in a database search with the PSSM derived from phage display results and, independently, in a pull-down experiment [73]. Binding of GABARAP to a clathrin peptide comprising the proposed interaction motif (residues 510 to 522) was investigated by NMR spectroscopy. Indeed, addition of CHC(510-522) lead to line broadening and reduction of peak intensities in $^{1}H^{15}N$-HSQC spectra of native GABARAP, indicating a direct interaction. As with the K1 and CRT(178-188) ligands, the pattern of affected GABARAP residues suggested that the two hydrophobic pockets constitute the major binding sites.

In accordance with the striking similarity in the modes of GABARAP binding to these ligands, we found that calreticulin is able to displace the heavy chain of clathrin from the complex.

Clathrin is an important player in the endocytosis of membrane proteins, such as the $GABA_A$ receptor. Since GABARAP is able to interact with both proteins, it seems reasonable to assume that it might be involved in the regulation of $GABA_A$ receptor endocytosis and thus in the control of receptor numbers at the postsynaptic membrane of neurons.

One of the earliest reports of physiological GABARAP interaction partners concerned NSF [36]. As a key component of the membrane fusion machinery, this protein is critically involved in cellular trafficking of membranes and associated polypeptides. NSF belongs to the AAA (ATPases associated with various cellular activities) group within the superfamily of Walker-type ATPases. Enzymes of this class usually form ring-like oligomers; in the case of NSF, a hexamer is believed to be the physiological state. Each chain folds into three domains, an N-terminal substrate binding domain (N) which is followed by two ATPase domains (D1 and D2). Unfortunately, crystal structures are available for the isolated N and D2 domains [74,75], but not for the full-length protein including the D1 domain. This is important because the relative orientation of the D1 and D2 domain rings (parallel vs. antiparallel) is still controversial [76]. In an attempt to investigate the GABARAP-NSF interaction *in silico*, we first built a homology model of hexameric NSF [77], using the structure of the related ATPase p97/VCP [78] as an additional template.

When we switched to an antiparallel orientation of the ATPase domains, our model revealed that a few hydrophobic side chains at the beginning of the D2 domain became exposed to the solvent; this site was chosen as an attractor for docking of the GABARAP molecule. Intriguingly, a reasonable result was found only if input coordinates of GABARAP were derived from complexes (such as with the K1 and CRT(178-188) peptides), whereas attempts with unliganded structures of GABARAP or GATE-16 were unsuccessful, indicating that conformational changes similar to those observed previously upon peptide binding are also required for interaction of Atg8-like proteins with NSF.

In the resulting model, the interface features a hydrophobic core which is flanked by polar contacts. Besides the apolar side chains residing in the NSF D2 domain, which are mainly in contact with hp1 residues of GABARAP, the interaction involves additional amino acids in both ATPase domains, leading to a relatively large surface area (950 $Å^2$) buried upon complex formation. The important role of the hydrophobic surface of GABARAP in NSF binding was verified in a pull-down experiment. Here we could demonstrate that a peptide containing the GABARAP binding motif was able to displace NSF from immobilised GABARAP, whereas a control peptide was inactive.

Based on the proximity of bound GABARAP to the D1 ATP binding site in our model, we have speculated that it may regulate ATP binding and/or hydrolysis. It is important to note that for GATE-16, a stimulating activity on the ATPase activity of NSF has been known for more than a decade [27]. Besides such direct effects on the enzymatic activity of NSF, association with lipid-conjugated Atg8 proteins may be required for anchoring this molecular machine to membranes. In support of this hypothesis, suppression of GABARAP lipidation has been reported to indeed alter the subcellular localization of NSF [79].

The results of the investigations outlined above, using ligands ranging from small molecule compounds to medium-sized proteins, have lead to the notion that GABARAP interactions with a wide variety of binding partners usually conform to a common theme. This paradigm involves the bipartite hydrophobic site on the surface of GABARAP, which usually accommodates a linear motif of the type *fn* x/[-] x/[-] Ω x/[-] x/[-] Φ *fc* (following the convention outlined in [80]). In this notation, Ω denotes an aromatic side chain, Φ can be any hydrophobic residue. The four remaining positions vary considerably, but at least one of these is usually acidic, thus complementing the basic side chains located in the vicinity of the hydrophobic patches.

In view of the significant conservation of the respective protein surface, it came at no surprise that these rules were found to be valid, with slight modifications, for other Atg8 proteins. The characteristic hydrophobic motif found in yeast Atg8 ligands has been named AIM (Atg8 family interacting motif [81]) whereas for mammalian homologues the term LIR (LC3 interacting region [82]) is preferred.

Figure 3 visualises the hydrophobic properties and electrostatic potential on the ligand binding surfaces of GABARAP and other family members. While the fundamental characteristics of the site are well conserved, these molecules do exhibit differences in detail, which is consistent with their overlapping yet non-identical spectrum of binding partners. Several examples illustrating this concept are discussed below. We are currently extending our phage display-based ligand screening to cover other major family members, aiming at a detailed understanding of their local preferences, which can be correlated with data on relative affinities for native binding partners.

It is important to note that virtually all structure-related investigations on Atg8-like proteins have been performed with soluble variants of these molecules, whereas the biologically active species are the lipid-conjugated forms attached to membranes. In order to investigate the effects of lipidation, we have used nanodiscs as a model membrane system [83]. Nanodiscs consist of a lipid bilayer patch which is laterally shielded by an apolipoprotein A-I-derived scaffold protein. As judged by $^{1}H^{15}N$-HSQC experiments, chemical coupling of the GABARAP C-terminus to nanodisc lipids does not change the overall structure of the molecule. In particular, the interaction surface comprising hp1 and hp2, which is located opposite to the membrane attachment site, retains its ligand binding capacity. While these observations support the general utility of soluble Atg8 proteins in biochemical and biophysical studies, it seems conceivable that both conformational dynamics and interaction propensities of these molecules may be affected by membrane attachment to varying extents.

5. Interaction of Atg8-like proteins with the autophagic machinery

Many interaction partners reported in the early years of research on mammalian Atg8 homologues did not bear obvious relation to the autophagy pathway. In the meantime, however, evidence has accumulated that (1) Atg8 proteins are involved in numerous contacts with core autophagy components, and (2) these interactions are usually critically dependent on the hydrophobic motif described above.

5.1. Conjugation enzymes

First and foremost, Atg8 proteins obviously need to interact with the components of their conjugation machinery, i.e. the cysteine protease Atg4, the E1-like Atg7, the E2-like Atg3 and

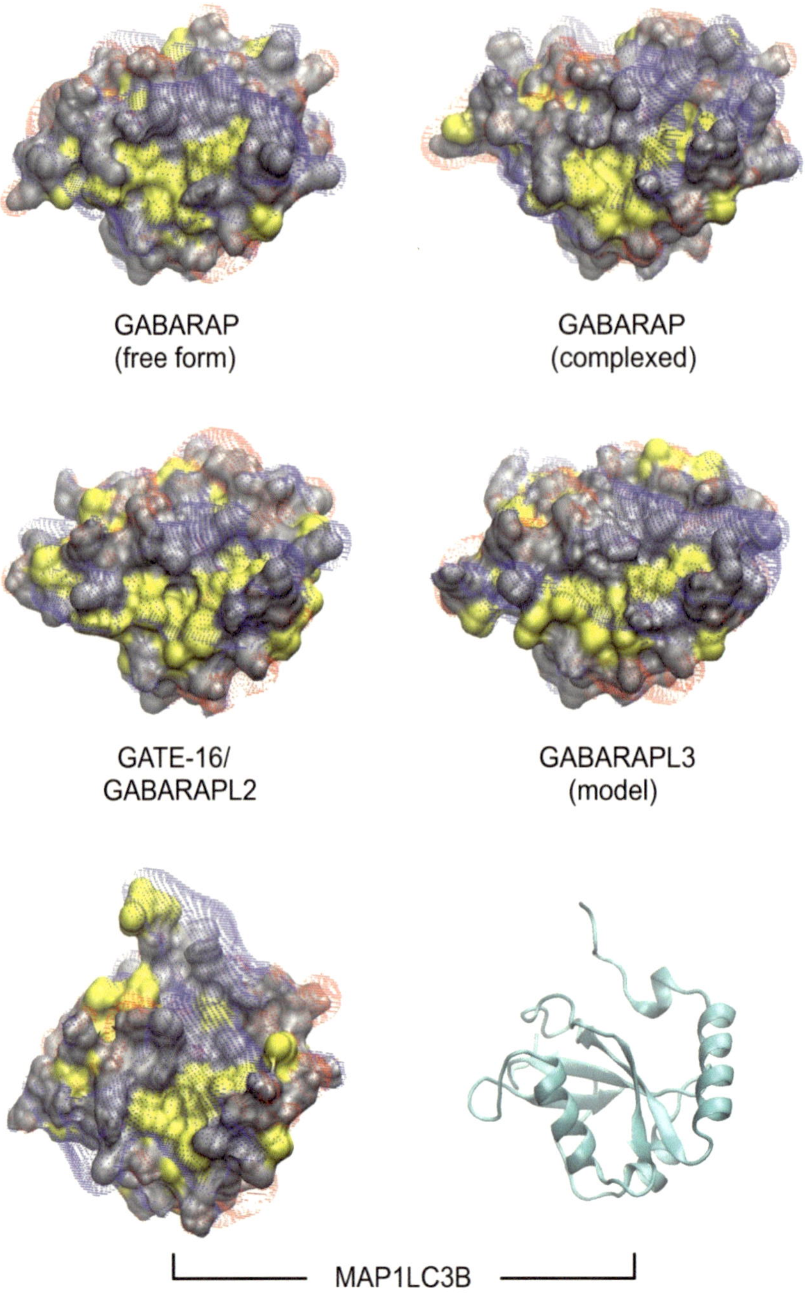

Figure 3. Surface properties of human Atg8 family proteins (LIR interaction sites face-on). Molecular surfaces were calculated with MSMS [84] and visualised in VMD [85]; hydrophobic patches are highlighted in yellow. Moreover, the electrostatic potential propagating into the solvent (calculated with ABPS [86], assuming dielectric constants of 1.0 and 80.0 in the protein and solvent regions, respectively) is contoured in blue (2 k_BT/e) and red (-2 k_BT/e). For ease of orientation, MAP1LC3B is additionally shown as ribbon diagram. Coordinates were derived from the following PDB entries: free GABARAP, 1KOT [48] model 2; complexed GABARAP, 3DOW [64]; GATE-16, 1EO6 [41]; MAP1LC3B, 1V49 [38]. The GABARAPL3 structure was built on the SWISS-MODEL server [87], using a GABARAP crystal structure (1GNU [50]) as template.

maybe the Atg5-Atg12-Atg16 complex acting as an E3 analogue. The molecular details of Atg8 binding to Atg4 are exemplified by several variants of the mammalian Atg4B-LC3B complex investigated by X-ray crystallography [88]. In these structures, the β-grasp domain of LC3B is involved in extensive polar and hydrophobic contacts with Atg4B, while its C-terminal tail adopts an extended conformation, reaching out into the catalytic centre of the protease. With respect to the free enzyme [89], the N-terminal segment of Atg4B as well as a regulatory loop, which occlude the entrance and exit of the catalytic groove, respectively, are moved aside upon complex formation. Despite its detachment from the catalytic domain, the N-terminus does not become disordered; via a canonical LIR motif (YDTL), it contacts the hydrophobic grooves of LC3 in a crystallographically equivalent complex. While experimental evidence supporting the significance of this interaction is not available at this time, it seems reasonable to assume that stoichiometric quantities of (possibly lipid-conjugated) Atg8 proteins may promote the hydrolytic activity of Atg4 on the phagophore membrane.

The mechanism of Atg8 processing by the activating enzyme Atg7 and its transfer to the conjugating enzyme Atg3 has been independently addressed in three seminal publications [90-92]. These groups have investigated different subcomplexes, mostly using X-ray crystallography. Combining biochemical and biophysical evidence, these data allow the structure of the full Atg7-Atg8-Atg3 assembly to be inferred. A particularly important finding concerns the functional organization of the Atg7 enzyme. While the initial adenylation of the Atg8 C-terminus and subsequent transfer to the catalytic cysteine takes place in the C-terminal domain of Atg7, the unique N-terminal domain of the enzyme, which is not found in canonical E1 proteins, has been found to recruit Atg3 (see below). Based on these observations, Atg7, Atg8 and Atg3 appear to form a complex with 2:2:2 stoichiometry. In this context, Atg7 dimerization may serve to bring Atg8 bound to the C-terminal domain of one protomer into physical proximity of Atg3 associated with the N-terminal domain of the second protomer, i.e. exchange of Atg8 between E1 and E2 components occurs via a *trans* mechanism.

Intriguingly, C-terminal truncation of Atg7 was found to severely reduce its affinity for Atg8 [91]. Sequence analysis reveals that this segment - in both yeast and human Atg7 - contains a tryptophan residue along with aliphatic and acidic side chains, but does not match the canonical AIM/LIR consensus. In accordance with these observations, NMR experiments indicated that a peptide corresponding to the C-terminal 30 amino acids of Atg7 bound to yeast Atg8 in an atypical manner, without assuming regular secondary structure [91]. While the orientation of the peptide in this assembly seems difficult to reconcile with the X-ray structure of the Atg7-Atg8 complex, available evidence suggests that the hydrophobic pockets of Atg8 do play a crucial role for interaction with the E1-E2 complex.

The crystal structure of yeast Atg3 has been determined, as well [93]. Overall, its fold is reminiscent of canonical E2 enzymes, but is distinguished by two large insertions. One of these is an acidic segment which is disordered in the crystal and has been implicated in the association with the Atg7 N-terminal domain. The second insertion forms a long helical extension followed by a disordered loop which might be involved in binding Atg8, as evidenced by deletion experiments. Subsequent studies confirmed that the WEDL sequence found in this region of yeast Atg3 indeed functions as an AIM [94]. Notably, however, the majority of this segment – including the hydrophobic motif – is conserved among various yeast species but is missing in higher eukaryotes. In accordance with this finding, the AIM of Atg3 appears to be required for the yeast-specific Cvt pathway, but not for starvation-induced autophagy.

5.2. Autophagic cargo adaptors

While autophagy has been initially described as a mechanism of bulk degradation, serving to replenish nutrient and energy resources under conditions of stress, accumulating evidence suggests that specific targeting for autophagic proteolysis plays a crucial role for cellular homeostasis. Indeed, autophagy has been recognised as the prevalent mechanism for turnover of long-lived proteins, and is the only available degradation pathway for large protein aggregates or complete organelles. Specificity is accomplished by a number of autophagy receptor (or adaptor) proteins which selectively bind certain types of targets and, at the same time, associate with Atg8-like proteins on the phagophore surface. In yeast cells, a precursor of aminopeptidase I (prApe1) and α-mannosidase (Ams1) are transported to the lytic compartment via the Cvt or Atg pathways [95], depending on nutrient availability. In this context, Atg19 is required to link the two hydrolases to Atg8 on the PAS or emerging phagophores [96]. Biochemical evidence and X-ray data revealed that the C-terminus of Atg19 contains an AIM (WEEL) which interacts with Atg8 in the canonical manner [97].

In mammalian cells, targets of selective autophagy are often tagged by polyubiquitin chains; examples include protein aggregates, dysfunctional organelles, and pathogens. Accordingly, the corresponding receptor proteins contain a ubiquitin binding domain in addition to a LIR motif (WTHL in p62, for instance), which mediates the association with Atg8 orthologues [82]. Again, the X-ray structure of LC3 with a p62 LIR peptide confirmed the expected binding mode of the ligand, which extends the central β-sheet of LC3 [97]. Similar to the Atg19-Atg8 complex, acidic residues adjacent to the core binding motif of p62 enhance its affinity for LC3. In humans, three additional autophagy receptors recognizing ubiquitinated targets have been identified; while NBR1 (neighbor of BRCA1 gene 1 [98]) is involved in similar functions as p62, NDP52 (nuclear dot protein of 52 kDa [99]) and optineurin [100] are required for the elimination of intracellular bacteria (discussed below).

Finally, the mitochondrial outer membrane protein Atg32 promotes mitophagy in yeast cells via direct association with Atg8 [101], and in mammalian erythrocytes, Nix (Nip-like protein x) has been ascribed a similar function [102].

In recent years, the role of autophagy in non-adaptive pathogen defence has attracted considerable attention. It is now well-established that removal of cytoplasmic bacteria is critically dependent on autophagy receptors, such as p62, optineurin and NDP52. In the context of this review, the latter two are particularly noteworthy since they illustrate two remarkable variations to the paradigm of Atg8 protein complexes. The optineurin sequence contains a LIR motif (FVEI) which is immediately preceded by a serine residue (S_{177}). Upon recruitment of the protein to ubiquitinated *Salmonella enterica*, S_{177} gets phosphorylated by TANK binding kinase 1 (TBK1), resulting in a significant enhancement in affinity for LC3 and thus improved efficiency of microbial clearance [100,103]. As expected, the effect of serine phosphorylation could be mimicked by substitution with acidic residues, thus confirming the concept of negative charges favouring LIR-mediated interactions. The presence of serine residues upstream of LIR motifs in other autophagy receptors suggests that this type of regulation may not be restricted to optineurin. NDP52, on the other hand, has a dual function in pathogen defence. Similar to p62 and optineurin, it is able to recognise polyubiquitin chains on the surface of cytosolic bacteria, but it is also recruited by Galectin-8 which acts as a sensor of endosomal damage [104]. The most intriguing property of this protein, however, is its clear preference for LC3C over all other members of the Atg8 family [55]. Indeed, the LC3C-NDP52 interaction appears to be essential for removal of cytosolic *Salmonella*, since in its absence other autophagy receptors are not efficiently recruited to the pathogen. This peculiar function is mirrored by some remarkable findings in the structure of the complex. While the interaction surface of LC3C is highly conserved with respect to other family members (enabling interaction with general autophagy receptors like p62), it appears to be skewed towards binding of the stunted LIR sequence of NDP52. This motif (LVV), which the authors term CLIR, leaves the hp1 site unoccupied, whereas the first and third residues together interact with a relatively flat hp2. Moreover, the CLIR β-strand is rotated with respect to the canonical orientation, allowing for optimised main chain hydrogen bonding [55].

5.3. Signalling components

A particularly interesting facet of the Atg8 family interactome concerns the autophagic initiator complex centred around the kinase Atg1/ULK, which is critical for the onset of autophagy under most circumstances. The first report proposing an interaction of GABARAP and GATE-16 with mammalian ULK1 dates back to 2000 [56]; these authors already mapped the binding site to the proline/serine rich domain of the kinase. Several recent contributions have provided more insight into the molecular details as well as the biological significance of this interaction in different organisms. Specifically, in yeast a fraction of cellular Atg1 was found to be included in autophagosomes, leading to degradation in the vacuole [7]. Subsequent investigations revealed that this pathway was dependent on Atg1 binding to Atg8; indeed, the interaction is mediated by a canonical AIM (YVVV) located in the proline/serine rich domain of Atg1 [105]. Similar results were reported for mammalian ULK1 [106]; here, the authors provided evidence that Atg13, owing to its affinity for Atg1, is co-transported to the degradative compartment. Finally, it has been shown that functional LIR motifs are, in fact, present in both mammalian Atg13 and FIP200, in addition to ULK1/2 [107]. It is interesting to note that all three proteins display a clear preference for GABARAP and its closest relatives over members of the LC3 subfamily. The authors of this study used synthetic peptide arrays for precise mapping of the interaction sites on ULK1/2, Atg13 and FIP200, yielding minimal LIR sequences DFVMV, DFVMI and DFETI, respectively.

Taken together, these reports have demonstrated that Atg1/ULK engages in a specific interaction with Atg8 proteins which is conserved during evolution from yeast to higher eukaryotes. In addition, they consistently found that membrane association of the kinase (mediated by Atg8) is required for efficient autophagosome formation. This has lead to the intriguing hypothesis that Atg8 proteins may form scaffolds supporting the organisation of the autophagic initiator complex [107]. On the other hand, their results differ in several respects, e.g.

concerning the significance of LIR motifs in additional components of the complex and the contribution of autophagy versus proteasomal degradation to cellular turnover of the kinase. These inconsistencies are likely to reflect species differences.

In order to define the importance of individual positions in the linear binding motifs found in ULK1, Atg13 and FIP200, Alemu et al. performed complete mutational analyses of the core sequences and their immediate environment. Combining these data with a compilation of published Atg8 binding sequences, they arrived at the consensus [D,E] [D,E,S,T] [W,F,Y] [D,E,L,I,V] x [I,L,V] [107]. This pattern is in excellent agreement with our results obtained by screening phage-displayed peptide libraries with GABARAP [61,63,73] and other family members (unpublished observations). In particular, it highlights the requirement for acidic side chains preceding the aromatic anchoring residue, which was confirmed by our investigations of GABARAP binding partners calreticulin [61], clathrin [73], and Bcl-2 (P. Ma et al., in revision).

Another previously unexpected interaction partner of Atg8 family proteins is ERK8 (extracellular signal-regulated kinase 8, also known as MAPK15). This MAP kinase is atypical in that it is not involved in a MAPKKK-MAPKK-MAPK cascade; instead, different stimuli, including starvation, appear to induce autophosphorylation of ERK8, resulting in kinase activation. Indeed, it was found to localise to autophagic membranes and to stimulate lipidation of Atg8-like proteins as well as autophagosome formation. These functions depend on the presence of a LIR motif (YQMI) in the C-terminal portion of the kinase [108]. In analogy to the Atg1/ULK complexes discussed above, this interaction might serve to recruit the MAP kinase to the surface of emerging phagophores or autophagosomes, where it may encounter potential substrates.

6. Linking autophagy to apoptosis signalling

Autophagy and apoptosis are recognised as fundamental cellular response programs supporting both normal development and adaptation to stress in multicellular organisms; their impact on the individual cell, however, is often antithetic: while apoptosis is, by definition, a process of controlled shut-down and removal of a cell, autophagy is directed towards sustaining its viability. The latter is accomplished by either removing potentially harmful structures, such as damaged organelles, aggregates, or pathogens, or by degrading dispensable material to compensate for nutrient or energy deprivation.

In recent years, evidence has accumulated in support of coordinated regulation of autophagy and apoptosis, particularly under conditions of stress. Several modulators acting in both processes have been identified, the most prominent being Bcl-2 (B-cell lymphoma 2). Besides its well-known function as an apoptosis inhibitor, which is largely based on sequestration of its pro-apoptotic siblings Bax and Bak on the mitochondrial outer membrane, it was found to interact with the autophagy regulator Beclin 1 (reviewed in [109]). As an activator of the class III PI3K Vps34, Beclin 1 is involved in the initiation of autophagy. Bcl-2 (as well as Bcl-xL) binds to the BH3 (Bcl-2 homology 3) region of Beclin 1, thus preventing it from promoting

Vps34 activity. In the presence of autophagic stimuli such as starvation, however, Beclin 1 is released from Bcl-2, resulting in activation of the PI3K and hence phagophore formation.

Research in our laboratory has added Atg8 homologues to the list of interaction partners of Bcl-2 family proteins, thus expanding the potential connections between autophagy and apoptosis pathways.

Among the matches returned in database searches with the PSSM for GABARAP binders was a sequence in the N-terminal part of the Nix protein [110]. Nix is a member of a pro-apoptotic subclass of Bcl-2 proteins characterised by the presence of BH1, BH2 and BH3 sequence motifs, as well as a C-terminal membrane anchor. Its interaction with GABARAP *in vitro* could indeed be confirmed by SPR measurements using either a synthetic peptide or the complete cytoplasmic portion of the molecule (NixΔC), yielding a dissociation constant of approx. 100 μM. Despite this relatively low affinity, the complex was detectable in co-immunoprecipitation experiments using cell extracts, and immunofluorescence staining revealed significant colocalization, predominantly in the perinuclear region. The residues of GABARAP involved in this interaction were probed by NMR spectroscopy. Titration of ^{15}N-GABARAP with NixΔC revealed alterations of numerous resonances in ^{1}H^{15}N-HSQC spectra. As expected, the residues most strongly affected by ligand binding mapped to the hydrophobic pockets hp1 and hp2, once again confirming the wide-spread adoption of this interaction paradigm.

Recent data have brought forward an interesting variation to this universal theme. During our investigation of the GABARAP-Nix complex, we set out to explore the influence of Bcl-2, which was anticipated to interact with Nix. Unexpectedly, Bcl-2 was found to exhibit significant GABARAP affinity as well, although a canonical binding motif was clearly absent (P. Ma et al., in revision). Due to the low solubility of native Bcl-2, we used a chimeric construct containing a segment of Bcl-xL in place of the long α1-α2 loop. Indeed, this modified Bcl-2 molecule bound to GABARAP with higher affinity than Nix (K_D = 25 μM). Mapping of the interacting residues via NMR spectroscopy revealed that Bcl-2 utilises an incomplete binding motif in its BH4 region to contact the hp1 site of GABARAP. Again, the significance of this complex *in vivo* is supported by co-immunoprecipitation and colocalization studies. Monitoring the cellular amounts of Atg8-like proteins by Western blotting, we found that Bcl-2 overexpression significantly decreased the fraction of lipidated GABARAP, whereas LC3 was unaffected. Concomitantly, the interaction of GABARAP with Atg4B was also reduced, indicating that Bcl-2 may sequester GABARAP, preventing it from entering the conjugation pathway.

It is interesting to note that Nix and Bcl-2 display distinct preferences for the members of the Atg8 family. In addition to all GABARAP-like proteins, Nix shows significant affinity for LC3A, but not LC3B [102]. Bcl-2, on the other hand, has a much more restricted spectrum of interactions, including GABARAP and its close relative GEC1, but not GATE-16 or LC3A/B.

7. Biological function and diversity

For about 15 years, Atg8 has been known to be essential for efficient autophagosome formation in yeast [Lang 1998]. In higher metazoans such as mammals, the interpretation of gene deletion experiments is complicated by the presence of several paralogues. Indeed, knockout mice lacking GABARAP do not show a marked phenotype [111], supporting the concept that significant functional redundancy might exist among Atg8 family members. In a series of elegant experiments, Weidberg et al. used siRNA pools targeting all members of a given subfamily. Their results indicated that the LC3 group (LC3A, LC3B, LC3C) is required for expansion of the phagophore membrane, whereas GABARAP and its relatives (GEC1 and GATE-16) are involved in a later stage, possibly closure of the phagophore to form a complete autophagosome [112].

What are the molecular foundations of this functional diversification? To answer this question, we need to consider the two fundamental activities ascribed to Atg8 family proteins in the context of autophagy. First of all, these molecules are well-established as binding partners of core autophagy components and adaptor proteins; as a result, they have been implicated not only in recruitment of specific cargo destined for degradation, but also in the organization of functional multi-protein complexes. By virtue of their reversible membrane association, they are perfectly suited for this type of regulated scaffolding function. As outlined above, the binding specificities may differ markedly between members of the two subfamilies, depending on the interaction partner. The components of the ULK1/2 complex, for instance, clearly prefer GABARAP-like over LC3-like proteins [107], and current evidence suggests that Atg1/ULK is not only involved in the initiation stage, but also exerts important functions on the phagophore membrane [105,106]. Conversely, the adaptor protein FYCO1 (FYVE and coiled-coil domain-containing protein 1), which links autophagosomes to the microtubule cytoskeleton, displays a strong preference for LC3 [113]. Among all Atg8/LC3 binding proteins identified to date, the most restricted scope has been found for NDP52. While most ubiquitin-directed autophagy receptors (p62, NBR1 and optineurin) interact with both GABARAP and LC3 subfamily proteins [82,98,100], NDP52 has been found to exclusively bind LC3C (discussed above). In general terms, these examples illustrate the great flexibility of the Atg8 interaction framework, which – despite its apparent simplicity – can be tuned to achieve any level of specificity. Indeed, LC3C appears to be unusual in several respects. While it has traditionally received less attention, recent evidence indicates that it exerts functions which are non-redundant with or even contrary to conventional LC3 proteins. In addition to its role in microbial clearance, it has been ascribed a unique anti-carcinogenic activity. Specifically, experiments with kidney cancer cells revealed that the von Hippel-Lindau (VHL) protein regulates LC3B and LC3C expression in opposite ways, with LC3B promoting tumour growth and LC3C suppressing it [114]. While it is tempting to speculate that this striking observation is related to differential recruitment of critical autophagy targets, the identity of these cargoes is still elusive.

A second activity of Atg8 proteins was reported by Nakatogawa et al. back in 2007. Using a reconstituted conjugation system, they showed that lipidated Atg8 is able to mediate tethering and even hemifusion of vesicle membranes [115]. The finding that (1) Atg8-PE, unlike the non-conjugated protein, forms oligomers, and (2) Atg8-decorated vesicles did not associate with bare ones led to the conclusion that the fusogenic effect was mediated by homotypic protein-protein interaction. More recently, the membrane tethering and fusion capabilities of the mammalian Atg8 orthologues have been investigated. Indeed, GATE-16 and LC3B, repre-

http://dx.doi.org/10.5772/55647

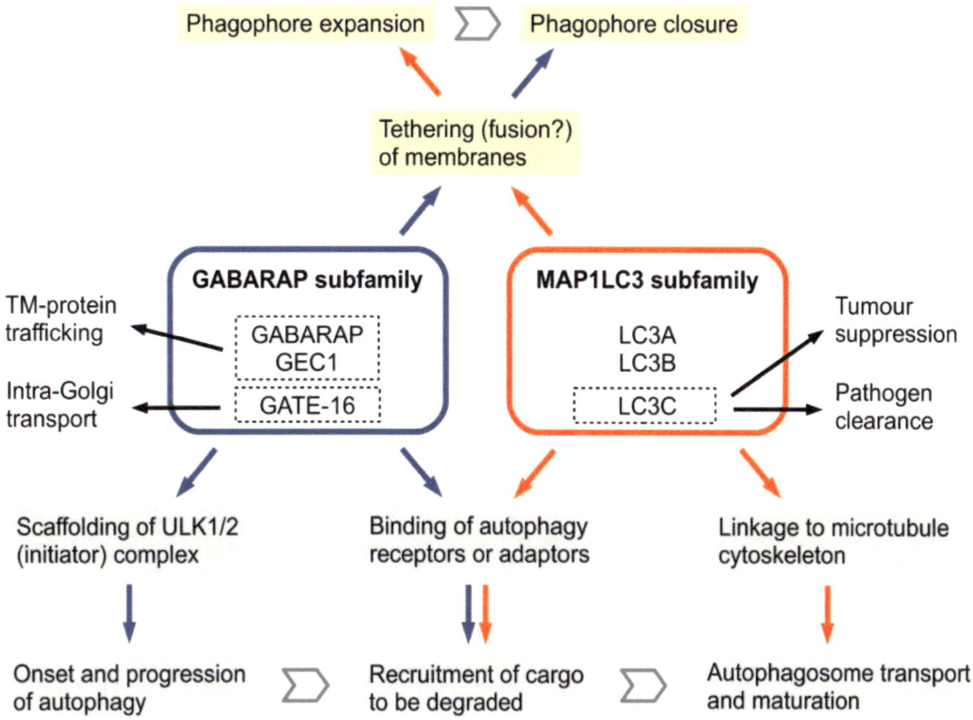

Figure 4. Schematic overview of the functions of GABARAP and LC3 subfamily proteins in mammalian cells. Activities which are essential for the autophagy pathway are highlighted by grey and yellow shading, signifying the involvement of protein-protein and protein-lipid interactions, respectively. Additional functions relevant for certain aspects of autophagy (in case of LC3C) or implicated in unrelated pathways (for GABARAP subfamily members) are indicated as well. See text for details.

senting the two subfamilies, were both confirmed to possess these activities when conjugated to liposomes [116]. Truncation experiments indicated that the short N-terminal α-helix ($\alpha 1$) plays an important role in fusogenicity. Intriguingly, the two Atg8 molecules tested in this study appear to employ different mechanisms to accomplish a similar effect: while in LC3B, two arginines (R_{10}, R_{11}) were found to be critical for membrane fusion, the same function was attributed to hydrophobic residues W_3 and M_4 in GATE-16. It is important to note that these authors postulated a direct protein-lipid interaction as an essential step in membrane fusion. The relevance of these findings is supported by transfection experiments: in both yeast and mammalian cells, expression of mutant Atg8 proteins which are defective in membrane tethering or fusion results in severely compromised autophagosome biogenesis. It should be noted, however, that the role of Atg8 in membrane fusion has been challenged by a series of elegant experiments, in which the effects of lipidated Atg8 and LC3 were separated from the bilayer-destabilizing properties of the PE lipid itself [117]. It turned out that with PE fractions found in cellular membranes, Atg8 proteins do possess membrane tethering activity, but are unable to initiate fusion. These observations are at variance with the results of Weidberg et al. [116], who used a similar lipid composition; the reasons for these discrepancies are currently unclear. Notably, accumulating evidence suggests that components of the conventional membrane fusion machinery, such as SNARE proteins and NSF, play an important role in autophagosome biogenesis [117,118]. In summary, these results support a model in which Atg8 and its homologues act as a "glue" promoting the aggregation of membrane structures, e.g. at the edge of an expanding phagophore, while the actual fusion process is mediated by a SNARE-based system.

Figure 4 summarises the concepts outlined above regarding the biological functions of the two Atg8 protein subfamilies found in mammalian cells.

8. Conclusions and outlook

During the past decade, the three-dimensional structures of all major members of the Atg8 family have been determined, and a wealth of biochemical and biophysical data on their functional properties has accumulated. Despite these great advances, our understanding of structure and biology of these proteins is far from complete. Three aspects will continue to require significant attention: First of all, the mechanistic details and functional implications of the growing number of molecular interactions identified for Atg8 proteins need to be defined more clearly. This includes what is probably the most ambiguous facet of these proteins: their proposed fusogenic potential and their relation to and interplay with the conventional membrane fusion machinery.

Second, recent reports have highlighted the involvement of previously unexpected regulatory mechanisms affecting Atg8 proteins. For instance, it appears that LC3A can be phosphorylated by protein kinase A at a site (S_{12}) which is conserved in other LC3 variants. Upon induction of autophagy, this residue is dephosphorylated, correlating with increased association with autophagic membranes [119]. Similarly, acetylation of Atg3 has been found to promote autophagy by increasing the processing of Atg8 in yeast [120].

Last but not least, while the diversification of the family in higher eukaryotes is likely to reflect both redundancy and functional specialization, the individual roles of these paralogues in autophagy are only beginning to emerge. The situation is confounded by the fact that several (maybe all) members of the family are involved in processes without direct relation to autophagy, implying an independent pattern of functional overlap. These issues need to be addressed by careful cell biological studies, involving knock-down of Atg8 proteins either individually or in groups, and monitoring of various cellular functions.

As the important implications of autophagy for human health and disease have become more and more obvious, therapeutic strategies targeting this pathway are now being developed. In this context, it should be noted that the functional differentiation of the mammalian Atg8 family may ultimately prove fortunate in terms of pharmacological interference, since it may allow targeting a specific function associated with a single member without globally disturbing the vital autophagy pathway.

Given the astonishing versatility of these proteins and the number of unresolved questions to be addressed, additional exciting twists are certainly to be expected.

Acknowledgements

The authors are aware that due to the extent of the research field covered in this review, the account of the literature is necessarily incomplete. We sincerely apologise to all colleagues whose contributions could not be considered. We are grateful to Dr. Felix Randow (MRC Laboratory of Molecular Biology, Cambridge, UK) and Dr. Masato Akutsu (Goethe University, Frankfurt, Germany) for providing the coordinates of the LC3C-NDP52 complex prior to public release. Research from our laboratory referred to in this article was supported by a grant of the Deutsche Forschungsgemeinschaft (DFG) to D.W. (Wi1472/5).

Author details

Oliver H. Weiergräber[1], Jeannine Mohrlüder[1] and Dieter Willbold[1,2*]

*Address all correspondence to: d.willbold@fz-juelich.de

1 Institute of Complex Systems, ICS-6 (Structural Biochemistry), Forschungszentrum Jülich, Jülich, Germany

2 Institut für Physikalische Biologie und BMFZ, Heinrich-Heine-Universität Düsseldorf, Düsseldorf, Germany

References

[1] Yang Z, Klionsky DJ. Eaten alive: a history of macroautophagy. Nature Cell Biology 2010;12(9) 814-822.

[2] Ravikumar B, Sarkar S, Davies JE, Futter M, Garcia-Arencibia M, Green-Thompson ZW, Jimenez-Sanchez M, Korolchuk VI, Lichtenberg M, Luo S, Massey DC, Menzies FM, Moreau K, Narayanan U, Renna M, Siddiqi FH, Underwood BR, Winslow AR, Rubinsztein DC. Regulation of mammalian autophagy in physiology and pathophysiology. Physiological Reviews 2010;90(4) 1383-1435.

[3] Nakatogawa H, Suzuki K, Kamada Y, Ohsumi Y. Dynamics and diversity in autophagy mechanisms: lessons from yeast. Nature Reviews Molecular Cell Biology 2009;10(7) 458-467.

[4] Williams RA, Woods KL, Juliano L, Mottram JC, Coombs GH. Characterization of unusual families of ATG8-like proteins and ATG12 in the protozoan parasite Leishmania major. Autophagy 2009;5(2) 159-172.

[5] Xie Z, Klionsky DJ. Autophagosome formation: core machinery and adaptations. Nature Cell Biology 2007;9(10) 1102-1109.

[6] Mizushima N, Yoshimori T, Ohsumi Y. The role of Atg proteins in autophagosome formation. Annual Review of Cell and Developmental Biology 2011;27 107-132.

[7] Suzuki K, Kubota Y, Sekito T, Ohsumi Y. Hierarchy of Atg proteins in pre-autophagosomal structure organization. Genes to Cells 2007;12(2):209-218.

[8] Cebollero E, van der Vaart A, Reggiori F. Understanding phosphatidylinositol-3-phosphate dynamics during autophagosome biogenesis. Autophagy. 2012;8(12) 1868-1870.

[9] Mizushima N, Noda T, Yoshimori T, Tanaka Y, Ishii T, George MD, Klionsky DJ, Ohsumi M, Ohsumi Y. A protein conjugation system essential for autophagy. Nature 1998;395(6700) 395-398.

[10] Fujioka Y, Noda NN, Nakatogawa H, Ohsumi Y, Inagaki F. Dimeric coiled-coil structure of Saccharomyces cerevisiae Atg16 and its functional significance in autophagy. The Journal of Biological Chemistry 2010;285(2) 1508-1515.

[11] Suzuki K, Kirisako T, Kamada Y, Mizushima N, Noda T, Ohsumi Y. The pre-autophagosomal structure organized by concerted functions of APG genes is essential for autophagosome formation. The EMBO Journal 2001;20(21) 5971-5981.

[12] Mizushima N, Yamamoto A, Hatano M, Kobayashi Y, Kabeya Y, Suzuki K, Tokuhisa T, Ohsumi Y, Yoshimori T. Dissection of autophagosome formation using Apg5-deficient mouse embryonic stem cells. The Journal of Cell Biology 2001;152(4) 657-668.

[13] Ichimura Y, Kirisako T, Takao T, Satomi Y, Shimonishi Y, Ishihara N, Mizushima N, Tanida I, Kominami E, Ohsumi M, Noda T, Ohsumi Y. A ubiquitin-like system mediates protein lipidation. Nature 2000;408(6811) 488-492.

[14] Tanida I, Komatsu M, Ueno T, Kominami E. GATE-16 and GABARAP are authentic modifiers mediated by Apg7 and Apg3. Biochemical and Biophysical Research Communications 2003;300(3) 637-644.

[15] Kirisako T, Ichimura Y, Okada H, Kabeya Y, Mizushima N, Yoshimori T, Ohsumi M, Takao T, Noda T, Ohsumi Y. The reversible modification regulates the membrane-binding state of Apg8/Aut7 essential for autophagy and the cytoplasm to vacuole targeting pathway. The Journal of Cell Biology 2000;151(2) 263-276.

[16] Kabeya Y, Mizushima N, Yamamoto A, Oshitani-Okamoto S, Ohsumi Y, Yoshimori T. LC3, GABARAP and GATE16 localize to autophagosomal membrane depending on form-II formation. Journal of Cell Science 2004;117(Pt 13) 2805-2812.

[17] Hanada T, Noda NN, Satomi Y, Ichimura Y, Fujioka Y, Takao T, Inagaki F, Ohsumi Y. The Atg12-Atg5 conjugate has a novel E3-like activity for protein lipidation in autophagy. The Journal of Biological Chemistry 2007;282(52) 37298-37302.

[18] Fujita N, Itoh T, Omori H, Fukuda M, Noda T, Yoshimori T. The Atg16L complex specifies the site of LC3 lipidation for membrane biogenesis in autophagy. Molecular Biology of the Cell 2008;19(5) 2092-2100.

[19] Kuznetsov SA, Gelfand VI. 18 kDa microtubule-associated protein: identification as a new light chain (LC-3) of microtubule-associated protein 1 (MAP-1). FEBS Letters 1987;212(1) 145-148.

[20] Mann SS, Hammarback JA. Molecular characterization of light chain 3. A microtubule binding subunit of MAP1A and MAP1B. The Journal of Biological Chemistry 1994;269(15) 11492-11497.

[21] Clark SL Jr. Cellular differentiation in the kidneys of newborn mice studies with the electron microscope. The Journal of Biophysical and Biochemical Cytology 1957;3(3) 349-362.

[22] Kabeya Y, Mizushima N, Ueno T, Yamamoto A, Kirisako T, Noda T, Kominami E, Ohsumi Y, Yoshimori T. LC3, a mammalian homologue of yeast Apg8p, is localized in autophagosome membranes after processing. The EMBO Journal 2000;19(21) 5720-5728.

[23] Lang T, Schaeffeler E, Bernreuther D, Bredschneider M, Wolf DH, Thumm M. Aut2p and Aut7p, two novel microtubule-associated proteins are essential for delivery of autophagic vesicles to the vacuole. The EMBO Journal 1998;17(13) 3597-3607.

[24] Shpilka T, Weidberg H, Pietrokovski S, Elazar Z. Atg8: an autophagy-related ubiquitin-like protein family. Genome Biology 2011;12(7) 226.

[25] Xin Y, Yu L, Chen Z, Zheng L, Fu Q, Jiang J, Zhang P, Gong R, Zhao S. Cloning, expression patterns, and chromosome localization of three human and two mouse homologues of $GABA_A$ receptor-associated protein. Genomics 2001;74(3) 408-413.

[26] Legesse-Miller A, Sagiv Y, Porat A, Elazar Z. Isolation and characterization of a novel low molecular weight protein involved in intra-Golgi traffic. The Journal of Biological Chemistry 1998;273(5) 3105-3109.

[27] Sagiv Y, Legesse-Miller A, Porat A, Elazar Z. GATE-16, a membrane transport modulator, interacts with NSF and the Golgi v-SNARE GOS-28. The EMBO Journal 2000;19(7) 1494-1504.

[28] Wang H, Bedford FK, Brandon NJ, Moss SJ, Olsen RW. $GABA_A$-receptor-associated protein links $GABA_A$ receptors and the cytoskeleton. Nature 1999;397(6714) 69-72.

[29] Leil TA, Chen ZW, Chang CS, Olsen RW. $GABA_A$ receptor-associated protein traffics $GABA_A$ receptors to the plasma membrane in neurons. The Journal of Neuroscience 2004;24(50) 11429-11438.

[30] Green F, O'Hare T, Blackwell A, Enns CA. Association of human transferrin receptor with GABARAP. FEBS Letters 2002;518(1-3) 101-106.

[31] Cook JL, Re RN, deHaro DL, Abadie JM, Peters M, Alam J. The trafficking protein GABARAP binds to and enhances plasma membrane expression and function of the angiotensin II type 1 receptor. Circulation Research 2008;102(12) 1539-1547.

[32] Laínez S, Valente P, Ontoria-Oviedo I, Estévez-Herrera J, Camprubí-Robles M, Ferrer-Montiel A, Planells-Cases R. $GABA_A$ receptor associated protein (GABARAP) modulates TRPV1 expression and channel function and desensitization. FASEB Journal 2010;24(6) 1958-1970.

[33] Chen C, Wang Y, Huang P, Liu-Chen LY. Effects of C-terminal modifications of GEC1 protein and γ-aminobutyric acid type A ($GABA_A$) receptor-associated protein (GABARAP), two microtubule-associated proteins, on κ opioid receptor expression. The Journal of Biological Chemistry 2011;286(17) 15106-15115.

[34] Mansuy V, Boireau W, Fraichard A, Schlick JL, Jouvenot M, Delage-Mourroux R. GEC1, a protein related to GABARAP, interacts with tubulin and $GABA_A$ receptor. Biochemical and Biophysical Research Communications 2004;325(2) 639-648.

[35] Chen C, Li JG, Chen Y, Huang P, Wang Y, Liu-Chen LY. GEC1 interacts with the κ opioid receptor and enhances expression of the receptor. The Journal of Biological Chemistry 2006;281(12) 7983-7993.

[36] Kittler JT, Rostaing P, Schiavo G, Fritschy JM, Olsen R, Triller A, Moss SJ. The subcellular distribution of GABARAP and its ability to interact with NSF suggest a role for this protein in the intracellular transport of $GABA_A$ receptors. Molecular and Cellular Neurosciences 2001;18(1) 13-25.

[37] Coyle JE, Qamar S, Rajashankar KR, Nikolov DB. Structure of GABARAP in two conformations: implications for $GABA_A$ receptor localization and tubulin binding. Neuron 2002;33(1) 63-74.

[38] Kouno T, Mizuguchi M, Tanida I, Ueno T, Kanematsu T, Mori Y, Shinoda H, Hirata M, Kominami E, Kawano K. Solution structure of microtubule-associated protein light chain 3 and identification of its functional subdomains. The Journal of Biological Chemistry 2005;280(26) 24610-24617.

[39] Fujita N, Hayashi-Nishino M, Fukumoto H, Omori H, Yamamoto A, Noda T, Yoshimori T. An Atg4B mutant hampers the lipidation of LC3 paralogues and causes defects in autophagosome closure. Molecular Biology of the Cell 2008;19(11) 4651-4659.

[40] Sou YS, Waguri S, Iwata J, Ueno T, Fujimura T, Hara T, Sawada N, Yamada A, Mizushima N, Uchiyama Y, Kominami E, Tanaka K, Komatsu M. The Atg8 conjugation system is indispensable for proper development of autophagic isolation membranes in mice. Molecular Biology of the Cell 2008;19(11) 4762-4775.

[41] Paz Y, Elazar Z, Fass D. Structure of GATE-16, membrane transport modulator and mammalian ortholog of autophagocytosis factor Aut7p. The Journal of Biological Chemistry 2000;275(33) 25445-25450.

[42] Vijay-Kumar S, Bugg CE, Cook WJ. Structure of ubiquitin refined at 1.8 Å resolution. Journal of Molecular Biology 1987;194(3) 531-544.

[43] Narasimhan J, Wang M, Fu Z, Klein JM, Haas AL, Kim JJ. Crystal structure of the interferon-induced ubiquitin-like protein ISG15. The Journal of Biological Chemistry 2005;280(29) 27356-27365.

[44] Ceccarelli DF, Song HK, Poy F, Schaller MD, Eck MJ. Crystal structure of the FERM domain of focal adhesion kinase. The Journal of Biological Chemistry 2006;281(1) 252-259.

[45] Kraulis PJ. MOLSCRIPT: A program to produce both detailed and schematic plots of protein structures. Journal of Applied Crystallography 1991;24(5) 946-950.

[46] Merritt EA, Bacon DJ. Raster3D: photorealistic molecular graphics. Methods in Enzymology 1997;277 505-524.

[47] Kabsch W, Sander C. Dictionary of protein secondary structure: pattern recognition of hydrogen-bonded and geometrical features. Biopolymers. 1983;22(12):2577-2637.

[48] Stangler T, Mayr LM, Willbold D. Solution structure of human $GABA_A$ receptor-associated protein GABARAP: implications for biological function and its regulation. The Journal of Biological Chemistry 2002;277(16) 13363-13366.

[49] Bavro VN, Sola M, Bracher A, Kneussel M, Betz H, Weissenhorn W. Crystal structure of the $GABA_A$-receptor-associated protein, GABARAP. EMBO Reports 2002;3(2) 183-189.

[50] Knight D, Harris R, McAlister MS, Phelan JP, Geddes S, Moss SJ, Driscoll PC, Keep NH. The X-ray crystal structure and putative ligand-derived peptide binding properties of γ-aminobutyric acid receptor type A receptor-associated protein. The Journal of Biological Chemistry 2002;277(7) 5556-5561.

[51] Kouno T, Miura K, Kanematsu T, Shirakawa M, Hirata M, Kawano K. ^{1}H, ^{13}C and ^{15}N resonance assignments of GABARAP, $GABA_A$ receptor associated protein. Journal of Biomolecular NMR 2002;22(1) 97-98.

[52] Pacheco V, Ma P, Thielmann Y, Hartmann R, Weiergräber OH, Mohrlüder J, Willbold D. Assessment of GABARAP self-association by its diffusion properties. Journal of Biomolecular NMR 2010;48(1) 49-58.

[53] Schwarten M, Stoldt M, Mohrlüder J, Willbold D. Solution structure of Atg8 reveals conformational polymorphism of the N-terminal domain. Biochemical and Biophysical Research Communications 2010;395(3) 426-431.

[54] Kumeta H, Watanabe M, Nakatogawa H, Yamaguchi M, Ogura K, Adachi W, Fujioka Y, Noda NN, Ohsumi Y, Inagaki F. The NMR structure of the autophagy-related protein Atg8. Journal of Biomolecular NMR. 2010;47(3) 237-241.

[55] von Muhlinen N, Akutsu M, Ravenhill BJ, Foeglein A, Bloor S, Rutherford TJ, Freund SM, Komander D, Randow F. LC3C, bound selectively by a noncanonical LIR motif in NDP52, is required for antibacterial autophagy. Molecular Cell. 2012;48(3) 329-342.

[56] Okazaki N, Yan J, Yuasa S, Ueno T, Kominami E, Masuho Y, Koga H, Muramatsu M. Interaction of the Unc-51-like kinase and microtubule-associated protein light chain 3 related proteins in the brain: possible role of vesicular transport in axonal elongation. Brain Research. Molecular Brain Research 2000;85(1-2) 1-12.

[57] Kanematsu T, Jang IS, Yamaguchi T, Nagahama H, Yoshimura K, Hidaka K, Matsuda M, Takeuchi H, Misumi Y, Nakayama K, Yamamoto T, Akaike N, Hirata M, Nakayama K. Role of the PLC-related, catalytically inactive protein p130 in $GABA_A$ receptor function. The EMBO Journal 2002;21(5) 1004-1011.

[58] Kittler JT, Arancibia-Carcamo IL, Moss SJ. Association of GRIP1 with a $GABA_A$ receptor associated protein suggests a role for GRIP1 at inhibitory synapses. Biochemical Pharmacology 2004;68(8) 1649-1654.

[59] Kneussel M, Haverkamp S, Fuhrmann JC, Wang H, Wassle H, Olsen RW, Betz H. The γ-aminobutyric acid type A receptor ($GABA_AR$)- associated protein GABARAP interacts with gephyrin but is not involved in receptor anchoring at the synapse. Proceedings of the National Academy of Sciences of the United States of America 2000;97(15) 8594-8599.

[60] Lee JH, Rho SB, Chun T. $GABA_A$ receptor-associated protein (GABARAP) induces apoptosis by interacting with DEAD (Asp-Glu-Ala-Asp/His) box polypeptide 47 (DDX 47). Biotechnology Letters 2005;27(9) 623-628.

[61] Mohrlüder J, Stangler T, Hoffmann Y, Wiesehan K, Mataruga A, Willbold D. Identification of calreticulin as a ligand of GABARAP by phage display screening of a peptide library. FEBS Journal 2007;274(21) 5543-5555.

[62] Thielmann Y, Mohrlüder J, Koenig BW, Stangler T, Hartmann R, Becker K, Höltje HD, Willbold D. An indole-binding site is a major determinant of the ligand specificity of the GABA type A receptor-associated protein GABARAP. Chembiochem 2008;9(11) 1767-1775.

[63] Weiergräber OH, Stangler T, Thielmann Y, Mohrlüder J, Wiesehan K, Willbold D. Ligand binding mode of $GABA_A$ receptor-associated protein. Journal of Molecular Biology 2008;381(5) 1320-1331.

[64] Thielmann Y, Weiergräber OH, Mohrlüder J, Willbold D. Structural framework of the GABARAP-calreticulin interface--implications for substrate binding to endoplasmic reticulum chaperones. FEBS Journal 2009;276(4) 1140-1152.

[65] Schrag JD, Bergeron JJ, Li Y, Borisova S, Hahn M, Thomas DY, Cygler M. The structure of calnexin, an ER chaperone involved in quality control of protein folding. Molecular Cell 2001;8(3) 633-644.

[66] Ellgaard L, Riek R, Herrmann T, Güntert P, Braun D, Helenius A, Wüthrich K. NMR structure of the calreticulin P-domain. Proceedings of the National Academy of Sciences of the United States of America 2001;98(6) 3133-3138.

[67] Gelebart P, Opas M, Michalak M. Calreticulin, a Ca^{2+}-binding chaperone of the endoplasmic reticulum. The International Journal of Biochemistry and Cell Biology 2005;37(2) 260-266.

[68] Afshar N, Black BE, Paschal BM. Retrotranslocation of the chaperone calreticulin from the endoplasmic reticulum lumen to the cytosol. Molecular and Cellular Biology 2005;25(20) 8844-8853.

[69] Fadel MP, Szewczenko-Pawlikowski M, Leclerc P, Dziak E, Symonds JM, Blaschuk O, Michalak M, Opas M. Calreticulin affects β-catenin associated pathways. The Journal of Biological Chemistry 2001;276(29) 27083-27089.

[70] Nakamura T, Hayashi T, Nasu-Nishimura Y, Sakaue F, Morishita Y, Okabe T, Ohwada S, Matsuura K, Akiyama T. PX-RICS mediates ER-to-Golgi transport of the N-cadherin/β-catenin complex. Genes & Development 2008;22(9) 1244-1256.

[71] Rojiani MV, Finlay BB, Gray V, Dedhar S. In vitro interaction of a polypeptide homologous to human Ro/SS-A antigen (calreticulin) with a highly conserved amino acid sequence in the cytoplasmic domain of integrin α subunits. Biochemistry 1991;30(41) 9859-9866.

[72] Kawaguchi SY, Hirano T. Integrin $\alpha3\beta1$ suppresses long-term potentiation at inhibitory synapses on the cerebellar Purkinje neuron. Molecular and Cellular Neurosciences 2006;31(3) 416-426.

[73] Mohrlüder J, Hoffmann Y, Stangler T, Hänel K, Willbold D. Identification of clathrin heavy chain as a direct interaction partner for the γ-aminobutyric acid type A receptor associated protein. Biochemistry 2007;46(50) 14537-14543.

[74] Yu RC, Jahn R, Brunger AT. NSF N-terminal domain crystal structure: models of NSF function. Molecular Cell 1999;4(1) 97-107.

[75] Yu RC, Hanson PI, Jahn R, Brünger AT. Structure of the ATP-dependent oligomerization domain of N-ethylmaleimide sensitive factor complexed with ATP. Nature Structural Biology 1998;5(9) 803-811.

[76] Furst J, Sutton RB, Chen J, Brunger AT, Grigorieff N. Electron cryomicroscopy structure of N-ethyl maleimide sensitive factor at 11 Å resolution. The EMBO Journal 2003;22(17) 4365-4374.

[77] Thielmann Y, Weiergräber OH, Ma P, Schwarten M, Mohrlüder J, Willbold D. Comparative modeling of human NSF reveals a possible binding mode of GABARAP and GATE-16. Proteins 2009;77(3) 637-646.

[78] Huyton T, Pye VE, Briggs LC, Flynn TC, Beuron F, Kondo H, Ma J, Zhang X, Freemont PS. The crystal structure of murine p97/VCP at 3.6 Ǻ. Journal of Structural Biology 2003;144(3) 337-348.

[79] Chen ZW, Chang CS, Leil TA, Olsen RW. C-terminal modification is required for GABARAP-mediated $GABA_A$ receptor trafficking. The Journal of Neuroscience 2007;27(25) 6655-6663.

[80] Aasland R, Abrams C, Ampe C, Ball LJ, Bedford MT, Cesareni G, Gimona M, Hurley JH, Jarchau T, Lehto VP, Lemmon MA, Linding R, Mayer BJ, Nagai M, Sudol M, Walter U, Winder SJ. Normalization of nomenclature for peptide motifs as ligands of modular protein domains. FEBS Letters 2002;513(1) 141-144.

[81] Noda NN, Ohsumi Y, Inagaki F. Atg8-family interacting motif crucial for selective autophagy. FEBS Letters 2010;584(7) 1379-1385.

[82] Pankiv S, Clausen TH, Lamark T, Brech A, Bruun JA, Outzen H, Øvervatn A, Bjørkøy G, Johansen T. p62/SQSTM1 binds directly to Atg8/LC3 to facilitate degradation of ubiquitinated protein aggregates by autophagy. The Journal of Biological Chemistry 2007;282(33) 24131-24145.

[83] Ma P, Mohrlüder J, Schwarten M, Stoldt M, Singh SK, Hartmann R, Pacheco V, Willbold D. Preparation of a functional GABARAP-lipid conjugate in nanodiscs and its investigation by solution NMR spectroscopy. Chembiochem 2010;11(14) 1967-1970.

[84] Sanner MF, Olson AJ, Spehner JC. Reduced surface: an efficient way to compute molecular surfaces. Biopolymers 1996;38(3) 305-320.

[85] Humphrey W, Dalke A, Schulten K. VMD: visual molecular dynamics. Journal of Molecular Graphics 1996;14(1):33-38.

[86] Baker NA, Sept D, Joseph S, Holst MJ, McCammon JA. Electrostatics of nanosystems: application to microtubules and the ribosome. Proceedings of the National Academy of Sciences of the United States of America 2001:98(18) 10037-10041.

[87] Arnold K, Bordoli L, Kopp J, Schwede T. The SWISS-MODEL Workspace: A web-based environment for protein structure homology modelling. Bioinformatics 2006:22(2) 195-201.

[88] Satoo K, Noda NN, Kumeta H, Fujioka Y, Mizushima N, Ohsumi Y, Inagaki F. The structure of Atg4B-LC3 complex reveals the mechanism of LC3 processing and delipidation during autophagy. The EMBO Journal 2009;28(9) 1341-1350.

[89] Sugawara K, Suzuki NN, Fujioka Y, Mizushima N, Ohsumi Y, Inagaki F. Structural basis for the specificity and catalysis of human Atg4B responsible for mammalian autophagy. The Journal of Biological Chemistry 2005;280(48) 40058-40065.

[90] Taherbhoy AM, Tait SW, Kaiser SE, Williams AH, Deng A, Nourse A, Hammel M, Kurinov I, Rock CO, Green DR, Schulman BA. Atg8 transfer from Atg7 to Atg3: a

distinctive E1-E2 architecture and mechanism in the autophagy pathway. Molecular Cell 2011;44(3) 451-461.

[91] Noda NN, Satoo K, Fujioka Y, Kumeta H, Ogura K, Nakatogawa H, Ohsumi Y, Inagaki F. Structural basis of Atg8 activation by a homodimeric E1, Atg7. Molecular Cell 2011;44(3) 462-475.

[92] Hong SB, Kim BW, Lee KE, Kim SW, Jeon H, Kim J, Song HK. Insights into noncanonical E1 enzyme activation from the structure of autophagic E1 Atg7 with Atg8. Nature Structural and Molecular Biology 2011;18(12) 1323-1330.

[93] Yamada Y, Suzuki NN, Hanada T, Ichimura Y, Kumeta H, Fujioka Y, Ohsumi Y, Inagaki F. The crystal structure of Atg3, an autophagy-related ubiquitin carrier protein (E2) enzyme that mediates Atg8 lipidation. The Journal of Biological Chemistry 2007;282(11) 8036-8043.

[94] Yamaguchi M, Noda NN, Nakatogawa H, Kumeta H, Ohsumi Y, Inagaki F. Autophagy-related protein 8 (Atg8) family interacting motif in Atg3 mediates the Atg3-Atg8 interaction and is crucial for the cytoplasm-to-vacuole targeting pathway. The Journal of Biological Chemistry 2010;285(38) 29599-29607.

[95] Baba M, Osumi M, Scott SV, Klionsky DJ, Ohsumi Y. Two distinct pathways for targeting proteins from the cytoplasm to the vacuole/lysosome. The Journal of Cell Biology 1997;139(7) 1687-1695.

[96] Scott SV, Guan J, Hutchins MU, Kim J, Klionsky DJ. Cvt19 is a receptor for the cytoplasm-to-vacuole targeting pathway. Molecular Cell 2001;7(6) 1131-1141.

[97] Noda NN, Kumeta H, Nakatogawa H, Satoo K, Adachi W, Ishii J, Fujioka Y, Ohsumi Y, Inagaki F. Structural basis of target recognition by Atg8/LC3 during selective autophagy. Genes to Cells 2008;13(12) 1211-1218.

[98] Kirkin V, Lamark T, Sou YS, Bjørkøy G, Nunn JL, Bruun JA, Shvets E, McEwan DG, Clausen TH, Wild P, Bilusic I, Theurillat JP, Øvervatn A, Ishii T, Elazar Z, Komatsu M, Dikic I, Johansen T. A role for NBR1 in autophagosomal degradation of ubiquitinated substrates. Molecular Cell 2009;33(4) 505-516.

[99] Thurston TL, Ryzhakov G, Bloor S, von Muhlinen N, Randow F. The TBK1 adaptor and autophagy receptor NDP52 restricts the proliferation of ubiquitin-coated bacteria. Nature Immunology 2009;10(11) 1215-1221.

[100] Wild P, Farhan H, McEwan DG, Wagner S, Rogov VV, Brady NR, Richter B, Korac J, Waidmann O, Choudhary C, Dötsch V, Bumann D, Dikic I. Phosphorylation of the autophagy receptor optineurin restricts Salmonella growth. Science 2011;333(6039) 228-233.

[101] Kondo-Okamoto N, Noda NN, Suzuki SW, Nakatogawa H, Takahashi I, Matsunami M, Hashimoto A, Inagaki F, Ohsumi Y, Okamoto K. Autophagy-related protein 32

acts as autophagic degron and directly initiates mitophagy. The Journal of Biological Chemistry 2012;287(13) 10631-10638.

[102] Novak I, Kirkin V, McEwan DG, Zhang J, Wild P, Rozenknop A, Rogov V, Löhr F, Popovic D, Occhipinti A, Reichert AS, Terzic J, Dötsch V, Ney PA, Dikic I. Nix is a selective autophagy receptor for mitochondrial clearance. EMBO Reports 2010;11(1) 45-51.

[103] Weidberg H, Elazar Z. TBK1 mediates crosstalk between the innate immune response and autophagy. Science Signaling 2011;4(187) pe39.

[104] Thurston TL, Wandel MP, von Muhlinen N, Foeglein A, Randow F. Galectin 8 targets damaged vesicles for autophagy to defend cells against bacterial invasion. Nature 2012;482(7385) 414-418.

[105] Nakatogawa H, Ohbayashi S, Sakoh-Nakatogawa M, Kakuta S, Suzuki SW, Kirisako H, Kondo-Kakuta C, Noda NN, Yamamoto H, Ohsumi Y. The autophagy-related protein kinase Atg1 interacts with the ubiquitin-like protein Atg8 via the Atg8 family interacting motif to facilitate autophagosome formation. The Journal of Biological Chemistry 2012;287(34) 28503-28507.

[106] Kraft C, Kijanska M, Kalie E, Siergiejuk E, Lee SS, Semplicio G, Stoffel I, Brezovich A, Verma M, Hansmann I, Ammerer G, Hofmann K, Tooze S, Peter M. Binding of the Atg1/ULK1 kinase to the ubiquitin-like protein Atg8 regulates autophagy. The EMBO Journal 2012;31(18) 3691-3703.

[107] Alemu EA, Lamark T, Torgersen KM, Birgisdottir AB, Bowitz Larsen K, Jain A, Olsvik H, Overvatn A, Kirkin V, Johansen T. ATG8 family proteins act as scaffolds for assembly of the ULK complex: sequence requirements for LC3-interacting region (LIR) motifs. The Journal of Biological Chemistry 2012;287(47) 39275-39290.

[108] Colecchia D, Strambi A, Sanzone S, Iavarone C, Rossi M, Dall'armi C, Piccioni F, Verrotti di Pianella A, Chiariello M. MAPK15/ERK8 stimulates autophagy by interacting with LC3 and GABARAP proteins. Autophagy 2012;8(12) 1724-1740.

[109] Levine B, Sinha S, Kroemer G. Bcl-2 family members: dual regulators of apoptosis and autophagy. Autophagy 2008;4(5) 600-606.

[110] Schwarten M, Mohrlüder J, Ma P, Stoldt M, Thielmann Y, Stangler T, Hersch N, Hoffmann B, Merkel R, Willbold D. Nix directly binds to GABARAP: a possible crosstalk between apoptosis and autophagy. Autophagy 2009;5(5) 690-698.

[111] O'Sullivan GA, Kneussel M, Elazar Z, Betz H. GABARAP is not essential for GABA receptor targeting to the synapse. The European Journal of Neuroscience 2005;22(10) 2644-2648.

[112] Weidberg H, Shvets E, Shpilka T, Shimron F, Shinder V, Elazar Z. LC3 and GATE-16/GABARAP subfamilies are both essential yet act differently in autophagosome biogenesis. The EMBO Journal 2010;29(11) 1792-1802.

[113] Pankiv S, Alemu EA, Brech A, Bruun JA, Lamark T, Overvatn A, Bjørkøy G, Johansen T. FYCO1 is a Rab7 effector that binds to LC3 and PI3P to mediate microtubule plus end-directed vesicle transport. The Journal of Cell Biology 2010;188(2) 253-269.

[114] Mikhaylova O, Stratton Y, Hall D, Kellner E, Ehmer B, Drew AF, Gallo CA, Plas DR, Biesiada J, Meller J, Czyzyk-Krzeska MF. VHL-regulated MiR-204 suppresses tumor growth through inhibition of LC3B-mediated autophagy in renal clear cell carcinoma. Cancer Cell 2012;21(4) 532-546.

[115] Nakatogawa H, Ichimura Y, Ohsumi Y. Atg8, a ubiquitin-like protein required for autophagosome formation, mediates membrane tethering and hemifusion. Cell 2007;130(1) 165-178.

[116] Weidberg H, Shpilka T, Shvets E, Abada A, Shimron F, Elazar Z. LC3 and GATE-16 N termini mediate membrane fusion processes required for autophagosome biogenesis. Developmental Cell 2011;20(4) 444-454.

[117] Nair U, Jotwani A, Geng J, Gammoh N, Richerson D, Yen WL, Griffith J, Nag S, Wang K, Moss T, Baba M, McNew JA, Jiang X, Reggiori F, Melia TJ, Klionsky DJ. SNARE proteins are required for macroautophagy. Cell 2011;146(2) 290-302.

[118] Moreau K, Ravikumar B, Renna M, Puri C, Rubinsztein DC. Autophagosome precursor maturation requires homotypic fusion. Cell 2011;146(2) 303-317.

[119] Cherra SJ 3rd, Kulich SM, Uechi G, Balasubramani M, Mountzouris J, Day BW, Chu CT. Regulation of the autophagy protein LC3 by phosphorylation. The Journal of Cell Biology 2010;190(4) 533-539.

[120] Yi C, Ma M, Ran L, Zheng J, Tong J, Zhu J, Ma C, Sun Y, Zhang S, Feng W, Zhu L, Le Y, Gong X, Yan X, Hong B, Jiang FJ, Xie Z, Miao D, Deng H, Yu L. Function and molecular mechanism of acetylation in autophagy regulation. Science 2012;336(6080) 474-477.

Chapter 3

Rab GTPases in Autophagy

Yuko Hirota, Keiko Fujimoto and Yoshitaka Tanaka

Additional information is available at the end of the chapter

http://dx.doi.org/10.5772/54578

1. Introduction

Rab proteins constitute a subfamily of small GTPases that play important roles in the spatio-temporal regulation of intracellular vesicle transport [1-3]. Rab GTPases represent a large family of small guanosine triphosphate (GTP)-binding proteins that comprise more than 60 known members. In mammalian cells, it is well established that different Rab proteins localize on distinct membrane-bound compartments, where they regulate multiple steps in membrane traffic, including vesicle budding, fusion, and movement, through cycling between an inactive guanosine diphosphate (GDP)-bound form and an active GTP-bound form [3]. Guanine nucleotide exchange factor (GEF) shifts GTPase from its inactive GDP-bound form to its active GTP-bound form, while GTPase activating domain protein (GAP) inactivates GTPase [Figure 1]. Many structural and biological studies have shown that specific amino acid mutation can make possible to keep Rab GTPase in its GDP-bound form or GTP-bound form; therefore, expression of GDP-bound form or GTP-bound form could imitate its function [3].

Autophagy is a degradation pathway that delivers cytoplasmic components and intracellular organelles at random and/or in a selective manner to lysosomes via doubled-membrane organelles called autophagosomes [4]. Although autophagy is induced by exposure of cells to nutrient- or growth factor-deprived medium, it also occurs at basal levels in most tissues and contributes to the routine turnover of cytoplasmic materials. So far, in the process of the formation of autophagosomes, many mammalian homologues of yeast *ATG* (autophagy-related) genes have been identified and extensively characterized, demonstrating that the molecular machinery of autophagy has been conserved from yeasts to mammals [5]. Analyses of the Atg proteins have identified two ubiquitin-like conjugation systems that are required for autophagosome formation [6]. Among these proteins, Atg5 and LC3 (a mammalian homolog of Atg8) have been analyzed in more detail. An Atg12-Atg5 conjugate is necessary

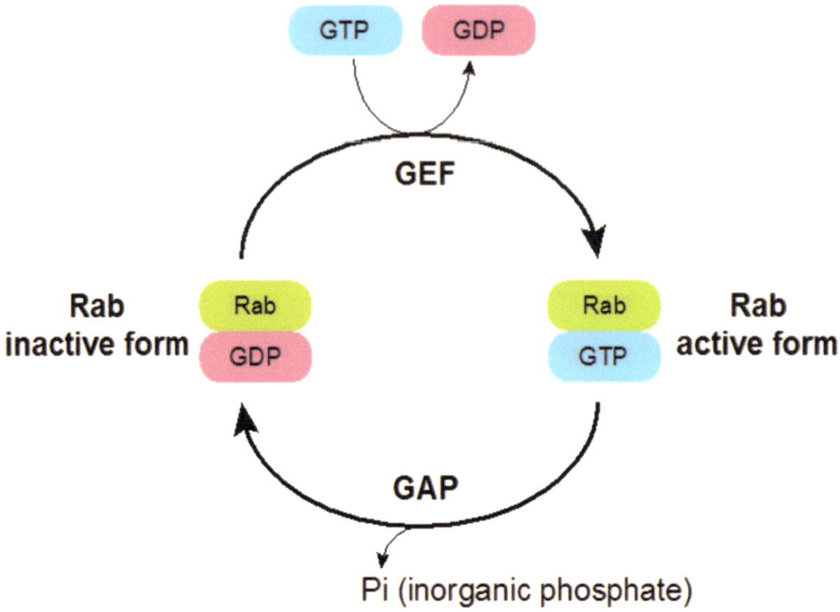

Figure 1. The Rab switch

for elongation of the isolation membrane on its outer side [7]. ProLC3 is processed by Atg4, a cystein protease, to a cytosolic form, LC3-I, exposing a carboxyl terminal glycine [8]. LC3-I is subsequently activated by the E1-like protein Atg7, and conjugates phosphatidylethanolamine to its C-terminal glycine via the E2-like enzyme Atg3, producing a membrane-bound form, LC3-II [9]. LC3-II localizes on the isolation membrane and the autophagosome membrane [10,11]. Because the amount of LC3-II correlates with the number of autophagosomes, detection of LC3-II by Western blotting can be used to measure autophagic activity. In addition, dot-like or ring-like staining of LC3 in immunofluorescence is widely utilized as a specific marker of the formation of autophagosomes.

Akin to the involvement of Rab proteins in vesicle transport processes, there is a growing body of evidence that many Rab proteins, such as Rab7, Rab9, Rab11, Rab24, Rab32, and Rab33, function in the formation and/or maturation of autophagic vacuoles. Each of these Rab proteins localizes to distinct intracellular compartments and thereby appears to be involved in a distinct step of autophagic flux. In this chapter, we focus on the roles of these Rab proteins in the regulation of the autophagic process.

2. Rab7 GTPase and autophagy

2.1. Rab7 GTPase and autophagosome maturation

Rab7 is a member of the Rab family, which is involved in vesicle transport from early endosomes to late endosomes/lysosomes as well as lysosome biogenesis [12, 13]. Rab7 is also implicated in the fusion between autophagosomes and lysosomes, i.e., autophagosome maturation [14-16]. Rab7 wild-type (WT) and active-forms of Rab7 (Rab7Q67L) are associated with ring-shaped vesicles labeled with the autofluorescent compound monodansylcadaverine (MDC), which is preferentially incorporated into mature autophagosomes and autolysosomes, and with LC3, which preferentially labels immature autophagosomes, indicating the association of Rab7 with autophagic vesicles [14]. On the other hand, overexpression of the inactive form of Rab7 (Rab7T22N) causes a marked increase in the size of MDC- and LC3-positive vesicles and the number of LC3-positive vesicles, but reduces the number of MDC-positive vesicles, indicating that the inactive-form of Rab7 impaired the fusion between autophagic vacuoles and lysosomes. Similar results were also obtained in cells depleted of Rab7 by RNAi [15]. Collectively, these results suggest that Rab7 is not essential for the initial step of autophagosome maturation, but is involved in the final step of the maturation of late autophagic vacuoles, possibly in the fusion with lysosomes [14, 15]. Interestingly, Rab7T22N that was diffusely localized in the cytoplasm under nutrient-rich conditions was redistributed to the membrane of MDC-positive vacuoles by amino acid starvation or by rapamycin treatment [14]. Thus, Rab7 is targeted to the autophagosomal membranes by a TOR (target of rapamycin)-kinase signal transduction mechanism in response to starvation.

SNARE proteins and the class C Vps (C-Vps) complex as well as Rab7 have been implicated in mammalian autophagy. In *Saccharomyces cerevisiae*, the fusion of autophagosomes and vacuoles is assumed to proceed in an identical manner to that of endocytic fusion, depending on SNARE proteins, Rab GTPase Ypt7, the yeast ortholog of mammalian Rab7, and its GEF, C-Vps tethering complex, all of which are known as regulators of the endocytic pathway [17-19]. Interestingly, while Rab7 and the C-Vps complex component Vps16 are essential for endocytic fusion with lysosomes, Rab7 but not Vps16 is required for complete autophagy flux in an autophagy induced by thapsigargin, an inhibitor of the sarco/ER Ca^{2+} ATPase [20]; therefore, autophagosomal-lysosomal fusion might be controlled by a molecular mechanism distinct from general endocytic fusion.

During autophagy, autophagosomes fuse with lysosomes to degrade materials within them by lysosomal hydrolases. So far, little is known about the fate of autolysosomes. Recently, it has been shown that mTOR regulates the termination of autophagy and reformation of lysosomes [21]. When cells are exposed to starvation, mTOR is inhibited, leading to the autophagy induction; however, prolonged starvation causes reactivation of mTOR and this reactivation generates proto-lysosomal tubules and vesicles from autolysosomal membranes to reform into functional lysosomes. Interestingly, the dissociation of Rab7 from autolysosomes is required for the reformation of lysosomes, and overexpression of the active form of Rab7 results in the accumulation of enlarged autolysosomes [21]; therefore, mTOR might regulate the reformation of lysosomes from autolysosomes via Rab7.

2.2. Rab7 GTPase and pathogen-containing autophagosome

Many pathogens are sequestered in phagosomes and fated to be degraded, since these phagosomes undergo a process of maturation, fusing with lysosomes [22, 23]; however, some pathogens reside in vacuoles that interact with other organelles, such as mitochondria, ER and

Golgi, while others escape from phagosomes or remain in vacuoles that neither acidify nor fuse with lysosomes [24]. In contrast, *Coxiella burnetii* bacteria, the agent of Q fever in humans and of coxiellosis in other animals, live and replicate in acidified compartments with phago-lysosomal characteristics. Lysosomal membrane proteins and enzymes are found in vacuoles containing *C. burnetii* [25]. In HeLa cells infected with *C. burnetii,* vacuoles containing these parasites and labeled with acidotropic probe LysoTracker were also labeled with MDC and LC3. Moreover, 3-methyladenine and wortmannin, known as reagents to inhibit the early stage of autophagosome formation, blocked the development of *Coxiella*-containing vacuoles [26]. These results suggest that *Coxiella*-containing vacuoles interact with the autophagic degradation pathway. Interestingly, exogenously expressed wild-type Rab7 and the active form of Rab7 colocalize with *Coxiella*-containing vacuoles, whereas the inactive form of Rab7 does not. This indicates that Rab7 associates with the biogenesis of *Coxiella*-containing vacuoles [26]. Also, the initial formation of Group A *streptococcus*-containing autophagosome-like vacuoles is prevented by expression of the inactive form of Rab7 or downregulation of Rab7 expression with RNAi, suggesting that Rab7 is required for the early stage of the formation of Group A *streptococcus*-containing autophagosome-like vacuoles [27].

2.3. Rab7 GTPase and interaction molecules

Rubicon (Run domain protein as Beclin 1 interacting and cysteine-rich containing) is a component of the class III phosphatidylinositol 3-kinase (PI3KC3) complex. PI3KC3 forms two protein subcomplexes that localize to autophagosomes or early endosomes and perform distinct functions. The autophagosomal subcomplex consists of the PI3KC3 core complex (hVps34, p150/Vps15, and Beclin 1) and Atg14L [28]. Atg14L is the targeting factor for this complex to the early stage of autophagosomes. The endosomal complex is composed of the PI3KC3 core complex, UV irradiation resistance-associated gene (UVRAG) and Rubicon. UVRAG activates PI3KC3 and is needed to mature autophagosomes and endosomes through its direct interaction [29]. UVRAG also interacts with C-Vps, and this interaction accelerates autophagosome recruitment and activation of Rab7, which facilitates autophagosome maturation [30]. On the other hand, Rubicon specifically interacts with Rab7 through the common C-terminal domain, called a regulator of G-protein signaling homology (RH) domain but not RUN domain (for RPIP8, UNC-14, and NESCA) to inhibit autophagosome maturation [31, 32]. The overexpressed active form of Rab7 competed with UVRAG for Rubicon binding much more efficiently than the inactive form of Rab7 [31]. Thus, Rubicon is a negative regulator of autophagosome maturation. Interestingly, Rubicon homologue, PLEKHM1, which contains an RH domain, specifically interacted with Rab7, and this interaction is important for their function [32]. In contrast to Rubicon, PLEKHM1 does not directly suppress autophagosome maturation. Rubicon, but not PLEKHM1, also interacted with the Beclin 1-PI3-kinase complex [32]. Rubicon functions to regulate the endocytic and autophagic pathways under the control of the association with Beclin 1-PI3-kinase complex or Rab7.

Phosphatidylinositol-3-phosphate (PI3P) is essential for autophagosome formation. Although the PI3P function in autophagy is unknown, it is already considered that effector proteins containing the FYVE (Fab1/YOTB/Vac1/EEA1) domain or PX (phox) domain are recruited to and activated on PI3P-enriched membranes. Recently, FYCO1 was identified as a novel protein interacting with LC3, Rab7, and PI3P [33]. FYCO1 interacts with Rab7 and PI3P via part of the coiled-coil domain and FYVE domain, respectively. Overexpression of FYCO1 redistributes LC3, Rab7, and ORP1L, a Rab7 effector protein, to the cell periphery in a microtubule-dependent manner [33]. In contrast, FYCO1 depletion leads to the accumulation of perinuclear clustering autophagosomes, indicating that FYCO1 binds to PI3P via its FYVE domain and functions as a Rab7 and LC3 effector molecule with microtubules plus end-directed transport.

3. Rab9 GTPase and autophagy

Rab9 GTPase resides in late endosomes, in which Rab7 localizes in a distinct microdomain, and plays a role in vesicle transport from late endosomes to the TGN [34]. Rab9 depletion using siRNA decreased the size of late endosomes and reduced the number of late endosomes/lysosomes, which were clustered in the perinuclear region [35], implying that Rab9 is associated with the maintenance of late endosomes/lysosomes.

Generally, it has been believed that Atg5 and Atg7 are essential for mammalian autophagy [36, 37]. In contrast, mouse embryonic fibroblasts deficient in Atg5 or Atg7 can still form autophagosomes and autophagic flux can function when exposed to autophagy-inducible stress conditions, and the lipidation of LC3 (autophagosome membrane-bound form) is also dispensable for this Atg5-/Atg7-independent autophagy [38]. Interestingly, in this alternative process of autophagy, but not in Atg5/Atg7-dependent conventional autophagy, the formation of autophagosomes seemed to be regulated in a Rab9-dependent manner by the fusion of isolation membranes with the TGN- and late endosome-derived vesicles [38]. In fact, the localization of Rab9 to autolysosomes was slightly increased with the active form of Rab9 (Rab9Q66L), but decreased with the dominant-negative form of Rab9 (Rab9S21N). Additionally, Rab9 silencing by siRNA decreased the number of autophagosomes but induced the accumulation of isolation membranes [38]. Thus, Rab9 plays a significant role in Atg5-/Atg7-independent autophagy.

4. Rab11 GTPase and autophagy

Rab11 has been shown to associate with perinuclear recycling endosomes and regulate transferrin recycling in CHO or BHK cells [39]; however, in K562, an erythroleukemic cell line, Rab11 localizes at MVBs, which are equivalent to late endosomes and are released into the extracellular space as so-called exosomes [40]. Overexpression of wild-type Rab11 and its active-form mutant produced large MVBs. Induction of autophagy by starvation or mTOR inhibitor rapamycin significantly increased the fusion between MVBs and autophagosomes [41]. This fusion was disturbed by the Ca^{2+} chelator BAPTA-AM and by the expression of the inactive form of Rab11 [41], indicating that the fusion of MVB with autophagosomes is a calcium- and Rab11 activity-dependent event.

Rab GTPase activity is controlled by GEF and GAP. Thirty-eight putative RabGAPs with a Tre-2/Bub2/Cdc16 (TBC) domain have been identified [43]. Recently, it was thought that RabGAP might be associated not only with the cellular endomembrane system but also with autophagy. In fact, TBC1D5 is identified as an interacting partner of LC3 and retromer complex and regulates the autophagy pathway and retrograde transport of cation-independent mannose 6-phosphate receptor from endosomes to the TGN [44]. Another RabGAP, TBC1D14, can bind a mammalian homologue of Atg1p ULK1, as can Rab11, and disrupts recycling endosome traffic [45]. Furthermore, under starvation conditions, TBC1D14 and Rab11 modulate the membrane transport from recycling endosomes to generate autophagosomes. TBC1D14 overexpression caused the tubulation of ULK1- and Rab11-positive recycling endosomes irrespective of nutrition conditions, impairing their function and preventing autophagosome formation [45]. However, the tubulation of recycling endosomes induced by the expression of TBC1D14 was dependent on Rab11 expression, since Rab11 depletion using siRNA gave rise to a loss of tubules and a diffuse distribution of TBC1D14 throughout the cytosol [45]. Amino acid-deprived starvation caused TBC1D14 relocation from recycling endosomes to Golgi, while the ULK1- and LC3-positive recycling endosome membrane was incorporated into the forming autophagosomes [45]. Thus, TBC1D14- and Rab11-dependent membrane transport from recycling endosomes participates in and controls starvation-induced autophagy.

5. Rab24 GTPase and autophagy

Rab24 is localized to perinuclear reticular structures that partially colocalize with marker proteins for ER, cis-Golgi, and ER-Golgi intermediate compartments [46]. Under starvation conditions, Rab24 relocated to large vesicles, where LC3 was localized. The appearance of these vesicles was enhanced in the presence of vinblastin, an agent that disrupts microtubules and prevents fusion of autophagosomes with lysosomes [46]. Interestingly, since no such distribution change was observed in cells expressing the mutant Rab24S67L that introduced the mutation into the GTP-binding motif, normally functioning Rab24 protein appears to be required for the formation of autophagosomes in response to starvation. *Coxiella burnetii* survives and replicates in MDC- and LC3-positive phagolysosomal compartments and Rab7 participates in the formation of *Coxiella*-containing vacuoles [26]. Overexpression of Rab24 or LC3 also accelerated the occurrence of *Coxiella*-containing vacuoles early after infection [47]. The expression of the Rab24 mutant (Rab24S67L), which does not localize to autophagosomes, significantly reduced the number and size of the phagolysosomal structures, although the inhibitory effect was not enduring but mutant expression delayed the generation of phagolysosomes [47]. Taken together, these results suggest that overexpression of proteins involved in the autophagic pathway, such as Rab24, increases the development of phagolysosomes for *Coxiella* replication.

Rab24 is also supposed to contribute to the degradation of aggregated proteins in rat cardiac myocytes. Glucose deprivation induced the formation of aggregates and aggresomes of polyubiquitinated proteins, and then they colocalized with exogenously expressing green fluorescent protein (GFP) tagged-LC3 and endogenous Rab24 [48]. Autophagy induced by glucose deprivation seemed to depend on the reactive oxygen species, because the treatment with N-acetylcysteine prevented aggresome formation and autophagy [48]. These results might imply that glucose deprivation induces oxidative stress, which is involved in aggresome formation and autophagy via Rab24 in cardiac myocytes.

6. Rab32 GTPase and autophagy

Mouse Rab32 and Rab38 operate in a functionally redundant manner in regulating skin melanocyte pigmentation and regulate post-Golgi trafficking of tyrosinase and tyrosinase-related protein 1, thereby suggesting their critical roles in melanosome maturation [49]. In *Xenopus* melanophores, Rab32 is involved in the regulation of melanosome transport by cAMP-dependent protein kinase A [50]. Although Rab32 is expressed in most human tissues [51], little is known about the physiological roles of Rab32 in tissues and cells other than melanocytes.

6.1. Rab32 GTPase and constitutive autophagy

Rab32 is supposed to localize to ER [52] and ER and mitochondria [53]. We showed previously that Rab32 participates in constitutive autophagy in HeLa cells derived from cervical cancer [52]. The expressed wild-type or GTP-bound active form of human Rab32 was primarily localized to the ER. Interestingly, overexpression of the wild-type or active form of Rab32 induced the formation of autophagic vacuoles containing LC3, the ER-resident protein calnexin and late endosomal/lysosomal membrane protein LAMP-2 even under nutrient-rich conditions. Moreover, the localization of Rab32 to ER was necessary for the formation of autophagosomes [52], because the expression of a mutant Rab32 deleted two cysteine residues that are essential for association with the membrane, impaired autophagy vacuole formation [50]. There is a long-standing debate concerning from where the autophagosomal membrane is derived. So far, two possibilities have been proposed: it arises from pre-existing organelles, such as the ER or Golgi, or from de novo formation [54]. Our findings mentioned above postulate, therefore, that Rab32 facilitates the formation of autophagic vacuoles whose membranes are derived from the ER. In addition, expression of the inactive form of Rab32 or depletion of Rab32 expression by siRNA caused the formation of p62 and ubiquitinated protein-accumulating aggresomes and prevented constitutive autophagy [52]. Thus, these results imply the physiological importance of Rab32 in the cellular clearance of aggregated proteins by basal constitutive autophagy.

As well as Rab7, Rab32 also seems to participate in phagosome maturation in pathogen-induced autophagic degradation by infection with *Salmonella enterica serovar Typhimurium* [55] or *Mycobacterium tuberculosis* [56], especially in the recruitment of lysosomal enzyme cathepsin D to phagosomes containing *M. tuberculosis* [56]. Rab32 and some other Rabs localized to *M. tuberculosis*-containing phagosomes transiently, and the expression of the inactive form of Rab32 showed the impairment of its recruitment to phagosomes [56], but had no effect on phagosomal fusion with lysosomes [55]. Therefore, these results imply that Rab32 regulates the recruitment of cathepsin D to the phagosomes.

6.2. Rab32 GTPase and interaction molecules

Recently, it has been reported that, in *Drosophila*, Rab32 colocalized with LysoTracker labeling lysosomes and GFP-Atg8 (LC3 homologue), indicating that Rab32 is localized at lysosomes and/or autophagosomes during programmed autophagy for metamorphosis to differentiate the fat body, salivary gland, and midgut [57]. Previously, Ma et al. reported that the Claret encoded by *claret*, a member of the granule group eye color genes [58], coprecipitated not only with Rab-RP1, a Rab GTPase encoded by *Drosophila lightoid*, but also with its human homologues, Rab32 and Rab38 [59]. Furthermore, the autophagosome formation was impaired in Rab32/*lightoid* mutants and Rab32 GEF/*claret* mutants, suggesting that Rab32 activity is required for the autophagic process of the fat body [57]. Previously, it has been suggested that autophagy impairment reduces lipid accumulation and impairs adipocytes differentiation in mice [60, 61]. In fact, downregulation of autophagy in *Drosophila* led to a decrease in the size of lipid droplets in *atg*-related genes in knocked down *Drosophila* fat body cells [57]; therefore, Rab32 appears to regulate lipid storage by controlling autophagy. Another report showed that Rab32 is upregulated in the epidermis and midgut during metamorphosis in *Helicoverpa armigera* [62], suggesting that Rab32 may participate in organeogenesis in insects.

In addition to GEF, a GAP for Rab32, RUTBC1, was identified [63]. RUTBC1 is a TBC domain-containing protein that binds to Rab9A in a nucleotide binding-state-dependent manner both in vitro and in vivo but has no GAP activity for Rab9A [63]; however, RUTBC1 acts as a GAP for Rab32 and Rab33B, and its TBC domain stimulates GTP hydrolysis [63]. Therefore, RUTBC1 may function in the autophagy process, as both Rab32 and Rab33B are suggested to be regulatory factors of autophagy.

6.3. Rab32 GTPase and disease

Recently, *RAB32* and *IL23R (interleukin receptor 23)* were identified as susceptibility genes for leprosy in a genome-wide association study [64], although *IL23R* was previously reported to be a gene involved in Crohn's disease [65]. Leprosy, also known as Hansen's disease, is a chronic granulomatous infectious disease caused by *Mycobacterium leprae*, which affects both peripheral nerves and mucosa of the upper respiratory tract. As Rab32 participates in regulating the recruitment of cathepsin D to phagosomes containing *M. tuberculosis* [56], referred to above, Rab32 may have function in the pathogenesis of leprosy, such as host defense against *M. leprae* infection.

7. Rab33 GTPase and autophagy

Rab33 has two isoforms, Rab33A and Rab33B. Rab33B is expressed ubiquitously, although Rab33A is expressed exclusively in the brain and cells of the immune system [66]. Rab33B is localized at the Golgi apparatus [67], although Rab33A also targets dense-core vesicles in neuroendocrine cells, not only in the Golgi [68]. It has been shown that both Rab33A and Rab33B interact with Atg16L, which associates with Atg5-12 complex on an isolation membrane for the duration of autophagosome formation and is essential for canonical autophagy [69]. These interactions occur in a GTP-dependent manner [70]. Expression of the active form of Rab33B induced the lipidation of LC3 even under nutrient-rich conditions, but this was not caused by the blockage of autophagic degradation [70]. Furthermore, this form of Rab33B inhibited constitutive autophagy, but not starvation-induced autophagy, as judged by the degradation of p62/SQSTM1, a substrate of autophagy [70]. The precise reason for this dysfunction is unknown, but it is possible to speculate that the active form of Rab33B recruits Atg5-12/16L complex to incorrect membranes because no obvious large LC3 dots were observed in cells expressing this mutant. Rab33B has another interaction partner, OATL1, a RabGAP, which can bind LC3 [71]. Rab33B seems to be a target substrate of OATL1 and to be involved in the fusion between autophagosomes and lysosomes. Overexpression of the wild-type or active form of Rab33B inhibits the fusion between autophagosomes and lysosomes, because OATL1 inactivates Rab33B as a GAP protein, and this inhibition leads to the blockade of autophagic flux. In fact, the expression of the active form of Rab33B increased the amount of LC3-II as previously reported [70], but did not show the colocalization of lysosomal membrane protein LAMP-1 with LC3-positive dots [71], suggesting that autophagosomes in the active form of Rab33B-expressing cells cannot efficiently fuse with lysosomes. Very recently, it was revealed that Atg16L interacts with Rab33A and that this interaction is required for the dense-core vesicle localization of Atg16L in neuroendocrine PC12 cells [72]. Knockdown of endogenous Atg16L in PC12 cells caused marked inhibition of hormone secretion independently of autophagic activity [72], indicating that Atg16L controls autophagy in all cell types as well as secretion from dense-core vesicles, presumably by acting as a Rab33A effector, in neuroendocrine cells.

8. Conclusion

Based on the results of a large body of research, regulation of autophagy by Rab proteins could be loosely classified into two types, (i) autophagosome formation or (ii) autophagosome maturation (i.e. the fusion of autophagosomes with lysosomes) [Figure 2]. We assumed that Rab7, Rab9, and Rab11 could be sorted into type (ii) profile, and Rab24, Rab32 and Rab33 into type (i). Autophagy indicates not only the sequestration of materials into autophagosomes but also the degradation by lysosomal enzymes. Therefore, Rab proteins associated with the fusion step of autophagosomes with lysosomes have a fatal role in autophagic flux, and Rabs also trigger autophagosome formation. Type (ii)-categorized Rab proteins are mostly localized at late endosomes and play a role in membrane traffic to lysosomes, recycling endosomes or TGN. It is conceivable that they are involved in autophagosome maturation, because late endosomes, TGN, recycling endosomes, or Golgi-derived vesicles fuse with lysosomes routinely [73]. So far, no clear result has shown that Rab24 and Rab33 directly associate with the formation of autophagosomes except for Rab32, which is associated with ER and/or mitochondria and is regarded as a key factor for the supply to autophagosome formation [52]. However, it could be plausible that Rab33B is involved in autophagosome formation, since no obvious large LC3 dots were observed in the active form of Rab33B-expressing cells, but LC3 lipidation was induced [70]. Although the precise mechanism of autophagy by Rab24 is unknown, Rab24 might participate in the early stage of autophagy since overexpression of Rab24 stimulates the sequestration of *Coxiella* into vacuoles early after infection [47]. Thus, these Rab proteins, Rab24, Rab33, and Rab32, could be involved in autophagosome formation.

Although significant progress has been made in understanding the mechanisms of the formation and maturation of autophagic vacuoles through the function of Rab proteins, the precise molecular mechanism of how they regulate or interact with the core autophagic complex including Atg proteins is still unclear. Moreover, temporal ordering, such as a signal pathway to initiate the restricted molecules for autophagy, and spatial regulation including the source of isolation membranes are important in the overall understanding of autophagy involving Rab GTPases.

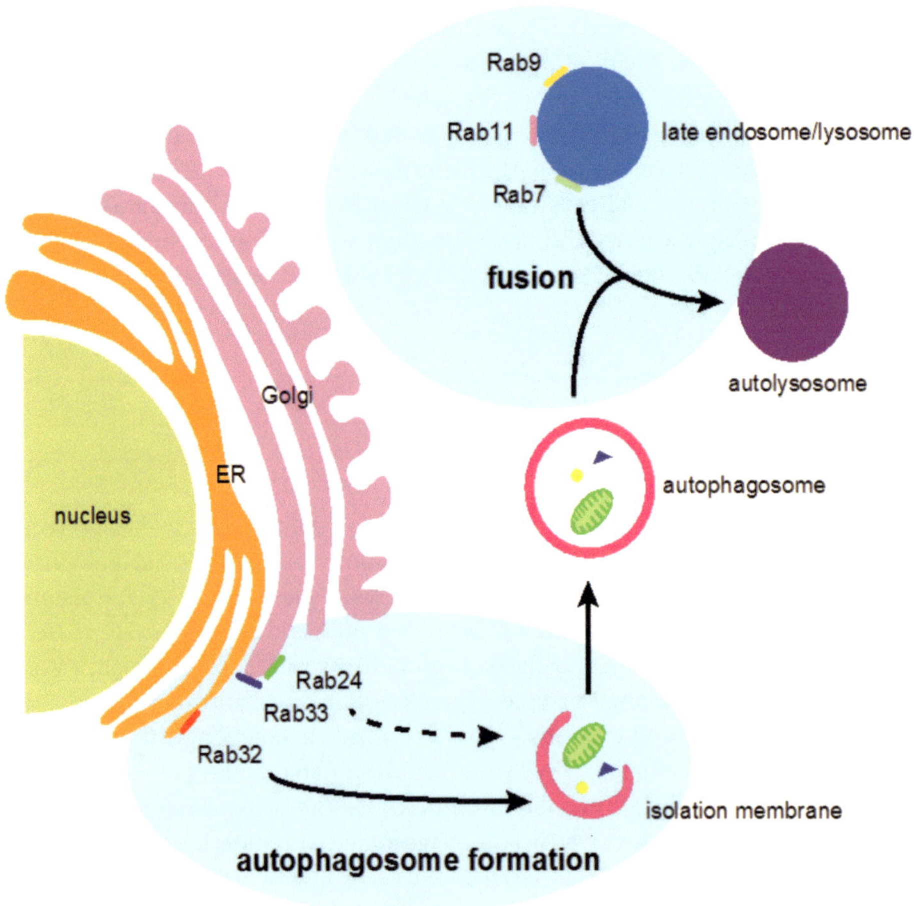

Figure 2. Localization and function of Rab GTPases for autophagy

http://dx.doi.org/10.5772/54578

Acknowledgements

This work was supported in part by a grant-in-aid from the Ministry of Education, Culture, Sports, Sciences and Technology of Japan.

Author details

Yuko Hirota[1], Keiko Fujimoto[2] and Yoshitaka Tanaka[2*]

*Address all correspondence to: ytanaka@phar.kyushu-u.ac.jp

1 Department of Clinical Chemistry and Laboratory Medicine, Graduate School of Medical Sciences, Kyushu University, Maidashi, Fukuoka, Japan

2 Division of Pharmaceutical Cell Biology, Graduate School of Pharmaceutical Sciences, Kyushu University, Maidashi, Fukuoka, Japan

References

[1] Klionsky DJ Emr SD. Autophagy as a regulated pathway of cellular degradation. Science 2000; 290(5497), 1717-1721.

[2] Mizushima N, Ohsumi Y, Yoshimori T. Autophagosome formation in mammalian cells. Cell Structure and Function 2002; 27(6), 421-429.

[3] Stenmark H. Rab GTPases as coordinators of vesicle traffic. Nature Reviews Molecular Cell Biology 2009; 10(8), 513-525.

[4] Weidberg H, Shvets E, Elazar Z. Biogenesis and cargo selectivity of autophagosomes. Annual Review of Biochemistry 2011; 80, 125-156.

[5] Nakatogawa H, Suzuki K, Kamada Y, Ohsumi Y. Dynamics and diversity in autophagy mechanisms: lessons from yeast. Nature Reviews Molecular Cell Biology 2009; 10(7), 458-467.

[6] Shpilka T, Mizushima N, Elazar Z. Ubiquitin-like proteins and autophagy at a glance. Journal of Cell Science 2012; 125. 2343-2348.

[7] Mizushima N, Yamamoto A, Hatano M, Kobayashi Y, Kabeya Y, Suzuki K, Tokuhisa T, Ohsumi Y, Yoshimori T. Dissection of autophagosome formation using Apg5-deficient mouse embryonic stem cells. Journal of Cell Biology 2001; 152(4), 657-668.

[8] Tanida I, Sou Y, Ezaki J, Minematsu-Ikeguchi N, Ueno T, Kominami E. HsAtg4B/HsApg4B/Autophagin-1 cleaves the carboxyl termini of three human Atg8 homo-

logues and delipidates microtubule-associated protein light chain 3- and $GABA_A$ receptor-associated protein-phospholipid conjugates. The Journal of Biological Chemistry 2004; 279(35), 36268-36276.

[9] Tanida I, Tanida-Miyake E, Komatsu M, Ueno T, Kominami E. Human Apg3p/Aut1p homologue Is an authentic E2 enzyme for multiple substrates, GATE-16, GABARAP, and MAP-LC3, and facilitates the conjugation of hApg12p to hApg5p. The Journal of Biological Chemistry 2002; 277(16), 13739-13744.

[10] Kabeya Y, Mizushima N, Ueno T, Yamamoto A, Kirisako T, Noda T, Kominami E, Ohsumi Y, Yoshimori T. LC3, a mammalian homologue of yeast Apg8p, is localized in autophagosome membranes after processing. The EMBO Journal 2000; 19(21), 5720-5728.

[11] Kabeya Y, Mizushima N, Yamamoto A, Oshitani-Okamoto S, Ohsumi Y, Yoshimori T. LC3, GABARAP and GATE16 localize to autophagosomal membrane depending on form-II formation. Journal of Cell Science 2004; 117, 2805-2812.

[12] Méresse S, Gorvel JP, Chavrier P. The rab7 GTPase resides on a vesicular compartment connected to lysosomes. Journal of Cell Science 1995; 108, 3349-3358.

[13] Feng Y, Press B, Wandinger-Ness A. Rab7 an important regulator of late endocytic membrane traffic. Journal of Cell Biology 1995; 131(6), 1435-1452.

[14] Gutierrez MG, Munafó DB, Berón W, Colombo MI. Rab7 is required for the normal progression of the autophagic pathway in mammalian cells. Journal of Cell Science 2004; 117, 2687-2697.

[15] Jager S, Bucci C, Tanida I, Ueno T, Kominami E, Saftig P, Eskelinen E-L. Role for Rab7 in maturation of late autophagic vacuoles. Journal of Cell Science 2004; 117, 4837-4848.

[16] Liang C, Lee JS, Inn KS, Gack MU, Li Q, Roberts EA, Vergne I, Deretic V, Feng P, Akazawa C, Jung JU. Beclin1-binding UVRAG targets the class C Vps complex to coordinate autophagosome maturation and endocytic trafficking. Nature Cell Biology 2008; 10(7), 776-787.

[17] Kirisako T, Baba M, Ishihara N, Miyazawa K, Ohsumi M, Yoshimori T, Noda T, Ohsumi Y. Formation process of autophagosome is traced with Apg8/Aut7p in yeast. Journal of Cell Biology 1999; 147(2), 435-446.

[18] Sato TK, Rehling P, Peterson MR, Emr SD. Class C Vps protein complex regulates vacuolar SNARE pairing and is required for vesicle docking/fusion. Molecular Cell 2000; 6(3), 661-671.

[19] Wurmser AE, Sato TK, Emr SD. New component of the vacuolar class C-Vps complex couples nucleotide exchange on the Ypt7 GTPase to SNARE-dependent docking and fusion. Journal of Cell Biology 2000; 151(3), 551-562.

[20] Ganley IG, Wong PM, Gammoh N, Jiang X. Distinct autophagosomal-lysosomal fusion mechanism revealed by thapsigargin-induced autophagy arrest. Molecular Cell 2011; 42(6), 731-743.

[21] Yu L, McPhee CK, Zheng L, Mardones GA, Rong Y, Peng J, Mi N, Zhao Y, Liu Z, Wan F, Hailey DW, Oorschot V, Klumperman J, Baehrecke EH, Lenardo MJ. Termination of autophagy and reformation of lysosomes regulated by mTOR. Nature 2010; 465(7300), 942-946.

[22] Alix E, Mukherjee S, Roy CR. Subversion of membrane transport pathways by vacuolar pathogens. Journal of Cell Biology 2011; 195(6), 943-952.

[23] Münz C. Enhancing immunity through autophagy. Annual Review of Immunology 2009; 27, 423-449.

[24] Birmingham CL, Canadien V, Kaniuk NA, Steinberg BE, Higgins DE, Brumell JH. Listeriolysin O allows Listeria monocytogenes replication in macrophage vacuoles. Nature 2008; 451(7176), 350-354.

[25] Heinzen RA, Hackstadt T, Samuel JE. Developmental Biology of Coxiella burnetii. Trends in Microbiology 1999; 7(4), 149-154.

[26] Berón W, Gutierrez MG, Rabinovitch M, Colombo MI. Coxiella burnetii localizes in a Rab7-labeled compartment with autophagic characteristics. Infection and Immunity 2002; 70(10), 5816-5821.

[27] Yamaguchi H, Nakagawa I, Yamamoto A, Amano A, Noda T, Yoshimori T. An initial step of GAS-containing autophagosome-like vacuoles formation requires Rab7. PLoS Pathogen 2009; 5(11), e1000670.

[28] Zhong Y, Wang QJ, Li X, Yan Y, Backer JM, Chait BT, Heintz N, Yue Z. Distinct regulation of autophagic activity by Atg14L and Rubicon associated with Beclin 1-phosphatidylinositol-3-kinase complex. Nature Cell Biology 2009; 11(4), 468-476.

[29] Matsunaga K, Saitoh T, Tabata K, Omori H, Satoh T, Kurotori N, Maejima I, Shirahama-Noda K, Ichimura T, Isobe T, Akira S, Noda T, Yoshimori T. Two Beclin 1-binding proteins, Atg14L and Rubicon, reciprocally regulate autophagy at different stages. Nature Cell Biology 2009; 11(4), 385-396.

[30] Liang C, Lee JS, Inn KS, Gack MU, Li Q, Roberts EA, Vergne I, Deretic V, Feng P, Akazawa C, Jung JU. Beclin 1-binding UVRAG targets the class C Vps complex to coordinate autophagosome maturation and endocytic trafficking. Nature Cell Biology 2008; 10(7), 776-787.

[31] Sun Q, Westphal W, Wong KN, Tan I, Zhong Q. Rubicon controls endosome maturation as a Rab7 effector. Proceedings of the National Academy of Sciences of the United States of America 2010; 107(45), 19338-19343.

[32] Tabata K, Matsunaga K, Sakane A, Sasaki T, Noda T, Yoshimori T. Rubicon and PLEKHM1 negatively regulate the endocytic/autophagic pathway via a novel Rab7-binding domain. Molecular Biology of the Cell 2010; 21(23), 4167-4172.

[33] Pankiv S, Alemu EA, Brech A, Bruun JA, Lamark T, Overvatn A, Bjørkøy G, Johansen T. FYCO1 is a Rab7 effector that binds to LC3 and PI3P to mediate microtubule plus end-directed vesicle transport. Journal of Cell Biology 2010; 188(2), 253-269.

[34] Barbero P, Bittova L, Pfeffer SR. Visualization of Rab9-mediated vesicle transport from endosomes to the trans-Golgi in living cells. Journal of Cell Biology 2002; 156(3), 511-518.

[35] Ganley IG, Carroll K, Bittova L, Pfeffer S. Rab9 GTPase regulates late endosome size and requires effector interaction for its stability. Molecular Biology of the Cell 2004; 15(12), 5420-5430.

[36] Kuma A, Hatano M, Matsui M, Yamamoto A, Nakaya H, Yoshimori T, Ohsumi Y, Tokuhisa T, Mizushima N. The role of autophagy during the early neonatal starvation period. Nature 2004; 432(7020), 1032-1036.

[37] Komatsu M, Waguri S, Ueno T, Iwata J, Murata S, Tanida I, Ezaki J, Mizushima N, Ohsumi Y, Uchiyama Y, Kominami E, Tanaka K, Chiba T. Impairment of starvation-induced and constitutive autophagy in Atg7-deficient mice. Journal of Cell Biology 2005; 169(3), 425-434.

[38] Nishida Y, Arakawa S, Fujitani K, Yamaguchi H, Mizuta T, Kanaseki T, Komatsu M, Otsu K, Tsujimoto Y, Shimizu S. Discovery of Atg5/Atg7-independent alternative macroautophagy. Nature 2009; 461(7264), 654-658.

[39] Ullrich O, Reinsch S, Urbé S, Zerial M, Parton RG. Rab11 regulates recycling through the pericentriolar recycling endosome. Journal of Cell Biology 1996; 135(4), 913-924.

[40] Savina A, Vidal M, Colombo MI. The exosome pathway in K562 cells is regulated by Rab11. Journal of Cell Science 2002; 115, 2505-2515.

[41] Fader CM, Sánchez D, Furlán M, Colombo MI. Induction of Autophagy promotes fusion of multivesicular bodies with autophagic vacuoles in K562 cells. Traffic 2008; 9(2), 230-250.

[42] Berg TO, Fengsrud M, Strømhaug PE, Berg T, Seglen PO. Isolation and characterization of rat liver amphisomes. Evidence for fusion of autophagosomes with both early and late endosomes. The Journal of Biological Chemistry 1998; 273(34), 21883-21892.

[43] Fuchs E, Haas AK, Spooner RA, Yoshimura S, Lord JM, Barr FA. Specific Rab GTPase-activating proteins define the Shiga toxin and epidermal growth factor uptake pathways. Journal of Cell Biology 2007; 177(6), 1133-1143.

[44] Popovic D, Akutsu M, Novak I, Harper JW, Behrends C, Dikic I. Rab GTPase-activating proteins in autophagy: regulation of endocytic and autophagy pathways by di-

rect binding to human ATG8 modifiers. Molecular and Cellular Biology 2012; 32(9), 1733-1744.

[45] Longatti A, Lamb CA, Razi M, Yoshimura S, Barr FA, Tooze SA. TBC1D14 regulates autophagosome formation via Rab11- and ULK1-positive recycling endosomes. Journal of Cell Biology 2012; 197(5), 659-675.

[46] Munafó DB, Colombo MI. Induction of autophagy causes dramatic changes in the subcellular distribution of GFP-Rab24. Traffic 2002; 3(7), 472-482.

[47] Gutierrez MG, Vázquez CL, Munafó DB, Zoppino FC, Berón W, Rabinovitch M, Colombo MI. Autophagy induction favours the generation and maturation of the Coxiella-replicative vacuoles. Cellular Microbiology 2005; 7(7), 981-993.

[48] Marambio P, Toro B, Sanhueza C, Troncoso R, Parra V, Verdejo H, García L, Quiroga C, Munafo D, Díaz-Elizondo J, Bravo R, González MJ, Diaz-Araya G, Pedrozo Z, Chiong M, Colombo MI, Lavandero S. Glucose deprivation causes oxidative stress and stimulates aggresome formation and autophagy in cultured cardiac myocytes. Biochimica et Biophysica Acta 2010; 1802(6), 509-518.

[49] Wasmeier C, Romano M, Plowright L, Bennet DC, Raposo G, Seabra MC. Rab38 and Rab32 control post-Golgi trafficking of melanogenic enzymes. Journal of Cell Biology 2006; 175(2), 271-281.

[50] Park M, Serpinskaya AS, Papalopulu N and Gelfand VI. (2007) Rab32 regulates melanosome transport in Xenopus melanophores by protein kinase A recruitment. Current Biology 2007; 17(23), 1-5.

[51] Bao, X., Faris, A.E., Jang, E.K. and Haslam, R.J. Molecular cloning, bacterial expression and properties of Rab31 and Rab32. European Journal of Biochemistry 2002; 269(1), 259-271.

[52] Hirota Y, Tanaka Y. A small GTPase, human Rab32, is required for the formation of autophagic vacuoles under basal conditions. Cellular and Molecular Life Sciences 2009; 66(17), 2913-2932.

[53] Bui M, Gilady SY, Fitzsimmons RE, Benson MD, Lynes EM, Gesson K, Alto NM, Strack S, Scott JD, Simmen T. Rab32 modulates apoptosis onset and mitochondria-associated membrane (MAM) properties. The Journal of Biological Chemistry 2010; 285(41), 31590-31602.

[54] Juhasz G, Neufeld TP. Autophagy: a forty-year search for a missing membrane source. PLoS Biology 2006; 4(2), e36.

[55] Smith AC, Heo WD, Braun V, Jiang X, Macrae C, Casanova JE, Scidmore MA, Grinstein S, Meyer T, Brumell JH. A network of Rab GTPases controls phagosome maturation and is modulated by Salmonella enterica serovar Typhimurium. Journal of Cell Biology 2007; 176(3), 263-268.

[56] Seto S, Tsujimura K, Koide Y. Rab GTPases regulating phagosome maturation are differentially recruited to mycobacterial phagosomes. Traffic 2011; 12(4), 407-420.

[57] Wang C, Liu Z, Huang X. Rab32 is important for autophagy and lipid storage in Drosophila. PLoS One 2012; 7(2), e32086.

[58] Yamamoto AH, Komma DJ, Shaffer CD, Pirrotta V, Endow SA. The claret locus in Drosophila encodes products required for eyecolor and for meiotic chromosome segregation. The EMBO Journal 1989; 8(12), 3543-3552.

[59] Ma J, Plesken H, Treisman JE, Edelman-Novemsky I, Ren M. Lightoid and Claret: a rab GTPase and its putative guanine nucleotide exchange factor in biogenesis of Drosophila eye pigment granules. Proceedings of the National Academy of Sciences of the United States of America 2004; 101(32), 11652-11657.

[60] Baerga R, Zhang Y, Chen PH, Goldman S, Jin S. Targeted deletion of autophagy-related 5 (atg5) impairs adipogenesis in a cellular model and in mice. Autophagy 2009; 5(8), 1118-1130.

[61] Zhang Y, Goldman S, Baerga R, Zhao Y, Komatsu M, Jin S. Adipose-specific deletion of autophagy-related gene 7 (atg7) in mice reveals a role in adipogenesis. Proceedings of the National Academy of Sciences of the United States of America 2009; 106(47), 19860-19865.

[62] Hou L, Wang JX, Zhao XF. Rab32 and the remodeling of the imaginal midgut in Helicoverpa armigera. Amino Acids 2011; 40(3), 953-961.

[63] Nottingham RM, Ganley IG, Barr FA, Lambright DG, Pfeffer SR. RUTBC1 protein, a Rab9A effector that activates GTP hydrolysis by Rab32 and Rab33B proteins. The Journal of Biological Chemistry 2011; 286(38), 33213-33222.

[64] Zhang F, Liu H, Chen S, Low H, Sun L, Cui Y, Chu T, Li Y, Fu X, Yu Y, Yu G, Shi B, Tian H, Liu D, Yu X, Li J, Lu N, Bao F, Yuan C, Liu J, Liu H, Zhang L, Sun Y, Chen M, Yang Q, Yang H, Yang R, Zhang L, Wang Q, Liu H, Zuo F, Zhang H, Khor CC, Hibberd ML, Yang S, Liu J, Zhang X. Identification of two new loci at IL23R and RAB32 that influence susceptibility to leprosy. Nature Geneticis 2011; 43(12), 1247-1251.

[65] Duerr RH, Taylor KD, Brant SR, Rioux JD, Silverberg MS, Daly MJ, Steinhart AH, Abraham C, Regueiro M, Griffiths A, Dassopoulos T, Bitton A, Yang H, Targan S, Datta LW, Kistner EO, Schumm LP, Lee AT, Gregersen PK, Barmada MM, Rotter JI, Nicolae DL, Cho JH. A genome-wide association study identifies IL23R as an inflammatory bowel disease gene. Science 2006; 314(5804), 1461-1463.

[66] Zheng JY, Koda T, Arimura Y, Kishi M, Kakinuma M. Structure and expression of the mouse S10 gene. Biochimica et Biophysica Acta 1997; 1351(1-2), 47-50.

[67] Zheng JY, Koda T, Fujiwara T, Kishi M, Ikehara Y, Kakinuma M. A novel Rab GTPase, Rab33B, is ubiquitously expressed and localized to the medial Golgi cisternae. Journal of Cell Science 1998; 111, 1061-1069.

[68] Tsuboi T, Fukuda M. Rab3A and Rab27A cooperatively regulate the docking step of dense-core vesicle exocytosis in PC12 cells. Journal of Cell Science 2006; 119, 2196-2203.

[69] Mizushima N, Kuma A, Kobayashi Y, Yamamoto A, Matsubae M, Takao T, Natsume T, Ohsumi Y, Yoshimori T. Mouse Apg16L, a novel WD-repeat protein, targets to the autophagic isolation membrane with the Apg12-Apg5 conjugate. Journal of Cell Science 2003; 116, 1679-1688.

[70] Itoh T, Fujita N, Kanno E, Yamamoto A, Yoshimori T, Fukuda M. Golgi-resident small GTPase Rab33B interacts with Atg16L and modulates autophagosome formation. Molecular Biology of the Cell 2008; 19(7), 2916-2925.

[71] Itoh T, Kanno E, Uemura T, Waguri S, Fukuda M. OATL1, a novel autophagosome-resident Rab33B-GAP, regulates autophagosomal maturation. Journal of Cell Biology 2011; 192(5), 839-853.

[72] Ishibashi K, Uemura T, Waguri S, Fukuda M. Atg16L1, an essential factor for canonical autophagy, participates in hormone secretion from PC12 cells independently of autophagic activity. Molecular Biology of the Cell 2012; 23(16), 3193-3202.

[73] Maxfield FR, McGraw TE. Endocytic recycling. Nature Reviews Molecular Cell Biology 2004; 5(2), 121-132.

Chapter 4

Flow Cytometric Measurement of Cell Organelle Autophagy

N. Panchal, S. Chikte, B.R. Wilbourn, U.C. Meier and G. Warnes

Additional information is available at the end of the chapter

http://dx.doi.org/10.5772/54711

1. Introduction

The term autophagy, (Type II Apoptosis) is derived from the Greek roots "auto" (self) and "phagy" (eat) and was first coined by De Duve in 1967 to epitomise this type of cell death. The mechanism of organelle autophagy in cells undergoing macro-autophagy (from here on termed autophagy) is poorly understood. Cytoplasm, misfolded protein aggregates, dysfunctional mitochondria and stressed endoplasmic reticulum (ER) are engulfed by the formation of a double membrane forming an autophagosome [1,2]. The formation of the autophagosome double membrane structure within the cytoplasm is thought to be formed from pre-existing membranes within the cell, although it is unknown whether the Golgi apparatus, endoplasmic reticulum (ER) or mitochondria are used preferentially to form the autophagosome structure [1,2]. During the formation of the autophagosome structure, organelles such as mitochondria, parts of the ER and Golgi apparatus are engulfed by the autophagosome with the final closure of the double membrane structure occurring next. This then fuses with nearby lysosomes, giving rise to an autolysosome, where the intracellular components are degraded by hydrolytic enzymes [1,3-5]. This process generates ATP, which may delay cell death if the cell is under nutrient depleted conditions leading to the survival of the cell. Thus it is unclear whether the process protects or causes diseases such as cancer and neurodegenerative disorders [6,7].

The process of autophagy is a cell survival mechanism that occurs when the cell is under stress, via external environmental pressures, including the lack of nutrients, or via the internal microenvironment of the cell, including the replacement of old and defective organelles such as mitochondria [8,9]. Autophagy is also induced by the formation and collection of misfolded proteins in the endoplasmic reticulum (ER) which causes ER-stress within the cell [10]. Prolonged adverse conditions results in the death of the cell by the autophagic process. The

ER is responsible for the folding and then delivery of proteins via the secretory pathway to a functional site. Misfolded proteins accumulate in the lumen of the ER due to high protein folding demand on the ER [10]. Only properly folded proteins are secreted with maintenance of the plasma membrane structure and ER folding capacity being under ER homeostatic control [10,11]. Once a threshold of misfolded protein accumulation has been reached, a signal activates the ER to nucleus signalling pathway or the Unfolded Protein Response (UPR) causing ER chaperone proteins to be synthesised which refold the misfolded proteins and translocate these proteins to the cytoplasm for degradation by the proteasome [10,11]. This process results in an increase of the ER capacity to fold proteins and maintain ER homeostasis. This increase in ER capacity is physically achieved by an increase in size of the ER at early stages of autophagy [12]. If the protein folding demand continues to increase, this will ultimately result in the phagy of the ER itself which can be caused by ER stress (induced *in vitro* by tunicamycin and dithiothreitol, (DTT) and also multiple mechanisms that induce autophagy *e.g.* drugs such as rapamycin and nutrient starvation [10].

Mitochondria are pivotal organelles in energy conversion. They act as the cell's power producers and are the site of cellular respiration, which ultimately generates fuel for the cell's activities. However, they are crucial for cell division, growth as well as cell death. As a major source of reactive oxygen species they consume cytosolic ATP when dysfunctional. It is, therefore pivotal for a cell to maintain of a cohort of healthy mitochondria. A sophisticated process called mitophagy, a selective process of autophagy, is responsible for the degradation of damaged organelles. In response to the triggers of mitophagy, mitochondria fragment, which sends out signals, which result in the engulfment by the autophagosomes [9,13]. Malfunctioning mitochondria are also generated through the process of aging or by having a high level of mutations in the mitochondrial DNA (mtDNA) induced by high levels of ROS. Mitochondrial DNA has 10-20 times more mutations than nuclear DNA [9]. Mitochondria reproduce by the process of fission, this produces two daughter mitochondria one of which hosts the damaged parts with reduced inner membrane potential of the original mitochondria but also a fully functional daughter [8]. The fission process can also stop excessive enlargement of mitochondria [8]. Conversely, mitochondrial fusion of damaged mitochondria dilutes the individual mitochondrion level of damaged macromolecules, and can prevent the early removal of such mitochondria [14]. Mitophagy is a physiological process which occurs during erythrocyte differentiation and during nutrient starvation [14]. The level of relative mitochondrial fusion and fission within cells maintains mitochondrial homoeostasis. Thus mitophagy provides a mechanism by which aged or ROS damaged mitochondria are ultimately removed resulting in the survival of the cell in question.

Different agents/stimuli induce autophagy via different signalling routes and thus may preferentially cause mitophagy or if the ER is stressed by such stimuli, ER phagy or ER enlargement may be detectable at specific time points during each treatment. To this end we have employed organelle mass dyes to measure the relative size of mitochondria and ER during nutrient starvation, rapamycin, and chloroquine treatments. We have previously developed and modified the technique employed by Ramdzan *et al* [15] to measure linear scaled fluorescent signals of MitoTracker Green and ER Tracker Green via a modification of the cell cycle analysis doublet discrimination technique to accurately measure mitochondrial & ER mass during such treatments [16]. These techniques will serve as an adjunct to the measurement of autophagy microtubule associated protein, LC3I and LC3II (or LC3B as referred to from here on) in the determination of the presence of autophagy within a population of cells.

Methods for monitoring autophagy started with the initial discovery of the process by the use of electron microscopy showing the presence of double and single membrane structures termed the autophagosome and autolysosome or autophagolysosome respectively [4,5]. Other techniques have also obviously centred upon the formation of the autophagic machinery by measurement of LC3B, such as by Western Blotting which can be used to quantitate the degree of autophagy in cells by measuring LC3B which is normally located in the cytoplasm in the form of LC3I but when incorporated into the autophagosome is cleaved and lipidated by phosphotidylethanolamine, to form LC3II [17-19]. Anti-LC3B antibody labelling of LC3B has also been employed to measure autophagosomes and autolysosomes by image and flow cytometric analysis [20-22]. The increase in number and intensity of fluorescently labelled LC3B autophagosomes-autolysosomes can be quantitated by time-consuming image analysis, whereas increase in MFI of LC3B antigen levels measured flow cytometrically makes the process significantly less burdensome [21,23]. Here we describe a protocol to determine the presence of autophagy by the determination of LC3B levels in normal growing cells and those undergoing autophagy.

Use of techniques to estimate the degree of loss of dysfunctional cell organelles has been more limited than those techniques investigating the development of the autophagic process. Previous studies have used ER Tracker and MitoTracker mass probes from Invitrogen to estimate ER and mitochondrial size based on median fluorescence intensity [16]. In experimental conditions different inducers of autophagy such as rapamycin, chloroquine, serum and total nutrient starvation may result in the different types of cell organelles being preferentially targeted by the autophagic process. Here we describe quantitative flow cytometric methods, which can be employed in the study of cell organelle-phagy. We show that a range of inducers cause the phagy of specific organelles and that this process is also cell type dependent. This article will highlight ways of monitoring the contribution of distinctive cell organelles in vitro to the autophagic process and highlight its diversity in different cell types.

2. Materials and methods

2.1. Cell lines

Jurkat T and K562 cell lines were grown in RMPI-1640 with L-glutamine (Cat No 21875-034, Invitrogen, Paisley, UK) supplemented with 10% foetal bovine serum (FBS, Cat No 10500-064, Invitrogen, Paisley, UK) and penicillin and streptomycin (Cat No 15140-122, Invitrogen, Paisley, UK) in the presence of 5% CO_2 at 37°C.

2.2. Induction of autophagy

Jurkat and K562 cells were grown in <1% FBS-RPMI, <1% FBS-Phosphate Buffered Saline (PBS) (PBS, Cat No 14190-094 Invitrogen, Paisley, UK) or treated with chloroquine, (CQ) at 50µM, (Cat No C6628-25G, Sigma, Poole, UK), rapamycin (80nM, Cat No PHZ1233, Invitrogen, Paisley, UK). Time points analysed were 48h, (n=3) as described in the section below.

2.3. Indirect immunofluorescence LC3B labeling

Jurkat and K562 cells with or without treatment were pelleted and resuspended in 100 µl of Solution A fixative for 15 min at room temperature (RT) (Cat No GAS-002A-1, Caltag UK). Cells were then washed in PBS buffer Cell pellets were then permeabilised with 0.25% Triton X-100 (Cat No X100-500ML, Sigma Chemicals, Poole, UK) for 15 min at RT. Cells were washed in PBS buffer. Anti-LC3B polyclonal antibody (0.25 µg) (Cat No L10382, Invitrogen, Paisley, UK) or rabbit immunoglobulin (0.25 µg) (Cat No I5006, Sigma Chemicals, Poole, UK) was used as an isotype control and incubated for 0.5h at RT. Cells were then washed in PBS buffer. Cells were then labelled with 0.125 µg of secondary fluorescent conjugate Alexa Flour 647 goat anti-rabbit IgG (Cat No A21244, Invitrogen, Paisley, UK) for 30 min at RT. Cells were then washed in PBS buffer and resuspended in 400 µl of PBS. Analysis of LC3B-Alexa Fluor647 signal was achieved by determining the MFI of the whole histogram signal for previously gated cells from a FSC versus SSC dot-plot and compared to the corresponding isotype control sample in an overlaid histogram. 10,000 events were collected by flow cytometry.

2.4. Organelle labelling

Jurkat and K562 cells with or without treatment were counted and adjusted to the same number per volume and loaded with 40nM MitoTracker Green (MTG) (Cat No M7514, Invitrogen, Paisley, UK) or 100nM ER Tracker Green (ERTG) (Cat No L7526, Invitrogen, Paisley, UK) by incubating cells with dyes for 15 or 30 minutes at 37°C respectively. Cells were then washed in PBS buffer and resuspended in 400 µl of PBS, in the presence of DNA viability dye, DAPI (200 ng/ml) (Cat No D9542, Sigma Chemicals, Poole, UK). Live cells were analysed for MTG or ERTG MFI levels by firstly gating on all cell material except small debris in the origin of a FSC versus SSC dot-plot. This data was then analysed on a DAPI versus FSC dot-plot with live cells being DAPI-ve. The 530/30nm channel on a BD FACS Canto II was set to linear and the width parameter activated. Doublet discrimination was then achieved by gating on single cells on a 530/30nm width and area plot. Median fluorescence Intensity (MFI) of MTG-Area or ERTG-Area signals from samples was then compared by histogram analysis of untreated and treated cells.

2.5. Flow cytometry

Single colour controls for MTG, or ERTG and DAPI were used to set compensations. MTG or ERTG were detected in the 530/30nm channel on the argon laser octagon (BD FACSCanto II); DAPI was detected in the 440/40nm channel on the violet diode trigon (BD FACSCanto II). No

http://dx.doi.org/10.5772/54711

compensation was required. LC3B-Alexa Fluor-647 was detected on the 660/20nm channel on the Red HeNe trigon (BD FACSCanto II).

Cells (10,000) were analysed on a Becton Dickinson FACSCanto II fitted with a 488nm Ti-Sapphire Argon laser, red HeNe 633nm laser and violet diode 405nm with FACSDiva Software ver 6.1.3. All data were analysed on FlowJo (ver 8.8.7, Treestar Inc, CA) in the form of list-mode data files version FCS 3.00 using the default bi-exponential transformation. Optical filters and mirrors in the BD FACSCanto II were fitted in 2008.

2.6. Statistics

Student t tests were performed in Microsoft Office Excel with P = >0.05 not significant (NS), P = <0.05*, P = <0.01**, P = <0.005***, P = <0.001****.

3. Results

Cells treated to induce autophagy were first shown to be autophagic by immunofluorescent labelling of LC3B which is located in the autophagosome of cells undergoing autophagy. These were compared to resting control cells and positivity determine by use of an rabbit immunoglobulin isotype control labelled with the same secondary Alexa Fluor-647 conjugate permitting the demonstration that LC3B levels have been up-regulated in cell samples undergoing autophagy treatments. Figure 1 shows the LC3B signals from Jurkat and K562 cells after 48h of treatment compared to resting control levels. Resting cells do show a detectable LC3B compared to isotype controls (Figure 1). Jurkat cells showed a high level of LC3B compared to resting cells when treated with 80nM rapamycin and grown in low serum conditions (<1% FBS/RPMI, Figure 1A). In contrast CQ showed a lower level (half) of LC3B up-regulation compared to cells grown in nutrient free (<1% FBS) conditions (Figure 1B). K562 cell up-regulation of LC3B was shown to be at a similar level of that of Jurkat cells, when treated with rapamycin and grown in low serum conditions (<1% FBS/RPMI, Figure 1C). CQ and nutrient free (<1% FBS) treatment of K562 cells showed a similar LC3B response to that of Jurkat cells (Figure 1D).

Rapamycin was shown to significantly up-regulate LC3B by 4-5 fold after 24 and 48h (P=<0.001, P=<0.005) in both cell lines (Figure 2A, B). In contrast low serum conditions caused a 5-6 fold increase in Jurkat LC3B only at 48h (P=<0.001, Figure 2A). Whilst, K562 responded to low serum treatment by a lower but significant 2-3 fold increase in LC3B levels at 24 and 48h (P=<0.005, P=<0.05, Figure 2B). Likewise nutrient depletion of K562 cells caused a 2 fold increase in LC3B at 24 and 48h (P=<0.01, P=NS, Figure 2B). Whilst nutrient depletion of Jurkat cells showed a significant 5-6 fold increase at 24h, reduced to a significant 2 fold increase above controls at 48h (P=<0.005, Figure 2A). CQ (50μM) in contrast to rapamycin induced a 50-100% increase in LC3B levels in Jurkat cells at 24 and 48h respectively (P=<0.05, P=NS, Figure 2A). Whilst K562 cells responded to CQ treatment (50μM) by doubling LC3B levels at 24h and falling back to control levels at 48h respectively (P=NS, Figure 2B).

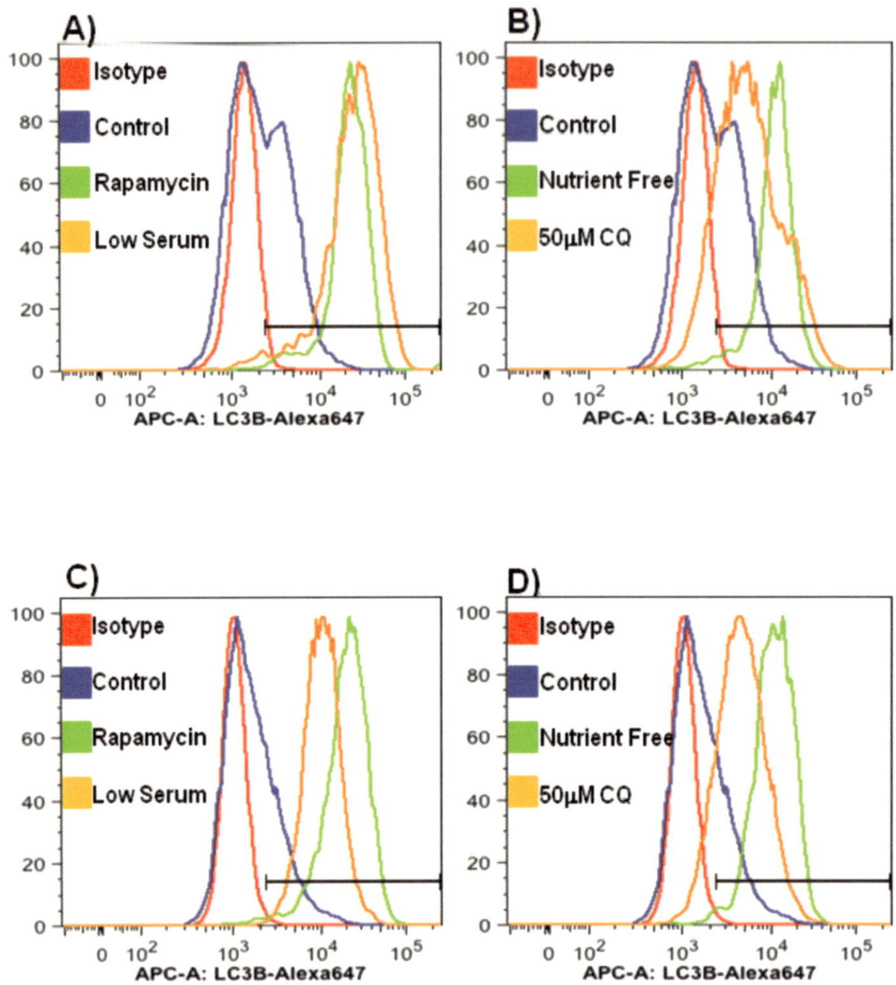

Figure 1. Demonstration of autophagy in Jurkat and K562 cells by measurement of LC3B levels by flow cytometry, see Materials & Methods. Resting Jurkat cell LC3B-Alexa Fluor-647 levels were compared after 48h treatment with 80nM rapamycin and low serum/RPMI LC3B levels A). Resting Jurkat cell LC3B-Alexa Fluor-647 levels were compared after 48h treatment of low serum/nutrient free conditions and CQ (50µM) LC3B levels B). Resting K562 cell LC3B-Alexa Fluor-647 levels were compared after 48h treatment with 80nM rapamycin and low serum/RPMI LC3B levels C). Resting K562 cell LC3B-Alexa Fluor-647 levels were compared after 48h treatment of low serum/nutrient free conditions and CQ (50µM) LC3B levels D).

Mitophagy was determined by linear flow cytometric analysis of the MTG signal in Jurkat and K562 cells by the following procedure. After gating on cells from a FSC v SSC dot-plot (Figure 3A), live cells were gated upon from a FSC v DAPI dot-plot (Figure 3B), DAPI negative cells being alive. Doublet discrimination was determined from an MTG Area v Width parameter plot (Figure 3C) and MFI of control and test samples compared on an overlaid histogram (Figure 3D). In this example the Jurkat cell control gave an MFI of 142,000 compared to a 50µM CQ test showing an MFI of 100,000 or a 30% reduction in the mitochondrial mass of live Jurkat cells.

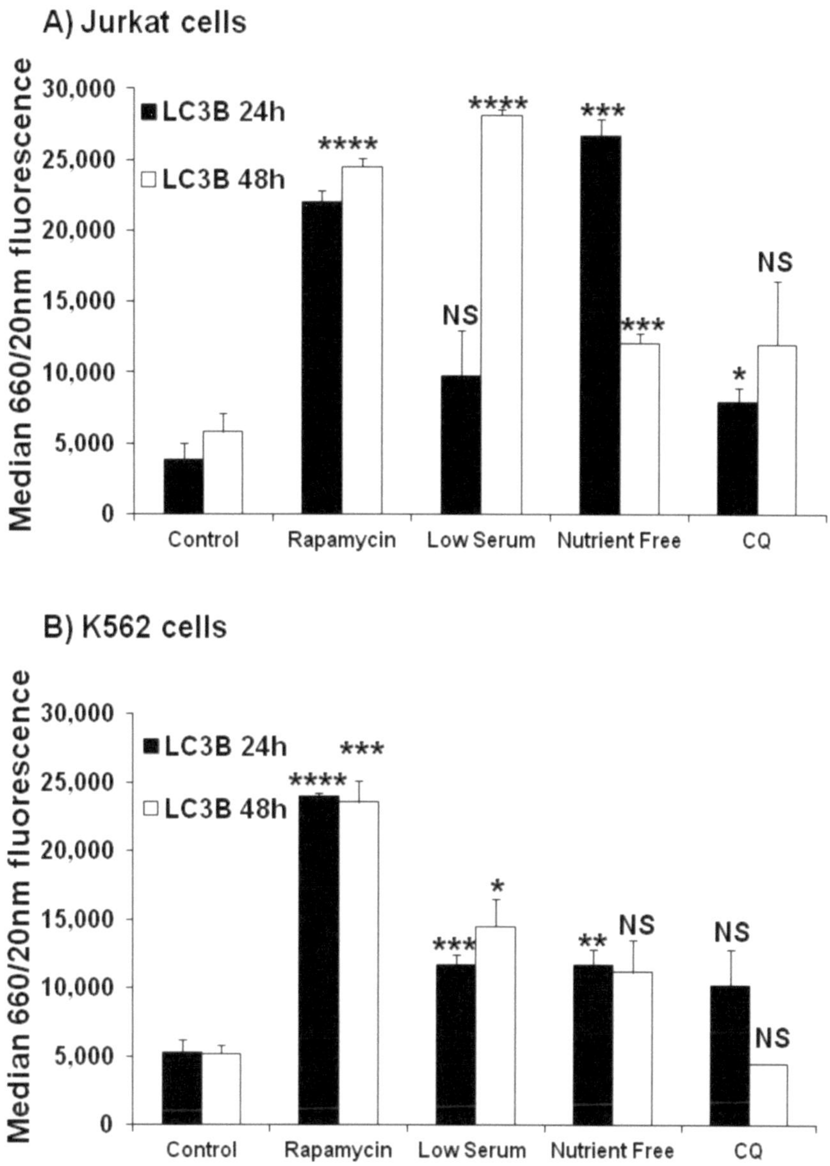

Figure 2. Demonstration of autophagy in Jurkat and K562 cells by measurement of MFI LC3B levels by flow cytometry, see Materials & Methods. Resting Jurkat cell MFI LC3B-Alexa Fluor-647 levels were compared after 24 and 48h treatment with 80nM rapamycin, low serum/RPMI, low serum/nutrient free conditions and CQ (50μM) LC3B levels A). Resting K562 cell MFI LC3B-Alexa Fluor-647 levels were compared after 24 and 48h treatment with 80nM rapamycin, low serum/RPMI, low serum/nutrient free conditions and CQ (50μM) LC3B levels A). T-test statistical analysis NS-not significant, * P=<0.05, ** P=<0.01, *** P=<0.005, **** P=<0.001.

Likewise ER-phagy was determined by linear flow cytometric analysis of the ERTG signal in Jurkat and K562 cells by the following procedure. After gating on cells from a FSC v SSC dot-plot (Figure 4A), live cells were gated upon from a FSC v DAPI dot-plot (Figure 4B), DAPI negative cells being alive. Doublet discrimination was determined from an ERTG Area v Width parameter plot (Figure 4C) and MFI of control and test samples compared on an overlaid histogram (Figure 4D). In this example the K562 cell control gave an MFI of 159,000 compared to a 40nM rapamycin test showing an MFI of 262,000 or a 65% increase in the ER mass of live K562 cells.

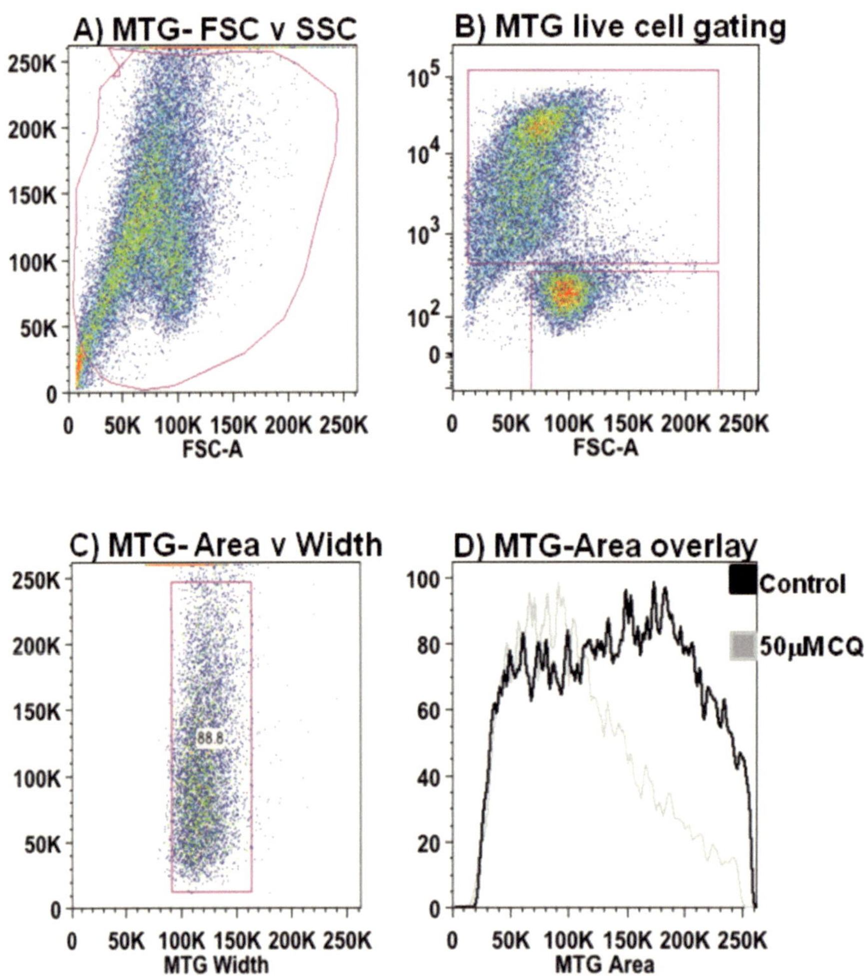

Figure 3. Jurkat cells were untreated (control) or treated with 50μM CQ for 48h. These cell cultures were then counted and adjusted to be the same number per volume and loaded with 40nM MTG for 15 minutes at 37°C. Cells were washed and adjusted to 0.5ml volume containing DAPI at 200ng/ml. Cells were then analysed on a BD Canto II according to Materials & Methods. Cells were gated upon except the small debris in the origin of a FSC versus SSC dot-plot (A). There data were then analysed on a DAPI versus FSC dot-plot with live cells being DAPI-ve (B). The 530/30nm channel on a BD FACS Canto II was set to linear and the width parameter activated. Doublet discrimination was then achieved by gating on single cells on a 530/30nm width and area plot (C). The MFI of MTG-Area signals from samples were then compared by histogram analysis of untreated, MFI 142,000 and CQ treated cells, MFI 100,000 (D).

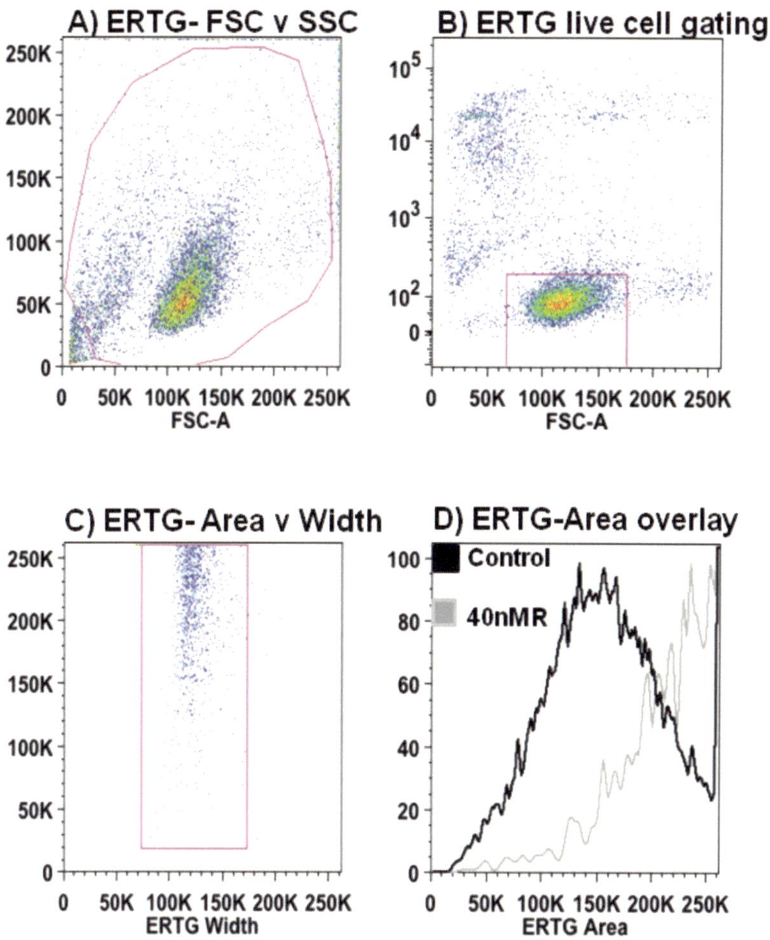

Figure 4. K562 cells were untreated (control) or treated with 40nM rapamycin (R) for 48h. These cell cultures were then counted and adjusted to be the same number per volume and loaded with 100nM ERTG for 30 minutes at 37°C. Cells were washed and adjusted to 0.5ml volume containing DAPI at 200ng/ml. Cells were then analysed on a BD Canto II according to Materials & Methods. Cells were gated upon except the small debris in the origin of a FSC versus SSC dot-plot (A). These data were then analysed on a DAPI versus FSC dot-plot with live cells being DAPI-ve (B). The 530/30nm channel on a BD FACS Canto II was set to linear and the width parameter activated. Doublet discrimination was then achieved by gating on single cells on a 530/30nm width and area plot (C). The MFI of ERTG-Area signals from samples were then compared by histogram analysis of untreated, MFI 159,000 and rapamycin treated cells, MFI 262,000 (D).

Induction of autophagy by the rapamycin mTOR signalling pathway induced a detectable mitophagy (34% reduction in mitochondrial mass) in Jurkat cells (Figure 5A). Whilst K562 cells displayed an ER stress response in that the ER mass was increased by 50% with rapamycin treatment without any mitophagy (Figure 5B). Like rapamycin, chloroquine induction of autophagy in Jurkat cells displayed a similar level of mitophagy (29%), whilst there was no significant organelle phagy displayed by K562 cells (Figure 5A, B). In contrast to rapamycin treatment of Jurkat cells, low serum/RPMI treatment caused a 19% increase in mitochondrial mass as opposed to a mitophagy. Under the same conditions Jurkat cells significantly increased

(37%, *P*<0.01) ER mass indicating a high level of misfolded proteins within the Jurkat cells (Figure 5A). However K562 cells responded in a similar manner to rapamycin treatment when treated with low serum/RPMI by displaying an increase in ER mass (43%) indicating again an autophagic response to a high level of misfolded proteins (Figure 5B). Nutrient depravation in the presence of low serum (<1%) showed the opposite response to low serum/RPMI treatment of cells in that a significant phagy of ER was observed in Jurkat cells (70%, *P*<0.05, Figure 5A). Whilst, K562 displayed both a significant ER phagy (42%) and mitophagy (36%), when undergoing nutrient depravation (*P*<0.05, Figure 5B).

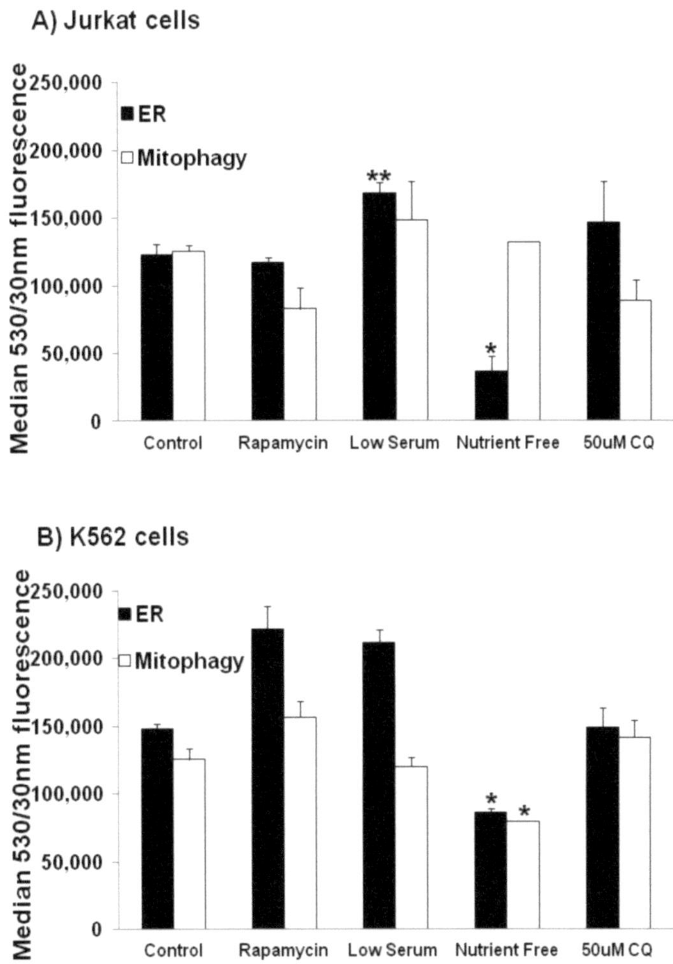

Figure 5. Jurkat (A) and K562 (B) cells were untreated (control) or treated 80nM rapamycin, low serum (<1% FBS) RPMI, nutrient free PBS (<1% FBS) or with 50μM CQ for 48h. These cell cultures were then counted and adjusted to be the same number per volume and loaded with 40nM MTG or 100nM ERTG for 15 or 30 minutes at 37°C respectively. Cells were washed and adjusted to 0.5ml volume containing DAPI at 200ng/ml. The MFI of MTG or ERTG-Area signals from samples were then compared by histogram analysis of untreated or treated cells. T-test statistical analysis NS-not significant, * P=<0.05, ** P=<0.01, *** P=<0.005, **** P=<0.001.

http://dx.doi.org/10.5772/54711

4. Discussion

Modulation of autophagy may be important in preventing or treating neurodegenerative conditions and cancers. It is, therefore, pivotal to understand the underlying mechanisms. Numerous new techniques have been developed to show that cells are undergoing the autophagic process; these include LC3B semi-quantification by image and flow cytometric analysis by the use of GFP tagged proteins or antibody labelling [17-22,24-30]. Here we demonstrate by flow cytometric measurement of anti-LC3B-Alexa Fluor-647 signals that rapamycin significantly up-regulated LC3B in Jurkat and K562 cells, whilst CQ at 50μM, induced in comparison a small degree of detectable up-regulated LC3B in both cell lines employed in this study. Nutrient depletion studies, including reduced serum <1% and lack of nutrients induced a low level of LC3B up-regulation in K562 cells compared to the high level (similar to rapamycin) detected in Jurkat cells.

Different drugs or treatments induce autophagy via different signalling routes in different cell types resulting in varying types and degrees of organelle phagy. Rapamycin although induced a similar LC3B up-regulation in the two cell lines employed in this study resulted in different organelle phagy responses. Jurkat cells showed a mitophagy whilst K562 cells did not. This was in contrast to the lack of an effect of rapamycin upon Jurkat ER mass, whilst K562 cells showed an increase in ER mass, indicating protein mis-folding within the ER, even though rapamycin caused little cell death over the 48h period studied [10,11]. This increase in ER size after 48h of rapamycin treatment may ultimately result in a measurable ER-phagy at a later time point.

Chloroquine induced apoptosis, death, low level autophagy (as indicated by mild LC3B up-regulation) and mitophagy as indicated by the reduction in mitochondrial mass in Jurkat cells, whilst no organelles were targeted in K562 cells [31]. Thus both rapamycin and chloroquine act upon Jurkat cells to produce a mitophagy with no significant affect upon ER mass; whilst K562 cell response to these autophagy inducers was to show an increase in ER mass.

Nutrient starvation commonly employed by removing serum from media in the study of autophagy caused protein mis-folding in both cell lines as indicated by an increase in ER mass and also of mitochondria in Jurkat cells [22]. In contrast total nutrient starvation which again caused varying degrees of up-regulation of LC3B levels in both cell lines as well as resulting in a large reduction in cell numbers caused a decrease in ER mass in both cell lines and a mitophagy in K562 cells. This again indicated that drug induction of autophagy may result in different organelle phagy in different cell types.

Thus the combination of measuring the induction of autophagy via LC3B flow cytometric measurements and the technique of organelle mass semi-quantification gives the investigator more information about the autophagic process occurring within the cell not only in terms of autophagic flux signals but also the degree and type of organelle phagy occurring during the autophagic process. This technique of organelle mass semi-quantification by flow cytometry on live cells permits researchers in the field to measure not only the degree of autophagy but also live cell functions such as mitochondrial membrane potential during the autophagic process giving an insight to the more precise mechanism of action of the wide variety of stimuli that can be employed in the study of the autophagic process. This area warrants further study as it holds new therapeutic and diagnostic potential.

Author details

N. Panchal[1], S. Chikte[1], B.R. Wilbourn[1], U.C. Meier[2] and G. Warnes[1]

1 Flow Cytometry Core Facility, The Blizard Institute, Barts and The London School of Medicine and Dentistry, Queen Mary London University, UK

2 Centre for Neuroscience and Trauma, The Blizard Institute, Barts and The London School of Medicine and Dentistry, Queen Mary London University, London, UK

References

[1] Tooze S, Yoshimori T. The origin of the autophagosomal membrane. Nature Cell Biology 2010;12:831-835.

[2] Mehrpour M, Esclatine A, Beau I, Codogno P. Overview of macroautophagy regulation in mammalian cells. Cell Research 2010;20:748-762.

[3] Yang Z, Klionsky D. Eaten alive: a history of macroautophagy. Nature Cell Biology 2010;12:814-822.

[4] Ashford TP, Porter K. Cytoplasmic components in hepatic cell lysosomes. J Cell Biol 1962;12:198-202.

[5] Deter RL, Duve CD. Influence of glucagon, an inducer of cellular autophagy, on some physical properties of rat liver lysosomes. J Cell Biol 1967;33:437-449.

[6] Shintani T, Klionsky D. Autophagy in health and disease: A double-edged sword. Science 2004;306:990-995.

[7] Rosello A, Warnes G, Meier CU. Cell death pathways and autophagy in the central nervous system and its involvement in neurodegeneration, immunity and CNS infection: to die or not to die - that is the question. Clin Exp Imm 2012;168:52-57.

[8] Terman A, Gustafsson B, Brunk UT. Autophagy, organelles and ageing. The Journal of Pathology 2007;211:134-143.

[9] Kim I, Rodriguezenriquez S, Lemasters J. Selective degradation of mitochondria by mitophagy. Archives of Biochemistry and Biophysics 2007;462:245-253.

[10] Yorimitsu T, Klionsky DJ. Eating the endoplasmic reticulum: quality control by autophagy. Trends in Cell Biology 2007;17:279-285.

[11] Mandl J, Mészáros T, Bánhegyi G, Hunyady L, Csala M. Endoplasmic reticulum: nutrient sensor in physiology and pathology. Trends in Endocrinology & Metabolism 2009;20:194-201.

[12] Bernales S, McDonald KL, Walter P. Autophagy counterbalances endoplasmic reticulum expansion during the unfolded protein response. PLoS Biology 2006;4:e423.

[13] Kanki T, Klionsky DJ. The molecular mechanism of mitochondria autophagy in yeast. Molecular Microbiology 2010;75:795-800.

[14] Twig G, Shirihai OS. The interplay between mitochondrial dynamics and mitophagy. Antioxidants & Redox Signaling 2011;14:1939-1951.

[15] Ramdzan YM, Polling S, Chia CPZ, Ng IHW, Ormsby AR, Croft NP, Purcell AW, Bogoyevitch MA, Ng DCH, Gleeson PA and others. Tracking protein aggregation and mislocalization in cells with flow cytometry. Nat Meth 2012;9:467-470.

[16] Jia W, He YW. Temporal regulation of intracellualr organelle homeostasis in T lymphocytes by autophagy. The Journal of Immunology 2011;186:5313-5322.

[17] Kabeya Y, Mizushima N, Ueno T, Yamamoto A, Kirisako T, Noda T, Kominami E, Ohsumi Y, T. Y. LC3, a mammalian homologue of yeast Apg8p, is localized in autophagosome membranes after processing. EMBO 2000;19:5720-5728.

[18] Barth S, Glick D, Macleod K. Autophagy: assays and artifacts. The Journal of Pathology 2010;221:117-124.

[19] Hansen TE, Johansen T. Following autophagy step by step. BMC Biology 2011;9:1-4.

[20] Geng Y, Kohil L, Klocke BJ, Roth KA. Chloroquine-induced autophagic vacuole accumulation and cell death in glioma cells is p53 independent. Neuro-Oncology 2010;12:473-481.

[21] Thomas S, Thurn KT, Biçaku E, Marchion DC, Münster N. Addition of a histone deacetylase inhibitor redirects tamoxifen-treated breast cancer cells into apoptosis, which is opposed by the induction of autophagy. Breast Cancer Research and Treatment 2011;130:437-447.

[22] Phadwal K, Alegre-Abarrategui J, Watson AS, Pike L, Anbalagan S, Hammond EM, Wade-Martins R, McMichael A, Klenerman P, Simon AK. A novel method for autophagy detection in primary cells: Impaired levels of macroautophagy in immunosenescent T cells. Autophagy 2012;8:677-689.

[23] Chen Y, McMillan-Ward E, Kong J, Israels SJ, Gibson S. Oxidative stress induces autophagic cell death independent of apoptosis in transformed and cancer cells. Cell Death and Differentiation 2007;15:171-182.

[24] Shvets E, Fass E, Elazar Z. Utlizing flow cytometry to monitor autophagy in living mammalian cells. Autophagy 2008;4:621-628.

[25] Kimura S, Noda T, Yoshimori T. Dissection of the autophagosome maturation process by a novel reporter protein, tandem fluorescent-tagged LC3. Autophagy 2007;3:452-460.

[26] Wu YT, Tan HL, Huang Q, Kim YS, Pan N, Ong WY, Liu ZG, Ong CN, Shen HM. Autophagy plays a protective role during zVAD-induced necrotic cell death. Autophagy 2008;4:457-466.

[27] Boya P, Gonzalez-Polo RA, Poncet D, Andreau K, Vieira HLA, Roumier T, Perfettini JL, Kroemer G. Mitochondrial membrane permeabilization is a critical step of lysosome-initiated apoptosis induced by hydroxychloroquine. Oncogene 2003;22:3927-3936.

[28] Boya P, Gonzalez-Polo RA, Casares N, Perfettini JL, Dessen P, Larochette N, Metivier D, Meley D, Souquere S, Yoshimori T and others. Inhibition of macroautophagy triggers apoptosis. Molecular and Cellular Biology 2005;25:1025-1040.

[29] Byun JY, Yoon CH, An S, Park IC, Kang CM, Kim MJ, SJ. L. The Rac1/MKK7/JNK pathway signals upregulation of Atg5 and subsequent autophagic cell death in response to oncogenic Ras. Carcinogenesis 2009;30:1880-1888.

[30] Mellen MA, dela Rosa EJ, Boya P. Autophagy is not universally required for phosphatidyl-serine exposure and apoptotic cell engulfment during neural development. Autophagy 2009;5:964-972.

[31] Cao C, Subhawong T, Albert JM, Kim KW, Geng L, Sekhar KR, Gi YJ, B. L. Inhibition of mammalian target of rapamycin or apoptotic pathway induces autophagy and radiosensitizes PTEN null prostate cancer cells. Cancer Research 2006;66:10040-10047.

Section 2

Consequences of Autophagy Deficits

Chapter 5

Autophagy, the "Master" Regulator of Cellular Quality Control: What Happens when Autophagy Fails?

A. Raquel Esteves, Catarina R. Oliveira and
Sandra Morais Cardoso

Additional information is available at the end of the chapter

http://dx.doi.org/10.5772/55196

1. Introduction

Autophagy an evolutionarily conserved process, is activated in response to nutrient deprivation, as well as to endogenous and exogenous stresses. It is currently known that the basic mechanism of autophagy has been well conserved during evolution since diverse organisms, including yeast, flies, and mammals, all carry a similar set of autophagy-related genes (ATGs), although there are some significant differences between yeast and higher eukaryotes. As a result, autophagy is a highly regulated process known to contribute to cellular cleaning through the removal of intracellular components in lysosomes showing therefore an important role in cellular quality control. Autophagy ensures that proteins damaged or incorrectly synthesized are removed from the cells by degradation, preventing thus the devastating cellular consequences associated with accumulation of malfunctioning proteins inside cells. Moreover autophagy works as a recycling system where it mediates the breakdown of proteins that are no longer needed into essential components (aminoacids, free fatty acids, sugars), which can then be used in the synthesis of new proteins. This has extreme importance in conditions of nutritional stress or starvation. At optimal physiological conditions in the absence of stressors, basal level of autophagy assures maintenance of cell homeostasis through regular turnover of proteins, lipids, and organelles. Therefore autophagy acts a cautious controller of cellular homeostasis. In addition, it also has an important role in cellular defence as in compromised situations it contributes to the proteolytic breakdown of components of invading pathogens and other types of biological cell aggressors. Autophagy may also modulate synaptic plasticity, which involves structural remodelling of nerve terminals and the trafficking and degradation of receptors and other synaptic proteins [1]. Finally, autophagy is also a key player in cellular adaptation since is able of changing very rapidly the rate of a particular protein's degradation allowing fast modulation of the proteome in response to stress. The lysosomal system and specifically the autophagic pathway is the principal mechanism for degradating longer-lived proteins and is the only system in cells capable of degrading organelles and protein aggregates/inclusions. Similar to what happens with the proteasome system, protein aggregates and certain organelles have been shown to be tagged with ubiquitination for selective removal by autophagy [2]. An adaptor molecule with an ubiquitin-binding domain engages the ubiquitinated structure and couples it to the pre-autophagosomal isolation membrane for subsequent sequestration. The exact types of ubiquitin motifs recognized by the proteasome and autophagy may differ and the degree to which ubiquitination drives autophagic protein turn-over relative to that by proteasomes is still unclear.

There are three basic forms of autophagy, namely, macroautophagy, microautophagy and chaperone-mediated autophagy (CMA), which primarily differ in the way in which cytosolic components are delivered to lysosomes [3]. Moreover, the different types of autophagy share a common endpoint, the lysosome, but differ in the substrates targeted, their regulation, and the conditions in which each of them is preferentially activated.

In macroautophagy and microautophagy there is the sequestration of cytoplasmic components into vesicular compartments. In the case of macroautophagy, vesicles are form *de novo* from a limiting membrane of non-lysosomal origin that closes and forms a double membrane organelle called autophagosome [4]. Because the autophagosome lacks any enzymes, the trapped contents are not degraded until the autophagosome fuses with a lysosome, which provides all the hydrolases required for degradation of cargo. Autophagosomes are then transported along microtubules to the perinuclear region of the cell (where lysosomes are clustered) to enhance the probability of autophagosome-lysosome fusion to form autophagolysosomes. Lysosomes are single membrane organelles dedicated to degradation of both intracellular and extracellular components. Lysosomal membranes possess vacuolar H^+ ATPases which are proton pumps that acidificate the autophagolysosome contents. This acidification is essential for the activation of lysosomal enzymes (including proteases, lipases, glycosidases, and nucleases), which are responsible for proteolysis of the components in the autophagolysosome and confer this organelle its high degradative capacity [5]. Although autophagy was initially considered to be a nonselective process, cargo-specific autophagy (the selective degradation of certain aggregated proteins and organelles) is now recognized to exist [6]. Numerous different types of cargo-specific autophagic processes have been described, like for instance mitophagy where damaged mitochondria are sequestered and degraded [7]. Macroautophagy occurs constitutively in cells and is markedly induced under stress conditions, such as starvation and has two major purposes: as a source to generate essential macromolecules and energy in conditions of nutritional scarcity, or as a mechanism for the removal of altered intracellular components [8].

In the case of microautophagy the engulfment of cargo occurs through invagination of the lysosomal membrane itself, to form vesicles that invaginate towards the lysosomal lumen [9]. Cytosolic regions are sequestered directly by the lysosome through invaginations or tubulations that "pinch off" from the membrane into the lysosomal lumen where they are rapidly degraded. Microautophagy participates in the continuous, slow "basal" turnover of cellular components in normal cellular conditions [10]. Furthermore microautophagy has been shown to be responsible for the selective removal of organelles when they are no longer needed [11]. The absence of mammalian homologs for the microautophagy yeast genes has made it difficult in gaining a better understanding of the pathophysiology of this process.

The third type of mammalian autophagy CMA distinguishes from the others because of its distinctive feature of selectivity. Unlike the other forms of autophagy, in which portions of the cytoplasm are typically engulfed, proteins degraded by CMA are identified and transported one-by-one by a cytosolic chaperone that delivers them to the surface of the lysosomes. First there is individual recognition of single proteins in the citosol, then the substrate proteins unfold and cross the lysosomal membrane. Only soluble proteins containing in their aminoacid sequence a pentapeptide motif related to KFERQ, are recognized by heat shock cognate protein of 70 kDa, hsc70, which mediates their delivery to the lysosomes for direct translocation across the lysosomal membrane [3, 12]. These pentapeptide motifs are recognized specifically by the cytosolic protein hsc70, the constitutively expressed member of the hsp70 family of cytosolic chaperones. hsc70 not only targets the CMA substract to the lysosomal membrane where it can interact with the CMA receptor but also facilitates substrate folding [13]. A second chaperone that also localizes on both sides of the lysosomal membrane is the heat shock protein of 90 kDa, hsp90. hsp90 is proposed to participate in substrate protein unfolding [14], and contributes to the stabilization of essential components of the translocation complex as they organize into a multimeric structure [15]. Additionally, the co-chaperone carboxyl terminus of Hsp70-interacting protein (CHIP) has been shown to mediate lysosomal degradation of proteins such as alpha-synuclein, a known CMA substrate [16], however, it is still undergoing investigation whether CHIP-mediated degradation occurs via CMA or via other forms of autophagy. The lysosome-associated membrane protein type 2A (LAMP-2A) acts as a receptor for substrates of this pathway [17]. This protein is one of the three splice variants encoded by the lamp2 gene that share identical lumenal regions and different transmembrane and cytosolic tails [18]. Once the cytosolic proteins destined for CMA degradation bind to the cytosolic tail of the single-span membrane protein LAMP-2A, they promote its assembly into a high molecular weight complex of about 700 kDa at the lysosomal membrane required for translocation of the CMA substrates into the lysosomal lumen [15]. Multimerization is a required step for substrate translocation. However, the translocation complex is not stable at the lysosomal membrane and, once the substrate reaches the lysosomal lumen, LAMP-2A dissociates into monomers. hsc70 mediates the disassembly generating monomeric forms of LAMP-2A that are required for substrate binding in order to sustain cycles of binding and uptake. Other proteins have been shown to modulate the stability of the CMA translocation complex, such as, glial fibrillary acidic protein (GFAP) and elongation factor 1 a (EF1a). One of the crucial functions of CMA is protein quality control via the selective degradation of damaged or altered proteins. Moreover proteins associated with several neurodegeneratives disorders have KFERQ-like sequences such as, alpha-synuclein, parkin, UCHL1 and Pink1, amyloid precursor protein (APP) and tau and huntingtin. These proteins can be degraded by CMA only when they are in soluble forms; once the insoluble inclusions are formed, they can only be degraded via other proteolytic pathways [19]. CMA is constitutively active in many cell types and similar to macroautophagy, CMA is maximally activated under stress conditions like nutritional stress or starvation and cellular stresses leading to protein damage [3]. Indeed, upon stressors that cause protein damage, such as oxidative stress, CMA is upregulated [20]. During starvation, macroautophagy is first activated, and if starvation persists, cells activate CMA, which selectively targets non-essential proteins for degradation to obtain the aminoacids required for the synthesis of essential proteins showing that macroautophagy and CMA act in a synchronized manner [21]. There are two lysosomal components that limit CMA activity: LAMP-2A and lys-hsc70. It is known that changes in the number of LAMP-2A molecules, as well as, in the membrane content of LAMP-2A, rapidly upregulate or downregulate CMA activity [17]. Levels of lys-hsc70 increase gradually with the increase in CMA activity, although the mechanisms modulating this increase are still poorly understood [14]. Furthermore, the activity of this autophagic pathway is also directly modulated by changes in other autophagic and proteolytic systems inside the cell. For examples, cells in culture respond to CMA blockage by upregulating macroautophagy [21]. Likewise, blockage of macroautophagy results in constitutive activation of CMA [22].

The identification of ATGs in yeast was a major breakthrough in the study of autophagy [23]. Until now more than 35 ATG genes have been found in yeast and many of them have orthologs in mammals to control the dynamic processes and different stages of autophagy. Indeed, autophagic activities are mediated by a complex molecular machinery including more than 35 ATG-related proteins and 50 lysosomal hydrolases. This dynamic macroautophagic process includes initiation, nucleation, elongation, maturation and degradation. Initiation involves formation of the phagophore, a cup-shaped membrane structure in the cytoplasm. Nucleation involves the recruitment and assembly of several proteins including Beclin-1 and Vps34, among others [24]. Vps34 has a phosphoinositide 3-kinase (PI3K) activity and produces phosphatidylinositol 3-phosphate, which recruits molecular components involved in subsequent vesicle elongation. Subsequently elongation is a critical step in forming the complete autophagosomes and is controlled by two ubiquitin-like conjugation systems. The first involves the Atg7 E1-like enzyme that mediates the covalent linkage of Atg12 to Atg5 followed by formation of a complex of Atg12-Atg5 with Atg16L, which dimerizes and associates with the phagophore membrane. In the second one LC3 has to be cleaved by Atg4 to generate the cytosolic LC3 (that arises from microtubular-associated protein-light chain) that can subsequently be conjugated to a phosphatidylethanolamine tag by Atg7 and the E2-like enzyme Atg3. The resulting lipidated form of LC3, LC3-II, is attached to both sides of the autophagosome membrane. LC3-II is clearly necessary for autophagosome formation and is commonly used as a marker for autophagy. Finally, maturation and degradation steps, involves fusion with endosomes-lysosomes to form autophagic vacuoles (AVs) and requires a number of lysosomal proteins necessary for the degradation of the lumenal contents [25]. The transmembrane Atg9 protein also contributes to autophagic vesicle nucleation and elongation, possibly by mediating the transport of lipid molecules. At this stage, LC3 on the outer autophagosome membrane is removed by Atg4 for reuse, while inner membrane LC3-II is digested along with the cargo. The rate-limiting step in the autophagic process is autophagosome formation.

Autophagy can be regulated by mTOR-dependent signaling pathways and mTOR-independent signaling pathways. The serine–threonine kinase mTOR is a master negative regulator of autophagy that acts by blocking the activity of the ULK1/Atg1 complex [26]. mTOR is inhibited when nutrients are scarce, when growth factor signaling is reduced, and under ATP depletion. When mTOR is inhibited and the repressive effect of mTOR on autophagosome formation disappears an increase in autophagosome biogenesis occurs [27]. In pathways that act independently of mTOR, when in nutrient-rich conditions, Beclin-1 is bound to the antiapoptotic protein Bcl-2. Under nutrient starvation, the stress-activated enzyme Jun N-terminal kinase 1 (JNK1) phosphorylates Bcl-2, which induces its dissociation from Beclin-1 [28]. Beclin can then interact with other members of the autophagic machinery and stimulate the induction of autophagy [29-30]. One of the mTOR-independent pathways can be mediated by the IP3 pathway and the Ca^{2+}–calpain–G-stimulatory protein alpha pathway [31]. Generation of IP3 induces the release of Ca^{2+} from stores in the endoplasmic reticulum and high levels of cytosolic Ca^{2+} inhibit autophagy by activating calpains. After cleavage by calpains, the Ca^{2+}–calpain–G-stimulatory protein alpha pathway becomes activated which, in turn, causes the production of more inhibitory cAMP [31]. On the other hand, decreased levels of IP3 signaling leads to reduced Ca^{2+} release from the endoplasmic reticulum and lower rates of mitochondrial Ca^{2+} uptake, causing a reduction in mitochondrial activity and ATP depletion, which results in 5'-AMP-activated protein kinase (AMPK) activation [32]. Activated AMPK directly phosphorylates ULK1, indirectly inactivating mTOR complex which leads to the induction of autophagy.

2. Autophagic failure at cellular level

Two of the main aggravating factors contributing to failure of autophagic pathways are believed to be oxidative stress and aging. Aging can lead to reductions in autophagosome formation and autophagosome-lysosome fusion [33]. Moreover leads to lysosomal alterations such as increased lysosome volume, decreased lysosomal stability, altered activity of hydrolases and intralysosomal accumulation of indigestible material such as lipofuscin [34]. These alterations are consistent with a decrease in autophagy, more specifically macroautophagy. Consequently, there is a reduction in the turnover of intracellular components as well as a reduction in the ability of cells to adapt to changes in the extracellular environment [35]. Ultimately this contributes to the intracellular accumulation of misfolded proteins in aged organisms [36]. Moreover, also a decline in CMA activity has been described in almost all cell types and tissues. This decrease is primarily due to a reduction in the levels of LAMP-2A at the lysosomal membrane [37]. In addition, the stability of LAMP-2A at the lysosomal membrane is markedly reduced with age, resulting in a decrease in the net content of this protein in lysosomes [38].

On the other hand, oxidative stress overloads the macroautophagic system, and oxidized proteins and damaged organelles engulfed by autophagosomes can become a source of ROS inside either autophagosomes or lysosomes. Subsequently ROS can react and damage lysosomal hydrolases and/or other components required for the lysosome/autophagosome fusion, resulting in the accumulation of undegraded products inside these cellular compartments, such as lipofuscin. Defects in basal autophagy could lead to altered neuronal homeostasis and degeneration through impaired utilization of nutrients or an imbalance of vesicular biogenesis and turnover. Alternatively, neuronal dysfunction might be a more direct result of failed protein degradation with resultant accumulation of ubiquitinated protein aggregates.

2.1. Lipofuscin accumulation

The most abundant pigment in the human brain is lipofuscin, a protein- and lipid-based pigment with broad distribution [39]. Lipofuscin is commonly considered to be a ubiquitous pigment within the brain, being recognized as a hallmark of aging [40]. Lipofuscin is a chemically and morphologically polymorphous and pigmented waste material that is formed exclusively in lysosomes where it can accumulate as well as in other lysosome-related vesicles due to incomplete digestion of engulfed components and subsequent intra-lysosomal oxidation [41]. Ivy and co-workers suggested that lipofuscin accumulation might result from age-related decrease in the activity of lysosomal enzymes, especially cysteine proteases, such as cathepsins B, H, and L [42]. However where some studies showed a decrease in lysosomal cysteine protease activity others did not because lipofuscin starts to accumulate immediately after birth and continues to do so more or less linearly through the life span [41]. Lipofuscin formation appears to depend on the rate of oxidative damage to proteins, the functionality of mitochondrial repair systems, the proteasomal system, and the functionality and effectiveness of the lysosomes. Major factors that contribute and enhance lipofuscin formation are increased autophagic activity, increased oxidative stress or decreased antioxidant defences, and high concentrations within the lysosomes of redox-active iron [43].

The cell recycles many of useful substrates by autophagic degradation to simple molecules, such as aminoacids, fatty acids and simple sugars, which may be reused after relocalization to the cytosol. Many of these macromolecules that are autophagocytosed contain iron, which is released from a variety of metalloproteins during their intralysosomal degradation [43]. Moreover ROS (mainly H_2O_2, which is produced in mitochondria by dismutation of $O^{2\bullet-}$) easily diffuse into lysosomes and by interacting with iron results in Fenton-type reactions which leads to excessive production of ROS and peroxidation of lysosomal protein content. Moreover the newly generated ROS are then able to interact with the proteins and lipids within the lysosome and form a complex array of Schiff bases and cross-linking of surrounding macromolecules culminating in lipofuscin formation [44]. This is supported by data that demonstrates that oxidative stress, high iron or decreased antioxidant systems stimulate lipofuscin formation whereas antioxidants and iron chelators attenuate lipofuscin formation [39]. Furthermore, oxidatively damaged mitochondria may contain some already peroxidized, undegradable macromolecules. In addition, such ineffective mitochondria are not only ferruginous but also may generate larger amounts of $O^{2\bullet-}$ than do functional mitochondria. Inside the lysosome, such production of $O^{2\bullet-}$ may continue for a while, as well as, iron-catalyzed oxidative modification of mitochondrial components. Moreover ATP synthase subunit c, a mitochondrial protein is a predominant component of lipofuscin from aged neurons [45]. Indeed, autophagocytosed mitochondria seem to be a major source for both macromolecular components of lipofuscin and the low mass iron that catalyzes the peroxidative reactions resulting in its formation. Therefore the primary constituents of lipofuscin are oxidatively modified protein residues, which are bridged into polymer complexes by acids, and lipid residues such as reactive aldehydes originating from the breakdown of triglycerides, free fatty acids, cholesterol and phospholipids, while carbohydrates form only a minor structural component [33]. Proteins within lipofuscin are linked by intramolecular and intermolecular cross-links; many of these cross-links are caused by nonproteineous compounds including oxidation products of other cellular components, such as 4-hydoxy-2-nonenal. Overall, the composition of lipofuscin is variable and dependent upon cell type, but in neurons, lipofuscin appears to be derived primarily from autophagocytosed mitochondria [46].

Hence, lipofuscin accumulation in post-mitotic tissues is both a hallmark of normal senescence and symptomatic of numerous age-related diseases including Alzheimer's disease (AD) and age-related macular degeneration. In addition, lysosomal storage diseases (neuronal ceroid lipofuscinosis, or Batten's disease) are also associated with the accumulation of lysosomal pigment displaying properties similar to that typical of lipofuscin. Lipofuscin complex is undegradable as the result of the excessive oxidation and cross-linking that occurs during its formation leading to the inability of the lysosome to degrade all incorporated materials. Lipofuscin inherently causes toxic effects, in part because of its ability to bind metals, such as iron, copper, zinc, manganese, and calcium, in a concentration up to 2% [47]. Being rich in heavy metals such as iron lipofuscin may jeopardize lysosomal stability under severe oxidative stress, causing enhanced lysosomal rupture and consequent apoptosis/necrosis. The only known mechanism allowing cells to get rid of lipofuscin is mitotic activity, which results in the dilution of the pigment. When lysosomes accumulate lipofuscin, lysosomal enzymes increasingly go to lipofuscin loaded lysosomes in an attempt to degrade the non-degradable material. Since the capacity to produce lysosomal enzymes for autophagy is not unlimited, the lack of lysosomal enzymes for autophagy leads to reduced ability to recycle other cellular organelles, such as mitochondria. Hence lipofuscin acts as a sink for newly produced lysosomal enzymes. If damaged mitochondria are not eliminated this subsequently results in a lower rate of ATP production and increased ROS production. Moreover, the release of the abundant iron from the aged intralysosomal compartment by free-radical-mediated membrane damage will also stimulate free radical production via Fenton chemistry, possibly leading to apoptotic cell death [46]. On the other hand, large numbers of lipofuscin-containing lysosomes, which also contain active hydrolases, may promote cellular damage if lysosomal membranes are destabilized by pathogenic factors (including oxidative stress), resulting in leak of hydrolytic enzymes into the cytosol. Because lipofuscin is separated from the rest of the cytoplasm by the lysosomal membrane, it cannot react directly with extralysosomal constituents. However, within lysosomes loaded lipofuscin iron may promote ROS production, sensitizing cells to oxidative injury through lysosomal destabilization. So, lipofuscin a yellowish-brown, autofluorescent, nondegradable polymeric substance that cells cannot get rid of it [48], reduce the lysosomal degradative capacity [34]. Interestingly, the rate of lipofuscin formation is inversely related to age and lysosomes containing lipofuscin have a reduced ability to fuse with autophagic structures [43].

Moreover, lipofuscin is able to decrease not only lysosomal degradation but also proteosomal degradation, perhaps as a result of binding proteasomal complexes in unsuccessful attempts of degradation [49]. Therefore, lipofuscin directly decreases cellular proteolysis by inhibition of the proteasomal turnover, resulting in reduced proteasomal activity [50]. Another reason for lipofuscin toxicity is the gradual filling of the cytoplasmic space over time, [51] resulting at first in decreased cellular functional capability, later in apoptotic cell death. In addition, also intracellular trafficking as well as cytoskeletal integrity may be compromised by the presence of large intracellular aggregates.

2.2. Defective autophagy

Defects in autophagy have been described to occur at very different levels within the cell and at very different diseases. Therefore understanding how autophagy step or steps are altered in specific disorders is a priority. There are several examples of defects that can arise in autophagy.

For example, initiation of autophagy may be compromised because of altered signaling through the insulin or mTOR pathways, which is tightly bound to activation of autophagy. Since autophagy can be cargo-specific which implicate cargo adaptors molecules that enable autophagosomes to identify specific substrates, alterations in the organelle-specific markers for degradation or in the autophagic machinery can impair autophagic turnover. For instance, if autophagosomes not recognize damaged mitochondria, defects in mitophagy occur which contributes to disease pathology, since removal of damaged mitochondria is impaired leading to high levels of ROS. Cells are then more susceptible to proapoptotic insults. Defects in cargo recognition occurs for instance in Huntington's disease (HD). HD is caused by gain-of-function mutations that confer neurotoxic effects on the ubiquitously expressed protein huntingtin [52]. This gain-of-function impairs the ability of autophagosomes to recognize certain cargoes, while mutant huntingtin is efficiently incorporated into autophagosomes, it may reduce autophagic sequestration of other cargoes [53].

In other conditions, autophagosomes form correctly and sequester relevant cargo but they fail to be cleared from the cytosol. Problems with clearance could result from alterations at very different levels. For example, problems with vesicular trafficking could indirectly interfere with the mobilization of autophagosomes toward the lysosomal compartment [54]. The cytoskeleton is extremely important in the trafficking of organelles and in fact has the role to maintain the spatial organisation for autophagy by conducting the trafficking of organelles involved in different interactions during autophagy. Therefore, autophagy is microtubule-dependent [55]. When microtubules are disrupted by colchicine, autophagosome–lysosomal fusion is also disrupted leading to an increase in the number of autophagic vacuoles (AVs) [56].

Pathogenic proteins can also interfere with the fusion step, which, although still not completely elucidated at the molecular level, is known to depend on different SNARE proteins, the actin cytoskeleton and the histone deacetylase 6 (HDAC6) [57]. Furthermore, if organelle traffic through the axon is "jammed" we will have organelle transport problems [58].

It is also possible that problems in maturation of autophagosomes and their fusion with lysosomes also occur. The maturation of autophagosomes and their fusion with lysosomes is dependent on the motor protein dynein, which mediates autophagosome movement along the cytoskeleton towards the microtubule-organizing center where the lysosomes are clustered. Loss of dynein function also results in the accumulation of autophagosomes and reduction of the clearance of intracytoplasmatic aggregation-prone proteins [54].

Furthermore, lysosomal defects also have a negative impact on clearance of autophagosomes. Autophagosomes can form properly and sequester the usual cargo but they are not eliminated through the lysosomal system. The reasons for lysosomal failure could be multiple. For instance in most lysosomal storage disorders (LSDs) the accumulation of undegraded products inside lysosomes limits their degradative capacity [59]. Undigestible material in lysosomes could build up or otherwise inhibit hydrolases, and could also dilute or divert the delivery of lysosomal hydrolases decreasing their efficiency. Interestingly, accumulation of β-amyloid (Aβ) 1-42, which has a high propensity to aggregate and therefore is less efficiently degraded, causes leakage of lysosomal enzymes into the cytosol [60-61].

Moreover the build up of AVs filled with undigestible material could inhibit secretory pathways interfering with nutrient uptake and response to growth factors or recovery from stress. Additionally, it could also inhibit organelle fusion or block the supply of aminoacids from autophagic protein breakdown, inducing cell starvation [62]. Additionally, conditions that alter lysosomal membrane stability, decreases lysosomal biogenesis or changes lysosomal pH inhibiting lysosomal proteolysis (because of the acidic pH requirements of their degradative enzymes) could also alter autophagosome clearance. Lysosomal hydrolases inhibition by defective acidification of the lysosomal lumen can be due to the inability to target to lysosomes one of the subunits of the proton pump that usually acidifies this compartment, or by direct inhibition of cathepsins that leads to reduced rates of autophagy. Changes in the lysosomal pH could be due to enhanced activity of the V-ATPase. The V-ATPase is a holoenzyme consisting of a membrane bound Vo and cytosolic V1 components, and both V1 and Vo are composed of multiple subunits. The Vo subunit a1 is required for acidification of degradative competent lysosomes [63]. Preservation of low pH is important for cargo release, lysosomal hydrolase and vesicle maturation, autophagy and neurotransmitter loading into synaptic vesicle [64].

Accelerated endocytosis also increases protein and lipid accumulation in endosomes and slows lysosomal degradation of endocytic cargoes [65], leading to lysosomal instability.

The process from protein sorting to endosomal-lysosomal fusion is maintained by the sequential interaction of four complexes termed the endosomal sorting complexes required for transport (ESCRT complexes). The four ESCRT complexes, numbered 0–III, are required for the degradation of aggregate-prone proteins by autophagy [66]. Mutations or depletions affecting ESCRT related genes results in deficient maturation of autophagosomes or in their inability to fuse with lysosomes and endosomes. This leads to autophagosomes accumulation without degradation of their cargo, as well as, ubiquitin-positive aggregates accumulation leading to neurodegeneration in many cases [67]. As proteasome is thought to be turned over by autophagy [68], blockade of proteasome turnover could disrupt additional degradative pathways.

More specifically alterations of CMA can also occur. Different pathogenic proteins have been shown to directly interfere with CMA activity [69]. For instance, mutant forms of alpha-synuclein fail to translocate to the lysosomal lumen whereas the wild-type protein binds to the lysosomal receptor and rapidly reaches the lumen for degradation. Consequently, mutant forms of alpha-synuclein block access of other cytosolic proteins to lysosomes via CMA by abnormally bind to the lysosomal receptor. Moreover, the accumulation at the surface of lysosomes of oligomeric forms of pathogenic proteins targeted via CMA destabilizes the lysosomal membrane and results in leakage of lysosomal enzymes into the cytosol, which often triggers cellular death.

2.2.1. *Consequences of defective autophagy*

Healthy cells harbour a high autophagic clearance capacity preventing the "traffic jams" of endosomes, autophagosomes, as well as, aggregated proteins and damaged organelles. Only unfolded monomer proteins can undergo degradation through any of these two systems: autophagy, specifically CMA and ubiquitin-proteasome system (UPS). Consequently, once organized in oligomers, protofibrils, and fibrils, proteins can only be removed by in-bulk degradation, such as via microautophagy or macroautophagy. Thus, oligomers and fibrils of particular proteins can block the proteolytic activity of the UPS and of CMA. These alterations in intracellular proteins can be due to exposure to intracellular or extracellular stressors, such as oxidative stress, endoplasmic reticulum stress, ultraviolet radiation and other toxic insults. In addition, genetic mutations can generate proteins that cannot fold properly or are prone to aggregation. This can lead to the generation of misfolded proteins or modified soluble proteins resulting from protein cross-linking and oligomerization. The most toxic forms of altered proteins are complex organized structures, such as fibrils or oligomers, although the mechanism by which they exert their cellular toxicity is still controversial. In the absence of a properly functioning quality control system, and as a last attempt to prevent toxicity, cells favor formation of protein aggregates rather than fibrils and oligomeric complexes. The maturation of misfolded or unfolded protein into protein aggregates can vary across different disorders, but generally protein aggregation results from proteins that fold into an abnormal conformation, leading to the formation of oligomeric intermediates [70]. These smaller aggregates, both structured and unstructured, continue to grow and multimerize into larger aggregates or inclusions. Larger cytoplasmic inclusions can evolve further and coalesce into an aggresome, a pericentriolar, membrane-free cytoplasmic inclusion formed specifically at the microtubule organizing center. It has been proposed that the aggresome is a protective structure, formed to sequester proteins that cannot be degraded by the proteasome and packaged for degradation by autophagy. Moreover they sequester toxic monomeric or oligomeric species diluting its toxicity and facilitating the removal of these toxic species. However, aggregates are not completely harmless, because they interfere with normal cellular tra ▪ cking and become a sink for still-functional proteins that usually get trapped in these aggregates [71]. Their prolonged presence in neurons is indicative of some failure in fundamental cellular processes.

Taking this into account protein aggregation occurs and enhances the complexity that each neurodegenerative disorders presents. Interestingly, despite the unique features of each neurodegenerative disease, protein aggregation also shares several common characteristics. Generally, the major component of the inclusions is often ubiquitously expressed, such as huntingtin, alpha-synuclein and tau. Moreover the inclusions are found throughout the brain, and they do not correlate only to the pattern of neurodegeneration that occurs in each neurodegenerative disorder [6].

In addition, accumulation of defective, no longer functional organelles is also deleterious for the cells [72]. Whole organelles including peroxisomes and mitochondria are degraded by macroautophagy. If AVs containing damaged mitochondria or peroxisomes are not degraded, damaged organelles will gradually accumulate [41, 73].Mitochondrial degradation by autophagy, known as mitophagy plays an important role in the regulation of mitochondrial function and remodeling. Mitophagy may also be important in attenuating apoptosis or necrosis, by clearance of damaged mitochondria. This could then prevent the release of cytochrome c, AIF (apoptosis-inducing factor) and other apoptotic factors that lead to cell death [74]. Clearly, autophagic removal of defective mitochondria is of crucial importance for cell survival. Inhibition of autophagy by 3-methyladenine in growth-arrested human fibroblasts, as a model of cell aging, results in the accumulation of lipofuscin-like material and of mitochondria, with a low-membrane potential [75]. Enlarged and structurally deteriorated mitochondria, showing swelling and disrupted *cristae* often result in the formation of amorphous material [76-77]. These mitochondria are defective in ATP production and produce increased amounts of ROS which are harmful for cells [78]. One would think that damaged mitochondria should be degraded, but their accumulation with age implies that they either acquire replicative advantage over normal mitochondria, or instead they accumulate due to a decrease in autophagic-lysosomal pathway.

Overall, autophagy is directly responsible for the maintenance of a proteome free of alterations [79]. Importantly, autophagy by mediating the removal of damaged organelles after stress restores organelles homeostasis being essential for the maintenance of cellular homeostasis and to guarantee cellular survival during stress. When quality control systems fail to accomplish its function is the basis for protein conformational disorders. In many late-onset neurodegenerative disorders, including Parkinson's disease (PD) and HD, there is accumulation of intracellular protein aggregates in the brain [6]. The elimination of these intracellular protein aggregates is often correlated with amelioration of symptoms of the disease [6]. Indeed, in mice with deficiencies in either Atg7[33] or Atg5[34] constitutive autophagy is required for the clearance of cytosolic aggregate-prone proteins from neurons [80].

3. Autophagy in disease

The broad array of physiological functions attributed to autophagy justifies why alterations in this catabolic process lead to cellular malfunctioning and often cell death. Autophagy is closely involved in the etiology of several human diseases contributing to its pathogenesis, including cancer, neurodegenerative diseases and metabolic disorders [81-82].

3.1. Autophagic dysfunction in neurodegenerative disorders

Alzheimer's disease

As AD progresses, either due to AD related genes or environmental/aging factors, several pathological changes of the lysosomal network occurs, such as deregulation of endocytosis and increased lysosomal biogenesis culminating in a progressive failure of lysosomal clearance mechanisms [83]. Enlargement of Rab5 and Rab7 positive endosomes is one of the earliest specific pathology reported in AD brain tissue which reflects a pathological acceleration of endocytosis. Interestingly, it develops in pyramidal neurons of the neocortex at a stage when plaques and tangles are restricted only to the hippocampus. Furthermore genes involved in endocytosis are up-regulated in AD and their corresponding proteins are abnormally recruited to endosomes promoting fusion and enlargement of early and late endosomes, which is a specific characteristic of AD and is not seen in normal aging brain. Acceleration of endosome pathology is also seen in individuals who inherit the ε4 allele of APOE, the major risk factor for late-onset AD [84-85]. Lipinski and co-workers recently reported that transcription of factors that promote autophagy are up-regulated in the brains of AD patients, while negative regulators of autophagy are down-regulated [86]. Indeed, cellular ultrastructural changes have been described in AD brain biopsies revealing a high level of AVs within dystrophic neurites [87]. AVs and lysosomes constitute more than 95% of the organelles in dystrophic neuritic swellings in AD. This means that autophagy initiation is up-regulated or its progression is either delayed or impaired. However the profuse and selective accumulation of AVs in dystrophic neurites indicates a defect in the clearance of AVs by lysosomes rather than an abnormally augment of autophagy. In the case of familial AD, for instance, presenilin 1 (a ubiquitous transmembrane protein involved in diverse biological roles) mutations hinder lysosome proteolysis and accelerate neuritic dystrophy which also supports a primary role for failure of proteolytic clearance. Presenilin 1 is required for lysosome acidification which is needed to activate cathepsins and other hydrolases that carry out digestion during autophagy. Mutations in Presenilin 1 result in impaired targeting of the a1 subunit of V0-ATPase from the endoplasmic reticulum to the lysosome. As V0-ATPase is required for acidification of the autolysosome contents, mutations in Presenilin 1 are proposed to be involved in the defective proteolysis of autophagic substrates in patients with AD [88]. Furthermore, Zhang and colleagues reported for the first time, a role for presenilins in regulating lysosomal biogenesis [89]. The role of impaired lysosomal degradation in the etiology of AD was underscored in a study of a transgenic mouse model of this disease. In these mice, deletion of cystatin B (an endogenous inhibitor of lysosomal cysteine proteases) stimulated the turnover of proteins by the lysosome, enhanced the clearance of the autophagic substrates (including Aβ), and rescued the deficient cognitive phenotype of the animals. Furthermore, in APP transgenic mouse models of AD, undigested autophagic substrates including LC3-II, p62, and ubiquitinated proteins accumulate in neuronal AVs [90]. This general failure to clear autophagy substrates affects clearance of various proteins relevant to AD pathogenesis, including the protein Aβ and tau promoting cell death [91]. These results indicate that mutant APP overexpression alone can lead to autophagic-lysosomal pathology. However the mechanism by which overexpression of mutant APP may lead to impaired autophagy and neuritic dystrophy, is not well understood. One possibility is that APP may affect the endosomal-lysosomal system. Strong overexpression of human Aβ42 in Drosophila neurons induces age-related accumulation of Aβ in AVs and neurotoxicity which is further enhanced by autophagy activation and is partially rescued by autophagy inhibition [92]. These authors propose that the structural integrity of post-fusion AVs may be compromised in Aβ42 affected neurons, leading to subcellular damage and loss of neuronal integrity in the Aβ42 flies. Moreover, expression of an APOEε4 allele, but not the APOEε3 allele, in a mouse AD model increases levels of intracellular Aβ in lysosomes, altering their function and causing neurodegeneration [93]. Interestingly, inhibition of Aβ aggregation rescues the autophagic deficits in the TgCRND8 mouse model of AD [94]. Autophagy sequesters and digests unneeded or damaged organelles, some of which are APP-rich [95]. Autophagosomes are enriched in APP as well as APP substrates and enzymes that are responsible for processing APP into Aβ. Under normal circunstances Aβ is subsequently degraded by lysosomes [96-97]. Therefore autophagosomes are a site of intracellular production of Aβ, thus upon their cellular accumulation amyloid deposition occurs [98].

Failure of the autophagic system also compromises the elimination of aggregate forms of tau, a protein that also accumulates in AD neurons [99]. In fact, for certain types of tau mutations, pathogenic tau could contribute to the failure of macroautophagy, due to the toxic effect that the still soluble forms of the protein exert in the membrane of lysosomes when they are delivered to this compartment by CMA. Furthermore, particular mutant forms of tau have been shown to abnormally interact with components of the lysosomal CMA translocation machinery [100]. More interestingly, stimulation of autophagy is neuroprotective in a mouse model of human tauopathy [101]. In addition to the defects in late stages of autophagy, evidence suggests that autophagy might be disrupted at the level of autophagosome formation in patients with AD. Compared with healthy individuals, the brains of patients with AD show reduced expression of Beclin-1, which could lead to an impairment in the initiation of autophagy [95]. Transgenic mice that expressed a mutated form of the human APP on a Beclin-1 haploinsufficient background had disrupted autophagy, as well as, increased intracellular Aβ accumulation and neurodegeneration, compared with mice that expressed the mutated human APP in the context of a normal Beclin-1 background [95]. Caspase 3-mediated cleavage of Beclin-1 occurs in the brains of patients with AD; thus, increased activity of this enzyme might contribute to the loss of Beclin-1 function in individuals with this disease [102].

Parkinson's disease

In PD the most common pathogenic protein, alpha-synuclein, is usually degraded by different autophagic and non-autophagic pathways. Soluble forms of the protein are substrates of both the UPS and of CMA [103]. However, macroautophagy is the only plausible way for the elimination of the pathogenic variants of this protein once they aggregate [104]. When alpha-synuclein aggregates can no longer be degraded by either the UPS or CMA, macroautophagy becomes the only proteolytic pathway able to remove these proteinaceous deposits from the neuronal cytosol. As in the other disorders, post-translational modifications and pathogenic forms of alpha-synuclein promote the formation of oligomeric species that interfere with the normal functioning of the UPS [105], CMA [103] and even macroautophagy at the level of autophagosome formation [106]. In fact, upregulation of wild-type alpha-synuclein leads to significant inhibition of macroautophagy [107] and mutant forms of alpha-synuclein A30P and A53T have been shown to be poorly degraded by CMA, because they are not translocated into the lysosome although they bind to the lysosomal membrane with high affinity. Furthermore, because of their high-a ▪ nity binding to the CMA receptor, they block the uptake and degradation of other CMA substrates, leading to a general CMA blockage [103]. Indeed, dopamine-modified alpha-synuclein, which is modified by non-covalent binding of oxidized dopamine, not only is poorly degraded by CMA but also block degradation of other substrates [108]. Subsequently, normal substrate proteins for CMA can no longer be turned over through this pathway and also end up accumulating inside the affected cells. Interestingly, inhibition of the UPS and CMA promote macroautophagy upregulation, [21] which maintains normal levels of protein degradation and removes the cytosolic toxic and aggregated alpha-synuclein [109]. AVs have also been described in melanized neurons of the substantia nigra in PD [110]. This accumulation is consistent with either an overproduction or impaired turnover of AVs. Additionally, LC3-II levels were increased in patients with diffuse Lewy Body Disease and PD, and LC3 co-localized with alpha-synuclein in most Lewy Bodies and Lewy neurites, suggesting an increase in macroautophagy, as well as, an attempt to clear alpha-synuclein pathology by up-regulating autophagic activation [111-113]. Furthermore, Alvarez-Erviti et al. revealed that CMA activity, as well as, LAMP-2A and hsc70 levels were significantly decreased in the substantia nigra from PD patients when compared to controls [114]. Likewise, cathepsin D immunoreactivity was significantly reduced in substantia nigra neurons of PD patients with an even greater decrease in alpha-synuclein inclusion-bearing cells [115].This suggests the presence of abundant and dysfunctional autophagosomes and lysosomes in PD and diffuse Lewy Body Disease. In several cell models, such as SK-SH5Y [116], BE-M17 [117], PC12 cells [118] , as well as, in primary mesencephalic [117] or cortical neurons [119], 1-methyl-4-phenylpyridinium (MPP^+) has been shown to perturb the autophagy flux, as mirrored by an evident increase in the number of LC3-positive autophagic vesicles, in association with apoptotic or nonapoptotic cell death. More interestingly, cells harboring PD patient's mitochondria or in mitochondrial DNA-depleted cells show compromise quality control autophagic response and defective clearance of the AVs [120]. The recently identified physiological role for Parkin and Pink1, two PD-causative genes, in mitophagy, suggests that alterations of this selective form of autophagy could also contribute to the pathogenesis of this disease [121]. PINK1 encodes the serine–threonine kinase and is localized to the outer mitochondrial membrane, and PARK2 encodes Parkin, an E3 ubiquitin ligase. Under basal conditions, Parkin is mainly found in the cytosol where its ubiquitin ligase activity is inhibited by an unknown mechanism. Loss of mitochondrial membrane potential drives the recruitment of Parkin to the mitochondria. In fact, under physiologic and pathologic stress, Parkin is selectively recruited from the cytosol to damaged mitochondria in a PINK-1-dependent manner where ubiquinates membrane proteins such as voltage-dependent anion channel 1 and mitofusins 1 and 2, mediating the elimination of the mitochondria by mitophagy [122-124].

Neurodegeneration in PINK-1 and Parkin-positive familial forms of PD may result from a defect in mitophagy, leading to the accumulation of damaged mitochondria and excessive ROS production. In fact, Parkin-deficient mice accumulate dysfunctional mitochondria and oxidative damage [125] while Drosophila flies with mutated Parkin are particularly sensitive to oxygen radical stress [126]. In addition, other PD causative gene, DJ-1 also plays a protective role in autophagy. Loss of DJ-1 has been shown to result in decreased basal autophagy [127]. Moreover, Irrcher and co-workers demonstrated that DJ-1 deficiency leads to altered autophagy in murine and human cells [128]. McCoy and Cookson also showed that loss of DJ-1 leads to mitochondrial phenotypes including reduced membrane potential, increased fragmentation and accumulation of autophagic markers [129]. Moreover, Lrrk2 gene which is mutated in certain dominant forms of familial PD was shown to modulate autophagy. For instance, G2019S mutation induces autophagy via MEK/ERK pathway and inhibition of this exacerbated autophagy reduces the sensitivity observed in G2019S mutant cells [130]. Furthermore, Lrrk2 was shown to co-localize with markers of the endosomal–lysosomal pathway in cases of diffuse Lewy Body Disease [131]. Loss of function of Lrrk2 mutants have been shown to decrease neuritic arbor and to cause the accumulation of autophagic vesicles and swollen lysosomes containing tau inclusion bodies [132]. However, the precise mechanism(s) by which LRRK2 regulates autophagy are still elusive. Interestingly, recent data suggest a mechanism involving late steps in autophagic-lysosomal clearance in a manner dependent on NAADP (nicotinic acid-adenine dinucleotide phosphate)-sensitive lysosomal Ca^{2+} channels [133]. Other connection between autophagy and PD has been established in studies where mutant forms of the ubiquitin carboxyl-terminal esterase L1 (UCHL-1) described in some familial forms of PD were shown to interact abnormally with CMA components, namely hsc70, hsp90, and LAMP-2A. The abnormally binding between UCHL-1 and these three CMA components was proposed to interfere with CMA activity [134].

3.2. Autophagic dysfunction in lysosomal storage disorders

Lysosomes are ubiquitous organelles that have crucial roles in cellular clearance. Individually rare, it is now known that about 1 in 7500 live-born infants in many populations will have a lysosomal disease [135]. The LSDs are a group of over 60 diseases, clinically characterized by a progressive phenotype involving multiple organs and tissues, including severe neurodegeneration [136]. Indeed, LSDs involving primary defects in lysosome function commonly exhibit prominent neurodegenerative phenotypes, including neuritic dystrophies closely resembling the ultrastructural morphologies of dystrophic neurites in AD and, in some of these disorders, neurofibrillary tangles as well as increased amyloidogenic processing of APP and diffuse Aβ deposits [137]. Walter and colleagues demonstrated in LSDs an accumulation of sphingolipids, as occurs in lysosomal lipid storage disorders and, decreases the lysosome-dependent degradation of APP C-terminal fragments that stimulates γ-secretase activity increasing the generation of both intracellular and secreted Aβ. Notably, primary fibroblasts from patients with different storage diseases show strong accumulation of potentially amyloidogenic APP C-terminal fragments [138]. LSDs are characterized by progressive accumulation of undigested metabolites such as glycogen, ceramide, heparin sulphate and glucocerebroside, among others, within the cell due to lysosomal dysfunction. Loss-of-function of lysosomal proteins causes accumulation of autophagic and endosomal substrates, and as a result lysosomal storage. Such proteins include lysosomal enzymes, lysosomal integral membrane proteins, and proteins involved in the post-translational modification and trafficking of lysosomal proteins. Defects in these proteins involved in lysosome regulation or function induce the accumulation of undigested molecules that can subsequently alter many cellular processes. These include lysosomal pH regulation, synaptic release, endocytosis, vesicle maturation, autophagy, exocytosis and Ca^{2+} homeostasis [139-140]. As a consequence, many tissues and organ systems are affected, including brain, viscera, bone and cartilage, with early onset central nervous system dysfunction predominating. Although individuals with LSDs can display early symptoms, the majority are clinically normal at birth which suggests that lysosomal dysfunction per se does not impact significantly the events of early brain development. Furthermore children typically meet early development milestones signifying that lysosomal storage does not affect neuronal function and maturation at early developmental stages. In patients with these disorders, dysfunction of a lysosomal enzyme (usually one of the many lysosomal hydrolases) causes impaired degradation of the enzyme's substrate, which accumulates within lysosomes [136]. An abundance of AVs is a common feature of cells from patients with LSDs. Lysosomal storage impairs autophagic delivery of bulk cytosolic contents to lysosomes. This can be attributed to a defect in autophagosome–lysosome fusion as a result of dysfunction of the lysosomal compartment. This, in turn, leads to a progressive accumulation of poly-ubiquitinated protein aggregates and of dysfunctional mitochondria, suggesting that neurodegeneration in LSDs may share mechanisms with some neurodegenerative disorders in which the accumulation of protein aggregates is a prominent feature [141]. This defect in autophagosome-lysosome fusion can result from an impairment of vesicular trafficking due to a microtubule-based transport deficiency. The motor protein dynein mediates the movement of mature autophagosomes towards lysosomes. Indeed, mutations affecting dynein impair the autophagic degradation of aggregate-prone proteins [54]. Another possible mechanism involves changes in the lipid composition of lysosomal membranes. In fact, lipids have been demonstrated to accumulate in a wide range of LSDs [142]. These lipids are the principal constituents of lipid-rafts. Since lipid-rafts play a critical role in membrane physiology influencing their plasticity, it is possible that the abnormal accumulation of lipids in LSDs leads to an increase in lipid-rafts that affects the dynamics of lysosomal membranes and specifically their ability to fuse with autophagosomes. Indeed, in conditions that are similar to what is found in Niemann Pick type C diseases-1 and-2, cholesterol accumulation in late endosomes perturb the intra-endosomal trafficking affecting most likely autophagosome maturation [143].

Impaired autophagy has been reported in several models of LSDs, including Pompe disease, Niemann-Pick disease, the neuronal ceroid lipofuscinoses, multiple sulphatase deficiency, and GM1-gangliosidosis. Additionally, lysosomal function is intimately linked to exocytosis (removal of cellular cargo by fusion of vesicle with the plasma membrane), and multiple LSDs, such as mucolipidosis type I, Niemann-Pick disease and sialidosis have been shown to have impaired exocytosis [144].

http://dx.doi.org/10.5772/55196

The first LSD in which an involvement of autophagy was reported was Danon disease (also called "glycogen storage disease due to LAMP2 deficiency" or "lysosomal glycogen storage disease with normal acid maltase activity") [145]. It was reported an accumulation of AVs in several tissues, particularly the muscle, from a mouse model of the disease [146]. The disease, extremely rare, is inherited as an X-linked trait and the disease phenotype is characterized by severe cardiomyopathy and variable skeletal muscle weakness often associated with mental retardation.

Another LSD is Multiple sulfatase deficiency which is attributed to deficiencies in the activity of sulfatase enzymes. The sulfatases are a family of enzymes that catalyze the hydrolysis of sulfate ester bonds in a wide variety of substrates, ranging from complex molecules, such as glycosaminoglycans, to sulfolipids and steroid sulfates. These enzymes can be divided, at least in mammals, into two main groups based on their subcellular localization: those found in lysosomes (acting at an acidic pH) and those found in the endoplasmic reticulum, the Golgi apparatus and at the cell surface (acting at a neutral pH) [147]. Sulfatases are activated upon a post-translational modification by sulfatase modifying factor 1. The gene encoding sulfatase modifying factor 1 in humans is mutated in this rare autosomal recessive disorder, in which the activity of all sulfatases is profoundly impaired [148]. As for other types of LSDs, the pathogenic mechanisms that lead from enzyme deficiency to cell death in Multiple sulfatase deficiency is not completely understood. An impairment of autophagy is postulated to play a major role in disease pathogenesis. In mouse models of multiple sulfatase deficiency, it was observed an accumulation of autophagosomes resulting from defective autophagosome-lysosome fusion which was demonstrated by the inefficient degradation of exogenous aggregate-prone proteins (expanded huntingtin and mutated alpha-synuclein) and defective organelles [59]. Moreover in chondrocytes from multiple sulfatase deficiency mice show a severe lysosomal storage defect and a defective autophagosome digestion leading to a defect in energy metabolism and to cell death [149]. Interestingly, the mechanisms underlying the impairment of the fusion between lysosomes and autophagosomes in multiple sulfatase deficiency seems to involve abnormalities of membrane lipid composition and SNARE protein distribution [150]. Overall, it seems that a global lysosomal dysfunction leads to the impaired autophagy observed in the pathogenesis of multiple sulfatase deficiency.

Niemann-Pick type C disease is caused by mutations in the NPC1 or NPC2 genes, [151] whose protein products are thought to act cooperatively in the efflux of cholesterol from late endosomes and lysosomes [152]. In the Niemann–Pick type C disease defects in lysosomal trafficking and biogenesis occur. Moreover, a marked accumulation of autophagosomes occurs in the brains of Niemann–Pick disease type C mice and in skin fibroblasts from Niemann–Pick disease type C patients [153-155]. Using human embryonic stem cell-derived neurons engineered to mimic the cholesterol lysosomal storage disease Niemann Pick type C, we have shown that excessive activation and impaired progression of the autophagic pathway lead to abnormal mitochondrial clearance [156]. NPC1 deficiency leads to both an induction of autophagy and an impairment of autophagic flux. The impairment in degradation of autophagic substrates may contribute to several aspects of NPC neuropathology, including the accumulation of ubiquitinated proteins and the generation of ROS.

Gaucher's disease is an autosomal recessive condition, due to deficient activity of lysosomal β-glucosylceramidase. The pathologically enlarged and often multinucleate 'storage' cell is a striking feature of Gaucher's disease. Similarly to Niemann-Pick type C disease, also models of Gaucher disease show both an induction of autophagy and an accumulation of autophagosomes and autophagic substrates [157-158].

In LSDs the primary alteration of the lysosomal compartment can indirectly affect CMA activity. However, there are two LSDs for which alterations in CMA are not merely a consequence of the lysosomal alteration, but rather are due to the possible functional association of the mutant protein with CMA, such as galactosialydosis and mucolipidosis type IV. The first one is caused by a mutation or loss of cathepsin A gene which is a protease involved in CMA regulation and contributes to lysosomal activity and stability. Moreover is also involved in the degradation of LAMP-2A [159]. Loss of cathepsin A enzymatic activity leads to slower LAMP-2A degradation resulting to an increase in CMA activity which contributes to an extreme weight loss that characterizes these patients. In the second LSD the mutated protein, the transient receptor potential mucolipin-1 (TRPML1) which is a endolysosomal cation channel, interacts with HSPA8 (HSC70) and DNAJB1 (HSP40), two components of the CMA molecular machinery leading to decreased CMA activity in response to serum removal [160]. Moreover it has been suggested that TRPML1 mediates Ca^{2+} efflux from late endosomes and lysosomes indicating that TRPML1 is a key regulator of membrane trafficking along the endosomal pathway [161]. In fact, in TRPML1-deficient cells the delivery of cargo from the cell surface to the lysosome and fusion of lysosomes with the plasma membrane is impaired [162-163].

In spite of all the above mentioned differences among diseases in most cases there is an impairment of autophagic flux, causing a secondary accumulation of autophagy substrates and on the other hand an increase in factors involved in autophagosome formation, such as Beclin1, as an attempt to compensate for the impaired autophagic flux. Accordingly, LSDs can be seen primarily as "autophagy disorders." [164].

3.3. Autophagic dysfunction in diabetes

In response to a variety of metabolic stressors such as nutrient starvation, growth factor withdrawal, high lipid content challenges or hypoxia macroautophagy and CMA can be induced [165]. Moreover autophagy is also capable of mobilizing energetically efficient molecules, such as lipids, glycogen and nucleic acids which can be used by TCA cycle, gluconeogenesis and glycolysis to produce ATP [165-166]. This capability of autophagy to maintain ATP production and support macromolecular synthesis makes it a pro-survival pathway of particular importance in organs with high energetic requirements, such as the heart or skeletal muscles. Therefore, alterations of this specific autophagic function can constitute the basis of some common metabolic disorders such as diabetes. Type II diabetes is caused, in most cases, by the inability of the body to buffer the free fatty acid concentration. This increases the redox pressure on the mitochondrial respiratory chain, increases ROS production, reduces mitochondrial function, and increases apoptosis of beta-cells [167]. Moreover it is characterized by insulin resistance and failure of beta-cells producing insulin. The most potent fatty acid causing insulin resistance is palmitate. As palmitate is the precursor of ceramide and sphingolipid biosynthesis, the sphingolipid pathway has been implicated in the etiology of insulin resistance [168]. Interestingly, ceramide has been shown to activate autophagy by upregulating beclin1 [169]. Moreover a decrease in intracellular aminoacid concentrations and a decrease in mTOR-dependent signaling may also contribute to the activation of autophagy by ceramide [170]. If in fact autophagy turns out to be up-regulated in type II diabetes, we could speculate that the onset of insulin resistance in elderly people could be an adaptive mechanism aiming to increase autophagy and helping to improve the ability to remove damaged organelles [171]. Autophagy is inhibited by the insulin -mTOR signaling pathway and can be activated either upon aminoacid depletion or by rapamycin administration. Insulin inhibits autophagy in two ways: first by activating mTOR in synergy with aminoacids, which results in the phosphorylation and inhibition of the protein kinase Atg1 (ULK1 in mammals); and second by protein kinase B-mediated phosphorylation and inhibition of the transcription factor FoxO3, which is responsible for the expression of ATG genes [172]. Because insulin inhibits autophagy in insulin-sensitive cells, such as beta-cells, autophagy might be increased. In addition, as a result of free fatty acid induced oxidative stress insulin resistance develops in response to over-feeding, which is one of the major causes of insulin resistance. Interestingly, ROS are known to be required to trigger autophagy, probably because they oxidize a critical cysteine residue in Atg4 [173]. A report from 2007 showed that oxidative stress induced by diabetes leads to ubiquitination and storage of proteins into cytoplasmatic aggregates that do not co-localize with insulin. Because accumulation of ubiquitinated protein may damage cells, such data suggests that autophagy may contribute to the regulation of beta-cell survival and death acting as a defense to cellular damage during diabetes [174]. To prove that a relationship between autophagy and beta-cell death indeed exists experiments were performed in a mouse model whose beta-cells were deficient in autophagic activity. In beta-cell specific Atg7 knockout mice it was reported by two different studies islet degeneration, decreased glucose tolerance, insulin secretion and accumulation of large ubiquitin-containing protein aggregates indicating that autophagy was impaired. This happened even with overexpression of LC3-binding protein p62, which is required for polyubiquitinated protein aggregates to be delivered to the autolysosome [175-176]. In beta-cells from diabetic animals the levels of autophagosomes was significantly increased in both diabetic *db/db* mice and in non-diabetic control mice fed with a high-fat diet. This can occur either as a result of impaired autophagosome/lysosome fusion, or impaired function of the lysosomal proton pump. Interestingly when Atg7(-/-) mice were fed with a high-fat diet their glucose tolerance decreased indicating once again the important role of autophagy in maintaining proper beta-cell function under stressful conditions. Ebato and colleagues also showed that free fatty acids which can cause peripheral insulin resistance associated with diabetes induced autophagy in beta-cells [176]. These findings suggest that basal autophagy is important for maintenance of normal islet architecture and function and protects beta-cells against cell damage [177]. Interestingly in autophagy-deficient beta-cells ubiquitinated proteins accumulated inside cells. Moreover also p62, a specific autophagy substrate also accumulated in autophagy-deficient beta-cells [178]. Furthermore, exposure of human neuroblastoma SH-SY5Y cells to sera from type 2 diabetic patients with neuropathy is associated with increased levels of autophagosomes [179]. Jung and co-workers showed that the presence of defective beta-cell mitochondria and endoplasmic reticulum presumably contributed to the reduced ability to produce insulin, indicating that decreased function and mass of beta-cells can result from mitochondrial dysfunction and endoplasmic reticulum stress, since both organelles are autophagic substrates [175].Moreover, deregulated autophagy may be involved in insulin resistance. In fact, mitochondrial impairment and defective endoplasmic stress response have been implicated in insulin resistance [180-182]. In pancreatic beta-cells, the endoplasmic reticulum is the crucial site for insulin biosynthesis. Consequently, perturbations to endoplasmic reticulum function of the beta-cell, such as those caused by high levels of free fatty acid and insulin resistance, can lead to an imbalance in protein homeostasis and ER stress, which has been recognized as an important mechanism for type 2 diabetes. Macroautophagy is activated as a novel signaling pathway in response to ER stress [183]. In the brain of young patients with poorly controlled type I diabetes mellitus and fatal diabetic ketoacidosis was demonstrated increased levels of macroautophagy-associated proteins as well as increased levels of the ER-associated GRP78. Therefore probably chronic metabolic instability and oxidative stress may cause alterations in the autophagy-lysosomal pathway [184].

Interestingly, insulin resistance, one of the major components of type II diabetes mellitus is a known risk factor for AD. Son and co-workers observed that insulin resistance promotes Aβ generation in the brain via altered insulin signal transduction, increased BACE1/β-secretase and γ-secretase activities, and accumulation of autophagosomes. These authors proposed that the insulin resistance that underlies the pathogenesis of type II diabetes mellitus might alter APP processing through autophagy activation, which might be involved in the pathogenesis of AD [185].

4. Targeting autophagy

The evidence indicating that induction of autophagy is cytoprotective in neurodegenerative disease models raises the possibility that this intracellular catabolic pathway may be exploited to clear toxic disease proteins and provide therapeutic benefit for patients. Consequently, novel approaches to manipulating autophagy in human patients are desirable.

Using a yeast screen, one recent study identified several small molecules capable of augmenting autophagy in mammalian cells and further demonstrated therapeutic benefit of these compounds in a Drosophila model of neurodegeneration [186].

Knockdown of autophagy blocks the protective effects of delayed aging, suggesting that neuroprotection associated with manipulation of aging pathways is autophagy dependent [97]. These results are consistent with the notion that protein aggregation neuronal dysfunction might occur coincidentally with age-related decline in cellular mechanisms to deal with misfolded protein species. Thus, the age-related onset of pathology in neurodegenerative conditions might be correlated with a decline in autophagic capacity beyond a critical threshold. However, additional studies will be required to define the precise relationship of aging, autophagy and neurodegeneration. In the case of some diseases autophagy is initially induced as a neuroprotective response in stressed or injured neurons, but is subsequently overwhelmed or impaired by disease-related factors. This could partly account for evidence that autophagy seems to be both induced and impaired in several major neurodegenerative diseases. Overexpression of HDAC6, a cytoplasmic deacetylase containing a ubiquitin-binding domain, was found to suppress neurodegeneration in a model of polyglutamine disease and to compensate for defects in the UPS by facilitating autophagic protein degradation [187]. These results suggest that HDAC6 functions at the intersection of the UPS and autophagy and identify HDAC6 as a promising target for pharmacological manipulation in neurodegeneration.

Pharmacological activation of autophagy can be achieved with rapamycin, a lipophilic macrolide antibiotic. Chronic low-level stimulation of autophagy through peripheral administration of rapamycin or other agents [188], or enhancing lysosomal proteolysis selectively [90, 189], can markedly diminish Aβ levels and amyloid load in APP transgenic mice, underscoring the importance of lysosomal clearance of Aβ. In both APP and triple transgenic mouse models of AD for instance peripheral administration of rapamycin significantly reduces Aβ deposition and tau pathology [188, 190-191]. Furthermore, by deleting an endogenous inhibitor of lysosomal cysteine proteases (cystatin B) in the TgCRND8 APP mouse model lysosomal pathology is rescued and abnormal autolysosomal accumulation of autophagy substrates including Aβ is decreased and learning and memory deficits also ameliorate [90]. In addition, inhibition of mTOR by rapamycin improves cognitive deficits and rescues Aβ pathology and intraneuronal neurofibrillary tangles by increasing autophagy [192]. In a cell model of HD treatment with rapamycin reduced cellular toxicity by reducing bolth levels of soluble mutant huntingtin and the formation of intracellular aggregates [193]. Likewise this was also replicated in vivo in both Drosophila and mouse models of HD [194]. Furthermore, treatment of animals with rapamycin ameliorated neuronal toxicity in models of both Frontotemporal Dementia, spinocerebellar ataxia type 3, and PD [195-199]. However rapamycin shows autophagy-independent functions, such as regulation of ribosome biogenesis and protein translation. Taking into account that rapamycin may not be a good choice for the treatment in humans, other compounds that are independent of mTOR pathway are being studied. Moreover, long-term massive degradation of intracellular components as a result of the upregulation of macroautophagy can have possible negative consequences. Current efforts are now focused on the identification of other methods that can activate macroautophagy.

The mood stabilizer lithium used in bipolar disorder induces autophagy independently of mTOR through the inhibition of inositol monophosphatase therefore depleting the intracellular signaling molecule IP3 [200-201]. Lithium enhances clearance of aggregate-prone proteins, such as mutant forms of huntingtin and alpha-synuclein [202]. Likewise, sodium valproate and carbamazepine are other mood stabilizing drugs that can also reduce the accumulation and toxic effects of aggregation-prone mutant proteins in cell models and protect against neurodegeneration in vivo. Enhanced clearance of mutant alpha-synuclein and neuroprotective effects through lithium were demonstrated in cell-culture models and mice, underscoring the potential of lithium and related molecules for further evaluation [203-204].

An interesting study has shown that lithium (when given in addition to riluzole), significantly delayed the onset of disability and death in human Amyotrophic lateral sclerosis patients, compared to those administered riluzole alone [205]. In addition, prolonged lithium treatment alleviates memory deficits and reduces Aβ production in AD mouse models [206]. On 1-methyl-4-phenyl-1,2,3,6-tetrahydropyridine (MPTP)-induced Parkinsonism mice valproate combined with lithium carbonate rescued dopaminergic neurons and ameliorated the loss of DOPAC, likely via activation of autophagic/lysosomal pathways [207].

The disaccharide trehalose is a stable disaccharide with unique physicochemical properties. This disaccharide directly acts as a chemical chaperone that can stabilize misfolded proteins but it also facilitates the clearance of aggregate-prone proteins. Trehalose was shown to augment autophagy and enhance the clearance of mutant forms of huntingtin and alpha-synuclein from neuronal cells in culture [208]. In APP(swe) mutant mice co-treatment with trehalose increases the expression of autophagic markers as well as the expression of chaperones and reduces the levels of Aβ peptide aggregates, tau plaques and levels of phospho-tau [209]. Moreover in PC12 cells overexpressing wild-type or A53T mutant alpha-synuclein trehalose promoted the clearance of A53T alpha-synuclein but not wild-type alpha-synuclein in PC12 cells, and increased LC3 and Lysotracker RED positive AVs by using lysotracker and LC3 staining [210].

These findings have opened the possibility of using autophagy modulators as therapeutic approaches for these types of pathology. Therefore the search for chemical autophagy modulators more selective and potency is ongoing. Nevertheless, there are certain limitations in the use of up-regulation of macroautophagy for anti-neurodegenerative purposes. One of the limitations is that they all act on early steps of macroautophagy enhancing autophagosome formation, but will not have a beneficial effect in those pathological conditions in which the autophagic defect is in steps past autophagosome formation. Moreover, it has been reported that not all protein aggregates are recognized by the macroautophagic machinery. Further studies are needed to address whether expression of other cellular components or specific post-translational modifications in the aggregated proteins could enhance their recognition by the autophagic systems. For example, pharmacological induction or inhibition of macroautophagy alters the rate of turnover of polyglutamine-expanded proteins, polyalanine-expanded proteins, as well as, wild-type and mutant forms of alpha-synuclein [193, 211].

A patent by Harvard and Cambridge Universities disclosed novel small molecule enhancers of autophagy, named SMERs (small molecule enhancers of rapamycin) [President and Fellows of Harvard College, Dana Farber Cancer Institute, Cambridge Enterprise Ltd. Regulating autophagy. WO2008122038 (2008)]. In mammalian cells, the three most active compounds: SMER10, SMER18 and SMER28, augmented autophagosome formation in an mTORC1-independent fashion and without affecting the levels of Beclin-1, Atg5, Atg7 and Atg12, or enhancing the conjugation of Atg12 to Atg5. Importantly, SMERs enhanced the clearance of aggregation-prone mutant forms of alpha-synuclein and huntingtin in mammalian cells [186]. In addition, in cell lines and primary neuronal cultures SMER28 reduced the levels of Aβ and APP C-terminal fragments [188]. In addition, drugs that involve the serine/threonine-protein kinase mammalian target of rapamycin (mTOR) are

http://dx.doi.org/10.5772/55196

being currently examined for the treatment of metabolic disorders, such as diabetes mellitus. However future investigations are necessary [212]. Drugs that take advantage of the potential protective effect of autophagy in ER stress, such as glucagon like peptide-1, will be a promising avenue of investigation [183, 213].

5. Concluding remarks

The progression of research in the field of autophagy has taken the understanding of autophagy from a means of survival during starvation to a possible means of treating pharmacologically diseases where autophagy holds an important role. Progress in understanding autophagy has emphasized the importance of autophagy in cellular homeostasis and quality control; prevention of neurodegeneration in several animal models, and the significance of its deficiency in the development and pathogenesis of several disorders. This may provide further insights into the role that autophagy has in disease and how it may be possibly use it as a therapeutic strategy.

Acknowledgements

Work in our laboratory is supported by funds from PTDC/SAU-NEU/102710/2008. AR Esteves is supported by Post-Doctoral Fellowship (SFRH/BPD/75044/2010) from Portuguese Foundation for Science and Technology (FCT-MCTES, Portugal). Sponsored by PEST-SAU/LA0001CNC.

Author details

A. Raquel Esteves[1*], Catarina R. Oliveira[1,2] and Sandra Morais Cardoso[1,2]

*Address all correspondence to: esteves.raquel@gmail.com

1 CNC – Center for Neuroscience and Cell Biology, University of Coimbra, Coimbra, Portugal

2 Institute of Biology, Faculty of Medicine, University of Coimbra, Coimbra, Portugal

References

[1] Boland B, Nixon RA. Neuronal macroautophagy: from development to degeneration. Mol Aspects Med. 2006 Oct-Dec;27(5-6):503-19.

[2] Youle RJ, Narendra DP. Mechanisms of mitophagy. Nat Rev Mol Cell Biol. 2011 Jan; 12(1):9-14.

[3] Cuervo AM. Chaperone-mediated autophagy: selectivity pays off. Trends Endocrinol Metab. 2010 Mar;21(3):142-50.

[4] Tanida I. Autophagosome formation and molecular mechanism of autophagy. Antioxid Redox Signal. 2011 Jun;14(11):2201-14.

[5] Yoshimori T, Yamamoto A, Moriyama Y, Futai M, Tashiro Y. Bafilomycin A1, a specific inhibitor of vacuolar-type H(+)-ATPase, inhibits acidification and protein degradation in lysosomes of cultured cells. J Biol Chem. 1991 Sep 15;266(26):17707-12.

[6] Yamamoto A, Simonsen A. The elimination of accumulated and aggregated proteins: a role for aggrephagy in neurodegeneration. Neurobiol Dis. 2011 Jul;43(1):17-28.

[7] Goldman SJ, Taylor R, Zhang Y, Jin S. Autophagy and the degradation of mitochondria. Mitochondrion. 2010 Jun;10(4):309-15.

[8] Mizushima N. The pleiotropic role of autophagy: from protein metabolism to bactericide. Cell Death Differ. 2005 Nov;12 Suppl 2:1535-41.

[9] Mortimore GE, Lardeux BR, Adams CE. Regulation of microautophagy and basal protein turnover in rat liver. Effects of short-term starvation. J Biol Chem. 1988 Feb 15;263(5):2506-12.

[10] Ahlberg J, Glaumann H. Uptake--microautophagy--and degradation of exogenous proteins by isolated rat liver lysosomes. Effects of pH, ATP, and inhibitors of proteolysis. Exp Mol Pathol. 1985 Feb;42(1):78-88.

[11] Farre JC, Subramani S. Peroxisome turnover by micropexophagy: an autophagy-related process. Trends Cell Biol. 2004 Sep;14(9):515-23.

[12] Kaushik S, Bandyopadhyay U, Sridhar S, Kiffin R, Martinez-Vicente M, Kon M, et al. Chaperone-mediated autophagy at a glance. J Cell Sci. 2011 Feb 15;124(Pt 4):495-9.

[13] Chiang HL, Terlecky SR, Plant CP, Dice JF. A role for a 70-kilodalton heat shock protein in lysosomal degradation of intracellular proteins. Science. 1989 Oct 20;246(4928):382-5.

[14] Agarraberes FA, Dice JF. A molecular chaperone complex at the lysosomal membrane is required for protein translocation. J Cell Sci. 2001 Jul;114(Pt 13):2491-9.

[15] Bandyopadhyay U, Kaushik S, Varticovski L, Cuervo AM. The chaperone-mediated autophagy receptor organizes in dynamic protein complexes at the lysosomal membrane. Mol Cell Biol. 2008 Sep;28(18):5747-63.

[16] Shin Y, Klucken J, Patterson C, Hyman BT, McLean PJ. The co-chaperone carboxyl terminus of Hsp70-interacting protein (CHIP) mediates alpha-synuclein degradation

decisions between proteasomal and lysosomal pathways. J Biol Chem. 2005 Jun 24;280(25):23727-34.

[17] Cuervo AM, Dice JF. A receptor for the selective uptake and degradation of proteins by lysosomes. Science. 1996 Jul 26;273(5274):501-3.

[18] Gough NR, Hatem CL, Fambrough DM. The family of LAMP-2 proteins arises by alternative splicing from a single gene: characterization of the avian LAMP-2 gene and identification of mammalian homologs of LAMP-2b and LAMP-2c. DNA Cell Biol. 1995 Oct;14(10):863-7.

[19] Wong E, Cuervo AM. Autophagy gone awry in neurodegenerative diseases. Nat Neurosci. 2010 Jul;13(7):805-11.

[20] Finn PF, Dice JF. Ketone bodies stimulate chaperone-mediated autophagy. J Biol Chem. 2005 Jul 8;280(27):25864-70.

[21] Massey AC, Kaushik S, Sovak G, Kiffin R, Cuervo AM. Consequences of the selective blockage of chaperone-mediated autophagy. Proc Natl Acad Sci U S A. 2006 Apr 11;103(15):5805-10.

[22] Kaushik S, Massey AC, Mizushima N, Cuervo AM. Constitutive activation of chaperone-mediated autophagy in cells with impaired macroautophagy. Mol Biol Cell. 2008 May;19(5):2179-92.

[23] Ohsumi Y, Mizushima N. Two ubiquitin-like conjugation systems essential for autophagy. Semin Cell Dev Biol. 2004 Apr;15(2):231-6.

[24] Matsunaga K, Saitoh T, Tabata K, Omori H, Satoh T, Kurotori N, et al. Two Beclin 1-binding proteins, Atg14L and Rubicon, reciprocally regulate autophagy at different stages. Nat Cell Biol. 2009 Apr;11(4):385-96.

[25] Yang Z, Klionsky DJ. Mammalian autophagy: core molecular machinery and signaling regulation. Curr Opin Cell Biol. 2010 Apr;22(2):124-31.

[26] Metcalf DJ, Garcia-Arencibia M, Hochfeld WE, Rubinsztein DC. Autophagy and misfolded proteins in neurodegeneration. Exp Neurol. 2012 Nov;238(1):22-8.

[27] Ravikumar B, Sarkar S, Davies JE, Futter M, Garcia-Arencibia M, Green-Thompson ZW, et al. Regulation of mammalian autophagy in physiology and pathophysiology. Physiol Rev. 2010 Oct;90(4):1383-435.

[28] Wei Y, Pattingre S, Sinha S, Bassik M, Levine B. JNK1-mediated phosphorylation of Bcl-2 regulates starvation-induced autophagy. Mol Cell. 2008 Jun 20;30(6):678-88.

[29] Cao Y, Klionsky DJ. Physiological functions of Atg6/Beclin 1: a unique autophagy-related protein. Cell Res. 2007 Oct;17(10):839-49.

[30] Ku B, Woo JS, Liang C, Lee KH, Jung JU, Oh BH. An insight into the mechanistic role of Beclin 1 and its inhibition by prosurvival Bcl-2 family proteins. Autophagy. 2008 May;4(4):519-20.

[31] Williams A, Sarkar S, Cuddon P, Ttofi EK, Saiki S, Siddiqi FH, et al. Novel targets for Huntington's disease in an mTOR-independent autophagy pathway. Nat Chem Biol. 2008 May;4(5):295-305.

[32] Cardenas C, Miller RA, Smith I, Bui T, Molgo J, Muller M, et al. Essential regulation of cell bioenergetics by constitutive InsP3 receptor Ca2+ transfer to mitochondria. Cell. 2010 Jul 23;142(2):270-83.

[33] Terman A, Brunk UT. Lipofuscin. Int J Biochem Cell Biol. 2004 Aug;36(8):1400-4.

[34] Terman A, Dalen H, Brunk UT. Ceroid/lipofuscin-loaded human fibroblasts show decreased survival time and diminished autophagocytosis during amino acid starvation. Exp Gerontol. 1999 Dec;34(8):943-57.

[35] Ward WF. Protein degradation in the aging organism. Prog Mol Subcell Biol. 2002;29:35-42.

[36] Kim I, Rodriguez-Enriquez S, Lemasters JJ. Selective degradation of mitochondria by mitophagy. Arch Biochem Biophys. 2007 Jun 15;462(2):245-53.

[37] Cuervo AM, Dice JF. Age-related decline in chaperone-mediated autophagy. J Biol Chem. 2000 Oct 6;275(40):31505-13.

[38] Kiffin R, Kaushik S, Zeng M, Bandyopadhyay U, Zhang C, Massey AC, et al. Altered dynamics of the lysosomal receptor for chaperone-mediated autophagy with age. J Cell Sci. 2007 Mar 1;120(Pt 5):782-91.

[39] Double KL, Dedov VN, Fedorow H, Kettle E, Halliday GM, Garner B, et al. The comparative biology of neuromelanin and lipofuscin in the human brain. Cell Mol Life Sci. 2008 Jun;65(11):1669-82.

[40] Nakano M, Oenzil F, Mizuno T, Gotoh S. Age-related changes in the lipofuscin accumulation of brain and heart. Gerontology. 1995;41 Suppl 2:69-79.

[41] Brunk UT, Terman A. Lipofuscin: mechanisms of age-related accumulation and influence on cell function. Free Radic Biol Med. 2002 Sep 1;33(5):611-9.

[42] Ivy GO, Kanai S, Ohta M, Smith G, Sato Y, Kobayashi M, et al. Lipofuscin-like substances accumulate rapidly in brain, retina and internal organs with cysteine protease inhibition. Adv Exp Med Biol. 1989;266:31-45; discussion -7.

[43] Terman A, Gustafsson B, Brunk UT. Autophagy, organelles and ageing. J Pathol. 2007 Jan;211(2):134-43.

[44] Szweda PA, Camouse M, Lundberg KC, Oberley TD, Szweda LI. Aging, lipofuscin formation, and free radical-mediated inhibition of cellular proteolytic systems. Ageing Res Rev. 2003 Oct;2(4):383-405.

[45] Elleder M, Sokolova J, Hrebicek M. Follow-up study of subunit c of mitochondrial ATP synthase (SCMAS) in Batten disease and in unrelated lysosomal disorders. Acta Neuropathol. 1997 Apr;93(4):379-90.

[46] Terman A, Brunk UT. Oxidative stress, accumulation of biological 'garbage', and aging. Antioxid Redox Signal. 2006 Jan-Feb;8(1-2):197-204.

[47] Jolly RD, Douglas BV, Davey PM, Roiri JE. Lipofuscin in bovine muscle and brain: a model for studying age pigment. Gerontology. 1995;41 Suppl 2:283-95.

[48] Terman A, Brunk UT. Lipofuscin: mechanisms of formation and increase with age. APMIS. 1998 Feb;106(2):265-76.

[49] Schmidtke G, Emch S, Groettrup M, Holzhutter HG. Evidence for the existence of a non-catalytic modifier site of peptide hydrolysis by the 20 S proteasome. J Biol Chem. 2000 Jul 21;275(29):22056-63.

[50] Giasson BI, Duda JE, Murray IV, Chen Q, Souza JM, Hurtig HI, et al. Oxidative damage linked to neurodegeneration by selective alpha-synuclein nitration in synucleinopathy lesions. Science. 2000 Nov 3;290(5493):985-9.

[51] Sitte N, Merker K, Grune T, von Zglinicki T. Lipofuscin accumulation in proliferating fibroblasts in vitro: an indicator of oxidative stress. Exp Gerontol. 2001 Mar;36(3):475-86.

[52] Imarisio S, Carmichael J, Korolchuk V, Chen CW, Saiki S, Rose C, et al. Huntington's disease: from pathology and genetics to potential therapies. Biochem J. 2008 Jun 1;412(2):191-209.

[53] Martinez-Vicente M, Talloczy Z, Wong E, Tang G, Koga H, Kaushik S, et al. Cargo recognition failure is responsible for inefficient autophagy in Huntington's disease. Nat Neurosci. 2010 May;13(5):567-76.

[54] Ravikumar B, Acevedo-Arozena A, Imarisio S, Berger Z, Vacher C, O'Kane CJ, et al. Dynein mutations impair autophagic clearance of aggregate-prone proteins. Nat Genet. 2005 Jul;37(7):771-6.

[55] Kochl R, Hu XW, Chan EY, Tooze SA. Microtubules facilitate autophagosome formation and fusion of autophagosomes with endosomes. Traffic. 2006 Feb;7(2):129-45.

[56] Webb JL, Ravikumar B, Rubinsztein DC. Microtubule disruption inhibits autophagosome-lysosome fusion: implications for studying the roles of aggresomes in polyglutamine diseases. Int J Biochem Cell Biol. 2004 Dec;36(12):2541-50.

[57] Lee JY, Koga H, Kawaguchi Y, Tang W, Wong E, Gao YS, et al. HDAC6 controls autophagosome maturation essential for ubiquitin-selective quality-control autophagy. EMBO J. 2010 Mar 3;29(5):969-80.

[58] Duncan JE, Goldstein LS. The genetics of axonal transport and axonal transport disorders. PLoS Genet. 2006 Sep 29;2(9):e124.

[59] Settembre C, Fraldi A, Jahreiss L, Spampanato C, Venturi C, Medina D, et al. A block of autophagy in lysosomal storage disorders. Hum Mol Genet. 2008 Jan 1;17(1): 119-29.

[60] Yang AJ, Chandswangbhuvana D, Margol L, Glabe CG. Loss of endosomal/lysosomal membrane impermeability is an early event in amyloid Abeta1-42 pathogenesis. J Neurosci Res. 1998 Jun 15;52(6):691-8.

[61] Glabe C. Intracellular mechanisms of amyloid accumulation and pathogenesis in Alzheimer's disease. J Mol Neurosci. 2001 Oct;17(2):137-45.

[62] Sulzer D, Bogulavsky J, Larsen KE, Behr G, Karatekin E, Kleinman MH, et al. Neuromelanin biosynthesis is driven by excess cytosolic catecholamines not accumulated by synaptic vesicles. Proc Natl Acad Sci U S A. 2000 Oct 24;97(22):11869-74.

[63] Williamson WR, Wang D, Haberman AS, Hiesinger PR. A dual function of V0-ATPase a1 provides an endolysosomal degradation mechanism in Drosophila melanogaster photoreceptors. J Cell Biol. 2010 May 31;189(5):885-99.

[64] Mijaljica D, Prescott M, Devenish RJ. V-ATPase engagement in autophagic processes. Autophagy. 2011 Jun;7(6):666-8.

[65] Cataldo AM, Mathews PM, Boiteau AB, Hassinger LC, Peterhoff CM, Jiang Y, et al. Down syndrome fibroblast model of Alzheimer-related endosome pathology: accelerated endocytosis promotes late endocytic defects. Am J Pathol. 2008 Aug;173(2): 370-84.

[66] Filimonenko M, Stuffers S, Raiborg C, Yamamoto A, Malerod L, Fisher EM, et al. Functional multivesicular bodies are required for autophagic clearance of protein aggregates associated with neurodegenerative disease. J Cell Biol. 2007 Nov 5;179(3): 485-500.

[67] Lee JA, Beigneux A, Ahmad ST, Young SG, Gao FB. ESCRT-III dysfunction causes autophagosome accumulation and neurodegeneration. Curr Biol. 2007 Sep 18;17(18): 1561-7.

[68] Cuervo AM, Palmer A, Rivett AJ, Knecht E. Degradation of proteasomes by lysosomes in rat liver. Eur J Biochem. 1995 Feb 1;227(3):792-800.

[69] Koga H, Cuervo AM. Chaperone-mediated autophagy dysfunction in the pathogenesis of neurodegeneration. Neurobiol Dis. 2011 Jul;43(1):29-37.

[70] Merlini G, Bellotti V, Andreola A, Palladini G, Obici L, Casarini S, et al. Protein aggregation. Clin Chem Lab Med. 2001 Nov;39(11):1065-75.

[71] Scheibel T, Buchner J. Protein aggregation as a cause for disease. Handb Exp Pharmacol. 2006(172):199-219.

[72] Tyedmers J, Mogk A, Bukau B. Cellular strategies for controlling protein aggregation. Nat Rev Mol Cell Biol. 2010 Nov;11(11):777-88.

[73] Brunk UT, Terman A. The mitochondrial-lysosomal axis theory of aging: accumulation of damaged mitochondria as a result of imperfect autophagocytosis. Eur J Biochem. 2002 Apr;269(8):1996-2002.

[74] Lee J, Giordano S, Zhang J. Autophagy, mitochondria and oxidative stress: cross-talk and redox signalling. Biochem J. 2012 Jan 15;441(2):523-40.

[75] Stroikin Y, Dalen H, Loof S, Terman A. Inhibition of autophagy with 3-methyladenine results in impaired turnover of lysosomes and accumulation of lipofuscin-like material. Eur J Cell Biol. 2004 Oct;83(10):583-90.

[76] Coleman R, Silbermann M, Gershon D, Reznick AZ. Giant mitochondria in the myocardium of aging and endurance-trained mice. Gerontology. 1987;33(1):34-9.

[77] Terman A, Dalen H, Eaton JW, Neuzil J, Brunk UT. Mitochondrial recycling and aging of cardiac myocytes: the role of autophagocytosis. Exp Gerontol. 2003 Aug;38(8): 863-76.

[78] Sohal RS, Sohal BH. Hydrogen peroxide release by mitochondria increases during aging. Mech Ageing Dev. 1991 Feb;57(2):187-202.

[79] Goldberg AL. Protein degradation and protection against misfolded or damaged proteins. Nature. 2003 Dec 18;426(6968):895-9.

[80] Komatsu M, Waguri S, Chiba T, Murata S, Iwata J, Tanida I, et al. Loss of autophagy in the central nervous system causes neurodegeneration in mice. Nature. 2006 Jun 15;441(7095):880-4.

[81] Mizushima N, Levine B, Cuervo AM, Klionsky DJ. Autophagy fights disease through cellular self-digestion. Nature. 2008 Feb 28;451(7182):1069-75.

[82] Meijer AJ, Codogno P. Autophagy: regulation and role in disease. Crit Rev Clin Lab Sci. 2009;46(4):210-40.

[83] Nixon RA, Cataldo AM. Lysosomal system pathways: genes to neurodegeneration in Alzheimer's disease. J Alzheimers Dis. 2006;9(3 Suppl):277-89.

[84] Ginsberg SD, Alldred MJ, Counts SE, Cataldo AM, Neve RL, Jiang Y, et al. Microarray analysis of hippocampal CA1 neurons implicates early endosomal dysfunction during Alzheimer's disease progression. Biol Psychiatry. 2010 Nov 15;68(10):885-93.

[85] Cataldo AM, Peterhoff CM, Troncoso JC, Gomez-Isla T, Hyman BT, Nixon RA. Endocytic pathway abnormalities precede amyloid beta deposition in sporadic Alzheim-

er's disease and Down syndrome: differential effects of APOE genotype and presenilin mutations. Am J Pathol. 2000 Jul;157(1):277-86.

[86] Lipinski MM, Zheng B, Lu T, Yan Z, Py BF, Ng A, et al. Genome-wide analysis reveals mechanisms modulating autophagy in normal brain aging and in Alzheimer's disease. Proc Natl Acad Sci U S A. 2010 Aug 10;107(32):14164-9.

[87] Nixon RA, Wegiel J, Kumar A, Yu WH, Peterhoff C, Cataldo A, et al. Extensive involvement of autophagy in Alzheimer disease: an immuno-electron microscopy study. J Neuropathol Exp Neurol. 2005 Feb;64(2):113-22.

[88] Lee JH, Yu WH, Kumar A, Lee S, Mohan PS, Peterhoff CM, et al. Lysosomal proteolysis and autophagy require presenilin 1 and are disrupted by Alzheimer-related PS1 mutations. Cell. 2010 Jun 25;141(7):1146-58.

[89] Zhang X, Garbett K, Veeraraghavalu K, Wilburn B, Gilmore R, Mirnics K, et al. A role for presenilins in autophagy revisited: normal acidification of lysosomes in cells lacking PSEN1 and PSEN2. J Neurosci. 2012 Jun 20;32(25):8633-48.

[90] Yang DS, Stavrides P, Mohan PS, Kaushik S, Kumar A, Ohno M, et al. Reversal of autophagy dysfunction in the TgCRND8 mouse model of Alzheimer's disease ameliorates amyloid pathologies and memory deficits. Brain. 2011 Jan;134(Pt 1):258-77.

[91] Yang DS, Kumar A, Stavrides P, Peterson J, Peterhoff CM, Pawlik M, et al. Neuronal apoptosis and autophagy cross talk in aging PS/APP mice, a model of Alzheimer's disease. Am J Pathol. 2008 Sep;173(3):665-81.

[92] Ling D, Song HJ, Garza D, Neufeld TP, Salvaterra PM. Abeta42-induced neurodegeneration via an age-dependent autophagic-lysosomal injury in Drosophila. PLoS One. 2009;4(1):e4201.

[93] Belinson H, Lev D, Masliah E, Michaelson DM. Activation of the amyloid cascade in apolipoprotein E4 transgenic mice induces lysosomal activation and neurodegeneration resulting in marked cognitive deficits. J Neurosci. 2008 Apr 30;28(18):4690-701.

[94] Lai AY, McLaurin J. Inhibition of amyloid-beta peptide aggregation rescues the autophagic deficits in the TgCRND8 mouse model of Alzheimer disease. Biochim Biophys Acta. 2012 Oct;1822(10):1629-37.

[95] Pickford F, Masliah E, Britschgi M, Lucin K, Narasimhan R, Jaeger PA, et al. The autophagy-related protein beclin 1 shows reduced expression in early Alzheimer disease and regulates amyloid beta accumulation in mice. J Clin Invest. 2008 Jun;118(6): 2190-9.

[96] Bahr BA, Bendiske J. The neuropathogenic contributions of lysosomal dysfunction. J Neurochem. 2002 Nov;83(3):481-9.

[97] Florez-McClure ML, Hohsfield LA, Fonte G, Bealor MT, Link CD. Decreased insulin-receptor signaling promotes the autophagic degradation of beta-amyloid peptide in C. elegans. Autophagy. 2007 Nov-Dec;3(6):569-80.

[98] Yu WH, Cuervo AM, Kumar A, Peterhoff CM, Schmidt SD, Lee JH, et al. Macroautophagy--a novel Beta-amyloid peptide-generating pathway activated in Alzheimer's disease. J Cell Biol. 2005 Oct 10;171(1):87-98.

[99] Wang Y, Mandelkow E. Degradation of tau protein by autophagy and proteasomal pathways. Biochem Soc Trans. 2012 Aug;40(4):644-52.

[100] Wang Y, Martinez-Vicente M, Kruger U, Kaushik S, Wong E, Mandelkow EM, et al. Tau fragmentation, aggregation and clearance: the dual role of lysosomal processing. Hum Mol Genet. 2009 Nov 1;18(21):4153-70.

[101] Schaeffer V, Lavenir I, Ozcelik S, Tolnay M, Winkler DT, Goedert M. Stimulation of autophagy reduces neurodegeneration in a mouse model of human tauopathy. Brain. 2012 Jul;135(Pt 7):2169-77.

[102] Rohn TT, Wirawan E, Brown RJ, Harris JR, Masliah E, Vandenabeele P. Depletion of Beclin-1 due to proteolytic cleavage by caspases in the Alzheimer's disease brain. Neurobiol Dis. 2011 Jul;43(1):68-78.

[103] Cuervo AM, Stefanis L, Fredenburg R, Lansbury PT, Sulzer D. Impaired degradation of mutant alpha-synuclein by chaperone-mediated autophagy. Science. 2004 Aug 27;305(5688):1292-5.

[104] Webb JL, Ravikumar B, Atkins J, Skepper JN, Rubinsztein DC. Alpha-Synuclein is degraded by both autophagy and the proteasome. J Biol Chem. 2003 Jul 4;278(27): 25009-13.

[105] Stefanis L, Larsen KE, Rideout HJ, Sulzer D, Greene LA. Expression of A53T mutant but not wild-type alpha-synuclein in PC12 cells induces alterations of the ubiquitin-dependent degradation system, loss of dopamine release, and autophagic cell death. J Neurosci. 2001 Dec 15;21(24):9549-60.

[106] Winslow AR, Chen CW, Corrochano S, Acevedo-Arozena A, Gordon DE, Peden AA, et al. alpha-Synuclein impairs macroautophagy: implications for Parkinson's disease. J Cell Biol. 2010 Sep 20;190(6):1023-37.

[107] Xilouri M, Vogiatzi T, Vekrellis K, Park D, Stefanis L. Abberant alpha-synuclein confers toxicity to neurons in part through inhibition of chaperone-mediated autophagy. PLoS One. 2009;4(5):e5515.

[108] Martinez-Vicente M, Talloczy Z, Kaushik S, Massey AC, Mazzulli J, Mosharov EV, et al. Dopamine-modified alpha-synuclein blocks chaperone-mediated autophagy. J Clin Invest. 2008 Feb;118(2):777-88.

[109] Rideout HJ, Lang-Rollin I, Stefanis L. Involvement of macroautophagy in the dissolution of neuronal inclusions. Int J Biochem Cell Biol. 2004 Dec;36(12):2551-62.

[110] Anglade P, Vyas S, Javoy-Agid F, Herrero MT, Michel PP, Marquez J, et al. Apoptosis and autophagy in nigral neurons of patients with Parkinson's disease. Histol Histopathol. 1997 Jan;12(1):25-31.

[111] Crews L, Spencer B, Desplats P, Patrick C, Paulino A, Rockenstein E, et al. Selective molecular alterations in the autophagy pathway in patients with Lewy body disease and in models of alpha-synucleinopathy. PLoS One. 2010;5(2):e9313.

[112] Higashi S, Moore DJ, Minegishi M, Kasanuki K, Fujishiro H, Kabuta T, et al. Localization of MAP1-LC3 in vulnerable neurons and Lewy bodies in brains of patients with dementia with Lewy bodies. J Neuropathol Exp Neurol. 2011 Apr;70(4):264-80.

[113] Tanji K, Mori F, Kakita A, Takahashi H, Wakabayashi K. Alteration of autophagosomal proteins (LC3, GABARAP and GATE-16) in Lewy body disease. Neurobiol Dis. 2011 Sep;43(3):690-7.

[114] Alvarez-Erviti L, Rodriguez-Oroz MC, Cooper JM, Caballero C, Ferrer I, Obeso JA, et al. Chaperone-mediated autophagy markers in Parkinson disease brains. Arch Neurol. 2010 Dec;67(12):1464-72.

[115] Chu Y, Dodiya H, Aebischer P, Olanow CW, Kordower JH. Alterations in lysosomal and proteasomal markers in Parkinson's disease: relationship to alpha-synuclein inclusions. Neurobiol Dis. 2009 Sep;35(3):385-98.

[116] Zhu JH, Horbinski C, Guo F, Watkins S, Uchiyama Y, Chu CT. Regulation of autophagy by extracellular signal-regulated protein kinases during 1-methyl-4-phenylpyridinium-induced cell death. Am J Pathol. 2007 Jan;170(1):75-86.

[117] Dehay B, Bove J, Rodriguez-Muela N, Perier C, Recasens A, Boya P, et al. Pathogenic lysosomal depletion in Parkinson's disease. J Neurosci. 2010 Sep 15;30(37):12535-44.

[118] Cai ZL, Shi JJ, Yang YP, Cao BY, Wang F, Huang JZ, et al. MPP+ impairs autophagic clearance of alpha-synuclein by impairing the activity of dynein. Neuroreport. 2009 Apr 22;20(6):569-73.

[119] Wong AS, Lee RH, Cheung AY, Yeung PK, Chung SK, Cheung ZH, et al. Cdk5-mediated phosphorylation of endophilin B1 is required for induced autophagy in models of Parkinson's disease. Nat Cell Biol. 2011 May;13(5):568-79.

[120] Arduino DM, Raquel Esteves A, Cortes L, Silva DF, Patel B, Grazina M, et al. Mitochondrial metabolism in Parkinson's disease impairs quality control autophagy by hampering microtubule-dependent traffic. Hum Mol Genet. 2012 Aug 29.

[121] Geisler S, Holmstrom KM, Treis A, Skujat D, Weber SS, Fiesel FC, et al. The PINK1/Parkin-mediated mitophagy is compromised by PD-associated mutations. Autophagy. 2010 Oct;6(7):871-8.

[122] Geisler S, Holmstrom KM, Skujat D, Fiesel FC, Rothfuss OC, Kahle PJ, et al. PINK1/Parkin-mediated mitophagy is dependent on VDAC1 and p62/SQSTM1. Nat Cell Biol. 2010 Feb;12(2):119-31.

[123] Narendra D, Tanaka A, Suen DF, Youle RJ. Parkin is recruited selectively to impaired mitochondria and promotes their autophagy. J Cell Biol. 2008 Dec 1;183(5):795-803.

[124] Matsuda N, Sato S, Shiba K, Okatsu K, Saisho K, Gautier CA, et al. PINK1 stabilized by mitochondrial depolarization recruits Parkin to damaged mitochondria and activates latent Parkin for mitophagy. J Cell Biol. 2010 Apr 19;189(2):211-21.

[125] Palacino JJ, Sagi D, Goldberg MS, Krauss S, Motz C, Wacker M, et al. Mitochondrial dysfunction and oxidative damage in parkin-deficient mice. J Biol Chem. 2004 Apr 30;279(18):18614-22.

[126] Pesah Y, Pham T, Burgess H, Middlebrooks B, Verstreken P, Zhou Y, et al. Drosophila parkin mutants have decreased mass and cell size and increased sensitivity to oxygen radical stress. Development. 2004 May;131(9):2183-94.

[127] Krebiehl G, Ruckerbauer S, Burbulla LF, Kieper N, Maurer B, Waak J, et al. Reduced basal autophagy and impaired mitochondrial dynamics due to loss of Parkinson's disease-associated protein DJ-1. PLoS One. 2010;5(2):e9367.

[128] Irrcher I, Aleyasin H, Seifert EL, Hewitt SJ, Chhabra S, Phillips M, et al. Loss of the Parkinson's disease-linked gene DJ-1 perturbs mitochondrial dynamics. Hum Mol Genet. 2010 Oct 1;19(19):3734-46.

[129] McCoy MK, Cookson MR. DJ-1 regulation of mitochondrial function and autophagy through oxidative stress. Autophagy. 2011 May;7(5):531-2.

[130] Bravo-San Pedro JM, Niso-Santano M, Gomez-Sanchez R, Pizarro-Estrella E, Aiastui-Pujana A, Gorostidi A, et al. The LRRK2 G2019S mutant exacerbates basal autophagy through activation of the MEK/ERK pathway. Cell Mol Life Sci. 2012 Jul 8.

[131] Higashi S, Moore DJ, Yamamoto R, Minegishi M, Sato K, Togo T, et al. Abnormal localization of leucine-rich repeat kinase 2 to the endosomal-lysosomal compartment in lewy body disease. J Neuropathol Exp Neurol. 2009 Sep;68(9):994-1005.

[132] MacLeod D, Dowman J, Hammond R, Leete T, Inoue K, Abeliovich A. The familial Parkinsonism gene LRRK2 regulates neurite process morphology. Neuron. 2006 Nov 22;52(4):587-93.

[133] Gomez-Suaga P, Luzon-Toro B, Churamani D, Zhang L, Bloor-Young D, Patel S, et al. Leucine-rich repeat kinase 2 regulates autophagy through a calcium-dependent pathway involving NAADP. Hum Mol Genet. 2012 Feb 1;21(3):511-25.

[134] Kabuta T, Furuta A, Aoki S, Furuta K, Wada K. Aberrant interaction between Parkinson disease-associated mutant UCH-L1 and the lysosomal receptor for chaperone-mediated autophagy. J Biol Chem. 2008 Aug 29;283(35):23731-8.

[135] Poupetova H, Ledvinova J, Berna L, Dvorakova L, Kozich V, Elleder M. The birth prevalence of lysosomal storage disorders in the Czech Republic: comparison with data in different populations. J Inherit Metab Dis. 2010 Aug;33(4):387-96.

[136] Raben N, Shea L, Hill V, Plotz P. Monitoring autophagy in lysosomal storage disorders. Methods Enzymol. 2009;453:417-49.

[137] Ohmi K, Kudo LC, Ryazantsev S, Zhao HZ, Karsten SL, Neufeld EF. Sanfilippo syndrome type B, a lysosomal storage disease, is also a tauopathy. Proc Natl Acad Sci U S A. 2009 May 19;106(20):8332-7.

[138] Tamboli IY, Hampel H, Tien NT, Tolksdorf K, Breiden B, Mathews PM, et al. Sphingolipid storage affects autophagic metabolism of the amyloid precursor protein and promotes Abeta generation. J Neurosci. 2011 Feb 2;31(5):1837-49.

[139] Vitner EB, Platt FM, Futerman AH. Common and uncommon pathogenic cascades in lysosomal storage diseases. J Biol Chem. 2010 Jul 2;285(27):20423-7.

[140] Bellettato CM, Scarpa M. Pathophysiology of neuropathic lysosomal storage disorders. J Inherit Metab Dis. 2010 Aug;33(4):347-62.

[141] Settembre C, Arteaga-Solis E, Ballabio A, Karsenty G. Self-eating in skeletal development: implications for lysosomal storage disorders. Autophagy. 2009 Feb;5(2):228-9.

[142] Futerman AH, van Meer G. The cell biology of lysosomal storage disorders. Nat Rev Mol Cell Biol. 2004 Jul;5(7):554-65.

[143] Sobo K, Le Blanc I, Luyet PP, Fivaz M, Ferguson C, Parton RG, et al. Late endosomal cholesterol accumulation leads to impaired intra-endosomal trafficking. PLoS One. 2007;2(9):e851.

[144] Schultz ML, Tecedor L, Chang M, Davidson BL. Clarifying lysosomal storage diseases. Trends Neurosci. 2011 Aug;34(8):401-10.

[145] Nishino I, Fu J, Tanji K, Yamada T, Shimojo S, Koori T, et al. Primary LAMP-2 deficiency causes X-linked vacuolar cardiomyopathy and myopathy (Danon disease). Nature. 2000 Aug 24;406(6798):906-10.

[146] Tanaka Y, Guhde G, Suter A, Eskelinen EL, Hartmann D, Lullmann-Rauch R, et al. Accumulation of autophagic vacuoles and cardiomyopathy in LAMP-2-deficient mice. Nature. 2000 Aug 24;406(6798):902-6.

[147] Diez-Roux G, Ballabio A. Sulfatases and human disease. Annu Rev Genomics Hum Genet. 2005;6:355-79.

[148] Ballabio A, Gieselmann V. Lysosomal disorders: from storage to cellular damage. Biochim Biophys Acta. 2009 Apr;1793(4):684-96.

[149] Settembre C, Arteaga-Solis E, McKee MD, de Pablo R, Al Awqati Q, Ballabio A, et al. Proteoglycan desulfation determines the efficiency of chondrocyte autophagy and

the extent of FGF signaling during endochondral ossification. Genes Dev. 2008 Oct 1;22(19):2645-50.

[150] Fraldi A, Annunziata F, Lombardi A, Kaiser HJ, Medina DL, Spampanato C, et al. Lysosomal fusion and SNARE function are impaired by cholesterol accumulation in lysosomal storage disorders. EMBO J. 2010 Nov 3;29(21):3607-20.

[151] Carstea ED, Morris JA, Coleman KG, Loftus SK, Zhang D, Cummings C, et al. Niemann-Pick C1 disease gene: homology to mediators of cholesterol homeostasis. Science. 1997 Jul 11;277(5323):228-31.

[152] Kwon HJ, Abi-Mosleh L, Wang ML, Deisenhofer J, Goldstein JL, Brown MS, et al. Structure of N-terminal domain of NPC1 reveals distinct subdomains for binding and transfer of cholesterol. Cell. 2009 Jun 26;137(7):1213-24.

[153] Pacheco CD, Kunkel R, Lieberman AP. Autophagy in Niemann-Pick C disease is dependent upon Beclin-1 and responsive to lipid trafficking defects. Hum Mol Genet. 2007 Jun 15;16(12):1495-503.

[154] Liao G, Yao Y, Liu J, Yu Z, Cheung S, Xie A, et al. Cholesterol accumulation is associated with lysosomal dysfunction and autophagic stress in Npc1 -/- mouse brain. Am J Pathol. 2007 Sep;171(3):962-75.

[155] Ko DC, Milenkovic L, Beier SM, Manuel H, Buchanan J, Scott MP. Cell-autonomous death of cerebellar purkinje neurons with autophagy in Niemann-Pick type C disease. PLoS Genet. 2005 Jul;1(1):81-95.

[156] Ordonez MP. Defective mitophagy in human Niemann-Pick Type C1 neurons is due to abnormal autophagy activation. Autophagy. 2012 May 31;8(7).

[157] Sun Y, Liou B, Ran H, Skelton MR, Williams MT, Vorhees CV, et al. Neuronopathic Gaucher disease in the mouse: viable combined selective saposin C deficiency and mutant glucocerebrosidase (V394L) mice with glucosylsphingosine and glucosylceramide accumulation and progressive neurological deficits. Hum Mol Genet. 2010 Mar 15;19(6):1088-97.

[158] Mazzulli JR, Xu YH, Sun Y, Knight AL, McLean PJ, Caldwell GA, et al. Gaucher disease glucocerebrosidase and alpha-synuclein form a bidirectional pathogenic loop in synucleinopathies. Cell. 2011 Jul 8;146(1):37-52.

[159] Cuervo AM, Mann L, Bonten EJ, d'Azzo A, Dice JF. Cathepsin A regulates chaperone-mediated autophagy through cleavage of the lysosomal receptor. EMBO J. 2003 Jan 2;22(1):47-59.

[160] Venugopal B, Mesires NT, Kennedy JC, Curcio-Morelli C, Laplante JM, Dice JF, et al. Chaperone-mediated autophagy is defective in mucolipidosis type IV. J Cell Physiol. 2009 May;219(2):344-53.

[161] Cheng X, Shen D, Samie M, Xu H. Mucolipins: Intracellular TRPML1-3 channels. FEBS Lett. 2010 May 17;584(10):2013-21.

[162] Vergarajauregui S, Connelly PS, Daniels MP, Puertollano R. Autophagic dysfunction in mucolipidosis type IV patients. Hum Mol Genet. 2008 Sep 1;17(17):2723-37.

[163] Medina DL, Fraldi A, Bouche V, Annunziata F, Mansueto G, Spampanato C, et al. Transcriptional activation of lysosomal exocytosis promotes cellular clearance. Dev Cell. 2011 Sep 13;21(3):421-30.

[164] Lieberman AP, Puertollano R, Raben N, Slaugenhaupt S, Walkley SU, Ballabio A. Autophagy in lysosomal storage disorders. Autophagy. 2012 May 1;8(5):719-30.

[165] Singh R, Cuervo AM. Autophagy in the cellular energetic balance. Cell Metab. 2011 May 4;13(5):495-504.

[166] Kotoulas OB, Kalamidas SA, Kondomerkos DJ. Glycogen autophagy in glucose homeostasis. Pathol Res Pract. 2006;202(9):631-8.

[167] Evans JL, Goldfine ID, Maddux BA, Grodsky GM. Oxidative stress and stress-activated signaling pathways: a unifying hypothesis of type 2 diabetes. Endocr Rev. 2002 Oct;23(5):599-622.

[168] Tagami S, Inokuchi Ji J, Kabayama K, Yoshimura H, Kitamura F, Uemura S, et al. Ganglioside GM3 participates in the pathological conditions of insulin resistance. J Biol Chem. 2002 Feb 1;277(5):3085-92.

[169] Scarlatti F, Bauvy C, Ventruti A, Sala G, Cluzeaud F, Vandewalle A, et al. Ceramide-mediated macroautophagy involves inhibition of protein kinase B and up-regulation of beclin 1. J Biol Chem. 2004 Apr 30;279(18):18384-91.

[170] Hyde R, Hajduch E, Powell DJ, Taylor PM, Hundal HS. Ceramide down-regulates System A amino acid transport and protein synthesis in rat skeletal muscle cells. FASEB J. 2005 Mar;19(3):461-3.

[171] Petersen KF, Befroy D, Dufour S, Dziura J, Ariyan C, Rothman DL, et al. Mitochondrial dysfunction in the elderly: possible role in insulin resistance. Science. 2003 May 16;300(5622):1140-2.

[172] Meijer AJ, Codogno P. Autophagy: a sweet process in diabetes. Cell Metab. 2008 Oct; 8(4):275-6.

[173] Scherz-Shouval R, Shvets E, Fass E, Shorer H, Gil L, Elazar Z. Reactive oxygen species are essential for autophagy and specifically regulate the activity of Atg4. EMBO J. 2007 Apr 4;26(7):1749-60.

[174] Kaniuk NA, Kiraly M, Bates H, Vranic M, Volchuk A, Brumell JH. Ubiquitinated-protein aggregates form in pancreatic beta-cells during diabetes-induced oxidative stress and are regulated by autophagy. Diabetes. 2007 Apr;56(4):930-9.

[175] Jung HS, Chung KW, Won Kim J, Kim J, Komatsu M, Tanaka K, et al. Loss of autophagy diminishes pancreatic beta cell mass and function with resultant hyperglycemia. Cell Metab. 2008 Oct;8(4):318-24.

[176] Ebato C, Uchida T, Arakawa M, Komatsu M, Ueno T, Komiya K, et al. Autophagy is important in islet homeostasis and compensatory increase of beta cell mass in response to high-fat diet. Cell Metab. 2008 Oct;8(4):325-32.

[177] Mizushima N, Klionsky DJ. Protein turnover via autophagy: implications for metabolism. Annu Rev Nutr. 2007;27:19-40.

[178] Bjorkoy G, Lamark T, Brech A, Outzen H, Perander M, Overvatn A, et al. p62/SQSTM1 forms protein aggregates degraded by autophagy and has a protective effect on huntingtin-induced cell death. J Cell Biol. 2005 Nov 21;171(4):603-14.

[179] Towns R, Kabeya Y, Yoshimori T, Guo C, Shangguan Y, Hong S, et al. Sera from patients with type 2 diabetes and neuropathy induce autophagy and colocalization with mitochondria in SY5Y cells. Autophagy. 2005 Oct-Dec;1(3):163-70.

[180] Petersen KF, Dufour S, Befroy D, Garcia R, Shulman GI. Impaired mitochondrial activity in the insulin-resistant offspring of patients with type 2 diabetes. N Engl J Med. 2004 Feb 12;350(7):664-71.

[181] Ozcan U, Cao Q, Yilmaz E, Lee AH, Iwakoshi NN, Ozdelen E, et al. Endoplasmic reticulum stress links obesity, insulin action, and type 2 diabetes. Science. 2004 Oct 15;306(5695):457-61.

[182] Ozcan U, Yilmaz E, Ozcan L, Furuhashi M, Vaillancourt E, Smith RO, et al. Chemical chaperones reduce ER stress and restore glucose homeostasis in a mouse model of type 2 diabetes. Science. 2006 Aug 25;313(5790):1137-40.

[183] Yin JJ, Li YB, Wang Y, Liu GD, Wang J, Zhu XO, et al. The role of autophagy in endoplasmic reticulum stress-induced pancreatic beta cell death. Autophagy. 2012 Feb 1;8(2):158-64.

[184] Hoffman WH, Shacka JJ, Andjelkovic AV. Autophagy in the brains of young patients with poorly controlled T1DM and fatal diabetic ketoacidosis. Exp Mol Pathol. 2012 Oct;93(2):273-80.

[185] Son SM, Song H, Byun J, Park KS, Jang HC, Park YJ, et al. Accumulation of autophagosomes contributes to enhanced amyloidogenic APP processing under insulin-resistant conditions. Autophagy. 2012 Aug 29;8(12).

[186] Sarkar S, Perlstein EO, Imarisio S, Pineau S, Cordenier A, Maglathlin RL, et al. Small molecules enhance autophagy and reduce toxicity in Huntington's disease models. Nat Chem Biol. 2007 Jun;3(6):331-8.

[187] Pandey UB, Nie Z, Batlevi Y, McCray BA, Ritson GP, Nedelsky NB, et al. HDAC6 rescues neurodegeneration and provides an essential link between autophagy and the UPS. Nature. 2007 Jun 14;447(7146):859-63.

[188] Tian Y, Bustos V, Flajolet M, Greengard P. A small-molecule enhancer of autophagy decreases levels of Abeta and APP-CTF via Atg5-dependent autophagy pathway. FASEB J. 2011 Jun;25(6):1934-42.

[189] Sun B, Zhou Y, Halabisky B, Lo I, Cho SH, Mueller-Steiner S, et al. Cystatin C-cathepsin B axis regulates amyloid beta levels and associated neuronal deficits in an animal model of Alzheimer's disease. Neuron. 2008 Oct 23;60(2):247-57.

[190] Caccamo A, Majumder S, Richardson A, Strong R, Oddo S. Molecular interplay between mammalian target of rapamycin (mTOR), amyloid-beta, and Tau: effects on cognitive impairments. J Biol Chem. 2010 Apr 23;285(17):13107-20.

[191] Spilman P, Podlutskaya N, Hart MJ, Debnath J, Gorostiza O, Bredesen D, et al. Inhibition of mTOR by rapamycin abolishes cognitive deficits and reduces amyloid-beta levels in a mouse model of Alzheimer's disease. PLoS One. 2010;5(4):e9979.

[192] Cai Z, Zhao B, Li K, Zhang L, Li C, Quazi SH, et al. Mammalian target of rapamycin: a valid therapeutic target through the autophagy pathway for Alzheimer's disease? J Neurosci Res. 2012 Jun;90(6):1105-18.

[193] Ravikumar B, Duden R, Rubinsztein DC. Aggregate-prone proteins with polyglutamine and polyalanine expansions are degraded by autophagy. Hum Mol Genet. 2002 May 1;11(9):1107-17.

[194] Ravikumar B, Vacher C, Berger Z, Davies JE, Luo S, Oroz LG, et al. Inhibition of mTOR induces autophagy and reduces toxicity of polyglutamine expansions in fly and mouse models of Huntington disease. Nat Genet. 2004 Jun;36(6):585-95.

[195] Berger Z, Ravikumar B, Menzies FM, Oroz LG, Underwood BR, Pangalos MN, et al. Rapamycin alleviates toxicity of different aggregate-prone proteins. Hum Mol Genet. 2006 Feb 1;15(3):433-42.

[196] Menzies FM, Huebener J, Renna M, Bonin M, Riess O, Rubinsztein DC. Autophagy induction reduces mutant ataxin-3 levels and toxicity in a mouse model of spinocerebellar ataxia type 3. Brain. 2010 Jan;133(Pt 1):93-104.

[197] Menzies FM, Rubinsztein DC. Broadening the therapeutic scope for rapamycin treatment. Autophagy. 2010 Feb;6(2):286-7.

[198] Liu K, Liu C, Shen L, Shi J, Zhang T, Zhou Y, et al. WITHDRAWN: Therapeutic effects of rapamycin on MPTP-induced Parkinsonism in mice. Neurochem Int. 2011 Jun 6.

[199] Malagelada C, Jin ZH, Jackson-Lewis V, Przedborski S, Greene LA. Rapamycin protects against neuron death in in vitro and in vivo models of Parkinson's disease. J Neurosci. 2010 Jan 20;30(3):1166-75.

[200] Arias E, Cuervo AM. Chaperone-mediated autophagy in protein quality control. Curr Opin Cell Biol. 2011 Apr;23(2):184-9.

[201] Forlenza OV, de Paula VJ, Machado-Vieira R, Diniz BS, Gattaz WF. Does lithium prevent Alzheimer's disease? Drugs Aging. 2012 May 1;29(5):335-42.

[202] Sarkar S, Floto RA, Berger Z, Imarisio S, Cordenier A, Pasco M, et al. Lithium induces autophagy by inhibiting inositol monophosphatase. J Cell Biol. 2005 Sep 26;170(7): 1101-11.

[203] Kim YH, Rane A, Lussier S, Andersen JK. Lithium protects against oxidative stress-mediated cell death in alpha-synuclein-overexpressing in vitro and in vivo models of Parkinson's disease. J Neurosci Res. 2011 Oct;89(10):1666-75.

[204] Xiong N, Jia M, Chen C, Xiong J, Zhang Z, Huang J, et al. Potential autophagy enhancers attenuate rotenone-induced toxicity in SH-SY5Y. Neuroscience. 2011 Dec 29;199:292-302.

[205] Fornai F, Longone P, Cafaro L, Kastsiuchenka O, Ferrucci M, Manca ML, et al. Lithium delays progression of amyotrophic lateral sclerosis. Proc Natl Acad Sci U S A. 2008 Feb 12;105(6):2052-7.

[206] Zhang X, Heng X, Li T, Li L, Yang D, Du Y, et al. Long-term treatment with lithium alleviates memory deficits and reduces amyloid-beta production in an aged Alzheimer's disease transgenic mouse model. J Alzheimers Dis. 2011;24(4):739-49.

[207] Li XZ, Chen XP, Zhao K, Bai LM, Zhang H, Zhou X. Therapeutic effects of valproate combined with lithium carbonate on MPTP-induced Parkinsonism in mice: possible mediation through enhanced autophagy. Int J Neurosci. 2012 Sep 17.

[208] Sarkar S, Davies JE, Huang Z, Tunnacliffe A, Rubinsztein DC. Trehalose, a novel mTOR-independent autophagy enhancer, accelerates the clearance of mutant huntingtin and alpha-synuclein. J Biol Chem. 2007 Feb 23;282(8):5641-52.

[209] Perucho J, Casarejos MJ, Gomez A, Solano RM, de Yebenes JG, Mena MA. Trehalose protects from aggravation of amyloid pathology induced by isoflurane anesthesia in APP(swe) mutant mice. Curr Alzheimer Res. 2012 Mar;9(3):334-43.

[210] Lan DM, Liu FT, Zhao J, Chen Y, Wu JJ, Ding ZT, et al. Effect of trehalose on PC12 cells overexpressing wild-type or A53T mutant alpha-synuclein. Neurochem Res. 2012 Sep;37(9):2025-32.

[211] Kabuta T, Suzuki Y, Wada K. Degradation of amyotrophic lateral sclerosis-linked mutant Cu,Zn-superoxide dismutase proteins by macroautophagy and the proteasome. J Biol Chem. 2006 Oct 13;281(41):30524-33.

[212] Maiese K, Chong ZZ, Shang YC, Wang S. Novel directions for diabetes mellitus drug discovery. Expert Opin Drug Discov. 2012 Oct 24.

[213] Jespersen MJ, Knop FK, Christensen M. GLP-1 agonists for type 2 diabetes: pharmacokinetic and toxicological considerations. Expert Opin Drug Metab Toxicol. 2012 Oct 24.

Chapter 6

Altering Autophagy: Mouse Models of Human Disease

Amber Hale, Dan Ledbetter, Thomas Gawriluk and
Edmund B. Rucker III

Additional information is available at the end of the chapter

http://dx.doi.org/10.5772/55258

1. Introduction

Since the advent of knockout technologies using mouse embryonic stem cells in the late 1980s, there has been an explosion of murine models to profile human diseases. The understanding of the genetic contribution to these diseases has been further enhanced with the incorporation of tissue-specific gene deletion strategies through the use of the Cre-lox and FLP-FRT site-specific recombination systems. Autophagy, a crucial regulator of cell energy homeostasis, is also a companion process to the ubiquitin-proteasome system to assist in the turnover of proteins. Two distinct types of mouse models have been engineered to characterize autophagy. The first type is based on the reporter model system to both detect and quantitate the *in vivo* levels of autophagy in all tissues and organs. The second type is based on genomic modification to perform global or tissue-specific gene deletions for generation of pathological disease conditions. A wide array of human diseases and conditions have been shown to be intimately linked to alterations in autophagy and include: 1) cancer, 2) heart disease, 3) neurodegenerative diseases (e.g. Alzheimer's and Parkinson's disease), 4) aging, 5) lysosomal storage disorders, 6) infectious disease and immunity (e.g. Crohn's disease), 7) muscle atrophy, 8) stroke, 9) type 2 diabetes, and 10) reproductive infertility. This article will address the role of autophagy in human disease progression by reviewing the strengths and weaknesses of current murine models, as well as discussing their utility as therapeutic models for disease prevention and amelioration.

2. Transgenic models for autophagy detection

GFP-LC3. The best characterized and most widely used detection model is the GFP-LC3 transgenic mouse generated by Mizushima and colleagues [1]. This robustly expressing transgenic mouse, in which LC3 is driven by a constitutive CAG promoter, displays punctate GFP fluorescence that corresponds to autophagosomes. With this transgenic model, quantitation of autophagosomes is feasible using a high resolution fluorescence microscopy. This reporter has been crossed into many of the knockout and floxed autophagy models generated in the field. For example, *Atg5* $^{-/-}$ mice are autophagy deficient and *Atg5* $^{-/-}$; GFP-LC3 mice do not exhibit the punctate fluorescence indicative of autophagosomes. Protocols for use are widely available and published references are helpful, for detailed information see [2]. This model is limited in that only autophagosome numbers, not autophagic flux, can be evaluated. In basal conditions, lysosomal degradation clears the autophagosome and contents from the cell, thus maintaining a "balance" of autophagosome formation and degradation. An accumulation of autophagosomes could either represent an increase in formation or a decrease in fusion events. Ferreting out these differences is relevant for proper data interpretation especially when using chemical autophagy inhibitors and inducers. Measuring autophagic flux *in vivo* has been problematic to date and the field is in need of an appropriate reporter model; currently, tandem fluorescent-tagged autophagy proteins are a valuable *in vitro* tool. An increase in GFP puncta (denoting labeled autophagosomes) could indicate an increase in autophagosome formation or a decrease in vesicular fusion with the lysosome. Due to its chemical nature, GFP is quenched by the low pH of the autolysosome. Since red fluorescent proteins are more pH stable, they have been utilized in assays designed to monitor autophagic flux. The need for better detection mechanisms with regard to cardiac autophagy has led to the generation of a double transgenic reporter. A cardiac muscle specific alpha myosin heavy chain (αMyHC) promoter was used to drive expression of a mCherry-LC3 construct. These mice were crossed with the GFP-LC3 model previously described to produce a double reporter which allowed for the monitoring of autophagic flux; autophagosomes were identified by GFP and mCherry co-localized puncta, while autolysosomes were tracked by mCherry-only puncta [3]. Although this model is cardiac-specific, a similar strategy could be used to target other tissues. Additionally, a dual labeled RFP-GFP-LC3 construct could be used to generate a ubiquitous transgenic model to quantitate autophagic flux.

GFP-GABARAP. GFP-GABARAP transgenic mice were originally generated to address the question of the role of GABARAP in podocytes. Since GABARAP was reported to be highly expressed in podocytes, a pCAG-GFP-GABARAP transgenic mouse was produced in order to examine subcellular localization in this specialized cell type. The expression level of GFP-GABARAP is low, yet visible, ameliorating many of the potential effects of highly expressing fluorescent proteins. In podocytes, GFP-GABARAP co-localized with p62 aggregates but not LC3II. Although it was shown that GABARAP was not the preferred Atg8 ortholog for conjugation in podocytes, this may prove to be a valuable reporter model in other tissues [4].

2.1. Cancer

As we will see as a common thread in many of the disease models featured in this chapter, there is no consensus in the literature as to whether autophagy is protective or maladaptive. The contribution of autophagy to cancer development is the most demonstrable example of the Janus-like role of autophagy in disease pathogenesis, and perhaps the most thoroughly investigated. Autophagy is hypothecated to contribute to tumor progression through two distinct mechanisms. Tumor growth is initially restricted when the centrally-located cells undergo nutrient deprivation, hypoxia, and if prolonged, subsequently undergo necrosis. This metabolic stress induces autophagy to maintain energy homeostasis and prevent necrosis until neo-vascularization occurs. Angiogenesis is potentially able to restore a nutrient supply throughout the tumor, promoting tumor survival and facilitating growth and metastasis. Additionally, transformed cells that harbor defects in apoptosis and autophagy pathways tumor cells undergo chronic necrosis, which foments vascularization of the necrotic area, thus promoting tumorigenesis. Conversely, intact autophagy can be cytoprotective by eliminating damaged or aged organelles and degrading aggregated or misfolded proteins, thus preventing accumulation of tumorigenic and/or cytotoxic cellular inclusions.

These two scenarios are manifested in mouse models of autophagy disruption. When *Beclin1* (*Becn1*) was heterozygously disrupted (*Becn1*$^{+/-}$), these mice exhibited an increase in tumor formation when compared to control littermates [5]. Disruptions in pro-autophagy genes *Becn1* [6], *UVRAG* [7], and *Bif-1*[8] resulted in increased frequencies of lymphomas and mammary neoplasias through an increase in genetic instability. *TSC1*/*TSC2*-deficient mice, which were unable to effectively suppress mTOR, had reduced autophagy and a subsequent increase in tumor occurrence [9, 10]. These models all bolster the stance that autophagy is tumor suppressive. The embryonic or neonatal lethality of many autophagy knockout models is prohibitive for studying diseases, such as cancer, that are associated with age. Tissue-specific knockout studies will provide more insights in elucidating gene-specific mechanisms for tumor suppression. Evidence is forthcoming from cancer cell lines and primary tumor profiling that autophagy is permissive for tumor growth and enhanced in primary tumors. In a pancreatic cancer cell line (8988T cells), inhibition of autophagy by Atg5 siRNA resulted in an inhibition of soft agar growth, a measure of tumorigenic potential [11]. In cell lines with activated *ras*, a strong oncogene, autophagy was necessary for the quick growth and high metabolic needs of tumor cells. These cancer cells were described as "addicted to autophagy," and the phenomenon was consistent amongst several Ras-activated cell lines (e.g. T14, H1299, and CHT116). These data suggest that in quickly growing tumors, autophagy inhibition may sensitize the cells to death thus enhancing available treatment options [12]. This study provides primarily *in vitro* evidence of a role for autophagy induction in supporting tumor growth.

2.2. Heart disease

Coronary heart disease (CHD) is the leading cause of death of men and women in the United States. CHD is caused by narrowing of the arteries that supply the heart. In mouse models a standard treatment is to induce an ischemic event by surgically occluding one or more coronary artery. This occlusion may be relieved at a later time (ischemia/reperfusion) or remain (permanent occlusion). Interestingly, autophagy is induced in both ischemia/reperfusion (I/R) events and permanent occlusion events but the outcomes are differential. During I/R events autophagy is activated, however in *Becn1*$^{+/-}$ mice damage was reduced indicating that induction of autophagy was maladaptive during reperfusion [13]. Conversely, autophagy appears to be protective during permanent occlusion events. Long term ischemia causes relief of mammalian target of rapamycin (mTOR) inhibition of autophagy, thus leading to autophagy induction. In concordance with this, expression of a dominant negative mTOR regulator resulted in a reduction of autophagy and subsequent increase in cardiac damage [14]. Two recent mouse models of heart disease provide evidence supporting a protective role of autophagy for the prevention of CHD [15]. Razani et al provided evidence that autophagy prevented cholesterol crystal-induced inflammation that normally can lead to atherosclerosis. Since high fat diets have been shown to inhibit autophagic flux, competent autophagic flux may be necessary for the prevention of CHD [16].

In addition to CAD, congenital mutations and other pathological conditions (e.g. type 2 diabetes) my result in cardiomyopathies. In a recent study, Choi et al showed that in a cardiomyopathy model resulting from a mutation in A-type lamins (A/C) resulted in active mTOR in cardiomyocytes, which inhibited autophagy activation [17]. This lack of autophagy activation was proposed to lead to an energy deficit and was detrimental to the survival of the cardiomyocytes and resulted in disease progression. Further evidence of the protective role of autophagy was obtained when autophagy was reactivated by the mTOR-inhibiting drug rapamycin. In mice mutant for A-type lamins, rapamycin treatment attenuated the cardiomyopathy phenotype. In another approach, a mouse model of diabetic cardiomyopathy, generated by diet alteration to include a surplus of saturated fatty acids, has revealed that autophagy was activated in response to pressure increase resultant from cardiac hypertrophy. Autophagy induction was measured by increased Becn1 and LC3B mRNA, as well as increased GFP-LC3 puncta in the diabetic hypertrophy model. In this system, isolated cardiomyocytes that were autophagy impaired did not develop hypertrophy, indicating that increased autophagy was required for hypertrophy. Furthermore, isolated cardiomyocytes treated with myristate led to an increase in Becn1 and Atg7 expression through a ceramide-dependent mechanism [18]. The role of autophagy is complex and seems to be highly context specific with regard to heart disease. It is apparent that the method by which autophagy is inhibited or induced and in which type of model differentially dictates whether autophagy will result in a positive or negative outcome.

2.3. Neurodegenerative diseases (e.g. Alzheimer's and Parkinson's disease)

Post-mitotic neurons rely heavily on basal autophagy to clear old, damaged organelles and potentially harmful protein aggregates. Several neurodegenerative disorders result from expansions of poly-glutamine or poly-alanine stretches, including Huntington's disease, Parkinson's disease, spinocerebral ataxias, and fronto-temporal dementia. Mutant proteins involved in these diseases have an increased propensity to aggregate and poison the cell; accordingly, the disease progression is directly related to the amount of protein aggregates formed in these patients. In normal physiology, neurons rely on both the proteasomal degradation system and autophagy to maintain protein and energy homeostasis. It has been shown that autophagy is used preferentially to remove large aggregated proteins or multi-protein plaques. In a rat model of Machado-Joseph disease (spinocerebral ataxia type 3) overexpression of Becn1 reduced both mutant ataxin-3 accumulation and ubiquitin-positive inclusions [19]. It seems that autophagy is used as a compensatory mechanism to relieve the toxic effects of these mutant protein aggregates in the cell, serving a protective role. Neuronal-specific conditional deletion of *FIP200* (*FIP200*$^{fl/fl}$; nestin-Cre) resulted in dystrophy (axonal swelling), neurodegeneration, accumulation of polyubiquitinated proteins, damaged mitochondria, and neuronal death. When compared to *Atg5*/*Atg7* neuronal conditional deletion models, the *FIP200*$^{-/-}$ phenotypes were more severe, present at an earlier age, and resulted in premature death. Unique to the *FIP200*$^{-/-}$ conditional knockout mice is the development of diffuse brain spongiosis, observed as early as 2 weeks of age, and associated with ubiquitin positive inclusions, indicative of impaired clearance of cytotoxic proteins by autophagy [20]. Targeted deletions of either *Atg5* or *Atg7* resulted in the accumulation of polyubiquitinated proteins and sensitized neurons to degeneration; supporting the hypothesis that autophagy is neuroprotective [21, 22]. Purkinje cells (PC) reside within the gray matter at the interface between the molecular and the granular layers of the cerebellar cortex and are important for signal integration, balance, and motor coordination. PC-specific conditional deletion models of *Atg7* (*Atg7*$^{fl/fl}$; pcp2-Cre) resulted in PC dystrophy and subsequent axon terminal degeneration. PC degeneration was followed by PC death and behavioral changes in the mutant mice [23]. In a similar experiment, deletion of *Atg5* in PCs (*Atg5*$^{fl/fl}$; pcp2-Cre) also resulted in axonal swelling and neurodegeneration [24]. Although both the nestin-Cre and pcp2-Cre have been employed as useful tools to investigate the result of tissue-specific autophagy loss, one important difference is that nestin-Cre is also expressed in some astrocyte populations as well as other CNS structures, while pcp2-Cre is restricted to PCs. This difference may alter the experimental designs, as nestin-Cre is more appropriate for diffuse CNS ablation studies and pcp2-Cre is more Purkinje cell-specific.

2.4. Aging

Autophagy and its association with aging have been explored in two distinct contexts: 1) the impact of autophagy on increasing lifespan or longevity, and 2) the role of autophagy in age-related disease states. Longevity-promoting regimens, including caloric restriction (CR) and inhibition of TOR with rapamycin, resveratrol or the natural polyamine spermidine, have been associated with autophagy induction [25]. CR can improve heart function through autophagy, as long-term CR preserved cardiac contractile function with improved cardiomyocyte function and lessened cardiac remodeling [26]. Rapamycin prolonged median and maximal lifespan of both male and female mice when fed beginning at 600 days of age; this supplementation led to a life span increase of 14% for females and 9% for males. In addition, rapamycin-treated mice beginning at 270 days of age also increased survival in both males and females [27]. Lifelong administration of rapamycin extended the lifespan of female 129/Sv mice, as 22.9% of rapamycin-treated mice survived the age of death of the last control mouse. Rapamycin also inhibited age-related weight gain, decreased aging rate, and delayed spontaneous cancer formation [28]. Although rapamycin and caloric restriction both increase the life span of mice, they probably do not occur through similar mechanisms. Dietary restricted mice (40% food restriction) and rapamycin-treated mice both exhibited increased levels of autophagy [29]. The fat mass was similar between control and rapamycin-treated mice, but lower for the caloric restricted mice. There were also striking differences in insulin sensitivity and expression of cell cycle and sirtuin genes in mice fed rapamycin compared with dietary restriction. Spermidine, a natural polyamine whose intracellular concentration declines during human aging, extended the lifespan of yeast, flies and worms, and human immune cells. In addition, spermidine administration potently inhibited oxidative stress in aging mice [30].

Basal autophagy helps to reduce the deleterious effects from oxidative stress, heat stress and cytoplasmic protein aggregates. During the aging process, basal autophagy levels gradually decline so that the cell is not equipped to deal with these stressors. Since many age-related diseases correlate with a decline in basal autophagy, a targeted therapeutic strategy would be to increase the levels of productive autophagy to reduce the severity of the disease. However, it is important to keep in mind that a tight regulation of basal autophagy levels is important, as too much autophagy could have a negative outcome. For example, in the *Zmpste24*-null progeroid mice, which model the human laminopathy Hutchinson-Gilford Progeria Syndrome (HPGS), there was an increase in autophagy instead of the anticipated reduction that occurs during normal aging. Although autophagy levels were similar to those found associated with caloric restriction and prolonged lifespan, in this instance autophagy was linked to having a potential role in the premature aging phenotype. However, these mice also have several metabolic alterations including changes in circulating hormones (e.g. leptin, insulin, adiponectin) and glucose levels that probably impact additional contributing cellular processes [31]. Progerin, the truncated form of lamin A protein, was found to co-localize with the autophagic adapter protein p62 and the autophagy linked FYVE protein, ALFY [32]. Moreover, rapamycin decreased progerin protein levels through autophagy induction, which rescued the progeria phenotype in HGPS fibroblasts [33]. Rapamycin-induced autophagy has therapeutic implications for other types of laminopathies as well. For example, *Lmna-null* (lamin A-deficient) mice exhibited skeletal muscle dystrophy and cardiac hypertrophy; these pathologies were improved through rapamycin administration [34]. An oxidative environment potentially plays a crucial role in the aging process, as *p62*$^{-/-}$ mice exhibited accelerated aging phenotypes and tissues displayed elevated oxidative stress due to defective mitochondrial electron transport [35]. Likewise, *Cisd2*-null mice exhibited nerve and muscle degeneration and a premature aging phenotype [36]. CISD2, the gene responsible for Wolfram syndrome 2 (WFS2), encodes for a mitochondrial protein involved in mammalian life-span control. Although mitochondrial degeneration was exacerbated with age with a concomitant elevation of autophagy, this elevation was most likely due to a cellular response of mitophagy to clear damaged mitochondria. In addition to induced-mutation mouse models, there have been several different naturally occurring strains of senescence-prone (9 lines) and senescence-resistant (3 lines) mice that have been developed since the 1970s at Kyoto University in Japan [37]. These mice have been important to model aging, senile dementia, and Alzheimer's disease. By 12 months of age, the senescence accelerated mouse prone 8 (SAMP8) mice demonstrated a decline in cognitive ability that corresponded to increased levels of ubiquitinated proteins and autophagic vacuoles (AV) in hippocampal neurons, and decreased expression levels of Becn1 [38]. In contract, the senescence-resistant strain did not show an accumulation of these autophagic vacuoles. In the future, it would be interesting to examine whether calorie restriction or rapamycin administration could reduce the accumulation of ubiquitinated proteins and improve learning and memory in the senescence-prone model.

2.5. Lysosomal storage disorders

The group of degenerative disorders included in umbrella term lysosomal storage disorders (LSDs) is a heterogeneous and emerging list of diseases that commonly present with an inability to metabolize a normal cellular substrate. The metabolic defect may reside with the ability of the lysosome to degrade the substrate, or a blockade of autophagic flux, most often inhibiting fusion of the autophagosome and lysosome. A reduction in autophagic flux may result in an increase in autophagosome like structures in the cytoplasm as well as uncharacteristically large autophagosome like structures being formed. Consequently, failure to eliminate/recycle the autophagosomal contents induces cellular stress and may result in death.

Pompe disease, the first LSD to be characterized, is caused by an inability to synthesize acid α-glucosidase (GAA), a lysosomal enzyme needed to breakdown glycogen. Pompe mice (GAA KO) phenocopied the human condition, and abnormal autophagosomal and autolysosomal structures were seen intracellularly. When Pompe mice were crossed with *Atg5/7* muscle-specific conditional knockouts to inhibit autophagy, mice metabolized glycogen more efficiently than the Pompe mice and had a more positive prognosis. In this model system, autophagy was contributing to the pathology of the disease and inhibition of autophagy was been shown to be a useful therapeutic intervention. A side effect of muscle-specific autophagy inhibition, i.e. muscular atrophy, will be discussed in a later section.

Multiple sulfatase deficiency (MSD) is a disease where affected individuals have a reduction in the activity of all sulfatases due to mutations in Sulfatase Modifying Factor 1 (SUMF1), an enzyme responsible for post-translational modification of all sulfatases. A *sumf1* knockout mouse model shared characteristic manifestations of MSD including: skeletal abnormalities, kyphosis, and growth retardation. Impaired autophagosome-lysosome fusion was implicated, as a build-up of undigested material was detrimental to cellular homeostasis and led to death. Though models have exhibited that MSD is accompanied by defective autophagic flux, whether autophagy is protective or detrimental to the pathogenesis of the disease is unclear.

Mucopolysaccharidosis type VI (MPS VI) is caused by a specific sulfatase deficiency (N-acetylgalactosamine-4-sulfate) and patients may be short in stature and suffer from joint stiffness and destruction, cardiac valve abnormalities and corneal clouding. In a rat model of MPS VI, an increased number of autophagic structures were identified by electron microscopy [39].

Niemann Pick type C disease is a metabolic disorder characterized by the accumulation of lipids in late endosomes/lysosomes. The vast majority of cases are due to mutations in the *NPC1* gene. *Npc1*$^{-/-}$ mice had higher levels of autophagy proteins (LC3II) than controls, and PCs were preferentially affected exhibiting an increase in autophagic vesicles by electron microscopy. *Npc*$^{-/-}$ mice had the ability to form autophagosomes but were defective in autophagosome-lysosome fusion, which resulted in a functional autophagic block and inability to metabolize cargo [40].

Mucolipidosis type IV disease (MLIV) is caused by mutations in the MCOLN1 gene which encodes a lysosomal cation channel. Affected patients suffer from psychomotor delays and multiple ophthalmic pathologies. The *mcoln1*$^{-/-}$ mouse model recapitulated most of the symptoms observed in patients with the exception of corneal clouding. In *mcoln1*$^{-/-}$ brains, lysosomal inclusions were observed in several anatomical areas and cell types [41]. Neurons had increased LC3-II expression and failed to clear LC3-II, once again indicating a functional autophagic block that led to the pathogenesis [42].

2.6. Infectious disease and immunity (e.g. Crohn's disease)

The innate immune system is the first line of defense against pathogens; it is evolutionarily more ancient than the adaptive immune system and is deployed quickly and effectively despite its lack of pathogen specificity or memory. Viruses, bacteria, and parasites can be eliminated in an autophagic process involved in innate immunity defense termed 'xenophagy.' Invading bacteria can generally be classified as vacuolar (e.g. *Salmonella*) or cytosolic (e.g. *Listeria, Shigella*). Cytosolic bacteria can undergo ubiquitin-dependent and ubiquitin-independent mechanisms for autophagosomal envelopment followed by translocation to lysosomes. Vacuolar bacteria can be routed into autophagosomes, or in the instance of *Mycobacteria,* autophagy proteins can resume the maturation of the vacuole and promoter fusion with the lysosome. The main recognition receptors that link detection and autophagy induction include the membrane TLRs (Toll-like receptors) and the cytoplasmic nucleotide-binding oligomerization domains (NOD)-like receptors (NLRs). The receptors can recognize the lipopolysaccharides (LPS) and peptidylglycans of Gram-negative bacteria. Microbial interference with autophagy can occur due to the adaptive nature of bacteria. For example, *Shigella flexneri* secretes the protein lcsB, which prevented ATG5-induced autophagy at the bacterial surface. *Yersinia pseudotuberculosis* resides within arrested autophagosomes in macrophages, since it can inhibit the fusion process with lysosomes [43, 44].

The more complicated adaptive immune system also relies on autophagy in many capacities. Studies inducing ablation of autophagy proteins have revealed an essential role for autophagy in maintaining normal numbers of B cells, T cells and hematopoietic stem cell survival and function. *Atg5*- and *Atg7*- deficient models have shown that autophagy was important for T cell survival and maintenance of mitochondria. An increase in mitochondrial mass has been correlated with T cell death in circulating T cells, indicating a potential mechanism of action [45]. Thymic epithelial cells have a high rate of basal autophagy compared to other cell types. During T cell selection thymic epithelial cells display decorations of "self" and "non-self" antigens, aiding in this process autophagy is proposed to facilitate ligand (MHC-II molecule) loading. When autophagy was depleted specifically in thymic epithelial cells, the mature T cell repertoire was diminished due to alterations in positive and negative T cell selection processes. Interestingly, severe colitis, patches of flakey skin, atrophy of uterus, absence of fat pads and enlargement of lymph nodes were observed in many cases. Inflammation was observed in the colon, uterus, lung and Harderian gland of recipient mice. These manifestations are indicative of autoimmune diseases and this model provides a clear linkage between autophagy and autoimmune/inflammatory diseases [46]. A B cell-specific ablation of *Atg5* was achieved by either a Cre-LoxP approach ($Atg5^{fl/fl}$; CD19-Cre) or by repopulating irradiated mice with progenitor cells derived from an $Atg5^{-/-}$ fetal liver. In these experiments, autophagy was found to be essential for the survival of pre-B cells (after the pro-B cell to pre-B cell transition).

Additionally, in peripheral circulation Atg5 was required to maintain normal numbers of B-1a B cell populations but not B-2 B cells [47].

Genome wide association studies of Crohn's Disease identified two autophagy associated genes, Atg16L and IRGM. A naturally occurring insertion/deletion mutation was identified in the 5'UTR (untranslated region) of IRGM (immunity-related GTPase family, M) which disrupted a transcription factor binding site [48]. In another study, a SNP in the coding region of IRGM was identified that affected a microRNA binding site [49]. These identified mutations suggested that IRGM expression level changes were associated with Crohn's disease in humans. A mouse knockout model of *Irgm1* (a.k.a. LRG-47) has been developed and does not display any overt phenotype, including development of the immune system. However, when challenged with infection, *Irgm*$^{-/-}$ mice were unable to control the replication of intracellular pathogens [50]. Unfortunately, the *Irgm1*-knockout mice have not been investigated specifically in the context of autophagy function in immunity to date. However in parallel with *in vitro* data, IRGM1 has recently been shown to induce autophagy in a mouse model of stroke. The promotion of autophagy, most likely at the level of LC3I to LC3II conversion, was generally protective [51].

The relationship between Crohn's disease and the autophagic process is more developed in terms of investigating the *Atg16L* risk allele association by using two *Atg16L* gene trap models and an intestinal epithelium-specific *Atg5* knockout (*Atg5*$^{fl/fl}$; villin-Cre). Both *Atg16L* gene trap models (HM1 and HM2) result in a hypomorphic expression of Atg16L protein. Interestingly, in the Atg16L mutants Paneth cells exhibited abnormal morphology including decreased granule number and disorganization of granules. Researchers concluded that autophagy was required to maintain fidelity of the Paneth cell granule exocytosis pathway. When challenged with infection, the *Atg16L* hypomorphs performed similarly to controls. In *Atg5*$^{fl/fl}$; villin-Cre ileum, abnormal Paneth cells were identified which paralleled to those identified in the *ATG16L* mutants. Human ileum samples from at risk patients were examined and also exhibited abnormal Paneth cell morphology [52]. It is noteworthy at this point to mention that genome wide association studies for ulcerative colitis did not identify either of these autophagy genes, nor others. This suggests that *Atg16L* and IRMG are specific for the physiopathology of Crohn's disease, not inflammatory bowel diseases generally.

2.7. Muscle atrophy

Muscle atrophy is a symptom of a multitude of pathological states including but not limited to fasting conditions, denervation, inactivity, cancer, cardiac failure, and diabetes. Autophagy has been shown to be active in muscular atrophy and other myopathies however due to the nature of the methods used it cannot be said with certainty whether autophagy is promoting atrophy or is activated as a cytoprotective mechanism and coincides with pathology.

In a mouse tissue-specific mouse model generated to investigate the effect of superoxide dismutase 1 (SOD1) ablation on skeletal muscle, it was found that mutants developed muscle atrophy, reduction in contractile force and abnormal mitochondria. Significant upregulation of the mitophagy (specific and selective form of autophagy wherein mitochondria are preferentially enveloped and degraded) gene *Bnip3*, as well as, autophagosome marker LC3 were detected by RT-PCR, most likely as a result of activation of transcription factor FoxO3. Oxidant accumulation in *SOD1*$^{-/-}$ mice resulted in muscular atrophy through autophagy; this phenotype was rescued by depletion of LC3 by siRNA knockdown suggesting that autophagy was the driving pathway of atrophy. These findings imply that autophagy inhibition is a potential therapeutic target for acute and chronic muscular atrophy [53].

Muscle-specific *Atg7* conditional knockout mice were autophagy incompetent and morphologically diverged from wild type control littermates beginning at 40 days of age. Muscles of the knockout mice exhibited degenerative changes and a decrease in myofiber size; these abnormal changes were concurrent with the loss of muscle contractile force which further decreased with increasing age. Even more telling may be the ultrastructural changes associated with loss of autophagy in the muscles, abnormally large mitochondria, centrally located nuclei and dilated sarcoplasmic reticulum all observed via electron microscopy [54].

As discussed previously, a muscle-specific *Atg5* knockout mouse was generated and bred onto a glycogen-degrading enzyme acid-alpha glucosidase knockout background (*GAA*-KO) to interrogate the nature autophagic degradation of glycogen in the pathogenesis of Pompe disease. This study provided additional evidence that autophagy functioned to prevent muscular wasting. Muscle-specific *GAA*$^{-/-}$; *Atg5*$^{-/-}$ mice developed progressive muscular weakness and eventual paralysis beginning earlier (2-3 months of age) and progressing more rapidly than autophagy-competent *GAA* KO mice. Ubiquitin-positive structures accumulated in both *GAA* KO and *GAA*$^{-/-}$; *Atg5*$^{-/-}$ mice with a differing distribution. In *GAA* KO myocytes, autophagic vesicles built up in the cell and ubiquitin-positive structures associated with the autophagosome. In *GAA*$^{-/-}$; *Atg5* $^{-/-}$ myocytes, ubiquitin-positive structures were distributed throughout the cell and appeared to associate with lysosomes, though were not membrane bound. These data indicated that the disruption of functional autophagy and accumulation of toxic ubiquitin-positive structures promoted muscular myopathy [55].

2.8. Stroke

Cerebral ischemia is achieved in mice most often by surgical intervention and results in global, restricted, or cerebral directed ischemia depending on the method selected. Furthermore in some models ischemia is reversed, allowing reperfusion of the cerebral tissue. Autophagy is induced by hypoxia/ischemia events; however it is unclear whether autophagy is protective or maladaptive in stroke models. Responsibility of much of the dissenting opinions may be attributed to the variety of techniques used to induce ischemia events, these have been reviewed at length by Hossmann in [56]. Neonatal mice subjected to hypoxic/ischemic (H/I) brain injury responded with a robust autophagic response in neurons and hippocampal neuron death. *Atg7*$^{-/-}$ neonates, which are autophagy incompetent, were protected from hippocampal neuron death when subjected to identical (H/I) brain injury as control mice, indicating a neurotoxic role for autophagy induction [57]. Conversely, in when wild type mice were subjected to H/I brain injury and reperfusion, damage was mitigated by intraperitoneal injection of NAD^+. NAD^+ administration inhibited autophagy induction. In the NAD^+ treated group, a reduction of autophagy was correlated with a decrease in neuronal damage. To further investigate this link, researchers subjected mice to H/I injury and treated them with 3-

methyladenine (3-MA), an autophagy inhibitor, and an amelioration of neuronal damage was observed. These data indicated that in adults, H/I brain injury followed by reperfusion autophagy was maladaptive; furthermore, inhibition of autophagy at the time of reperfusion was neuroprotective [58]. Moreover, in a rat ischemia model, inhibition of autophagy by Becn1-directed shRNA or 3-MA treatment led to a reduction in damage and neuronal loss in the ipsilateral thalamus. This study supported the hypothesis that autophagy induction increased damage when activated following an ischemia/reperfusion event [59]. The variance amongst the model systems could account for much of the disparity seen in outcomes. The age of the individual, duration of ischemic event, and presence or absence of reperfusion is all potential modulators autophagic response.

2.9. Type 2 diabetes

Type 2 diabetes (T2D) is a complex disease that manifests in tissues, especially adipose, muscles, and liver, becoming resistant to insulin signaling and causing hyperglycemia. Pancreatic β–cells initially respond by increasing their production of insulin, but prolonged insulin resistance results in atresia of β–cells and a marked reduction in insulin production. Due to the high metabolic demands placed on β–cells it follows that autophagy would be play an important role in the pathophysiology of this chronic disease. In wild type C57BL/6 mice β–cells, unlike most organs, autophagosomes are sparingly observed after a period of starvation. However, when mice were fed high fat diets for 12 weeks, autophagosomes were readily observed. These results were confirmed *in vitro* by treating INS-1 β–cells with free fatty acid (FFA), glucose, or tolbutamide. Cells treated with FFA had increased LC3II levels and observable autophagosomes while cells treated with glucose or tolbutamide (a drug regularly prescribed for T2D) did not show any significant change in LC3 conversion or autophagosome formation. To further investigate the link between autophagy and β–cells, a β–cell specific $Atg7^{-/-}$ mutant mice was generated ($Atg7^{fl/fl}$; Rip-Cre). As early as 4 weeks, and degenerating in an age-dependent manner, enlarged cells with pale staining cytoplasm were identified near the periphery of $Atg7^{fl/fl}$; Rip-Cre islets. Inclusion bodies were observed at a high frequency in the enlarged cells the presence of inclusion bodies increased with age, and deformed mitochondria were also observed in these enlarged cells. Resting blood glucose levels were higher and insulin secretion was reduced in $Atg7^{fl/fl}$; Rip-Cre mice when compared to control $Atg7^{fl/fl}$ mice: these differences were amplified when control and $Atg7^{fl/fl}$; Rip-Cre mice were fed high fat diets [60]. In a related model, Marsh et al generated β–cells with a defective secretory pathway in $Rab3A^{-/-}$ mice. Although increased intracellular insulin levels were expected, this was not observed. Increased autophagy of the peptide hormone maintained the levels of insulin and prevented accumulation of potentially toxic cellular substrates [61]. These experiments together suggested that β–cells depended on autophagy to clear damaged organelles and toxic intracellular protein aggregates.

2.10. Reproductive infertility

The role of autophagy during folliculogenesis is a comparatively new topic of study. In rats, LC3II expression was characterized in follicles of varying developmental stages.

Primordial follicles exhibited only weak expression, but antral follicles exhibited robust expression restricted to granulosa cells. Staining was not observed in the oocyte proper or theca cells in any stage follicle [62]. In mice, expression studies also indicate a role for autophagy during folliculogenesis. An expression profile of *Becn1* mRNA revealed significantly higher expression in primordial follicles than in other stages. Immunohistochemistry for BECN1 confirmed this result and consistent staining of granulose cells, theca cells and oocyte cytoplasm was seen in all follicular stages examined [63]. Interestingly, when *Becn1* or *Atg7* is ablated specifically in the female germline (MMTV-CreA; *Becn1*$^{fl/fl}$) or globally (*Atg7*$^{-/-}$), fewer primordial follicles were present in the perinatal ovary. These results indicated that autophagy may be vital for survival of the primordial follicle pool, and be active during folliculogenesis and follicular atresia. In males it is well accepted that autophagy is responsible for post fertilization paternal mitochondrial clearance to prevent paternal mitochondrial DNA transmission; however, the role of autophagy during spermatogenesis is a field in its nascence. It has been shown in *Arabidopsis* that *Becn1* was essential for pollen development [64]. Also, autophagy induction in stallion sperm, as measured by LC3I to LC3II conversation, was important for the survival of sperm post ejaculation [65]. In mice, an *Atg5* sperm-specific conditional deletion has been generated and males developed an infertility phenotype at approximately 15 weeks due to accumulation of abnormal structures in the seminiferous tubules as well as abnormal sperm. Tsukamoto et al. suggested that autophagy may be essential for normal spermiogenesis in mice in order to effect cytoplasmic reduction [66].

In conclusion, a variety of mouse models have been established and interrogated to understand the implications of autophagy in human disease. These genetic-based models are primarily either reliant upon: 1) the generation of an autophagy-defective mouse to characterize a given disease state, or 2) the characterization of autophagy within a pre-established murine model. As this review has shown, conditional knockout models have been extremely useful in disease profiling. The next wave of studies will invariably utilize inducible-based systems for conditional knockouts, genetic-based rescue experiments of disease models, or pharmaceutical-based modification of autophagy. A prevailing theme in the field is that autophagy can either be beneficial or deleterious depending on the disease and its progression state, a theme which must be addressed in designing and implementing appropriate treatment regimes.

Author details

Amber Hale, Dan Ledbetter, Thomas Gawriluk and Edmund B. Rucker III*

*Address all correspondence to: edmund.rucker@uky.edu

Department of Biology, University of Kentucky, SAD

References

[1] Mizushima N, Yamamoto A, Matsui M, Yoshimori T, Ohsumi Y. In vivo analysis of autophagy in response to nutrient starvation using transgenic mice expressing a fluorescent autophagosome marker. Molecular Biology of the Cell 2004;15(3)1101-11.

[2] Mizushima N, Kuma A. Autophagosomes in GFP-LC3 Transgenic Mice. Methods in Molecular Biology 2008;445119-24.

[3] Terada M, Nobori K, Munehisa Y, Kakizaki M, Ohba T, Takahashi Y, et al. Double transgenic mice crossed GFP-LC3 transgenic mice with alphaMyHC-mCherry-LC3 transgenic mice are a new and useful tool to examine the role of autophagy in the heart. Circulation Journal 2010;74(1)203-6.

[4] Takagi-Akiba M, Asanuma K, Tanida I, Tada N, Oliva Trejo JA, Nonaka K, et al. Doxorubicin-induced glomerulosclerosis with proteinuria in GFP-GABARAP transgenic mice. American Journal of Physiology - Renal Physiology 2012;302(3)F380-9.

[5] Qu XP, Yu J, Bhagat G, Furuya N, Hibshoosh H, Troxel A, et al. Promotion of tumorigenesis by heterozygous disruption of the beclin 1 autophagy gene. Journal of Clinical Investigation 2003;112(12)1809-20.

[6] Yue Z, Jin S, Yang C, Levine AJ, Heintz N. Beclin 1, an autophagy gene essential for early embryonic development, is a haploinsufficient tumor suppressor. Proceedings of the National Academy of Sciences 2003;100(25)15077-82.

[7] Liang C, Feng P, Ku B, Oh BH, Jung JU. UVRAG: a new player in autophagy and tumor cell growth. Autophagy 2007;3(1)69-71.

[8] Takahashi Y, Coppola D, Matsushita N, Cualing HD, Sun M, Sato Y, et al. Bif-1 interacts with Beclin 1 through UVRAG and regulates autophagy and tumorigenesis. Nature Cell Biology 2007;9(10)1142-51.

[9] Kobayashi T, Minowa O, Sugitani Y, Takai S, Mitani H, Kobayashi E, et al. A germline Tsc1 mutation causes tumor development and embryonic lethality that are similar, but not identical to, those caused by Tsc2 mutation in mice. Proceedings of the National Academy of Sciences 2001;98(15)8762-7.

[10] Onda H, Lueck A, Marks PW, Warren HB, Kwiatkowski DJ. Tsc2(+/-) mice develop tumors in multiple sites that express gelsolin and are influenced by genetic background. Journal of Clinical Investigation 1999;104(6)687-95.

[11] Yang S, Wang X, Contino G, Liesa M, Sahin E, Ying H, et al. Pancreatic cancers require autophagy for tumor growth. Genes & Development 2011;25(7)717-29.

[12] Guo JY, Chen HY, Mathew R, Fan J, Strohecker AM, Karsli-Uzunbas G, et al. Activated Ras requires autophagy to maintain oxidative metabolism and tumorigenesis. Genes & Development 2011;25(5)460-70.

[13] Matsui Y, Takagi H, Qu X, Abdellatif M, Sakoda H, Asano T, et al. Distinct roles of autophagy in the heart during ischemia and reperfusion: roles of AMP-activated protein kinase and Beclin 1 in mediating autophagy. Circulation Research 2007;100(6)914-22.

[14] Konstantinidis K, Whelan RS, Kitsis RN. Mechanisms of cell death in heart disease. Arteriosclerosis, Thrombosis, and Vascular Biology 2012;32(7)1552-62.

[15] Liao X, Sluimer JC, Wang Y, Subramanian M, Brown K, Pattison JS, et al. Macrophage autophagy plays a protective role in advanced atherosclerosis. Cell Metabolism 2012;15(4)545-53.

[16] Razani B, Feng C, Coleman T, Emanuel R, Wen H, Hwang S, et al. Autophagy links inflammasomes to atherosclerotic progression. Cell Metabolism 2012;15(4)534-44.

[17] Choi JC, Wu W, Muchir A, Iwata S, Homma S, Worman HJ. Dual specificity phosphatase 4 mediates cardiomyopathy caused by lamin A/C (LMNA) gene mutation. The Journal of Biological Chemistry 2012., 287(48), 404513-24

[18] Russo SB, Baicu CF, Van Laer A, Geng T, Kasiganesan H, Zile MR, et al. Ceramide synthase 5 mediates lipid-induced autophagy and hypertrophy in cardiomyocytes. Journal of Clinical Investigation 2012., 122(11), 3919-3930

[19] Nascimento-Ferreira I, Santos-Ferreira T, Sousa-Ferreira L, Auregan G, Onofre I, Alves S, et al. Overexpression of the autophagic beclin-1 protein clears mutant ataxin-3 and alleviates Machado-Joseph disease. Brain 2011; 134(Pt 5)1400-15.

[20] Liang CC, Wang C, Peng X, Gan B, Guan JL. Neural-specific deletion of FIP200 leads to cerebellar degeneration caused by increased neuronal death and axon degeneration. The Journal of Biological Chemistry 2010;285(5)3499-509.

[21] Hara T, Nakamura K, Matsui M, Yamamoto A, Nakahara Y, Suzuki-Migishima R, et al. Suppression of basal autophagy in neural cells causes neurodegenerative disease in mice. Nature 2006;441(7095)885-9.

[22] Komatsu M, Waguri S, Chiba T, Murata S, Iwata J, Tanida I, et al. Loss of autophagy in the central nervous system causes neurodegeneration in mice. Nature 2006;441(7095)880-4.

[23] Komatsu M, Wang QJ, Holstein GR, Friedrich VL, Jr., Iwata J, Kominami E, et al. Essential role for autophagy protein Atg7 in the maintenance of axonal homeostasis and the prevention of axonal degeneration. Proceedings of the National Academy of Sciences 2007;104(36)14489-94.

[24] Nishiyama J, Miura E, Mizushima N, Watanabe M, Yuzaki M. Aberrant membranes and double-membrane structures accumulate in the axons of Atg5-null Purkinje cells before neuronal death. Autophagy 2007;3(6)591-6.

[25] Madeo F, Tavernarakis N, Kroemer G. Can autophagy promote longevity? Nature Cell Biology 2010;12(9)842-6.

[26] Han X, Turdi S, Hu N, Guo R, Zhang Y, Ren J. Influence of long-term caloric restriction on myocardial and cardiomyocyte contractile function and autophagy in mice. The Journal of Nutritional Biochemistry 2012., 23(12), 1592-9

[27] Harrison DE, Strong R, Sharp ZD, Nelson JF, Astle CM, Flurkey K, et al. Rapamycin fed late in life extends lifespan in genetically heterogeneous mice. Nature 2009;460(7253)392-5.

[28] Anisimov VN, Zabezhinski MA, Popovich IG, Piskunova TS, Semenchenko AV, Tyndyk ML, et al. Rapamycin increases lifespan and inhibits spontaneous tumorigenesis in inbred female mice. Cell Cycle 2011;10(24)4230-6.

[29] Fok WC, Zhang Y, Salmon AB, Bhattacharya A, Gunda R, Jones D, et al. Short-Term Treatment With Rapamycin and Dietary Restriction Have Overlapping and Distinctive Effects in Young Mice. Journal of Gerontology: Biological Sciences 2012., 68(2), 108-16

[30] Eisenberg T, Knauer H, Schauer A, Buttner S, Ruckenstuhl C, Carmona-Gutierrez D, et al. Induction of autophagy by spermidine promotes longevity. Nature Cell Biology 2009;11(11)1305-14.

[31] Marino G, Ugalde AP, Salvador-Montoliu N, Varela I, Quiros PM, Cadinanos J, et al. Premature aging in mice activates a systemic metabolic response involving autophagy induction. Human Molecular Genetics 2008;17(14)2196-211.

[32] Tan JM, Wong ES, Kirkpatrick DS, Pletnikova O, Ko HS, Tay SP, et al. Lysine 63-linked ubiquitination promotes the formation and autophagic clearance of protein inclusions associated with neurodegenerative diseases. Human Molecular Genetics 2008;17(3)431-9.

[33] Cao K, Graziotto JJ, Blair CD, Mazzulli JR, Erdos MR, Krainc D, et al. Rapamycin reverses cellular phenotypes and enhances mutant protein clearance in Hutchinson-Gilford progeria syndrome cells. Science Translational Medicine 2011;3(89)89ra58.

[34] Ramos FJ, Chen SC, Garelick MG, Dai DF, Liao CY, Schreiber KH, et al. Rapamycin reverses elevated mTORC1 signaling in lamin A/C-deficient mice, rescues cardiac and skeletal muscle function, and extends survival. Science Translational Medicine 2012;4(144)144ra03.

[35] Kwon J, Han E, Bui CB, Shin W, Lee J, Lee S, et al. Assurance of mitochondrial integrity and mammalian longevity by the p62-Keap1-Nrf2-Nqo1 cascade. EMBO Reports 2012;13(2)150-6.

[36] Chen YF, Kao CH, Chen YT, Wang CH, Wu CY, Tsai CY, et al. Cisd2 deficiency drives premature aging and causes mitochondria-mediated defects in mice. Genes & Development 2009;23(10)1183-94.

[37] Takeda T, Higuchi K, Hosokawa M. Senescence-accelerated Mouse (SAM): With Special Reference to Development and Pathological Phenotypes. ILAR Journal 1997;38(3)109-18.

[38] Ma Q, Qiang J, Gu P, Wang Y, Geng Y, Wang M. Age-related autophagy alterations in the brain of senescence accelerated mouse prone 8 (SAMP8) mice. Experimental Gerontology 2011;46(7)533-41.

[39] Tessitore A, Pirozzi M, Auricchio A. Abnormal autophagy, ubiquitination, inflammation and apoptosis are dependent upon lysosomal storage and are useful biomarkers of mucopolysaccharidosis VI. Pathogenetics 2009; 2(1)4.

[40] Liao G, Yao Y, Liu J, Yu Z, Cheung S, Xie A, et al. Cholesterol accumulation is associated with lysosomal dysfunction and autophagic stress in Npc1 -/- mouse brain. The American Journal of Pathology 2007;171(3)962-75.

[41] Micsenyi MC, Dobrenis K, Stephney G, Pickel J, Vanier MT, Slaugenhaupt SA, et al. Neuropathology of the Mcoln1(-/-) knockout mouse model of mucolipidosis type IV. Journal of Neuropathology & Experimental Neurology 2009;68(2)125-35.

[42] Lieberman AP, Puertollano R, Raben N, Slaugenhaupt S, Walkley SU, Ballabio A. Autophagy in lysosomal storage disorders. Autophagy 2012;8(5)719-30.

[43] Deretic V. Autophagy in immunity and cell-autonomous defense against intracellular microbes. Immunological Reviews 2011;240(1)92-104.

[44] Knodler LA, Celli J. Eating the strangers within: host control of intracellular bacteria via xenophagy. Cell Microbiology 2011;13(9)1319-27.

[45] Stephenson LM, Miller BC, Ng A, Eisenberg J, Zhao Z, Cadwell K, et al. Identification of Atg5-dependent transcriptional changes and increases in mitochondrial mass in Atg5-deficient T lymphocytes. Autophagy 2009;5(5)625-35.

[46] Nedjic J, Aichinger M, Emmerich J, Mizushima N, Klein L. Autophagy in thymic epithelium shapes the T-cell repertoire and is essential for tolerance. Nature 2008;455(7211)396-400.

[47] Miller BC, Zhao Z, Stephenson LM, Cadwell K, Pua HH, Lee HK, et al. The autophagy gene ATG5 plays an essential role in B lymphocyte development. Autophagy 2008;4(3)309-14.

[48] Prescott NJ, Dominy KM, Kubo M, Lewis CM, Fisher SA, Redon R, et al. Independent and population-specific association of risk variants at the IRGM locus with Crohn's disease. Human Molecular Genetics 2010;19(9)1828-39.

[49] Brest P, Lapaquette P, Souidi M, Lebrigand K, Cesaro A, Vouret-Craviari V, et al. A synonymous variant in IRGM alters a binding site for miR-196 and causes deregulation of IRGM-dependent xenophagy in Crohn's disease. Nature Genetics 2011;43(3)242-5.

[50] Collazo CM, Yap GS, Sempowski GD, Lusby KC, Tessarollo L, Woude GF, et al. Inactivation of LRG-47 and IRG-47 reveals a family of interferon gamma-inducible genes with essential, pathogen-specific roles in resistance to infection. The Journal of Experimental Medicine 2001;194(2)181-8.

[51] He S, Wang C, Dong H, Xia F, Zhou H, Jiang X, et al. Immune-related GTPase M (IRGM1) regulates neuronal autophagy in a mouse model of stroke. Autophagy 2012; 8(11), 1621-1627.

[52] Cadwell K, Liu JY, Brown SL, Miyoshi H, Loh J, Lennerz JK, et al. A key role for autophagy and the autophagy gene Atg16l1 in mouse and human intestinal Paneth cells. Nature 2008;456(7219)259-63.

[53] Dobrowolny G, Aucello M, Rizzuto E, Beccafico S, Mammucari C, Boncompagni S, et al. Skeletal muscle is a primary target of SOD1G93A-mediated toxicity. Cell Metabolism 2008;8(5)425-36.

[54] Masiero E, Agatea L, Mammucari C, Blaauw B, Loro E, Komatsu M, et al. Autophagy is required to maintain muscle mass. Cell Metabolism 2009;10(6)507-15.

[55] Raben N, Hill V, Shea L, Takikita S, Baum R, Mizushima N, et al. Suppression of autophagy in skeletal muscle uncovers the accumulation of ubiquitinated proteins and their potential role in muscle damage in Pompe disease. Human Molecular Genetics 2008;17(24)3897-908.

[56] Hossmann KA. Cerebral ischemia: models, methods and outcomes. Neuropharmacology 2008;55(3)257-70.

[57] Koike M, Shibata M, Tadakoshi M, Gotoh K, Komatsu M, Waguri S, et al. Inhibition of autophagy prevents hippocampal pyramidal neuron death after hypoxic-ischemic injury. The American Journal of Pathology 2008;172(2)454-69.

[58] Zheng C, Han J, Xia W, Shi S, Liu J, Ying W. NAD(+) administration decreases ischemic brain damage partially by blocking autophagy in a mouse model of brain ischemia. Neuroscience Letters 2012;512(2)67-71.

[59] Xing S, Zhang Y, Li J, Zhang J, Li Y, Dang C, et al. Beclin 1 knockdown inhibits autophagic activation and prevents the secondary neurodegenerative damage in the ipsilateral thalamus following focal cerebral infarction. Autophagy 2012;8(1)63-76.

[60] Ebato C, Uchida T, Arakawa M, Komatsu M, Ueno T, Komiya K, et al. Autophagy is important in islet homeostasis and compensatory increase of beta cell mass in response to high-fat diet. Cell Metabolism 2008;8(4)325-32.

[61] Marsh BJ, Soden C, Alarcon C, Wicksteed BL, Yaekura K, Costin AJ, et al. Regulated autophagy controls hormone content in secretory-deficient pancreatic endocrine beta-cells. Molecular Endocrinology 2007; 21(9)2255-69.

[62] Choi JY, Jo MW, Lee EY, Yoon BK, Choi DS. The role of autophagy in follicular development and atresia in rat granulosa cells. Fertility and Sterility 2010;93(8)2532-7.

[63] Gawriluk TR, Hale AN, Flaws JA, Dillon CP, Green DR, Rucker EB, 3rd. Autophagy is a cell survival program for female germ cells in the murine ovary. Reproduction 2011;141(6)759-65.

[64] Qin G, Ma Z, Zhang L, Xing S, Hou X, Deng J, et al. Arabidopsis AtBECLIN 1/AtAtg6/AtVps30 is essential for pollen germination and plant development. Cell Research 2007;17(3)249-63.

[65] Gallardo Bolanos JM, Miro Moran A, Balao da Silva CM, Morillo Rodriguez A, Plaza Davila M, Aparicio IM, et al. Autophagy and apoptosis have a role in the survival or death of stallion spermatozoa during conservation in refrigeration. PLoS One 2012;7(1)e30688.

[66] Tsukamoto S, editor. The role of autophagy during spermatogenesis in mice. Reproductive Biotechnology; 2010.

Section 3

Autophagy in GNE Myopathy

Chapter 7

Autophagy in GNE Myopathy

Anna Cho and Satoru Noguchi

Additional information is available at the end of the chapter

http://dx.doi.org/10.5772/55223

1. Introduction

Muscle diseases represent specific muscle pathology. The characteristic features as hallmarks of diseases have been historically used to diagnose the patients. The "Rimmed vacuole (RV)" (Figures 1) is one of such characteristic features in certain groups of the diseases. This structure consists of the space (vacuole) and purple granules (rim) within myofibers, while the space is sometimes occupied with cytosolic contents indicating that the space is artificially produced during the staining process and the rims have the nature of this pathological hallmark. Ultrastructurally, as discussed later, many autophagic vacuoles and multi-lamellar bodies are observed in RVs.

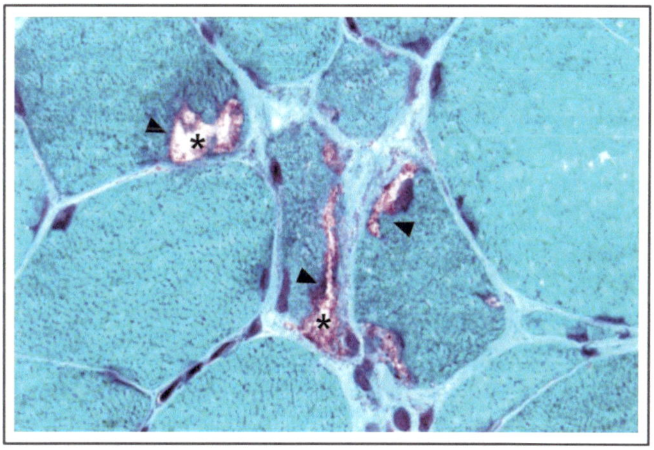

Figure 1. Rimmed vacuoles in a modified Gomori trichrome section. The purple granules (arrow heads) are surrounding vacuoles (asterisks).

Skeletal muscle represents 40~50% of the human body and is one of the most important sites for the control of metabolism. During catabolic conditions, muscle proteins are mobilized to provide alternative energy substrates for the other organs. The RV formation indicates dysfunction of autophagy and breakdown of energy homeostasis in diseased skeletal muscles. In addition, it also suggests the importance of autophagy in muscle functions. There is a group of muscle disease, generally referred to as autophagic vacuolar myopathies (AVM), which are characterized by the accumulation of autophagic vacuoles on skeletal muscle pathology.

In this review, we will give an outline of general knowledge and classification on AVMs and overview the molecular processes underlying autophagic vacuoles formation in rimmed vacuolar myopathies on the basis of our experimental evidences regarding GNE myopathy.

2. Autophagic vacuolar myopathies

Dysfunctional autophagy is associated with several neurodegenerative disorders [1-3]. As for muscle disorders, these are referred to as AVMs [4]. Since the autophagosomes are not observed in normal muscle fibers, autophagic vacuoles have been often recognized as pathologic hallmarks of numerous neuromuscular disorders. Two major categories in AVMs include lysosomal myopathies and rimmed vacuolar myopathies (Table 1) [4-6]. The former are associated with a primary defect in lysosomal proteins and the two best-described and genetically diagnosable AVMs, Pompe disease and Danon disease, are classified in this group. In contrast, autophagic vacuoles in rimmed vacuolar myopathies are secondarily caused by extra-lysosomal defects and usually observed at later stages of the disease. There are various kinds of rimmed vacuolar myopathies including sporadic inclusion body myositis (sIBM) and myofibrillar myopathies and most of them are clinically and etiologically heterogeneous disorders.

Disease	Causative Genes
Lysosomal Myopathy	
Acid maltase deficiency (Pompe disease)	*GAA*
Danon disease	*LAMP2*
X-linked myopathy with excessive autophagy (XMEA)	(identified)
Myopathy with rimmed vacuoles (RVs)	
Inclusion body myositis (sIBM)	?
Myofibrillar myopathy	*DES CRYAB MYOT ZASP etc*
GNE myopathy	*GNE*
Inclusion body myopathy, Paget's disease of bone, and frontotemporal dementia (IBMPFD)	*VCP*
Other myopathies often showing RVs	
Oculopharyngeal muscular dystrophy (OPMD)	*PABPN1*
Marinesco-Sjögren syndrome	*SIL1*
Myotonic dystrophy	*DMPK*

Table 1. Lists of autophagic vacuolar myopathies.

http://dx.doi.org/10.5772/55223

2.1. Lysosomal myopathy

Two best known disorders in AVMs are associated with primary lysosomal protein defects, namely, Pompe disease [7] and Danon disease [8]. The former is caused by a deficiency of lysosomal enzymes within the vacuoles, whereas the latter is caused by a deficiency of lysosomal membrane structural protein [9].

2.1.1. Pompe disease

Pompe disease [7], also referred to as glycogen storage disease type II and acid maltase deficiency, is the best characterized lysosomal myopathy caused by a deficiency of acid α-glucosidase (GAA, also known as acid maltase). This enzyme defect results in lysosomal glycogen accumulation in multiple tissues and cell types, with skeletal and cardiac muscle cells the most seriously affected [10, 11]. The classic infantile form is a rapidly progressive disease with hypotonia, generalized muscle weakness, and hypertrophic cardiomyopathy, usually leading to death from cardiorespiratory failure or respiratory infection in the first year of life [12]. But enzyme replacement therapy with recombinant human GAA is now available, which can dramatically improve the clinical features and life expectancy of the infantile Pompe disease patients [13-15]. The late-onset type shows less progressive clinical characteristics and absence of severe cardiomyopathy; these phenotypical differences are related to residual enzyme activity [16]. The *GAA* gene is located on human chromosome 17q25.2-25.3 and more than 200 pathogenic sequence variations have been characterized up to date [17].

On muscle pathology, cytoplasmic vacuoles are so remarkable and large that these occupy most of the space in many muscle fibers (Figure 2). The vacuoles contain amorphous materials that are presumably glycogen because of the strong reactivity with periodic acid Schiff stain. Acid phosphatase staining also shows strong signals in these vacuoles, indicating high lysosomal content [6]. In terms of pathomechanism, the failure of the lysosomal degradation of glycogen leads to the accumulation of autophagic vacuoles, which may cause cellular dysfunction and abnormal cytoskeletal organization [18].

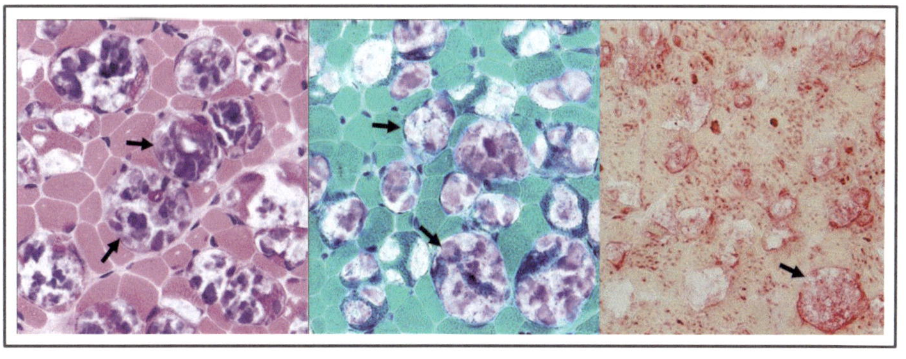

Figure 2. Pathologic findings in Pompe disease. Hematoxylin and eosin (left) and modified Gomori trichrome (middle) sections show pathognomonic vacuolar structures (arrows) in myofibers. These vacuolar structures are strongly stained by acid phosphatase (right).

2.1.2. Danon disease

Danon disease is an X-linked disorder caused by the primary deficiency of lysosome-associated membrane protein-2 (LAMP-2) [9]. Characteristic clinical features include skeletal myopathy, cardiomyopathy, and mental retardation. Male patients usually manifest the disease in their teens and die before their 30s from cardiac problems. LAMP-2 deficiency causes accumulation of autophagic vacuoles in a variety of tissues, including skeletal and cardiac muscles [5]. As LAMP-2 is required for the maturation of early autophagic vacuoles by fusion with endosomes and lysosomes, deficiency of LAMP-2 leads to a failure in the normal progression of autophagic maturation [6].

Muscle biopsy from the patients with Danon disease show scattered small basophilic granules in myofibers and lysosomal acid phosphatase activity is increased in these granules (Figure 3). Large vacuolar structures having sarcolemmal features with acetylcholine esterase activity are surrounding those lysosomal granules and these structures are known as autophagic vacuoles with sarcolemmal features (AVSF) [19]. This characteristic pathology in Danon disease (AVSF) is also seen in a number of diseases including X-linked myopathy with excessive autophagy (XMEA) [20], infantile autophagic vacuolar myopathy [21], and X-linked congenital autophagic vacuolar myopathy [22]. The list of this group of autophagic vacuolar myopathy is rapidly expanding [5] and they are expected to be related with lysosomal function because the pathologic features are quite similar to those in Danon disease.

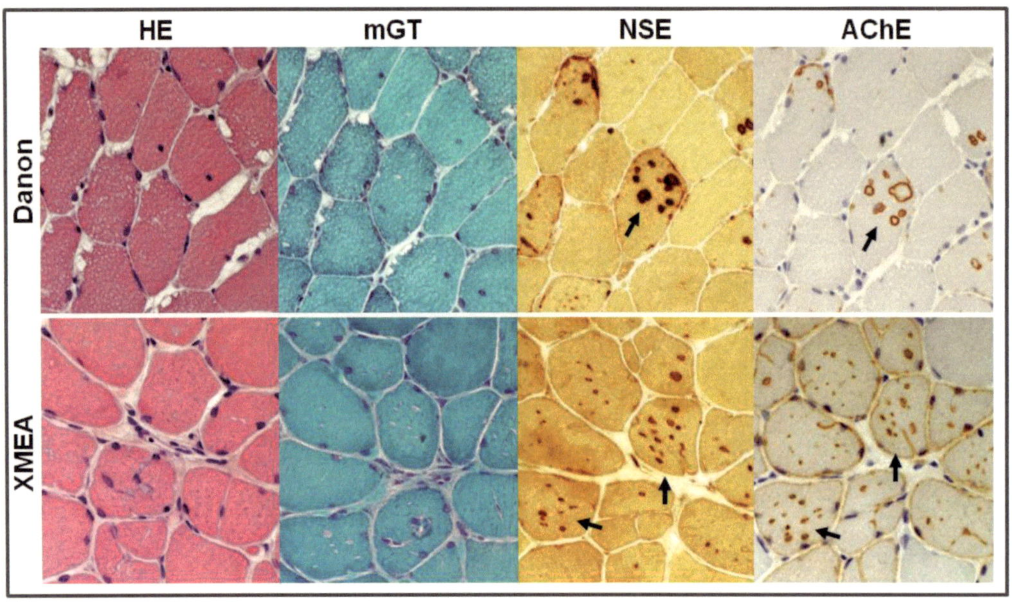

Figure 3. Pathologic findings in Danon disease and XMEA. Many AVSFs (arrows) are showing acetylcholine esterase and nonspecific esterase positivity. HE-hematoxylin and eosin; mGT-modified Gomori trichrome; NSE-nonspecific esterase; AChE-acetylcholine esterase.

2.2. Myopathy with rimmed vacuoles

Rimmed vacuolar myopathies comprise more various and heterogeneous disorders. The most common disease in this group is sIBM, which has been traditionally considered as an inflammatory myopathy. Myofibrillar myopathy, a group of chronic myopathies with a similar pathologic phenotype, is caused by several different genes. And VCP myopathy and GNE myopathy are well known single gene disorders which can be classified as hereditary inclusion body myopathies (hIBM). In addition, it is not uncommon that rimmed vacuoles are appreciated in numerous chronic myopathies which are not classically classified as rimmed vacuolar myopathies.

2.2.1. Inclusion Body Myositis (sIBM)

sIBM is the most common muscle disease in elderly patients [23-25]. Clinically, general progressive muscle weakness starts after age 50 years. The quadriceps muscle and finger flexors are usually affected early on. sIBM Patients may become unable to perform daily living activities and require assistive devices within 10 years of symptom onset. Muscle biopsy characteristically reveals rimmed vacuolar muscle fibers with endomysial T-cell inflammatory infiltrates. Although there still remains controversy whether sIBM is an autoimmune inflammatory myopathy or a primary degenerative myopathy with secondary inflammation, it is becoming more likely that abnormal myoproteostasis and muscle fiber degeneration with aging play primary pathogenic roles in this disorder [26].

Askanas and Engel [27] indicated that several phenomena observed in the degeneration of sIBM muscle fibers are similar to the neuronal degenerative processes occurring in both Alzheimer's disease and Parkinson's disease. Abnormal accumulations of various pathogenic proteins, posttranslational modifications of the accumulated proteins, abnormal protein disposal, and impaired autophagy and 26S proteasome function are common intracellular features of neurodegenerative disorders and thus suggest that sIBM is, like neurodegenerative diseases, a complex degenerative disorder caused by protein misfolding and associated with multiprotein aggregation [28].

2.2.2. Myofibrillar myopathies

RVs are often appreciated in large numbers of myofibrillar myopathies [29-31], which is a group of hereditary myopathies pathologically characterized as markedly disorganized myofibrils with cytoplasmic inclusion. Clinical symptoms of myofibrillar myopathies are very variable. The onset age rages from infancy to the eighth decade. Some patients show limb girdle muscle involvement, whereas others show distal myopathy. Cardiomyopathy is often involved and even can be seen in patients with no obvious skeletal muscle weakness. Seven disease-related genes have been identified (*DES*, *CRYAB*, *MYOT*, *ZASP*, *FLNC*, *BAG3*, and *FHL1*) up to date and all of them encode proteins closely related to Z-line. Electron microscopy findings imply that disintegration of myofibrils near Z-line causes accumulation of filamentous material and aggregation of membranous organelles and glycogen, leading to the entrapment of dislocated membranous organelles in autophagic vacuoles [31].

In the cardiomyocytes-restricted *CRYAB* over-expressed mice, autophagic activity is increased in response to protein aggregates and blunting autophagy *in vivo* dramatically worsen the disease progression [32]. Although myofibrillar myopathy includes various genetically and clinically heterogeneous disorders, accumulation of misfolded proteins is considered as a common pathological pattern and autophagy in myofibrillar myopathy is now becoming to be considered as an adaptive response.

2.2.3. Inclusion Body Myopathy, Paget's disease of the bone, and Frontotemporal Dementia (IBMPFD); Valosin-Containing Protein (VCP) myopathy

Inclusion body myopathy (IBM) with Paget's disease of bone (PDB) and frontotemporal dementia (FTD), now called IBMPFD or valosin-containing protein (VCP) myopathy, is a progressive autosomal dominant disorder caused by mutations in the *VCP* [33]. It is a rare multisystem degenerative disorder with three variably penetrated phenotypic features [34]. 90% of patients develop muscle weakness with a mean onset of 45 years of age and 50% of patients have osteolytic lesions consistent with PDB at the same mean age. About 30% patients develop a typical FTD manifested by apparent language and behavior dysfunction at fifties [35]. Other phenotypic features have been reported as well, including dilated cardiomyopathy, cataracts and sensory-motor neuropathy [36]. Muscle biopsy shows degenerating fibers with RVs and sarcoplasmic inclusions. While molecular pathogenesis in IBMPFD is unknown, the extensive accumulation of ubiquitin conjugates in affected tissues suggests impairment of protein degradation pathways in this disease. In addition, impaired maturation of ubiquitin-containing autophagosomes in cells expressing *VCP* mutants imply that defective autophagy also contributes to the pathogenesis of IBMPFD [37].

2.2.4. Other myopathies often related with rimmed vacuoles

Although they are not pathognomonic, RVs are often accompanied in various chronic myopathic conditions including Marinesco-Sjögren syndrome and oculopharyngeal muscular dystrophy (OPMD). It is interesting that these clinically and genetically different disorders are sharing a similar pathologic feature in skeletal muscles.

Marinesco-Sjögren syndrome is an autosomal recessive disorder clinically characterized by cerebellar ataxia, cataracts from infancy, mental retardation, and myopathy [38]. Loss of function mutations in *SIL1*, which encodes a nucleotide exchange factor for the Hsp70 chaperone BiP, was identified as a causative gene. Previous ultra-structural study showed that myofibrillar degeneration with autophagic phenomenon is prominent in Marinesco-Sjögren syndrome muscles [39]. In addition, it was demonstrated that increased ER stress and altered protein folding lead to neurodegeneration in *SIL1* knock-out mice [40], from which we can infer similar pathogenic process may occur in skeletal muscles of Marinesco-Sjögren syndrome.

OPDM is known to be caused by repeat expansion mutations in *PABPN1* [41]. It has recently become evident that autophagy has an important role in the pathogenesis of repeat expansion disease [42]. The role of autophagy has been extensively studied especially in the polyglutamine diseases such as Huntington's disease and spinocerebellar ataxia. Most of research evidences suggest that autophagy has up-regulated for the degeneration of misfolded proteins and usually is neuroprotective in those disorders.

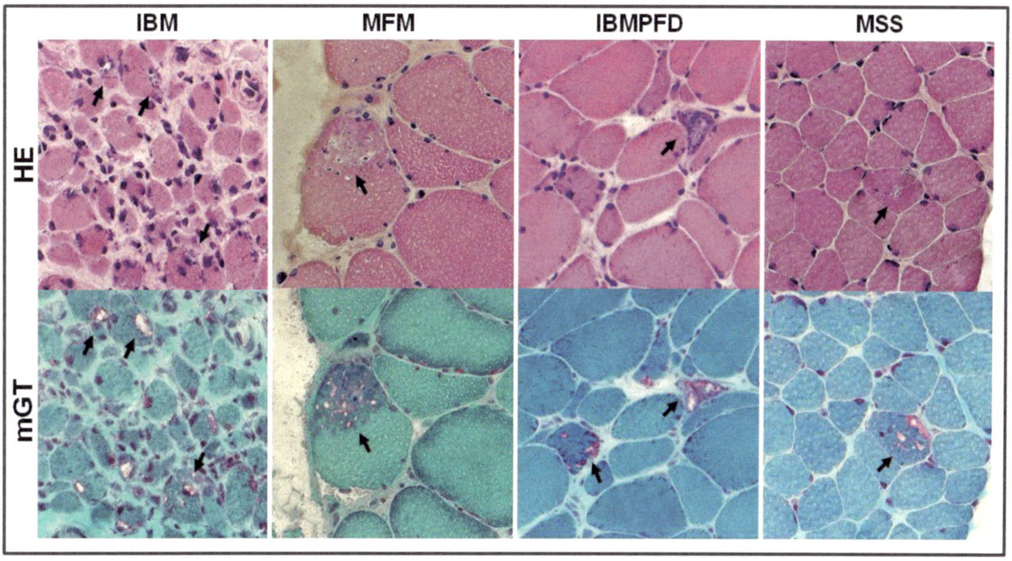

Figure 4. Pathologic findings in myopathies with rimmed vacuoles. Clinically and etiologically different disorders are showing same pathologic features (RVs; arrows) in skeletal muscles. HE-hematoxylin and eosin; mGT-modified Gomori trichrome; IBM-inclusion body myositis; MFM-myofibrillar myopathy; IBMPFD-inclusion body myopathy with Paget's disease of bone and frontotemporal dementia; MSS-Marinesco-Sjögren syndrome.

3. GNE myopathy

GNE myopathy is one of the well described rimmed vacuolar myopathies. It is an autosomal recessive myopathy originally reported in 1981 by Nonaka et al. [43, 44], and thus is also referred as distal myopathy with rimmed vacuoles (DMRV) or Nonaka myopathy. In 1984, Argov and Yarom [45] reported a similar disorder among Iranian Jews with the title of 'rimmed vacuolar myopathy sparing quadriceps'. And the term 'quadriceps sparing myopathy' and 'hereditary inclusion body myopathy (hIBM)' has also been used to refer this disease. Since these two disorders are thought to be identical and caused by GNE mutations [46], it would be better to harmonized the naming of this disease. The experts have recently designated the disease to be called "GNE myopathy".

Among the AVMs secondarily caused by extra-lysosomal defects, GNE myopathy has some advantages for the pathomechanism research. It is a single gene disorder with a homogeneous phenotype and the model mice which evidently display similar features of a human GNE myopathy have been generated [47]. Regardless of the upstream causes, autophagy in myopathies is thought to be mainly attributed to abnormal lysosomal function regarding their effects on myofiber breakdown in common and RVs appreciated in various kinds of myopathies is known to share similar histological and biochemical features. Thus, a comprehensive review of achievements and addressed issues in GNE myopathy research can broaden our understanding of this common pathomechanism of autophagy in rimmed vacuolar myopathies.

3.1. Clinical and pathologic features of GNE myopathy

Clinically, GNE myopathy is an early adult-onset progressive myopathy that affects the tibialis anterior muscle preferentially but spares quadriceps femoris muscles. The symptoms of distal limb muscle weakness start to affect the patient from the second or third decades, and most of the patients become wheelchair-bound between twenties and sixties with a median time to loss of ambulation of 17 years after disease onset [48]. Although the tibialis anterior muscle is most significantly affected, gastrocnemius, hamstrings, paraspinal, and sternocleidomastoid muscles are also involved from an early stage. Cardiac and respiratory muscles are less involved. Serum creatine kinase (CK) levels are normal to mildly elevated [49].

Muscle pathology (Figure 5) is characterized by the presence of RVs predominantly in atrophic fibers, which are occasionally aggregated and form small groups. These RVs are actually clusters of autophagic vacuoles and multi-lamellar bodies. They often contain congophilic amyloid material and deposits that are immunoreactive to β–amyloid and its precursor protein, ubiquitin, and tau protein. Ultrastructurally, the filamentous inclusions measuring 15-20 nm in diameter are seen in both cytoplasm and nucleus with the presence of autophagic vacuoles. Necrotic and regenerating fibers can be rarely seen in GNE myopathy [4, 43, 44, 49].

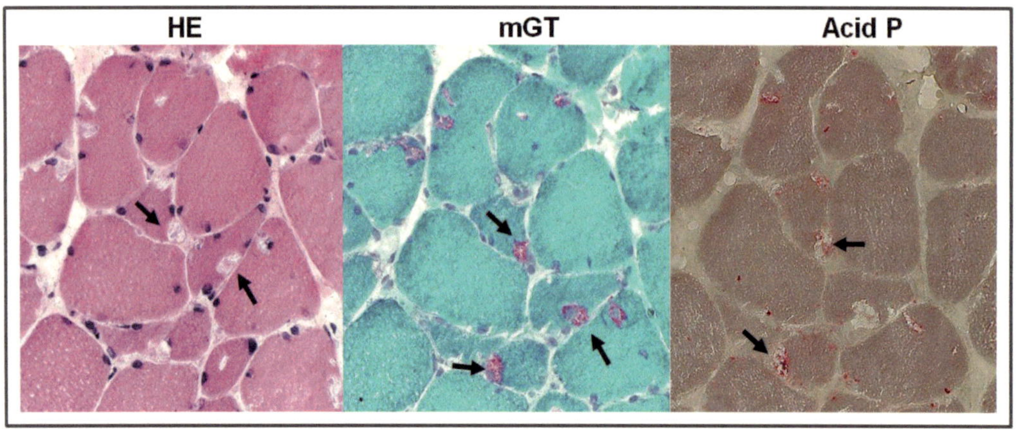

Figure 5. Pathologic findings in GNE myopathy. Rimmed vacuoles (arrows) are predominantly present in atrophic fibers. HE-hematoxylin and eosin; mGT-modified Gomori trichrome; Acid P-acid phoaphatase.

3.2. Molecular pathomechanism of GNE myopathy

GNE myopathy is caused by mutations in the gene encoding a key enzyme in sialic acid biosynthesis, UDP-*N*-acetylglucosamine 2-epimerase/*N*-acetylmannosamine kinase (GNE) [50]. Previous studies have shown that predominant *GNE* mutations exist in certain populations, such as

http://dx.doi.org/10.5772/55223

the V572L mutation in Japanese patients and the M712T mutation in Middle Eastern Jews [46, 50-52]. But it is now evident that GNE myopathy is not restricted to people of Japanese and Jewish ancestry, but they are widely distributed throughout all ethnic groups [53-56]. Recent study showed that homozygosity for V572L (the most common mutation in Japanese population) resulted in more severe phenotypes with earlier symptom onset and faster disease progression, implying the existence of genotype/phenotype correlation in GNE myopathy [48].

After the identification of *GNE* mutations, we found that GNE enzymatic activity measured *in vitro* in cells transfected with mutated *GNE* constructs is decreased. And we demonstrated that the levels of sialic acid in primary cultured cells from GNE myopathy patients are reduced and can be corrected by the addition of free sialic acid [57]. But the mechanism how the hyposialylation makes the pathognomonic pathologic change in skeletal muscle of GNE myopathy has still remained unknown. To answer the question, we developed a model mouse for GNE myopathy.

4. Animal model: A *Gne* knock out mouse expressing human *GNE* D176V mutation

Since the null mutation in *GNE* is known to be embryonically lethal [58], we adopted a different strategy to generate Gne$^{-/-}$hGNED176VTg, a mouse model for GNE myopathy [47, 59]. Our model harbored a transgene of D176V mutated human *GNE* (one of the most prevalent mutations among Japanese GNE myopathy patients) but is knocked-out of endogenous *GNE*, resulted in that only mutated GNE proteins were highly expressed and the endogenous one was disrupted. These mice exhibited marked hyposialylation in serum, muscle and other organs and reproduced similar myopathic phenotypes seen in the skeletal muscles of human GNE myopathy patients.

The muscle weakness, decreased whole muscle mass and reduced contractile power appeared in an age-related manner [60]. After 20 weeks of age, the GNE myopathy mice started to show physiologic muscle weakness, observed as impaired motor performance and reduced force generation of the skeletal muscle. This reduction of the force might be attributed to muscle atrophy, as specific twitch and tetanic forces per cross-section area are maintained at this age. The reduction in gross size of the skeletal muscle is accompanied by an increase in the number of small angular fibers. After 30 weeks of age, specific force generation in the gastrocnemius and tibialis anterior muscles was notably reduced, while myofiber size variation became more remarkable. Intracellular deposition of amyloid and other various proteins was appreciated in the gastrocnemius muscle at this age. After 40 weeks, the muscle force generation worsened, as reflected by increased twitch/tetanic ratio, which might be due to the appearance of the characteristic RV and accumulation of autophagic vacuoles [61]. With these results, the GNE$^{-/-}$hGNED176VTg mouse is the only existing pathogenic model for GNE myopathy up to date.

Muscle pathology of GNE myopathy mice reveals RVs after 40 weeks of age (Figure 6). Intense acid phosphatase staining and expression of lysosomal-associated membrane proteins (LAMPs) and LC3 imply that autophagic process is activated in skeletal muscles of the model mice [47]. Inclusion bodies are also appreciated with expression of various protein markers.

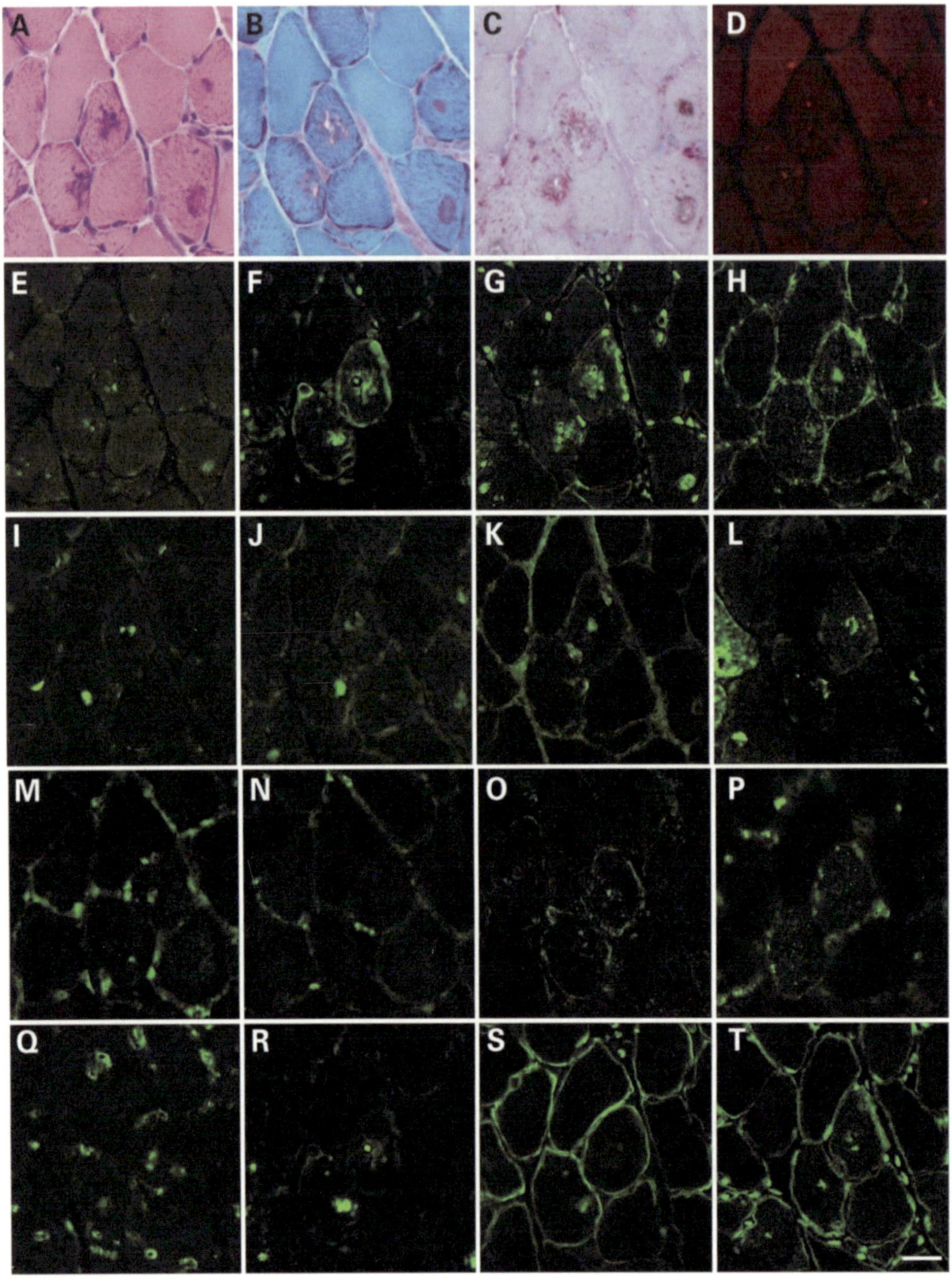

Figure 6. Pathologic findings in the GNE myopathy model mouse. (A) Hematoxylin and eosin sections showing fibers with RVs and cytoplasmic inclusions. (B) modified Gomori trichrome; (C) Acid phosphatase; (D) Congo red. Immunoreactivity to lysosomal proteins confirm the presence of autophagy in fibers with RVs. (E) LAMP-1; (F) LAMP-2; (G) LC3. Intracellular deposition of amyloid is seen in vacuolated or nonvacuolated fibers. (H) BACE2; (I) AβPP; (J) β-amyloid 1-42; (K) β -amyloid 1-40; (L) β -amyloid oligomeric. Neurofilament deposition is observed in the myofibers. (M) SM-31; (N) SM-310. (O) phosphorylated tau; (P) ubiquitin; (Q) Grp94. Sarcolemmal proteins are deposited within the vicinity of RVs. (R) α-dystroglycan (S) β-dystroglycan (T) α-sarcoglycan. Bar-20μm. (Reproduced from [47])

http://dx.doi.org/10.5772/55223

4.1. Autophagy in a mouse model of GNE myopathy

The characteristic RVs are observed after 40 weeks in the GNE myopathy model mouse. Like human muscle pathologic findings, these vacuoles have high acid phosphatase activity and strongly stained by various lysosomal antibodies (Figure 6) [47]. Ultrastructurally, the RVs contain multi-lamellar bodies, electron-dense bodies, and heterogeneous cytoplasmic debris which are surrounded by double membranes, indicating these are autophagic vacuoles (Figure 7). In the near areas, several vacuoles have a single limiting membrane and some cellular debris have no membrane, indicating degradated or ruptured vacuoles. Interestingly, filamentous or granular deposits considered as amyloid often appear with the autophagic vacuoles. And these probable amyloid deposits are also observed in the normal areas, which may suggest that the deposition of protein precede the accumulation of autophagic vacuoles [61].

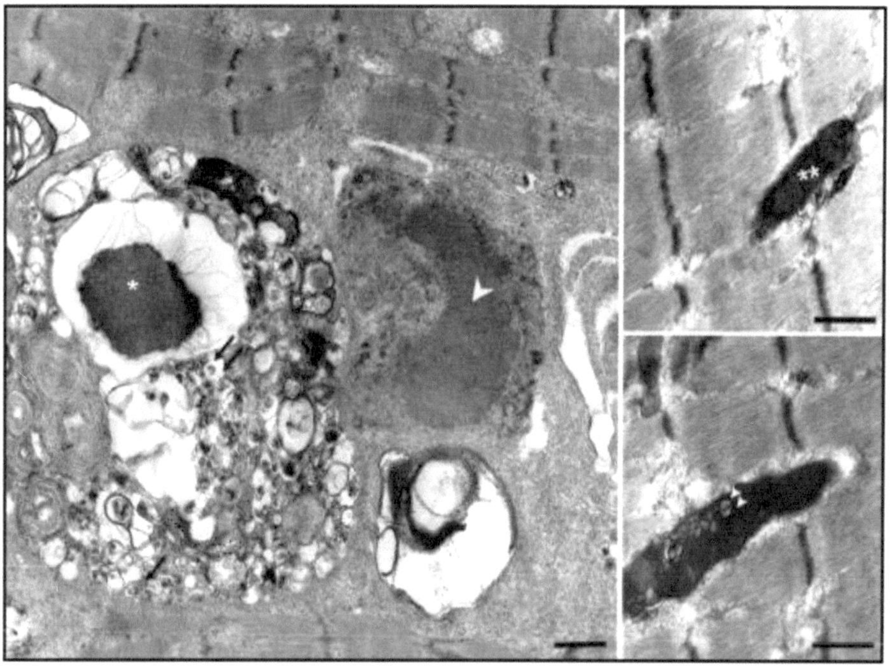

Figure 7. Electron microscopy findings in the GNE myopathy mouse model. Various cellular debris are enclosed by nacent (arrows) and degenerative (double arrows) autophagic vacuoles. Large osmiophilic doposits (asterisk) can be also seen. Dense ovoid bodies (double asterisk) are seen with autophagic vacuoles (double arrowheads) suggesting that these deposits predate RVs formation. Bars-1μm. (Reproduced from [61])

5. Prophylactic treatment with sialic acid metabolites in the GNE myopathy model mice

A possibility of the development of therapy for GNE myopathy was demonstrated in our model mice. As the addition of sialic acid metabolites has been shown to recover overall hyposialylation in cells [57], we have challenged in administering sialic acid compounds *in vivo*. We administered 2-epimerase/N-acetylmannosamine (ManNAc) to the GNE myopathy model mice from the preclinical age (5~6 weeks) continuously until the mice reached the age when all myopathic symptoms appear (54~57 weeks). ManNAc was added to drinking water and given in three doses: 20mg (low dose), 200mg (medium dose), and 2000mg (high dose)/kg body weight of mice in a day. During the treatment period, survival rate was remarkably improved as compared with control-treated mice at all three doses (Figure 8). At the end of the treatment, the phenotypes of the mice were evaluated and compared with placebo group and non-affected littermates. At all doses, motor performance and physiological contractile properties of skeletal muscles were remarkably improved. Sialic acid levels in the blood and tissues were elevated, and more importantly, the levels of sialic acid in the muscle were recovered to an almost normal level after treatment, providing evidence that prophylactic oral administration of ManNAc to the the model mice was notably effective. Then we also examined the effect of oral *N*-acetylneuraminic acid (NeuAc) and sialyllactose together with minimum dose of ManNAc (20 mg/kg bodyweight/day) on GNE myopathy mice starting at the preclinical age of 10~20 weeks. Treatment was also continued up to 54~57 weeks of age and similar beneficial effects on motor performance and force generation of skeletal muscles were obtained [62].

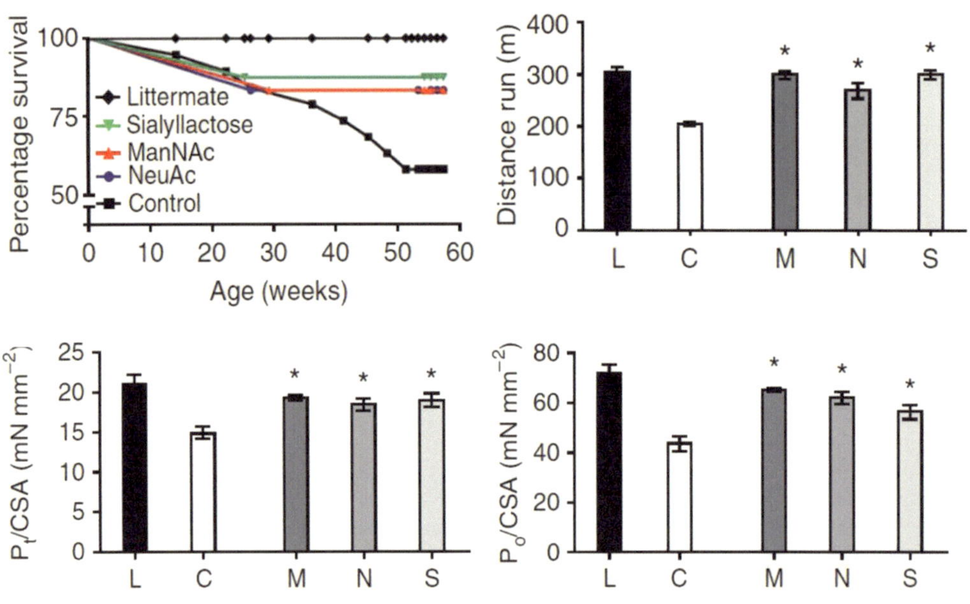

Figure 8. Favorable effects of sialic acid metabolites administration in GNE myopathy model mice. Survival curves (left upper), treadmill performance test (right upper), and contractile properties of specific isometric force (P_t/CSA; left lower) and specific tetanic force (P_0/CSA; right lower). (Reproduced from [62])

Sialic acid metabolites administration also led to a marked change in the muscle pathology of GNE myopathy mice (Figure 9) [62]. Although all mice in the control-treated group showed

http://dx.doi.org/10.5772/55223

RVs in the gastrocnemius muscles, only a few in the treatment groups showed RVs. As RVs are autophagic in nature, we checked for acid phosphatase staining and found decreased staining. The expression of LC3 and Lamp2, which are markers for autophagosomal structures, were not observed in the muscle sections of ManNAc treated mice, except for one mouse that had few RVs in the muscle. The amounts of LC3-I and LC3-II, used as an index to analyze autophagic induction in tissues, were lower after treatment. Treatment also increased muscle cross-sectional area (CSA) and diminished congophilic, amyloid-positive and tau-positive inclusions.

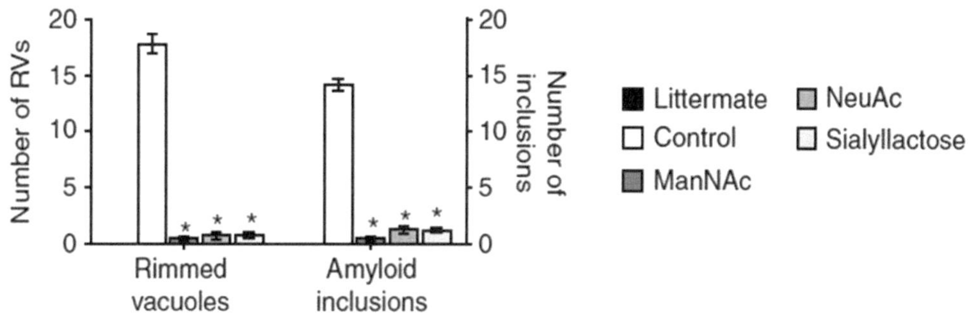

Figure 9. Pathologic improvement after treatment with sialic acid metabolites in GNE myopathy model mice. (Reproduced from [62])

Our successful prophylactic treatment results on the model mice supports the current concept that hyposialylation is one of main factors contributing the pathogenesis of GNE myopathy. This concept suggests that GNE myopathy is a potentially treatable disease and A phase I clinical trial for human patients using oral sialic acid therapy is recently underway in Japan.

6. Future issue 1 — Hypothesized pathway from hyposialylation to RVs formation

Although it has been already demonstrated that hyposialylation is a key factor in the pathomechanism of GNE myopathy and sialic acid metabolites administration can prevent the development of myopathic phenotype in GNE myopathy model mouse, there still remains an unexplained link between hyposialylation due to GNE mutations and the pathognomonic findings in the muscle. However, since the the model mice exhibit hyposialylation and intracellular amyloid deposition before the characteristic RVs appear, we can appreciate that the dysfunctional autophagy is a downstream phenomenon to hyposialylation and amyloid deposition in GNE myopathy.

In normal macroautophgy process, cytoplasm and organelles are enclosed by an isolated membrane (phagophore) to form an autophagosome. The outer membrane of the autophagosome fuses with the lysosome, and the internal material is degraded in the Autolysosome [63]. However, in hyposialylated condition, the autophagy does not progress normally. As hyposialylation can lead to abnormal protein configuration or misfolding, an excessive amount of misfolded glycoproteins which were not degraded in the ER may cause abnormal autophagy in GNE myopathy (Figure 10). Hyposialylation may also lead to abnormal protein processing which can induce abnormal protein deposits in cytoplasm. In addition, there are several reports that suggested oxidative stress is involved in the upstream pathways to amyloid deposition and/or RVs formation. One previous report showed that autophagic vacuoles were associated to be a site of amyloidogenic amyloid precursor protein processing and intra-lysosomal amyloid-β accumulation was induced by oxidative stress [64]. And another report demonstrated that reactive oxygen species may contribute to the formation of autophagosomes [65]. An experimental result implying a biologic function of sialic acid as an oxygen radical scavenger suggests hyposialylation can directly contribute to the increase of oxidative stress and support the above concept [66].

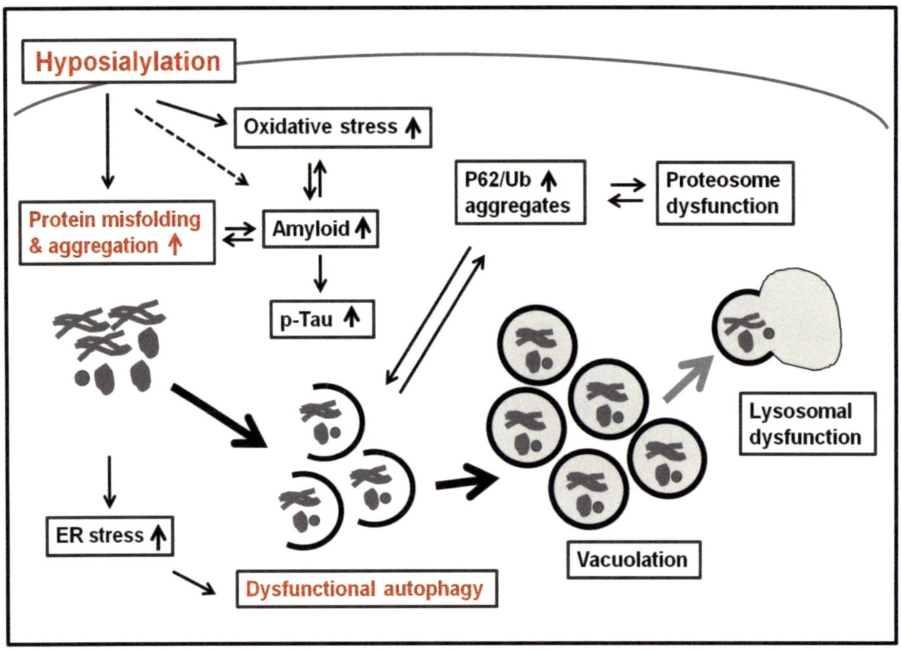

Figure 10. Hypothesized mechanism of dysfunctional autophagy in GNE myopathy.

7. Future issue 2 — The common molecular processes underlying autophagic vacuoles formation in rimmed vacuolar myopathies

Regarding the pathomechanism of rimmed vacuolar myopathies, it is interesting to figure out how various kinds of myopathies from different etiologies can share the similar pathology and undergo similar pathogenic process. One of the most impressive finding is that the accumulation of misfolded proteins and subsequent activation of autophagy are observed in all kinds of rimmed vacuolar myopathies commonly and constantly, which bring us that intra-myofiber accumulation of conformationally modified proteins plays a primary pathologic role in these disorders. However, up-regulation of autophagy alone may not be enough to form RVs if the pathways are operating properly. The primary role of autophagy is to protect cells under stress conditions and it is widely accepted that in most of neurodegenerative diseases, activation of autophagic process is an adaptive response against disease-related stress conditions. The fact that not all myofibrillar myopathy muscles show RVs suggests that the other pathways are necessary to complete RV formation.

Dysfunctional autophagy is another important common feature in rimmed vacuolar myopathies. It was already demonstrated that autophagy is impaired in sIBM, IBMPFD, GNE myopathy, and Marinesco-Sjögren syndrome [26, 37, 40, 61]. Regardless of upstream process, autophagosomes are proliferated and enlarged with lysosomal dysfunction and macroautophagy disregulation, and it might be contribute or worsen the abnormal accumulation of various proteins such as amyloids, p-tau, α-synuclein, and p62. This two major common process, up-regulated and dysfunctional autophagy, possibly develop characteristic RVs in skeletal muscle pathology.

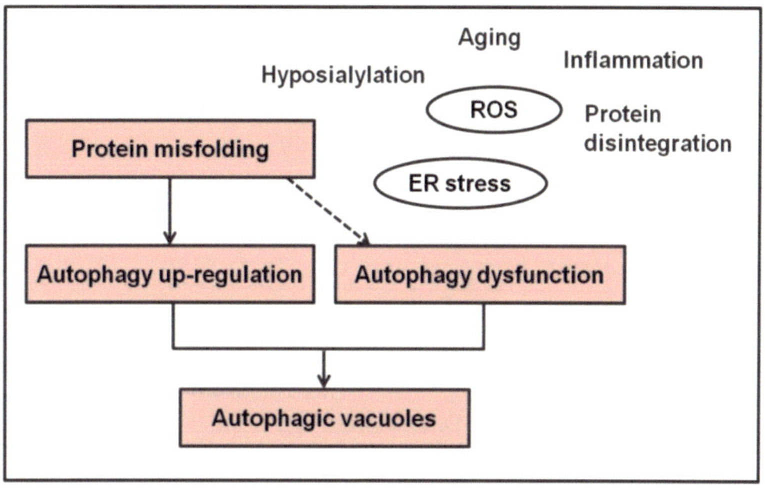

Figure 11. Hypothesized common mechanism of rimmed vacuoles formation in various AVMs.

8. Conclusion

Herein we presented the current knowledge on AVMs and recent approaches to the pathogenesis of rimmed vacuolar myopathies. With the model mice, we proved that hyposialylation is a key factor in the pathomechanism of GNE myopathy. And we also provided evidences that prophylactic treatment with sialic acid metabolites prevents the myopathic phenotype and substantially reduced the number of RVs in the GNE myopathy mice. Since rimmed vacuolar myopathies have revealed to share common pathways regarding the autophagic vacuoles formation irrespective of heterogeneous clinical phenotype and etiology, our experimental achievement can broaden the general understanding on the common pathomechanism of AVMs.

Acknowledgements

This work is supported by JSPS KAKENHI Grant Number 23390236, by Comprehensive Research on Disability Health and Welfare from the Ministry of Health and Labour, and by Intramural Research Grant (22-5) for Neurological and Psychiatric Disorders of NCNP.

Author details

Anna Cho and Satoru Noguchi*

Department of Neuromuscular Research, National Institute of Neuroscience, NCNP, Tokyo, Japan

References

[1] Wong, E, & Cuervo, A. M. Autophagy gone awry in neurodegenerative diseases. Nature neuroscience (2010). , 13(7), 805-811.

[2] Mizushima, N, Levine, B, Cuervo, A. M, & Klionsky, D. J. Autophagy fights disease through cellular self-digestion. Nature (2008). , 451(7182), 1069-1075.

[3] Winslow, A. R, & Rubinsztein, D. C. Autophagy in neurodegeneration and development. Biochimica et biophysica acta (2008). , 1782(12), 723-729.

[4] Nishino, I. Autophagic vacuolar myopathies. Current neurology and neuroscience reports (2003). , 3(1), 64-69.

[5] Nishino, I. Autophagic vacuolar myopathy. Seminars in pediatric neurology (2006). , 13(2), 90-95.

[6] Malicdan, M. C, & Nishino, I. Autophagy in lysosomal myopathies. Brain Pathol (2012). , 22(1), 82-88.

[7] Smith, J, Zellweger, H, & Afifi, A. K. Muscular form of glycogenosis, type II (Pompe). Neurology (1967). , 17(6), 537-549.

[8] Danon, M. J, & Oh, S. J. DiMauro S, Manaligod JR, Eastwood A, Naidu S, et al. Lysosomal glycogen storage disease with normal acid maltase. Neurology (1981). , 31(1), 51-57.

[9] Nishino, I, Fu, J, Tanji, K, Yamada, T, Shimojo, S, Koori, T, et al. Primary LAMP-2 deficiency causes X-linked vacuolar cardiomyopathy and myopathy (Danon disease). Nature (2000). , 406(6798), 906-910.

[10] Kishnani, P. S, & Howell, R. R. Pompe disease in infants and children. The Journal of pediatrics (2004). Suppl) S, 35-43.

[11] Van Der Ploeg, A. T, & Reuser, A. J. Pompe's disease. Lancet (2008). , 372(9646), 1342-1353.

[12] van den Hout HM, Hop W, van Diggelen OP, Smeitink JA, Smit GP, Poll-The BT, et al. The natural course of infantile Pompe's disease: 20 original cases compared with 133 cases from the literature. Pediatrics (2003). , 112(2), 332-340.

[13] Van Der Beek, N. A, Hagemans, M. L, Van Der Ploeg, A. T, Reuser, A. J, & Van Doorn, P. A. Pompe disease (glycogen storage disease type II): clinical features and enzyme replacement therapy. Acta neurologica Belgica (2006). , 106(2), 82-86.

[14] Kishnani, P. S, Corzo, D, Nicolino, M, Byrne, B, Mandel, H, Hwu, W. L, et al. Recombinant human acid [alpha]-glucosidase: major clinical benefits in infantile-onset Pompe disease. Neurology (2007). , 68(2), 99-109.

[15] Chen, L. R, Chen, C. A, Chiu, S. N, Chien, Y. H, Lee, N. C, Lin, M. T, et al. Reversal of cardiac dysfunction after enzyme replacement in patients with infantile-onset Pompe disease. The Journal of pediatrics (2009). e272., 155(2), 271-275.

[16] Winkel, L. P, Hagemans, M. L, Van Doorn, P. A, Loonen, M. C, Hop, W. J, Reuser, A. J, et al. The natural course of non-classic Pompe's disease; a review of 225 published cases. Journal of neurology (2005). , 252(8), 875-884.

[17] Kroos, M, Hoogeveen-westerveld, M, Michelakakis, H, Pomponio, R, Van Der Ploeg, A, Halley, D, et al. Update of the pompe disease mutation database with 60 novel GAA sequence variants and additional studies on the functional effect of 34 previously reported variants. Human mutation (2012). , 33(8), 1161-1165.

[18] Fukuda, T, Ewan, L, Bauer, M, Mattaliano, R. J, Zaal, K, Ralston, E, et al. Dysfunction of endocytic and autophagic pathways in a lysosomal storage disease. Annals of neurology (2006). , 59(4), 700-708.

[19] Sugie, K, Noguchi, S, Kozuka, Y, Arikawa-hirasawa, E, Tanaka, M, Yan, C, et al. Autophagic vacuoles with sarcolemmal features delineate Danon disease and related myopathies. Journal of neuropathology and experimental neurology (2005). , 64(6), 513-522.

[20] Kalimo, H, Savontaus, M. L, Lang, H, Paljarvi, L, Sonninen, V, Dean, P. B, et al. X-linked myopathy with excessive autophagy: a new hereditary muscle disease. Annals of neurology (1988). , 23(3), 258-265.

[21] Yamamoto, A, Morisawa, Y, Verloes, A, Murakami, N, Hirano, M, Nonaka, I, et al. Infantile autophagic vacuolar myopathy is distinct from Danon disease. Neurology (2001). , 57(5), 903-905.

[22] Yan, C, Tanaka, M, Sugie, K, Nobutoki, T, Woo, M, Murase, N, et al. A new congenital form of X-linked autophagic vacuolar myopathy. Neurology (2005). , 65(7), 1132-1134.

[23] Amato, A. A, & Barohn, R. J. Inclusion body myositis: old and new concepts. Journal of neurology, neurosurgery, and psychiatry (2009). , 80(11), 1186-1193.

[24] Griggs, R. C, & Askanas, V. DiMauro S, Engel A, Karpati G, Mendell JR, et al. Inclusion body myositis and myopathies. Annals of neurology (1995). , 38(5), 705-713.

[25] Amato, A. A, Gronseth, G. S, Jackson, C. E, Wolfe, G. I, Katz, J. S, Bryan, W. W, et al. Inclusion body myositis: clinical and pathological boundaries. Annals of neurology (1996). , 40(4), 581-586.

[26] Askanas, V, Engel, W. K, & Nogalska, A. Pathogenic considerations in sporadic inclusion-body myositis, a degenerative muscle disease associated with aging and abnormalities of myoproteostasis. Journal of neuropathology and experimental neurology (2012). , 71(8), 680-693.

[27] Askanas, V, & Engel, W. K. Inclusion-body myositis: muscle-fiber molecular pathology and possible pathogenic significance of its similarity to Alzheimer's and Parkinson's disease brains. Acta neuropathologica (2008). , 116(6), 583-595.

[28] Askanas, V, & Engel, W. K. Inclusion-body myositis: a myodegenerative conformational disorder associated with Abeta, protein misfolding, and proteasome inhibition. Neurology (2006). Suppl 1) S, 39-48.

[29] Hayashi, Y. K. Myofibrillar myopathy]. Brain and nerve = Shinkei kenkyu no shinpo (2011). , 63(11), 1179-1188.

[30] Selcen, D. Myofibrillar myopathies. Neuromuscular disorders : NMD (2011). , 21(3), 161-171.

[31] Selcen, D, Ohno, K, & Engel, A. G. Myofibrillar myopathy: clinical, morphological and genetic studies in 63 patients. Brain : a journal of neurology (2004). Pt 2) 439-451.

[32] Tannous, P, Zhu, H, Johnstone, J. L, Shelton, J. M, Rajasekaran, N. S, Benjamin, I. J, et al. Autophagy is an adaptive response in desmin-related cardiomyopathy. Proceedings of the National Academy of Sciences of the United States of America (2008). , 105(28), 9745-9750.

[33] Watts, G. D, Wymer, J, Kovach, M. J, Mehta, S. G, Mumm, S, Darvish, D, et al. Inclusion body myopathy associated with Paget disease of bone and frontotemporal de-

mentia is caused by mutant valosin-containing protein. Nature genetics (2004). , 36(4), 377-381.

[34] Weihl, C. C, Pestronk, A, & Kimonis, V. E. Valosin-containing protein disease: inclusion body myopathy with Paget's disease of the bone and fronto-temporal dementia. Neuromuscular disorders : NMD (2009). , 19(5), 308-315.

[35] Kimonis, V. E, & Watts, G. D. Autosomal dominant inclusion body myopathy, Paget disease of bone, and frontotemporal dementia. Alzheimer disease and associated disorders (2005). Suppl 1 S, 44-47.

[36] Guyant-marechal, L, Laquerriere, A, Duyckaerts, C, Dumanchin, C, Bou, J, Dugny, F, et al. Valosin-containing protein gene mutations: clinical and neuropathologic features. Neurology (2006). , 67(4), 644-651.

[37] Tresse, E, Salomons, F. A, Vesa, J, Bott, L. C, Kimonis, V, Yao, T. P, et al. VCP/97 is essential for maturation of ubiquitin-containing autophagosomes and this function is impaired by mutations that cause IBMPFD. Autophagy (2010).

[38] Sjogren, T. Hereditary congenital spinocerebellar ataxia accompanied by congenital cataract and oligophrenia; a genetic and clinical investigation. Confinia neurologica (1950). , 10(5), 293-308.

[39] Goto, Y, Komiyama, A, Tanabe, Y, Katafuchi, Y, Ohtaki, E, & Nonaka, I. Myopathy in Marinesco-Sjogren syndrome: an ultrastructural study. Acta neuropathologica (1990). , 80(2), 123-128.

[40] Zhao, L, Rosales, C, Seburn, K, Ron, D, & Ackerman, S. L. Alteration of the unfolded protein response modifies neurodegeneration in a mouse model of Marinesco-Sjogren syndrome. Human molecular genetics (2010). , 19(1), 25-35.

[41] Brais, B, Bouchard, J. P, Xie, Y. G, Rochefort, D. L, Chretien, N, Tome, F. M, et al. Short GCG expansions in the PABP2 gene cause oculopharyngeal muscular dystrophy. Nature genetics (1998). , 18(2), 164-167.

[42] La Spada AR, Taylor JP.Repeat expansion disease: progress and puzzles in disease pathogenesis. Nature reviews Genetics (2010). , 11(4), 247-258.

[43] Nonaka, I, Sunohara, N, Ishiura, S, & Satoyoshi, E. Familial distal myopathy with rimmed vacuole and lamellar (myeloid) body formation. Journal of the neurological sciences (1981). , 51(1), 141-155.

[44] Nonaka, I, Murakami, N, Suzuki, Y, & Kawai, M. Distal myopathy with rimmed vacuoles. Neuromuscular disorders : NMD (1998). , 8(5), 333-337.

[45] Argov, Z, & Yarom, R. Rimmed vacuole myopathy" sparing the quadriceps. A unique disorder in Iranian Jews. Journal of the neurological sciences (1984). , 64(1), 33-43.

[46] Nishino, I, Noguchi, S, Murayama, K, Driss, A, Sugie, K, Oya, Y, et al. Distal myopathy with rimmed vacuoles is allelic to hereditary inclusion body myopathy. Neurology (2002)., 59(11), 1689-1693.

[47] Malicdan, M. C, Noguchi, S, Nonaka, I, Hayashi, Y. K, & Nishino, I. A Gne knockout mouse expressing human GNE D176V mutation develops features similar to distal myopathy with rimmed vacuoles or hereditary inclusion body myopathy. Human molecular genetics (2007)., 16(22), 2669-2682.

[48] Mori-yoshimura, M, Monma, K, Suzuki, N, Aoki, M, Kumamoto, T, Tanaka, K, et al. Heterozygous UDP-GlcNAc epimerase and N-acetylmannosamine kinase domain mutations in the GNE gene result in a less severe GNE myopathy phenotype compared to homozygous N-acetylmannosamine kinase domain mutations. Journal of the neurological sciences (2012)., 2.

[49] Nonaka, I, Noguchi, S, & Nishino, I. Distal myopathy with rimmed vacuoles and hereditary inclusion body myopathy. Current neurology and neuroscience reports (2005)., 5(1), 61-65.

[50] Eisenberg, I, Avidan, N, Potikha, T, Hochner, H, Chen, M, Olender, T, et al. The UDP-N-acetylglucosamine 2-epimerase/N-acetylmannosamine kinase gene is mutated in recessive hereditary inclusion body myopathy. Nature genetics (2001)., 29(1), 83-87.

[51] Arai, A, Tanaka, K, Ikeuchi, T, Igarashi, S, Kobayashi, H, Asaka, T, et al. A novel mutation in the GNE gene and a linkage disequilibrium in Japanese pedigrees. Annals of neurology (2002)., 52(4), 516-519.

[52] Tomimitsu, H, Shimizu, J, Ishikawa, K, Ohkoshi, N, Kanazawa, I, & Mizusawa, H. Distal myopathy with rimmed vacuoles (DMRV): new GNE mutations and splice variant. Neurology (2004)., 62(9), 1607-1610.

[53] Broccolini, A, Ricci, E, Cassandrini, D, Gliubizzi, C, Bruno, C, Tonoli, E, et al. Novel GNE mutations in Italian families with autosomal recessive hereditary inclusion-body myopathy. Human mutation (2004).

[54] Kim, B. J, Ki, C. S, Kim, J. W, Sung, D. H, Choi, Y. C, & Kim, S. H. Mutation analysis of the GNE gene in Korean patients with distal myopathy with rimmed vacuoles. Journal of human genetics (2006)., 51(2), 137-140.

[55] Liewluck, T, Pho-iam, T, Limwongse, C, Thongnoppakhun, W, Boonyapisit, K, Raksadawan, N, et al. Mutation analysis of the GNE gene in distal myopathy with rimmed vacuoles (DMRV) patients in Thailand. Muscle & nerve (2006)., 34(6), 775-778.

[56] Li, H, Chen, Q, Liu, F, Zhang, X, Liu, T, Li, W, et al. Clinical and molecular genetic analysis in Chinese patients with distal myopathy with rimmed vacuoles. Journal of human genetics (2011)., 56(4), 335-338.

[57] Noguchi, S, Keira, Y, Murayama, K, Ogawa, M, Fujita, M, Kawahara, G, et al. Reduction of UDP-N-acetylglucosamine 2-epimerase/N-acetylmannosamine kinase activity

and sialylation in distal myopathy with rimmed vacuoles. The Journal of biological chemistry (2004). , 279(12), 11402-11407.

[58] Schwarzkopf, M, Knobeloch, K. P, Rohde, E, Hinderlich, S, Wiechens, N, Lucka, L, et al. Sialylation is essential for early development in mice. Proceedings of the National Academy of Sciences of the United States of America (2002). , 99(8), 5267-5270.

[59] Malicdan, M. C, Noguchi, S, & Nishino, I. A preclinical trial of sialic acid metabolites on distal myopathy with rimmed vacuoles/hereditary inclusion body myopathy, a sugar-deficient myopathy: a review. Therapeutic advances in neurological disorders (2010). , 3(2), 127-135.

[60] Malicdan, M. C, Noguchi, S, Hayashi, Y. K, & Nishino, I. Muscle weakness correlates with muscle atrophy and precedes the development of inclusion body or rimmed vacuoles in the mouse model of DMRV/hIBM. Physiological genomics (2008). , 35(1), 106-115.

[61] Malicdan, M. C, Noguchi, S, & Nishino, I. Autophagy in a mouse model of distal myopathy with rimmed vacuoles or hereditary inclusion body myopathy. Autophagy (2007). , 3(4), 396-398.

[62] Malicdan, M. C, Noguchi, S, Hayashi, Y. K, Nonaka, I, & Nishino, I. Prophylactic treatment with sialic acid metabolites precludes the development of the myopathic phenotype in the DMRV-hIBM mouse model. Nature medicine (2009). , 15(6), 690-695.

[63] Mizushima, N, & Komatsu, M. Autophagy: renovation of cells and tissues. Cell (2011). , 147(4), 728-741.

[64] Yu, W. H, Cuervo, A. M, Kumar, A, Peterhoff, C. M, Schmidt, S. D, Lee, J. H, et al. Macroautophagy--a novel Beta-amyloid peptide-generating pathway activated in Alzheimer's disease. The Journal of cell biology (2005). , 171(1), 87-98.

[65] Scherz-shouval, R, Shvets, E, Fass, E, Shorer, H, Gil, L, & Elazar, Z. Reactive oxygen species are essential for autophagy and specifically regulate the activity of Atg4. The EMBO journal (2007). , 26(7), 1749-1760.

[66] Iijima, R, Takahashi, H, Namme, R, Ikegami, S, & Yamazaki, M. Novel biological function of sialic acid (N-acetylneuraminic acid) as a hydrogen peroxide scavenger. FEBS letters (2004).

Section 4

Autophagy and the Liver

Chapter 8

Autophagy and the Liver

Ricky H. Bhogal and Simon C. Afford

Additional information is available at the end of the chapter

http://dx.doi.org/10.5772/55585

1. Introduction

Autophagy is a cellular process that involves lysosomal degradation and recycling of intracellular organelles and proteins to maintain energy homeostasis during times of cellular stress [1]. It also serves to remove damaged cellular components such as mitochondria and long-lived proteins. Autophagy is catabolic mechanism and although hepatic autophagy performs the standard functions of degrading damaged organelles/aggregated proteins and regulating cell death it also regulates lipid accumulation within the liver. Autophagy can be divided into three distinct sub-groups that are discussed below. This chapter focuses upon the role of autophagy in a variety of liver diseases including hepatocellular carcinoma (HCC) and viral hepatitis. The increased understanding of the cellular machinery regulating autophagy within the liver may foster the development of therapeutic strategies that will ultimately help treat liver disease.

As stated above autophagy involves lysosomal-dependent degradation of long-lived proteins and cellular organelles [2]. In rodent models of nutrient starvation autophagy was responsible for degradation of 35% of total liver protein within 24 hours [3] illustrating the role of autophagy in liver homeostasis and energy conservation. Conversely inhibition of autophagy in hepatocytes led to a 4-fold increase in liver mass because of failure to degrade a variety of cellular components [4]. Therefore autophagy plays an important role maintaining liver function under basal conditions but may also be manipulated by pathological processes to cause liver disease. The precise role of autophagy within the liver is detailed below. This chapter is not exhaustive with regard to the role of autophagy in liver disease but draws upon the current understanding of autophagy role in selected liver diseases.

2. Autophagic pathways

Three distinct types of autophagy have been described in eukaryotic cells; macroautophagy (referred to hereafter as autophagy), chaperone-mediated autophagy (CMA), and microautophagy [5]. In autophagy, a portion of cytosol is engulfed by a double-membrane structure, termed an autophagosome, that fuses with a lysosome whose enzymes degrade the cellular constituents sequestered in the autophagosome [6]. The regulation of this process is highly complex and controlled by the co-ordinated actions of the evolutionarily conserved autophagy-related genes (Atgs). Over 32 Atg genes have been identified in yeast and humans [7, 8]. The source of the double membrane is controversial, but it might be derived from the endoplasmic reticulum (ER), mitochondria, or plasma membrane [9]. The double membrane of the autophagosome is formed and elongated by as yet unclear mechanisms, but a number of multiprotein complexes are known to mediate these processes [10]. CMA allows the direct lysosomal import of unfolded, soluble proteins that contain a particular pentapeptide motif. In the third form a autophagy, microautophagy, cytoplasmic material is directly engulfed into the lysosome at the surface of the lysosme by membrane rearrangement. Despite being three separate mechanisms each type of autophagy involves engulfment of a part of the cytosol and lysosomal dependent degradation within a double membrane. Each of these three autophagic processes is discussed below.

2.1. Autophagy (Macroautophagy)

32 Atg genes have been identified thus far that regulate autophagy of which 16 Atg genes are required for all types of autophagy [11]. The major cellular pathways regulating autophagy aside for the Atg proteins include the inhibitory mammalian target of rapamycin (mTOR) and class III phosphatidylinositol 3-kinase (PI3K). Thus the Atg proteins, PI3K and mTOR pathways all co-ordinate a highly complex cellular signaling pathways to regulate autophagy [8].

The formation of the autophagosome can be divided in to several steps. The description the follow is a simplified pathway and in reality the process is likely to be much more complex. The autophagy process is induced by the ULK1 kinase complex [12] and later class III PI3K complex is involved in vesicle nucleation [13]. This is followed by membrane expansion that involves various Atg proteins including the Atg2-Atg18 complex, Atg12-Atg5-Atg16 conjugation system and Atg8-phosphatidylethanolamine (Atg8-PE) conjugation system [14]. The precise role of each molecular regulator is beyond the scope of this chapter although the process is discussed in detail below. The reader is referred to recent excellent reviews for more detailed reviews of the whole process [8, 10, 15].

In general, the Atg1–Atg13–Atg17 complex recruits and organizes other proteins for the developing autophagosome [16]. Activation of the Atg1–Atg13–Atg17 complex leads to organization of the Atg6/beclin-1–Vps34 complex on the lipid membrane [12]. Vps34 produces phosphatidylinositol 3-phosphate, which can recruit other proteins to the complex [17]. Vps34 is the target of the widely used pharmacologic inhibitor of autophagy 3-methyladenine (3-MA) [18, 19]. Importantly, beclin-1 is an important interface between the autophagic and cell death pathways, because the anti-apoptotic proteins Bcl-2 and Bcl-X_L bind beclin-1 to inhibit autophagy [20]. The regulation of this interaction is complex but includes its disruption by c-Jun N-terminal kinase 1–mediated phosphorylation of Bcl-2 [21].

Following the above, autophagosome formation and elongation involves 2 ubiquitin-like conjugation processes that generate membrane-bound protein complexes. In the first, Atg7 and Atg10 mediate the conjugation of Atg12 to Atg5 [22], which subsequently interact with Atg16 [23]. The Atg12–Atg5 complex associates with the membrane and then dissociates upon completion of the autophagosome. The second critical conjugation reaction involves Atg8 or microtubule-associated protein 1 light chain 3 (LC3). LC3 is constitutively cleaved by Atg4 to produce LC3-I. With a signal to induce autophagy, Atg7 and Atg3 mediate the conjugation of LC3-I to the membrane lipid PE to form LC3-II [24]. LC3-II associates with the autophagosomal membrane, where the lipidated protein can mediate membrane elongation and closure. LC3-II is degraded late in the autophagic pathway, after autophagosome fusion with a lysosome [25]. The formation of these autophagic vacuoles can be detected experimentally by labeling cells with the specific autophagic marker monodansylcadaverine (MDC).

Once formed, autophagosomes traffic along microtubules by a dynein-dependent mechanism to reach perinuclear lysosomes located near the microtubule-organising center. Another method to monitor the induction of autophagy is therefore to detect perinuclear, LC3-positive aggregates by immunofluorescence. Before they fuse with lysosomes, autophagosomes can fuse with early and late endosomes to form an amphisome. This process allows for a point of convergence between the pathways of autophagy and endocytosis [10]. Autophagosomes dock and fuse to form an autophagolysosome or autolysosome by a process that has not been well defined in mammalian cells. The term "autophagic vacuole" has been used for autophagosomes, amphisomes, and autolysosomes, which can be indistinguishable experimentally. In yeast, fusion is mediated by soluble N-ethylmaleimide–sensitive factor attachment protein receptors [26]. Whereas in mouse hepatocytes the soluble N-ethylmaleimide–sensitive factor attachment protein receptor protein vti1b mediates autophagosome-endosome fusion [27]. Other factors that regulate autophagosome fusion include the guanosine triphosphate binding protein Rab7 [28]. Autophagosome-lysosome fusion allows mixing of their contents and degradation of the cargo of the autophagosome by lysosomal acid hydrolases, which include proteases, nucleases, lipases, glycosidases, and phosphatases. The degraded products are then transported back to the cytosol for recycling. Constitutive levels of autophagy are required for cell survival and function in most organs, including the liver. Autophagy therefore has vital cellular functions even when not activated. The overall process of autophagy is summarised in Figure 1.

The formation of autophagy is commenced by ULK1 kinase complex. This process can be inhibited by mTOR. The process of membrane development involves beclin-1 a process that can be inhibited by Bcl2. Following this various Atg conjugation systems are involved in membrane elongation before the double membrane autophagic vacuoles fuse with the lysosome to form the autophagolysome and degrade the autophagic vacuole cargo for recycling.

2.2. Chaperone Mediated Autophagy (CMA)

In CMA, soluble proteins with a specific pentapeptide motif are recognised by the chaperone protein Hsc70 for translocation to the lysosome, where binding to the lysosome-associated membrane protein type 2A (LAMP-2A) receptor leads to protein internalization and degradation [29]. Similar to autophagy, CMA is constitutively active and increases with cellular

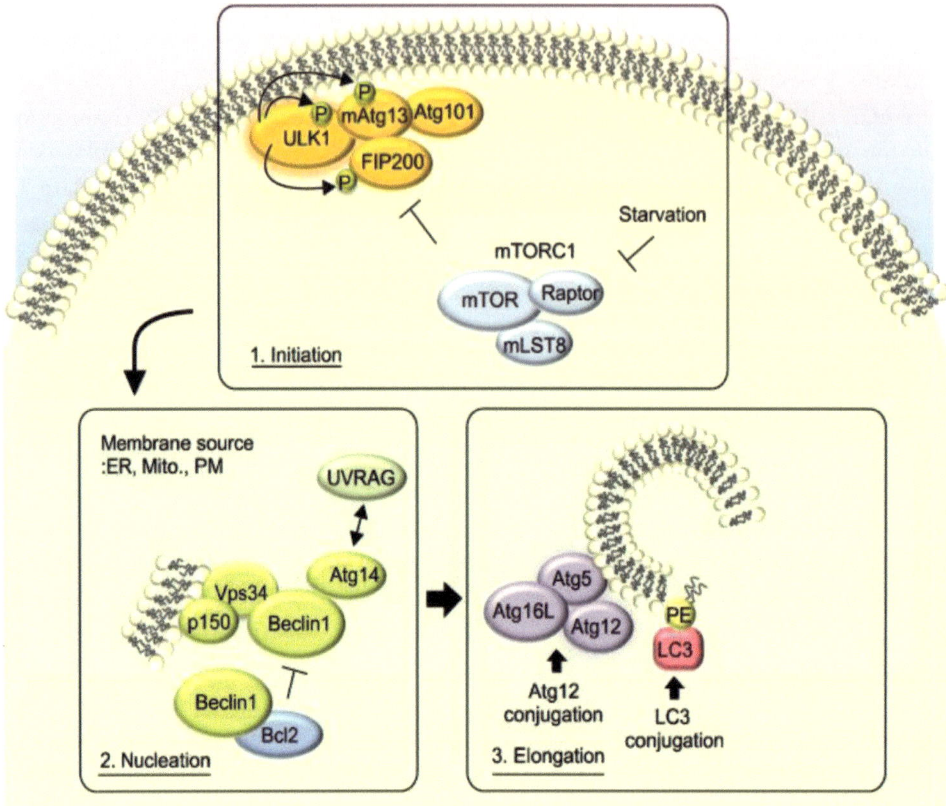

Figure 1. The Process of Autophagy.

stresses. CMA function in the liver has not been well studied [30], although CMA has been shown to mediate hepatocyte resistance to oxidant stress [31]. In addition few studies attempt to distinguish between CMA and autophagy. However, although autophagy and CMA are distinct pathways, they are likely to interact and a reduction in one pathway may lead to activation of the other. This interaction can complicate interpretation of the effects of inhibiting either one [32]. It is likely that both autophagy and CMA serve similar purposes within the cell.

2.3. Microautophagy

Microautophagy is a non-inducible, lysosomal internalisation of cellular constituents that occurs by invagination of the lysosomal membrane. Much of the data concerning microautophagy has focused upon the role of microautophagy of the mitochondria termed mitophagy [33]. Mitochondria are the central executioners for apoptosis and many times damaged mitochondria trigger cell death. If damaged mitochondria are limited to a fraction of mitochondria within the capacity of autophagic removal, cell death will be avoided. Nevertheless, cells may eventually die if the number of damaged mitochondria is too overwhelming to be removed by autophagy. Thus, autophagy may increase the threshold for the death stimuli. Accumulating evidence demonstrates that pharmacologic or genetic inhibition of autophagy greatly enhanced cell death [34]. The trafficking of mitochondria to lysosomal structures has long been known to occur in liver, and the livers of mice with a conditional knockout of Atg7 develop massive hepatomegaly marked by the accumulation of deformed mitochondria [4]. The process of lysosomal removal of mitochondria by autophagy was first characterized in hepatocyte models by Lemasters *et al* [2]. Before removal of dysfunctional mitochondria, mitochondria undergo the mitochondrial permeability transition (MPT), in which inner mitochondrial membrane pores open, leading to adenosine triphosphate (ATP) depletion from the uncoupling of oxidative phosphorylation and outer membrane rupture with release of pro-apoptotic factors. The MPT induces mitophagy, because the MPT inhibitor cyclosporine blocks this process. Indeed after liver injury LC3-positive structures appear in hepatocytes in the area of injury [2]. These findings indicate that a mechanism exists for the selective targeting of MPT-damaged mitochondria for autophagic degradation. Recent studies in non-hepatic cells have implicated Atg32 [35], the Bcl-2 family member Nix [36], and the ubiquitin ligase Parkin [37] in this process, but further studies are required to determine whether these proteins mediate selective mitophagy in hepatocytes.

Mitochondrial damage that induces the MPT leads to death of hepatocytes [38], and these findings provide a mechanism by which insufficient or impaired microautophagy can promote hepatocyte cell death. These processes may become even more prominent during ischaemia and hypoxia that is often encountered during liver disease. Selective mitophagy might be a mechanism to protect against hepatocyte death, because removal of mitochondria that have undergone the MPT could prevent mitochondrial oxidative stress, ATP depletion, or the release of pro-apoptotic factors. The contribution of mitophagy to the response of the hepatocyte to various death stimuli needs to be more carefully examined.

Aside from mitophagy, microautophagy can also selectively occur in the endoplasmic reticulum [39] and perixosomes [40]. These latter two processes will not be discussed further in this chapter.

3. Autophagy in liver diseases

As one of the most metabolically active organs, the liver plays a central role in regulating the overall organisms energy balance by controlling carbohydrate and lipid metabolism. The liver functions as a major buffering system to maintain the homeostasis of macro- and micronutrients to allow other tissues to function normally under physiological stress. Liver-targeted autophagy deficiency results in accumulation of protein aggregates, damaged mitochondria, steatosis and liver injury. These findings support a pro-survival and cyto-protective role of autophagy in maintaining protein, lipid and organelle quality control by eliminating damaged proteins and organelles as well as excessive lipid droplets in the liver during stress. In addition, accumulating evidence now indicates that autophagy is also involved in hepatocyte cell death, steatohepatitis, hepatitis virus infection and HCC. The role of autophagy in a variety of liver diseases is discussed below.

3.1. Autophagy and liver Ischaemia Reperfusion Injury (IRI)

Liver IRI occurs during many clinical scenarios within the liver. It involves a period of oxygen and nutrient deprivation due to the lack of blood (ischaemia) followed by the re-introduction of blood to the liver (reperfusion). IRI is well known to have detrimental effects upon the liver and can induce considerable liver injury [41]. IRI is an obligatory part of liver transplantation but also occurs as part of the clinical scenario involved in haemorrhagic shock, abdominal trauma and liver resectional surgery. The ischaemia associated with these processes induces the formation of reactive oxygen species (ROS) that can mediate liver parenchymal injury [41, 42]. The main cellular target for the IRI is the hepatocyte. The restoration of blood flow although of undoubted benefit to the liver can perpetuate the accumulation of ROS and further accentuate liver damage [42, 43]. The precise role of autophagy within the context of oxidative stress within the liver remains controversial unlike the roles of apoptosis and necrosis that have firmly established as detrimental [43]. Recent evidence suggests that autophagy is primarily a cyto-protective mechanism during liver IRI *in vitro* at least [19].

A variety of experimental approaches have been used to assess the role of autophagy within the liver during IRI. *In vitro* studies have shown that hepatocytes up-regulate autophagy in a ROS-PI3K-Atg protein dependent manner to protect against primarily apoptotic cell death principally by inducing mitophagy [19]. It must be stressed however that these isolated studies have focused only upon autophagy within hepatocytes and the role of autophagy in cholangiocytes, endothelial cell and hepatic stellate cells remains to be established. These observations have also been extended to suggest that autophagy also limits necrotic cell death during ischaemia [44]. Various cellular pathways have been shown to be important in autophagy mediated hepato-protection during ischaemia including calcium-calmodulin dependent protein kinase IV [45], PI3K [19] and AMPK [46]. The emerging consensus *in vitro* appears to be that autophagy is a cyto-protective mechanism although this remains contentious [47].

In vivo autophagy appears to protect the liver against ischaemia and oxidative stress [47, 48] although studies report both increase and decrease in autophagy after liver ischaemia [49-51]. These observations are likely to reflect the method used to assess autophagy in liver tissue. For instance an increase in Atg protein levels does not necessarily equate to autophagy induction.

Very few studies have assessed the effect of autophagy in human livers. Domart *et al* study does provide some data as to the role of autophagy in patients undergoing liver surgery [52]. In this study the surgical technique of ischaemic preconditioning (IPC) was used to assess whether it was hepatoprotective. IPC involves short periods of total liver ischaemia followed by reperfusion with the premise being that this increases the resistance of the liver to oxidative stress. In this study, two liver biopsies were taken, one prior to ischemia required by liver resection and another after liver reperfusion. Although overall the study did not show any overall benefit from IPC a subgroup of patients who underwent IPC for 10 minutes followed by 10 minutes of reperfusion before the prolonged ischemia required by liver resection showed a significant increase in liver cell autophagy [52]. This suggests that in this context, autophagy enhancement could allow for decreasing liver cell death.

Studies assessing autophagy during and after liver transplantation give contradictory results [44, 53, 54]. The explanation for such discrepancies may be the solution for cold preservation used. Indeed, a decrease in autophagy was observed in a study using a histidine–tryptophan–ketoglutarate cold-storage solution for 24 h cold preservation, while the contrary was reported when using the University of Wisconsin (UW) cold-storage solution. Importantly, UW cold storage solution does not contain amino acids. It is well demonstrated that amino acid depletion rapidly induces autophagy [55] and that anoxia decreases autophagy protein level. This induction of autophagy due to the absence of amino acids, may explain not only the apparent discrepancy between these studies but also the protection of the liver obtained with preservation solution such as the UW solution. Indeed, hypoxia/reoxygenation induces mitochondrial dysfunction [42]. Due to the decrease in autophagy proteins induced by anoxia/reoxygenation, autophagy fails to remove dysfunctional mitochondria, so that the mitochondria laden with ROS and calcium undergo the MPT, which in turn leads to uncoupling of oxidative phosphorylation, energetic failure, ATP depletion, and ultimately cell death. In case of associated nutrient depletion, autophagy is enhanced and facilitates autophagy of damaged mitochondria, leading to cell survival. This hypothesis is supported by the beneficial effect on liver tolerance to IRI of several strategies aiming at increasing autophagy in murine models [48, 49, 56]. It is striking to notice that many of the studies that suggest that inhibiting autophagy could ameliorate liver tolerance to ischemia used non-specific inhibitors of autophagy known to also have autophagy independent activities. This demands the development of specific autophagy inhibitors that will allow the dissection of autophagic pathways.

The role of autophagy in liver repair and regeneration will not be considered in this chapter but autophagy level decreases following partial hepatectomy suggesting a shift from the physiological steady state between anabolism and catabolism to the positive balance which is required for the compensatory growth of the liver after partial hepatectomy [57]. Finally no studies have yet specifically evaluated the autophagic pathway in liver sinusoidal endothelial cells or cholangiocytes. This is a limitation in understanding the effect of autophagy in liver IRI since these cells are also sensitive to ischemia and lesions to these cells are a key event in this context.

3.2. Autophagy and Alcoholic Liver Disease (ALD)

At the cellular level chronic alcohol abuse can result in mitochondrial damage, inhibition of insulin signaling, steatosis, apoptotic and necrotic cell death, all of which can be regulated by autophagy. Indeed, ethanol exposure is known to induces autophagy in primary cultured mouse hepatocytes [58] and in hepatoma cells expressing alcohol dehydrogenase (ADH) and cytochrome P450 2E1 (Cyp2E1) [59]. Indeed ethanol treated hepatocytes demonstrate increased autophagosome number when compared to controls as assessed by electron microscopy and suppression of autophagy exacerbates alcohol-induced liver injury [58]. In ALD cell death may still occur in the presence of cyto-protective autophagy as in alcoholic patients have other pathophysiological conditions such as diabetes, hyperinsulinemia, obesity and hepatitis C virus (HCV) that impair autophagy.

The majority of ethanol is metabolized in the liver and individuals who abuse alcohol by routinely drinking 50-60 g of ethanol per day are at risk for developing ALD [60]. ALD can ultimately progress to HCC. The pathogenesis of liver disease from alcohol abuse comes from the interaction of several factors, including the generation of oxidants and reactive metabolites from ethanol oxidation, which, in turn, causes other metabolic derangements [61]. Chronic alcohol consumption leads to liver steatosis and protein accumulation within the liver [62]. Furthermore chronic ethanol consumption slows down the catabolism of long-lived proteins in rodent livers [63] with reduction in the number of autophagosomes in hepatocytes [64]. This protein accumulation contributes to the formation of Mallory Denk (M-D) bodies in liver cells the hallmark of ALD [65]. Moreover recent evidence indicates that autophagy can degrade these insoluble complexes such as M-D bodies but clearance is hampered by an ethanol-elicited suppression of autophagy.

Several mechanisms have been suggested to contribute to ethanol-induced autophagy within the liver, including ROS production, ethanol metabolism and the suppression of mTOR although the precise mechanisms remain the subject of controversy. Oxidants such as ROS derived from ethanol metabolites such as acetaldehyde and malondialdehyde-acetaldehyde may impair autophagy [66]. Furthermore ethanol-induced ROS production can induce the expression of cytochrome Cyp2E1 and damage mitochondria [59]. ROS can also activate autophagy by the inactivation of Atg4B [67]. However, whether ethanol-induced ROS also inactivate Atg4B to promote autophagy in hepatocytes remains unknown. It is important that ROS may regulate the activity of many Atg proteins.

Chronic ethanol exposure may induce lysosomal fragility in an analogous manner to iron-induced oxidative stress [68]. Furthermore It is worth noting that lysosomes seem to exhibit differential sensitivity to ethanol levels in the serum [69] but even low levels of serum ethanol appear to be impair autophagy [70]. Ethanol-induced suppression of autophagy may result from alterations in hepatic amino acid pool sizes, especially those of leucine, which have been deemed regulatory amino acids and suppressors of autophagy. L-leucine is a potent autophagic suppressor reducing autophagy in chronic ethanol administration in rats when compared to control [71]. Thus, the association of an ethanol-induced reduction in autophagy with higher levels of intrahepatic leucine may partially explain autophagic suppression in the ethanol-fed state. Leucine accumulation could reflect a reduced ability of the liver to synthesize proteins, which indeed occurs in ethanol-fed animals [71].

Autophagic suppression by ethanol is well documented to disrupt protein trafficking in the liver. As discussed above autophagy requires the action of cytoskeletal elements to aid autophagosome formation and subsequent fusion with other vesicular bodies [72]. Disruption of vesicular movement within the hepatocyte by ethanol treatment occurs by mechanisms that are independent of the molecular motors, dyenin and kinesin, although there is evidence for alterations in the protein, dynamin [73]. Trafficking of exogenous proteins into the hepatocyte by endocytosis and the intracellular delivery of proteases to lysosomes are both inhibited by ethanol consumption. Furthermore, the anti-secretory properties of ethanol in the liver are well documented. Studies with liver slices and cultured cells indicate that ethanol metabolism is required for disruption of these protein trafficking events [74]. *In vitro* investigations also revealed that acetaldehyde, the initial product of ethanol oxidation, inhibits the polymerization of tubulin to form microtubules, indicating that the reactive metabolite may impair protein trafficking by forming adducts with tubulin subunits, thereby blocking their polymerization into microtubules [75].

Acute and chronic ethanol treatment also suppress Akt function in the liver, through the upregulation of the PTEN (phosphatase and tensin homolog) phosphatase [76]. Suppression of mTOR leads to the activation of the downstream ULK1 complex to trigger autophagy. It remains to be seen whether alcohol can affect the ULK1 complex. In addition to mTOR, ethanol-induced autophagy also requires the activation of beclin 1/VPS34 PI3K complex because 3MA, a PI-3 kinase inhibitor, suppresses acute ethanol-induced autophagy. Furthermore, ethanol has been shown to suppress proteasome activity, induce ER stress and activate JNK in hepatocytes, and all of these mechanisms have been shown to induce autophagy in non-hepatocyte models. Whether proteasome inhibition, ER stress and JNK activation play a role in ethanol-induced autophagy in hepatocytes needs to be further studied.

In addition to the accumulation of potentially toxic proteins, ethanol also causes mitochondrial damage [77, 78]. It is crucial that such damaged mitochondria be removed by mitophagy. While there is no firm evidence of ethanol-elicited suppression of mitophagy, the detection of increased numbers of damaged mitochondria in livers of ethanol-fed animals provides circumstantial evidence of mitophagy inhibition. Autophagy protects against ethanol-induced toxicity in livers of mice. Reagents that modify autophagy might be developed as therapeutics for patients with ALD [58].

Finally it has been speculated that ethanol may inhibit autophagy because chronic ethanol consumption reduces AMPK activity in the liver. However the role of AMPK in autophagy is still controversial although AMPK agonists such as metformin significantly protect against ethanol induced liver injury.

3.3. Autophagy and viral hepatitis

Besides the physiological function of autophagy in maintaining cellular homeostasis detailed above, autophagy is a newly recognized facet of the innate and adaptive immune system. Hepatotrophic viruses such as hepatitis C virus (HCV) and hepatitis B virus (HBV) have developed strategies to subvert and manipulate autophagy for their own survival benefit [79]. The role of HCV and HBV will be considered separately below.

3.3.1. HCV

Several studies have assessed the autophagic pathway in hepatocytes infected with HCV both *in vitro* and in liver biopsies from chronic HCV patients [80]. Using various experimental approaches including LC3, Atg5 or Beclin-1 immunoblotting, electron microscopy or GFP-LC3 immunofluorescence, studies consistently demonstrated an accumulation of autophagic vacuoles in HCV-infected hepatocytes [81]. Importantly, other viruses such as vesicular stomatitis virus and mutant herpes simplex virus 1 can be captured and eliminated by the autophagic pathway [82] but HCV has evolved to avoid and subvert autophagy using multiple strategies [83]. HCV appears to avoid its recognition by the cellular autophagic machinery [84]. This is based upon studies showing no or rare co-localization of HCV proteins with autophagic vacuoles.

HCV prevents the maturation of the autophagosome into an autolysosome [85]. For instance in HCV infected hepatocytes there is an increase in the number of autophagic vacuole without enhancement in autophagic protein degradation. Secondly, there is an absence of co-localization of lysosomes with autophagic vacuoles in HCV-infected cells in contrast to nutrient starved cells. Thirdly, there is a reduction in the number of autophagic vacuoles following HCV elimination and finally the absence of increase in the number of late autophagic vesicles in hepatocytes from chronic HCV patients as compared to controls, while a strong augmentation in the number of autophagic vesicles is observed. This may be related to a lack of fusion between autophagosome and lysosome.

HCV also utilizes functions or components of autophagy to enhance its intracellular replication [86]. Indeed, it has been recently shown that autophagy proteins are required for translation and/or delivery of incoming HCV RNA to the cell translation apparatus [87]. However, autophagy proteins are not needed for the translation of progeny HCV once replication is established since down-regulation of autophagy proteins 10 days after transduction had no effect on HCV replication. Therefore it is suggested that by remodelling endoplasmic reticulum membranes, the autophagy proteins or autophagic vesicles might provide an initial membranous support for translation of incoming RNA, prior to accumulation of viral proteins and the eventual establishment of virus-induced cellular modifications [88]. Alternatively, autophagy proteins might contribute directly or indirectly to the cytoplasmic transport of the incoming RNA to cellular factors or sites that are required for its translation. Importantly, autophagy proteins are required neither for HCV entry nor for HCV secretion. Altogether, these data explain the apparent contrast between the results of some *in vitro* studies reporting the implication of autophagy proteins in HCV replication and the absence of correlation between the number of autophagic vacuoles or the LC3-II level and the HCV load in chronic hepatitis C patients: autophagy proteins are required only for initial steps of HCV replication, but not once replication is established.

Notably, cytosolic RNA-sensing protein kinase PKR and eIF2-α phosphorylation regulate virus- and starvation-induced autophagy [89]. It is tempting to speculate that recognition of the incoming HCV RNA by RNA-sensing molecules induces autophagy and hence, favours its initial translation. Alternatively, constitutive basal autophagic vesicle formation might be required for this initial HCV RNA translation. The above observations suggest that autophagy proteins are pro-viral factors for HCV and can be manipulated to facilitate HCV infection of the liver.

3.3.2. HBV

HBV also induces autophagosomes in hepatocytes, as demonstrated both *in vitro* in several liver derived cell lines [90] and *in vivo* in the liver of transgenic mouse lines harboring HBV

DNA [91]. Importantly, this induction was also observed in the liver of an HBV-infected patients but not of a non-infected patients [92]. In contrast to HCV, HBV enhances the autophagic flux, as late autophagic vacuoles could be detected in mouse hepatocytes using electron microscopy and given the existence of an extensive co-localization of lysosome-associated membrane protein 1 (LAMP1) with GFP–LC3 puncta [93]. However, without being able to provide the reason for it, no significant increase in protein degradation was observed in HBV DNA-transfected cells.

An HBV-encoding protein, HBx, plays a crucial role in this HBV-induced autophagy. Indeed, transfection of Huh7.5 cells with an HBV unable to express HBx did not enhance autophagy [94]. Moreover, expression of HBx alone was sufficient to induce autophagy; similar results were obtained *in vivo* in transgenic mice [91]. This effect of HBx is due, at least partly, to its ability to bind to class III PI3K, a regulatory molecule that controls autophagy. Although conflicting, HBx may also up-regulate the transcription of beclin-1 thus sensitizes the cells to starvation-induced autophagy. Whether the role of HBx is confined to short nutrient starvation conditions or also exists in normal conditions remains controversial. If, in the same way as HCV, HBV subverts autophagy, the strategy applied is somewhat different. Autophagy enhances HBV replication mostly at the step of viral DNA replication, slightly at the step of RNA transcription, and not at other levels. How autophagy may enhance HBV DNA replication remains unresolved.

The question whether HBV could be engulfed in autophagic vacuoles is not fully elucidated despite the observation that HBV core/e antigens and surface antigens partially co-localized with autophagic vacuoles. Immuno-electron microscopy studies would be required to address this issue. However, as HBV seems to benefit from autophagy proteins and as the autophagic protein degradation rate is not increased, this hypothesis seems unlikely and autophagic vacuoles may rather serve as the sites for viral DNA replication and morphogenesis.

3.4. Autophagy and Hepatocellular Carcinoma (HCC)

The incidence of HCC is increasing worldwide primarily through the increase of cirrhosis secondary to viral hepatitis. The role of autophagy in the development of cancer has been comprehensively reviewed recently [95]. Autophagy is generally thought to be an anti-tumour mechanism in cells. The tumor suppressor role of autophagy is not yet clear but may involve limiting chromosomal instability, restricting oxidative stress and reducing intratumoral necrosis and local inflammation. The role of autophagy in the regulation of neoplasia has originated from observation where demonstrating mono-allelically deletion of Beclin-1 in 40–75% of cases of human breast, ovarian, and prostate cancer [96]. Moreover, the regulation of autophagy overlaps closely with signaling pathways that regulate tumorigenesis.

Studies assessing autophagy in HCC have clearly demonstrated *in vitro*, in mice and in patients that, in this context, autophagy is a tumor suppressor mechanism. This follows the general paradigm that autophagy is a cyto-protective mechanism. Murine models with heterozygous disruption of Beclin-1 have a high frequency of spontaneous HCC [97]. Moreover, crossing beclin-1 +/− mice, with mice, that transgenically express the HBV large-envelope polypeptide under the transcriptional control of the mouse albumin promoter, resulted in the acceleration of the development of hepatitis B virus-induced small-cell dysplasia [98]. This maybe the prelude to the development of HCC. Expression of several autophagic genes such as Atg5, Atg7 and Beclin-1 and their corresponding autophagic activity is decreased in HCC cell lines compared to that in a normal hepatic cell line [99]. Similarly, Beclin-1 mRNA and protein levels are lower in HCC tissue samples than in adjacent non-tumor tissues from the same patients [100].

The most aggressive malignant HCC cell lines and HCC tissues with recurrent disease display much lower autophagic levels than less aggressive cell lines or tissues, especially when the anti-apoptotic B-cell leukemia/lymphoma (Bcl)-xL protein is over-expressed [101]. Interestingly, in a tissue microarray study consisting of 300 HCC patients who underwent curative resection, the expression of Beclin-1 was significantly correlated with disease-free survival and overall survival only in the Bcl-xL+ patients. Multivariate analyses revealed that Beclin-1 expression was an independent predictor for disease-free survival and overall survival in Bcl-xL+ patients [101]. In addition, there was a significant correlation between Beclin-1 expression and tumor differentiation in Bcl-xL+ but not in Bcl-xL− HCC patients. These data suggest that autophagy defect synergizes with altered apoptotic activity and facilitates tumor progression and poor prognosis of HCC. The role of other Atg proteins in the development and progression of HCC remains to be established but on the basis of the data with Beclin-1 is would appear that these proteins would also have anti-neoplastic effect.

The mechanisms responsible for this low autophagy protein level are not elucidated. However, a recent study has demonstrated that HAb18G/CD147, a transmembrane glycoprotein highly expressed in HCC, contributes to this decreased autophagic level in HCC through the class I phosphatidylinositol 3-kinase–Akt pathway upregulation [102]. Other oncoproteins such as the Bcl-2 family proteins may also be implicated in HCC, like in other cancers. Stimulation of hypoxia-inducible factors (HIFs) due to hypoxic stress within HCC may also contribute to autophagy modulation [103]. However many of these signaling pathways have not been conclusively shown to be involved in HCC progression. It does remain an attractive notion that these pathways are modulated by HCC leading to decrease levels of autophagy and increased susceptibility to HCC. Indeed pharmacological therapies that inhibit autophagy improves survival of patient undergoing liver transplantation for HCC when compared to non-HCC recipients suggesting the specificity of its beneficial impact for cancer patients [104]. In addition ROS accumulation, and DNA damage also facilitates the development and progression of HCC [105]. In studies carried out using tetrandrine, a calcium channel blocker, it regulated the expression of Atg7, which then promoted tetrandrine-induced autophagy [106]. In vitro and in vivo tetrandrine caused the accumulation of ROS and induced cell autophagy in a tumor xenograft model. Therefore, these findings suggest that tetrandrine is a potent autophagy agonist and may be a promising clinical chemotherapeutic agent [106]. It was further demonstrated that oroxylin A-triggered autophagy contributed to cell death using over-expression of Atg5 and Atg7 and inhibition of autophagy by siBeclin 1 and 3-methyladenine (3-MA) [99]. In vivo study, oroxylin A inhibited xenograft tumor growth and induced obvious autophagy in tumors. These findings define and support a novel function of autophagy in promoting death of HCC cells [99]. Furthermore miR-375 inhibits autophagy by reducing expression of Atg7 and impairs viability of HCC cells under hypoxic conditions in culture and in mice. miRNAs that inhibit autophagy of cancer cells might be developed as therapeutics [107].

These data provide potential therapeutic targets to modulate the development of HCC. However until the precise regulatory role of autophagy in HCC is established these treatment cannot be used in the clinic.

3.5. Autophagy and other liver diseases

The role of autophagy in other liver disease has yet to be firmly established. However limited studies have been conducted in some liver diseases. In primary biliary cirrhosis (PBC) autophagy is specifically seen in the damaged small bile ducts along with cellular senescence. The inhibition of autophagy suppressed cellular senescence in cultured cells suggesting that autophagy may mediate the process of biliary epithelial senescence and be involved in the pathogenesis of bile duct lesions seen in PBC [108]. Furthermore the expression of LC3 was seen in coarse vesicles in the cytoplasm of bile ductular cells and significantly more frequently in PBC of both early and advanced stages when compared to control livers. Autophagy is frequently seen in bile ductular cells in ductular reactions (DRs) in PBC. Since cellular senescence of bile ductular cells is rather frequent in the advanced stage of PBC, autophagy may precede cellular senescence of bile ductular cells in DRs in PBC [109]. The aggregation of p62 is specifically increased in the damage bile ducts in PBC and may reflect dysfunctional autophagy, followed by cellular senescence in the pathogenesis of bile duct lesions in PBC [110]. Whether the same occurs in other biliary diseases such as primary sclerosing cholangitis is not known.

Recent reports have suggested a role for autophagy in alpha-1-antitrypsin (AT) deficiency where a mutant protein activates autophagy [111]. Autophagy is thought to be involved in the degradation of the mutant protein and hence defective autophagy may contribute to the hepatic fibrosis seen in AT [111].

Furthermore the role of autophagy in NAFLD, NASH and metabolic liver disease remains the subject of on-going research.

4. Conclusion

This chapter has outlined the extensive role played by autophagy within the liver and its role in various liver diseases. In general autophagy appears to be primarily a cyto-protective mechanism within the liver and especially hepatocytes. Autophagy induction appears to protect the liver from IRI and appears to negatively regulate neoplasia and reduce the effects of alcohol. However the autophagy cell machinery can be used by viral infections and biliary disease to aid disease propagation within the liver. It is important that aside from disease pathogenesis autophagy may have a role in liver repair and regeneration that has not been considered here.

Acknowledgements

The authors would like to thank the Centre for Liver Research, University of Birmingham, The Department of HpB and Liver Transplantation Surgery and National Institute for Health Research, University Hospitals of Birmingham for continued help with research endeavours.

Author details

Ricky H. Bhogal and Simon C. Afford

Centre for Liver Research, Institute for Biomedical Research, University of Birmingham, Edgbaston, Birmingham, UK

References

[1] Levine, B. and D.J. Klionsky, Development by self-digestion: molecular mechanisms and biological functions of autophagy. Dev Cell, 2004. 6(4): p. 463-77.

[2] Kim, I., S. Rodriguez-Enriquez, and J.J. Lemasters, Selective degradation of mitochondria by mitophagy. Arch Biochem Biophys, 2007. 462(2): p. 245-53.

[3] Cuervo, A.M., et al., Activation of a selective pathway of lysosomal proteolysis in rat liver by prolonged starvation. Am J Physiol, 1995. 269(5 Pt 1): p. C1200-8.

[4] Komatsu, M., et al., Impairment of starvation-induced and constitutive autophagy in Atg7-deficient mice. J Cell Biol, 2005. 169(3): p. 425-34.

[5] Klionsky, D.J., Autophagy. Curr Biol, 2005. 15(8): p. R282-3.

[6] Mehrpour, M., et al., Autophagy in health and disease. 1. Regulation and significance of autophagy: an overview. Am J Physiol Cell Physiol, 2010. 298(4): p. C776-85.

[7] Mizushima, N. and B. Levine, Autophagy in mammalian development and differentiation. Nat Cell Biol, 2010. 12(9): p. 823-30.

[8] Pyo, J.O., J. Nah, and Y.K. Jung, Molecules and their functions in autophagy. Exp Mol Med, 2012. 44(2): p. 73-80.

[9] Hamasaki, M. and T. Yoshimori, Where do they come from? Insights into autophagosome formation. FEBS Lett, 2010. 584(7): p. 1296-301.

[10] Das, G., B.V. Shravage, and E.H. Baehrecke, Regulation and function of autophagy during cell survival and cell death. Cold Spring Harb Perspect Biol, 2012. 4(6).

[11] Longatti, A. and S.A. Tooze, Vesicular trafficking and autophagosome formation. Cell Death Differ, 2009. 16(7): p. 956-65.

[12] Mizushima, N., The role of the Atg1/ULK1 complex in autophagy regulation. Curr Opin Cell Biol, 2010. 22(2): p. 132-9.

[13] Thi, E.P., U. Lambertz, and N.E. Reiner, Class IA phosphatidylinositol 3-kinase p110alpha regulates phagosome maturation. PLoS One, 2012. 7(8): p. e43668.

[14] Mizushima, N., Autophagy. FEBS Lett, 2010. 584(7): p. 1279.

[15] Jing, K. and K. Lim, Why is autophagy important in human diseases? Exp Mol Med, 2012. 44(2): p. 69-72.

[16] Sarbassov, D.D., S.M. Ali, and D.M. Sabatini, Growing roles for the mTOR pathway. Curr Opin Cell Biol, 2005. 17(6): p. 596-603.

[17] Liang, X.H., et al., Induction of autophagy and inhibition of tumorigenesis by beclin 1. Nature, 1999. 402(6762): p. 672-6.

[18] Seglen, P.O. and P.B. Gordon, 3-Methyladenine: specific inhibitor of autophagic/lysosomal protein degradation in isolated rat hepatocytes. Proc Natl Acad Sci U S A, 1982. 79(6): p. 1889-92.

[19] Bhogal, R.H., et al., Autophagy: a cyto-protective mechanism which prevents primary human hepatocyte apoptosis during oxidative stress. Autophagy, 2012. 8(4): p. 545-58.

[20] Levine, B., S. Sinha, and G. Kroemer, Bcl-2 family members: dual regulators of apoptosis and autophagy. Autophagy, 2008. 4(5): p. 600-6.

[21] Wei, Y., et al., JNK1-mediated phosphorylation of Bcl-2 regulates starvation-induced autophagy. Mol Cell, 2008. 30(6): p. 678-88.

[22] Mizushima, N., et al., A protein conjugation system essential for autophagy. Nature, 1998. 395(6700): p. 395-8.

[23] Ohsumi, Y. and N. Mizushima, Two ubiquitin-like conjugation systems essential for autophagy. Semin Cell Dev Biol, 2004. 15(2): p. 231-6.

[24] Ichimura, Y., et al., A ubiquitin-like system mediates protein lipidation. Nature, 2000. 408(6811): p. 488-92.

[25] Tanida, I., et al., Lysosomal turnover, but not a cellular level, of endogenous LC3 is a marker for autophagy. Autophagy, 2005. 1(2): p. 84-91.

[26] Sudhof, T.C. and J.E. Rothman, Membrane fusion: grappling with SNARE and SM proteins. Science, 2009. 323(5913): p. 474-7.

[27] Atlashkin, V., et al., Deletion of the SNARE vti1b in mice results in the loss of a single SNARE partner, syntaxin 8. Mol Cell Biol, 2003. 23(15): p. 5198-207.

[28] Jager, S., et al., Role for Rab7 in maturation of late autophagic vacuoles. J Cell Sci, 2004. 117(Pt 20): p. 4837-48.

[29] Kaushik, S. and A.M. Cuervo, Chaperone-mediated autophagy: a unique way to enter the lysosome world. Trends Cell Biol, 2012. 22(8): p. 407-17.

[30] Amir, M. and M.J. Czaja, Autophagy in nonalcoholic steatohepatitis. Expert Rev Gastroenterol Hepatol, 2011. 5(2): p. 159-66.

[31] Wang, Y., et al., Macroautophagy and chaperone-mediated autophagy are required for hepatocyte resistance to oxidant stress. Hepatology, 2010. 52(1): p. 266-77.

[32] Wang, Y., et al., Loss of macroautophagy promotes or prevents fibroblast apoptosis depending on the death stimulus. J Biol Chem, 2008. 283(8): p. 4766-77.

[33] Gomes, L.C. and L. Scorrano, Mitochondrial morphology in mitophagy and macroautophagy. Biochim Biophys Acta, 2013. 1833(1): p. 205-12.

[34] Glick, D., S. Barth, and K.F. Macleod, Autophagy: cellular and molecular mechanisms. J Pathol, 2010. 221(1): p. 3-12.

[35] Okamoto, K., N. Kondo-Okamoto, and Y. Ohsumi, Mitochondria-anchored receptor Atg32 mediates degradation of mitochondria via selective autophagy. Dev Cell, 2009. 17(1): p. 87-97.

[36] Novak, I., et al., Nix is a selective autophagy receptor for mitochondrial clearance. EMBO Rep, 2010. 11(1): p. 45-51.

[37] Lim, K.L., et al., Mitochondrial dynamics and Parkinson's disease: focus on parkin. Antioxid Redox Signal, 2012. 16(9): p. 935-49.

[38] Lemasters, J.J., et al., Role of mitochondrial inner membrane permeabilization in necrotic cell death, apoptosis, and autophagy. Antioxid Redox Signal, 2002. 4(5): p. 769-81.

[39] Appenzeller-Herzog, C. and M.N. Hall, Bidirectional crosstalk between endoplasmic reticulum stress and mTOR signaling. Trends Cell Biol, 2012. 22(5): p. 274-82.

[40] Li, W.W., J. Li, and J.K. Bao, Microautophagy: lesser-known self-eating. Cell Mol Life Sci, 2012. 69(7): p. 1125-36.

[41] Vardanian, A.J., R.W. Busuttil, and J.W. Kupiec-Weglinski, Molecular mediators of liver ischemia and reperfusion injury: a brief review. Mol Med, 2008. 14(5-6): p. 337-45.

[42] Bhogal, R.H., et al., Reactive oxygen species mediate human hepatocyte injury during hypoxia/reoxygenation. Liver Transpl, 2010. 16(11): p. 1303-13.

[43] Malhi, H. and G.J. Gores, Cellular and molecular mechanisms of liver injury. Gastroenterology, 2008. 134(6): p. 1641-54.

[44] Degli Esposti, D., et al., Ischemic preconditioning induces autophagy and limits necrosis in human recipients of fatty liver grafts, decreasing the incidence of rejection episodes. Cell Death Dis, 2011. 2: p. e111.

[45] Evankovich, J., et al., Calcium/calmodulin-dependent protein kinase IV limits organ damage in hepatic ischemia-reperfusion injury through induction of autophagy. Am J Physiol Gastrointest Liver Physiol, 2012. 303(2): p. G189-98.

[46] Padrissa-Altes, S., et al., Ubiquitin-proteasome system inhibitors and AMPK regulation in hepatic cold ischaemia and reperfusion injury: possible mechanisms. Clin Sci (Lond), 2012. 123(2): p. 93-8.

[47] Wang, D., et al., The role of AKT1 and autophagy in the protective effect of hydrogen sulphide against hepatic ischemia/reperfusion injury in mice. Autophagy, 2012. 8(6): p. 954-62.

[48] Cardinal, J., P. Pan, and A. Tsung, Protective role of cisplatin in ischemic liver injury through induction of autophagy. Autophagy, 2009. 5(8): p. 1211-2.

[49] Shin, T., et al., Activation of peroxisome proliferator-activated receptor-gamma during hepatic ischemia is age-dependent. J Surg Res, 2008. 147(2): p. 200-5.

[50] Kim, J.S., et al., Impaired autophagy: A mechanism of mitochondrial dysfunction in anoxic rat hepatocytes. Hepatology, 2008. 47(5): p. 1725-36.

[51] Cardinal, J., et al., Cisplatin prevents high mobility group box 1 release and is protective in a murine model of hepatic ischemia/reperfusion injury. Hepatology, 2009. 50(2): p. 565-74.

[52] Domart, M.C., et al., Concurrent induction of necrosis, apoptosis, and autophagy in ischemic preconditioned human livers formerly treated by chemotherapy. J Hepatol, 2009. 51(5): p. 881-9.

[53] Gotoh, K., et al., Participation of autophagy in the initiation of graft dysfunction after rat liver transplantation. Autophagy, 2009. 5(3): p. 351-60.

[54] Minor, T., et al., Hypothermic reconditioning by gaseous oxygen improves survival after liver transplantation in the pig. Am J Transplant, 2011. 11(12): p. 2627-34.

[55] Munafo, D.B. and M.I. Colombo, A novel assay to study autophagy: regulation of autophagosome vacuole size by amino acid deprivation. J Cell Sci, 2001. 114(Pt 20): p. 3619-29.

[56] Minor, T., et al., Impaired autophagic clearance after cold preservation of fatty livers correlates with tissue necrosis upon reperfusion and is reversed by hypothermic reconditioning. Liver Transpl, 2009. 15(7): p. 798-805.

[57] Jackson, L.N., et al., PI3K/Akt activation is critical for early hepatic regeneration after partial hepatectomy. Am J Physiol Gastrointest Liver Physiol, 2008. 294(6): p. G1401-10.

[58] Ding, W.X., et al., Autophagy reduces acute ethanol-induced hepatotoxicity and steatosis in mice. Gastroenterology, 2010. 139(5): p. 1740-52.

[59] Wu, D., et al., Alcohol steatosis and cytotoxicity: the role of cytochrome P4502E1 and autophagy. Free Radic Biol Med, 2012. 53(6): p. 1346-57.

[60] Zakhari, S. and T.K. Li, Determinants of alcohol use and abuse: Impact of quantity and frequency patterns on liver disease. Hepatology, 2007. 46(6): p. 2032-9.

[61] Zhou, Z., et al., A critical involvement of oxidative stress in acute alcohol-induced hepatic TNF-alpha production. Am J Pathol, 2003. 163(3): p. 1137-46.

[62] Baraona, E., et al., Pathogenesis of alcohol-induced accumulation of protein in the liver. J Clin Invest, 1977. 60(3): p. 546-54.

[63] Donohue, T.M., Jr., R.K. Zetterman, and D.J. Tuma, Effect of chronic ethanol administration on protein catabolism in rat liver. Alcohol Clin Exp Res, 1989. 13(1): p. 49-57.

[64] Kharbanda, K.K., et al., Ethanol consumption alters trafficking of lysosomal enzymes and affects the processing of procathepsin L in rat liver. Biochim Biophys Acta, 1996. 1291(1): p. 45-52.

[65] Jensen, K. and C. Gluud, The Mallory body: theories on development and pathological significance (Part 2 of a literature survey). Hepatology, 1994. 20(5): p. 1330-42.

[66] Donohue, T.M., Jr., et al., Role of the proteasome in ethanol-induced liver pathology. Alcohol Clin Exp Res, 2007. 31(9): p. 1446-59.

[67] Scherz-Shouval, R., et al., Reactive oxygen species are essential for autophagy and specifically regulate the activity of Atg4. EMBO J, 2007. 26(7): p. 1749-60.

[68] Britton, R.S., Metal-induced hepatotoxicity. Semin Liver Dis, 1996. 16(1): p. 3-12.

[69] Donohue, T.M., Jr., et al., L-Buthionine (S,R) sulfoximine depletes hepatic glutathione but protects against ethanol-induced liver injury. Alcohol Clin Exp Res, 2007. 31(6): p. 1053-60.

[70] Donohue, T.M., et al., Lysosomal leakage and lack of adaptation of hepatoprotective enzyme contribute to enhanced susceptibility to ethanol-induced liver injury in female rats. Alcohol Clin Exp Res, 2007. 31(11): p. 1944-52.

[71] Bernal, C.A., J.A. Vazquez, and S.A. Adibi, Leucine metabolism during chronic ethanol consumption. Metabolism, 1993. 42(9): p. 1084-6.

[72] Kochl, R., et al., Microtubules facilitate autophagosome formation and fusion of autophagosomes with endosomes. Traffic, 2006. 7(2): p. 129-45.

[73] Torok, N., et al., Vesicle movement in rat hepatocytes is reduced by ethanol exposure: alterations in microtubule-based motor enzymes. Gastroenterology, 1997. 113(6): p. 1938-48.

[74] Clemens, D.L., et al., Impairment of the asialoglycoprotein receptor by ethanol oxidation. Biochem Pharmacol, 1996. 52(10): p. 1499-505.

[75] Shepard, B.D. and P.L. Tuma, Alcohol-induced alterations of the hepatocyte cytoskeleton. World J Gastroenterol, 2010. 16(11): p. 1358-65.

[76] He, J., S. de la Monte, and J.R. Wands, Acute ethanol exposure inhibits insulin signaling in the liver. Hepatology, 2007. 46(6): p. 1791-800.

[77] Venkatraman, A., et al., Chronic alcohol consumption increases the sensitivity of rat liver mitochondrial respiration to inhibition by nitric oxide. Hepatology, 2003. 38(1): p. 141-7.

[78] Donohue, T.M., Jr., Autophagy and ethanol-induced liver injury. World J Gastroenterol, 2009. 15(10): p. 1178-85.

[79] Alavian, S.M., et al., Virus-triggered autophagy in viral hepatitis - possible novel strategies for drug development. J Viral Hepat, 2011. 18(12): p. 821-30.

[80] Rautou, P.E., et al., Changes in autophagic response in patients with chronic hepatitis C virus infection. Am J Pathol, 2011. 178(6): p. 2708-15.

[81] Ait-Goughoulte, M., et al., Hepatitis C virus genotype 1a growth and induction of autophagy. J Virol, 2008. 82(5): p. 2241-9.

[82] Iwasaki, A., Role of autophagy in innate viral recognition. Autophagy, 2007. 3(4): p. 354-6.

[83] Shrivastava, S., et al., Hepatitis C virus upregulates Beclin1 for induction of autophagy and activates mTOR signaling. J Virol, 2012. 86(16): p. 8705-12.

[84] Estrabaud, E., S. De Muynck, and T. Asselah, Activation of unfolded protein response and autophagy during HCV infection modulates innate immune response. J Hepatol, 2011. 55(5): p. 1150-3.

[85] Gregoire, I.P., C. Rabourdin-Combe, and M. Faure, Autophagy and RNA virus interactomes reveal IRGM as a common target. Autophagy, 2012. 8(7).

[86] Sir, D., et al., Replication of hepatitis C virus RNA on autophagosomal membranes. J Biol Chem, 2012. 287(22): p. 18036-43.

[87] Dreux, M., et al., The autophagy machinery is required to initiate hepatitis C virus replication. Proc Natl Acad Sci U S A, 2009. 106(33): p. 14046-51.

[88] Dreux, M. and F.V. Chisari, Impact of the autophagy machinery on hepatitis C virus infection. Viruses, 2011. 3(8): p. 1342-57.

[89] He, C. and D.J. Klionsky, Regulation mechanisms and signaling pathways of autophagy. Annu Rev Genet, 2009. 43: p. 67-93.

[90] Wang, P., et al., HBx induces HepG-2 cells autophagy through PI3K/Akt-mTOR pathway. Mol Cell Biochem, 2013. 372(1-2): p. 161-8.

[91] Tian, Y., et al., Autophagy required for hepatitis B virus replication in transgenic mice. J Virol, 2011. 85(24): p. 13453-6.

[92] Sir, D., D.K. Ann, and J.H. Ou, Autophagy by hepatitis B virus and for hepatitis B virus. Autophagy, 2010. 6(4).

[93] Kim, I. and J.J. Lemasters, Mitochondrial degradation by autophagy (mitophagy) in GFP-LC3 transgenic hepatocytes during nutrient deprivation. Am J Physiol Cell Physiol, 2011. 300(2): p. C308-17.

[94] Li, J., et al., Subversion of cellular autophagy machinery by hepatitis B virus for viral envelopment. J Virol, 2011. 85(13): p. 6319-33.

[95] White, E., Deconvoluting the context-dependent role for autophagy in cancer. Nat Rev Cancer, 2012. 12(6): p. 401-10.

[96] Cao, Y. and D.J. Klionsky, Physiological functions of Atg6/Beclin 1: a unique autophagy-related protein. Cell Res, 2007. 17(10): p. 839-49.

[97] Takamura, A., et al., Autophagy-deficient mice develop multiple liver tumors. Genes Dev, 2011. 25(8): p. 795-800.

[98] Qu, X., et al., Promotion of tumorigenesis by heterozygous disruption of the beclin 1 autophagy gene. J Clin Invest, 2003. 112(12): p. 1809-20.

[99] Zou, M., et al., Beclin 1-mediated autophagy in hepatocellular carcinoma cells: implication in anticancer efficiency of oroxylin A via inhibition of mTOR signaling. Cell Signal, 2012. 24(8): p. 1722-32.

[100] Kotsafti, A., et al., Autophagy and apoptosis-related genes in chronic liver disease and hepatocellular carcinoma. BMC Gastroenterol, 2012. 12: p. 118.

[101] Ding, Z.B., et al., Association of autophagy defect with a malignant phenotype and poor prognosis of hepatocellular carcinoma. Cancer Res, 2008. 68(22): p. 9167-75.

[102] Gou, X., et al., HAb18G/CD147 inhibits starvation-induced autophagy in human hepatoma cell SMMC7721 with an involvement of Beclin 1 down-regulation. Cancer Sci, 2009. 100(5): p. 837-43.

[103] Menrad, H., et al., Roles of hypoxia-inducible factor-1alpha (HIF-1alpha) versus HIF-2alpha in the survival of hepatocellular tumor spheroids. Hepatology, 2010. 51(6): p. 2183-92.

[104] Menon, K.V., A.R. Hakeem, and N.D. Heaton, Meta-analysis: recurrence and survival following the use of sirolimus in liver transplantation for hepatocellular carcinoma. Aliment Pharmacol Ther, 2013. 37(4): p. 411-9.

[105] Lin, H., F. Hua, and Z.W. Hu, Autophagic flux, supported by toll-like receptor 2 activity, defends against the carcinogenesis of hepatocellular carcinoma. Autophagy, 2012. 8(12).

[106] Gong, K., et al., Autophagy-related gene 7 (ATG7) and reactive oxygen species/ extracellular signal-regulated kinase regulate tetrandrine-induced autophagy in human hepatocellular carcinoma. J Biol Chem, 2012. 287(42): p. 35576-88.

[107] Chang, Y., et al., miR-375 inhibits autophagy and reduces viability of hepatocellular carcinoma cells under hypoxic conditions. Gastroenterology, 2012. 143(1): p. 177-87 e8.

[108] Sasaki, M., et al., Autophagy mediates the process of cellular senescence characterizing bile duct damages in primary biliary cirrhosis. Lab Invest, 2010. 90(6): p. 835-43.

[109] Sasaki, M., et al., Autophagy may precede cellular senescence of bile ductular cells in ductular reaction in primary biliary cirrhosis. Dig Dis Sci, 2012. 57(3): p. 660-6.

[110] Sasaki, M., et al., A possible involvement of p62/sequestosome-1 in the process of biliary epithelial autophagy and senescence in primary biliary cirrhosis. Liver Int, 2012. 32(3): p. 487-99.

[111] Hidvegi, T., et al., The role of autophagy in alpha-1-antitrypsin deficiency. Methods Enzymol, 2011. 499: p. 33-54.

Section 5

Autophagy in Cancer

Chapter 9

Role of Autophagy in Cancer and Tumor Progression

Bassam Janji, Elodie Viry, Joanna Baginska,
Kris Van Moer and Guy Berchem

Additional information is available at the end of the chapter

http://dx.doi.org/10.5772/55388

1. Introduction

Autophagy, which is constitutively executed at basal level in all cells, promotes cellular homeostasis by regulating organelles and proteins turnover. In tumor cells, autophagy is activated in response to various cellular stresses, including nutrient and growth factor starvation, as well as hypoxia [1]. It is now well established that autophagy can act as tumor suppressor and tumor promoter. The different roles of autophagy in cancer cells seem to depend on tumor type, stage, and genetic context. Indeed, autophagy clearly suppresses the initiation and development of tumors, however, it is considered as a key survival pathway in response to stress, and many established tumors require autophagy to survive. In this section, we will summarize the different mechanisms involved in the activation of autophagy in tumor and discuss recent reports about the dual role of autophagy in carcinogenesis and tumor progression.

1.1. Role of autophagy in tumor suppression

Several lines of indirect evidence indicate that autophagy acts as a tumor suppressor. Indeed in various cases, oncogenic transformations, such as activation of the PI3K/Akt pathway *via* activating *PI3K* mutations, *AKT* amplifications, or *PTEN* loss, are correlated, with a decreased autophagy through mTOR activation [2]. Moreover, the amplification of the apoptosis inhibitor Bcl-2 has been reported in some circumstances to inhibit autophagy through its binding to Beclin1 [3, 4]. The involvement of p53 in the regulation of autophagy seems to be more complex. Indeed, the activation of p53 by nutrient deprivation or genotoxic stress leads to the activation of autophagy through the inhibition of mTOR or by the activation of DRAM (damage-regulated autophagy modulator) [5-7]. However, consistent with the role of autophagy as tumor suppressor, the functional loss of p53 was expected to decrease autophagy or to suppress basal autophagy. The later effect seems to depend on the cytoplasmic, not the nuclear, pool of p53 [8].

Beside the indirect evidences outlined above, there are more direct ones supporting the tumor suppressing properties of autophagy. Thus, the autophagy execution protein Beclin1 is a haplo-insufficient tumor suppressor protein. Monoallelic deletions of *BECLIN1* are found in sporadic human breast and ovarian carcinomas [9], and heterozygous deletion of *BECLIN1* predisposes mice to a variety of tumors including mammary neoplastic lesions, lung adeno-carcinomas, hepatocellular carcinomas, and B cell lymphomas [10]. These results suggest that intact autophagy may be constraining tumor initiation [11]. Similarly, homozygote deletion of *ATG5* was shown to predisposed mice specifically to liver tumors with high penetrance [12].

The tumor suppressive functions of autophagy have been extensively investigated. Below we will provide mechanistic insights into the tumor-suppressive functions of autophagy.

1.1.1. Autophagy inhibits necrosis and inflammation

During the last decade, strong evidence supported that the inflammatory microenvironment plays a major role in tumor development. Indeed, chronic inflammation is a common future of early cancer development. In this regards, it has been proposed that autophagy can modulate those inflammatory reactions through different mechanisms, as autophagy-deficient tumors display an increased level of necrosis and inflammation.

First, it has been reported that activation of autophagy in tumor cells can inhibit necrotic cell death. Unlike apoptotic cell death, cells dying by necrosis stimulate a robust inflammatory response *in vivo* [13]. In 2006, a major study from the E. White's group reported that impairment of both apoptosis and autophagy promotes necrotic cell death, *in vitro* and *in vivo*, associated with an inflammatory response and an accelerated tumor growth [14]. These results suggest that autophagy takes part in the regulation of necrosis-induced cell death and thus in the subsequent inflammation.

Several studies have confirmed that autophagy is able to prevent the two forms of necrotic cell death (i) necroptosis and (ii) poly-(ADP-ribose) polymerase (PARP)-mediated cell death. Necroptosis is a form of caspase-independent cell death mediated by cell death ligands (*i.e.* TNF-α and FasL) [15, 16]. For example, Wu *et al.* have shown that autophagy is essential to overcome zVAD-induced necroptosis in L929 cells. Activation of PI3K-Akt-mTOR pathway, well-known as inhibitor of autophagy, is able to sensitize L929 cells to zVAD-induced necroptosis, while amino-acid and serum starvation offers some protection to these cells [17]. PARP-mediated cell death is another form of programmed-necrotic cell death mainly triggered by DNA damage. The cytoprotective role of autophagy in PARP-mediat-ed necrosis was illustrated in a recent study of Muñoz-Gámez *et al.*. They have reported that DNA damages induced by doxorubicin in fibroblasts lead to PARP-1 activation and autophagy induction which protects cells against necrosis. By specific knocking down of the autophagic genes *ATG5* or *BECLIN1*, the authors were able to sensitize cells to doxorubicin-induced necrotic cell death [18].

Autophagy acts also through different mechanisms to decrease inflammation. Autophagy is essential for the maintenance of intracellular ATP level, which in turn is required for the secretion of lysophosphatidylcholine (LPC). Secretion of LPC is associated with the acute phase of the inflammatory response and is involved in the development of chronic inflammation. It also has been shown that autophagy-deficient cells fail to generate phosphatidylserine on the outer membrane surface – an important anti-inflammatory pro-apoptotic marker. This explains how defect in autophagy can stimulate inflammatory response subsequently to insufficient clearance of dead cells [19]. Accumulation of p62 in autophagy-deficient cells activates the pro-inflammatory transcription factor NF-κB and the stress-responsive transcription factor NRF2, thus favoring inflammation and tissue injury [20]. Transcription factors of NF-κB family regulate the expression of a broad range of genes involved in the development, the proliferation, and the survival of tumor cells. Moreover, these transcription factors are important in regulation of inflammation and innate and adaptive immune responses [21]. Activation of NF-κB is mediated by the IκB kinase (IKK) complexes. It has been shown that IKK complexes are targets for degradation by autophagy when the heat shock protein 90 (Hsp90) function is inhibited [22]. Another mechanism of regulation of NF-κB by autophagy is mediated by the Kelch-like ECH-associated protein 1 (Keap1). Keap1 interacts with the kinase domain of IKKβ through its C-terminal domain. This domain is also required for the binding of Keap1 to the transcription factor NRF2, which controls expression of certain antioxidant genes. In response to tumor necrosis factor (TNF), Keap1 negatively regulates activation of NF-κB through inhibition of the IKKβ phosphorylation and induction of IKKβ degradation by autophagy pathway [23]. The E3 ubiquitin ligase Ro52 is another signaling molecule that targets IKKβ for degradation through the autophagy pathway. In response to distinct stimuli, specific interactions of Hsp90, Keap1 and Ro52 with IKKs regulate NF-κB activity through their ability to activate or repress the degradation of IKKs by autophagy [24]. It has been shown that the crosstalk between NF-κB and autophagy regulates inflammasome activity leading to the modulation of the activation of caspase-1 and subsequently the secretion of potent pro-inflammatory cytokines [25]. Overall, it appears that autophagy exerts a significant impact on the regulation of inflammation, and is an important modulator of cancer pathogenesis.

1.1.2. Autophagy prevents oxidative stress and genomic instability

Over the last years, the link between autophagy and suppression of cancer development has been confirmed by several *in vivo* studies using genetically engineered mice [26]. As mentioned above, autophagy-defective mice with targeted deletion of the essential autophagy gene *BECLIN1* showed an increased susceptibility to develop cancer [10] [27]. It appears that the involvement of autophagy in the management of oxidative stress and in the maintenance of the genomic integrity is related to its antitumorigenic activity. Indeed, Mathew *et al.* demonstrated that autophagy can limit DNA damage, chromosomal instability and aneuploidy, which may explain its antitumorigenic activity [28]. Several studies suggested that the ubiquitin- and LC3-binding protein p62 may play a determinant role [29] [30]. Indeed, the inability of autophagy-deficient cells to degrade p62 leads to the aberrant accumulation of this protein, which is sufficient to promote tumorigenesis [30]. Recently, two groups demonstrated that p62 activates the transcription factor NRF2 through the direct inhibition of Keap1 [31] [32]. However, the role of NRF2 in DNA damage promotion is not clearly understood. In addition, p62 may act as an important NF-κB modulator in tumorigenesis [33]. This study highlights that the increase in DNA damage in autophagy-deficient cells was associated with high levels of damaged mitochondria and reactive oxygen species (ROS), accumulation of endoplasmic reticulum (ER) chaperones and protein disulfide isomerases. DNA alterations were suppressed by ROS scavengers, confirming the essential role of autophagy in oxidative stress management and, subsequently in protein quality control [30]. On one hand, excessive ROS exposure can directly alter the function of multiple cellular macromolecules by oxidation (*e.g.* nucleic acids, lipids, proteins). On the other hand, oxidative stress is closely linked to mitochondria dysfunction. Since autophagy is the only process allowing the mitochondrial turnover (so-called mitophagy), preventing accumulation of damaged mitochondria highly reduces the risk of oxidative stress. Moreover, mitochondria produce the bulk of ATP required for vital cellular functions (*e.g.* DNA replication, mitosis, transcription). In this regard, the ability of autophagy to control proteins/organelles quality and to maintain cellular energy homeostasis highlights its antitumorigenic activity [34]. As an illustration of this concept, the presence of damaged proteins, which are crucial in DNA replication, mitosis or centrosome function, may favor DNA damage in autophagy-deficient cells. Moreover, default in ATP production following a dysfunction in mitochondrial clearance, may also alter DNA replication or repair by leading to the arrest of the replication forks and to the generation of breakage/fusion/bridge cycles responsible for gene amplification [35]. Finally, the implication of autophagy the "normal" protein turnover may also influence the occurrence of DNA damage. For example, cell cycle progression is driven by the periodic activity of certain proteins (*e.g.* Cyclin-dependent kinases (CDKs), Cyclins, CDKs inhibitors). It is possible that a deregulation in "normal" protein turnover in autophagy-deficient cells may alter the correct sequence of the cell cycle progression [35]. Taken together, it has become clear that autophagy helps normal cells to overcome several types of stresses (*e.g.* metabolic, oncogenic), that directly limits their oncogenic transformation. By contrast, this management of cellular stresses is also observed in cancer cells, and leads in this case to cancer promotion (see section 1.2.1.) [36].

Autophagy is also able to mitigate the accumulation of genomic alteration by inducing the mitotic senescence transition. Senescence is an irreversible cell cycle arrest associated with an active metabolism, which can limit the proliferation of abnormal cells. Young *et al.* reported an accumulation of autophagosomes in Ras-induced IMR90 senescent fibroblasts suggesting that autophagy is required for tumor senescence. In addition, knockdown of *ATG5/7* delayed the senescent phenotype, while induction of autophagy clearly enhanced the protein turnover that contributed to synthesis of pro-senescence cytokines (*e.g.* IL-6, IL-8) [37]. This study suggests that autophagy not only facilitates the entry into senescence but also reinforce the senescent phenotype of cells.

1.1.3. Autophagy contributes to tumor cell death

The induction of autophagic cell death has been proposed as a possible tumor suppression mechanism. This statement is based on the observation that apoptosis occurs concomitantly with features of autophagy [38] and that prolonged stress and progressive autophagy can lead to cell death [1].

Autophagic cell death was first described in 1973 based on the morphological features as a modality of cell death with the presence of autophagosomes and was subsequently named as type II cell death, together with apoptosis (type I) and necrosis (type III) [39]. The relevance of autophagic cell death in development has been established in lower eukaryotes and invertebrates like *Dictyostelium discoideum* and *Drosophila melanogaster* [40, 41]. There is clear evidence that mammalian development does not require autophagy as newborn mice lacking essential autophagy genes show no anatomical or histological defects and no impairment of the cell death [42]. This evidence is supported by the fact that in cultured mammalian cells (human or murine), autophagy genes depletion rather induces apoptosis than protects cell against death induced by different stresses [43, 44]. The role of autophagy in cell death induction is not clear, and needs further investigation. So far, the more convincing evidence showing autophagic cell death in mammals is from neuronal cells. Following insulin starvation, hippocampal neural stem cells undergo autophagic cell death, while suppression of autophagy by *ATG7* knock-down blocks the cell death. In this study, autophagic cell death occurs only in cells with functional apoptosis and is caspase-independent [45]. In cancer cells, recent study showed a novel anti-cancer function of the cytosolic protein FoxO1 which is able to induce the autophagic cell death, but it remains unclear if apoptosis is involved in FoxO1-mediated autophagy [46]. At present, most of experiments showing autophagic cell death in mammalian cells were mainly conducted under *in vitro* cell culture conditions and in cells with defective apoptosis machinery. It has been shown that DAPK (death associated protein kinase) plays an important role in regulation of both autophagy and apoptosis. Indeed, DAPK induces autophagy by phosphorylation of Beclin1, and is associated with the induction of apoptosis. However this type of DAPK-dependent autophagic death is caspase dependent, and it remains to be elucidated whether DAPK-mediated cell death is a real autophagic cell death, or whether autophagy only assists in the apoptosis execution phase [47]. It has been proposed by Kroemer *et al.* that cells rather die *with* autophagy, and not *by* autophagy as they showed that none of 1,400 compounds, evaluated for their ability to induce autophagic puncta and increase autophagic flux, killed tumor cells through the induction of autophagy [48]. Moreover a careful determination of the autophagic flux is needed to differentiate autophagic cell death from other forms of non-apoptotic programmed-cell death, such as necroptosis. These examples illustrate that autophagy may be involved in lethal signaling although the role of autophagy itself in cell killing remains unclear. Thus, further studies are required before the exact role and the precise mechanism of autophagic cell death will be known.

1.1.4. Autophagy modulates the anti-tumor immune response

The immune system plays an important role in controlling cancer progression. It is now well established that immune cells can mediate the destruction of mutated, aberrant or over-expressing self-antigens tumor cells. However, evasion of immune-mediated killing has recently been recognized as an universal hallmark of cancer [49]. It has become increasingly clear that hypoxic tumor microenvironment plays a crucial role in the control of immune protection [50]. On one hand, tumor cells have evolved to utilize hypoxic stress to their own advantage by activating key biochemical and cellular pathways that are important for tumor progression, survival, and metastasis. Autophagy is one of these pathways activated under hypoxia that may be exploited to modulate the responsiveness of tumor cells to immune system. On the other hand, immune cells that infiltrate tumor microenvironment also encounter hypoxia, resulting in hypoxia-induced autophagy. It is now clearly established that autophagy impacts on the immune system as this process is crucial for immune cell proliferation as well as for their effector functions such as antigen presentation and T-cell-mediated killing of tumor cells [51]. In the subsequent section we will discuss the role of autophagy activation in both tumor and immune cells in the context of cancer immune response. Indeed, understanding how tumor cells evade effective immunosurveillance represents a major challenge in the field of tumor immunotherapy.

1.1.4.1. Role of autophagy in immune cells

Despite the inhospitable hypoxic microenvironment, multiple cell types within the innate and adaptive immune system are capable to recognize and eliminate tumor cells. This was attributed to the ability of immune cells to adjust their metabolic dependency once they have reached the tumor and enhance their survival by activating autophagy. Here we will discuss how autophagy impacts specific immune subsets.

Autophagy in neutrophils

The effect of autophagy induction by hypoxia was investigated in neutrophils as this type of immune cells are the first to migrate to the inflammatory site of the tumor where they promote inflammation and activate macrophages and dendritic cells (DCs) [52]. Neutrophils display high glycolytic rate making them resistant to hypoxia. Autophagy activation in neutrophils has been reported to mediate neutrophil cell death. This will decrease inflammation and ultimately lead to limit tumor growth under these circumstances [53].

Autophagy in antigen presenting cells (APCs)

In contrast to neutrophils, APCs such as macrophages and dendritic cells (DCs) must metabolically adapt to hypoxia through stabilization of hypoxia-inducible factor-1α (HIF-1α) to induce the expression of glucose transporters and glycolytic enzymes as well as limiting oxygen consuming oxidative phosphorylation [54]. As a consequence of hypoxia, macrophages and DCs have decreased phagocytosis, reduced migratory capacity, and increased production of proangiogenic and proinflamatory cytokines. While, hypoxia is involved in dampening APC activity, autophagy may contribute to survival of APCs under these conditions. It has been proposed that culturing DCs under hypoxia resulted in the stabilization of HIF-1α which initiates BNIP3 expression and promotes survival of mature DCs, possibly due to induction of autophagy [55]. It has been proposed that autophagy induction in APCs infiltrating tumor occurs *via* different signaling mechanisms such as toll-like receptor (TLR) [56, 57] and TLR4/HMGB1 [51] signaling pathways. Based on the data outlined above, we could assume that autophagy plays a role in cell death of neutrophils which may serve as an anti-inflammatory mechanism in hypoxic tumors. However, autophagy in tumor-infiltrating APCs is involved

http://dx.doi.org/10.5772/55388

in survival, likely by liberation of nutrients required to support the energy demands of activated cells and is important for the cell's antigen presentation capabilities [58, 59]. DCs also use autophagy to promote cross-presentation of tumor antigens on major histocompatibility complex (MHC) class I complexes for cytotoxic T-Lymphocyte (CTL) activation [60] and to facilitate antigen expression on MHC class II molecules for T-helper (Th) cell activation [59, 61]. Considering the fact that autophagy was shown to be important for the process of antigen presentation, it may be involved in positive effects of APC presence within tumors such as activation of T cells through improved MHC expression. Thus, inhibiting autophagy likely dampens cancer immune response.

Autophagy in T lymphocytes

The effect of autophagy on the activity of T cells was also investigated. Indeed, autophagy is activated in these cells upon T cell receptor engagement in both CD4+ and CD8+ subtypes [62-64]. Targeting autophagy by silencing ATG5 or ATG7 during T cell receptor stimulation leads to a significant decrease in cellular proliferation, highlighting the importance of autophagy during T cell activation [63, 64]. Evidence has been recently provided showing that autophagy is upregulated at the immunological synapse during DC and T cell contact. Suppression of autophagy in DCs resulted in hyper-stable contacts between the DC and CD4+ T cells and increased T-cell activation [65]. Autophagy is upregulated in Th2 CD4+ T cells compared with Th1 CD4+ T cells and was shown to be important for the survival of a Th2 cell line upon growth factor withdrawal [66]. In addition, cells cultured under Th1-polarizing conditions rely more heavily on autophagy for survival compared to the Th17 subset. These findings indicate that the role of autophagy is dependent on the cell type and stimuli and that blocking autophagy can skew the balance of immune subsets [67]. Once T cells mature and traffic to the periphery, autophagy is required for survival [63, 67-70]. The role of autophagy in promoting mature T-cell survival has been attributed to autophagy degrading essential components of the apoptotic cell death machinery [67] and maintaining mitochondrial turnover [68-70]. In addition, it has been demonstrated that activated CD4+ T cells exhibit reduced cytokine secretion, adenosine triphosphate (ATP) production, fatty acid utilization, and glycolytic activity when autophagy is inhibited [64]. These findings support the notion that autophagy is required for cellular function by providing metabolism through the liberation of biosynthetic precursors. It has been shown that during sustained growth factor withdrawal, autophagy supplies the metabolites necessary to generate ATP production in bone marrow hematopoietic cells [71] supporting the hypothesis that immune cells use autophagy to generate metabolites required for cell survival. More recently, it has been shown that autophagy is involved in the liberation of the ubiquitous protein puromycin-sensitive aminopeptidase epitope, thereby creating a CTL epitope that mimic tumor-associated antigens [72].

1.1.4.2. Role of autophagy in tumor cells

Autophagy has been found activated in many tumors and its inhibition can lead to either increased death or increased survival, depending on tissue type, tumor grade and any concomitant therapy used [73, 74]. The role of autophagy induction in the anti-tumor immune response has recently received widespread attention. We have investigated the role of autophagy induction under hypoxia in tumor response to CTL-mediated lysis. Using non-small cell lung carcinoma and their autologous CTL, we clearly showed that the activation of autophagy under hypoxia in tumor cells is associated with resistance to CTL-mediated lysis (Figure 1).

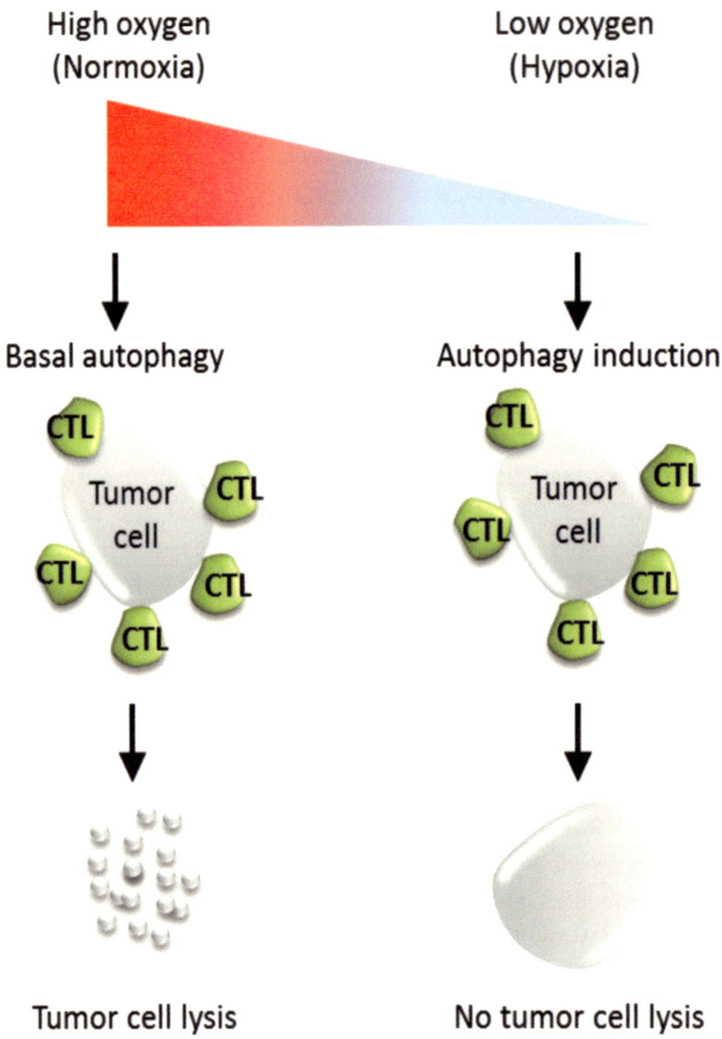

Figure 1. Effect of hypoxia-induced autophagy in CTL-mediated tumor cell killing

Targeting autophagy in hypoxic tumor cells restores CTL-mediated killing [75]. The mechanism by which hypoxia-induced autophagy leads to tumor resistance to CTL was investigated. We provided evidence that hypoxia-inducible factor (HIF)-1α and autophagy coordinately operate to induce and stabilize a survival pathway involving the activated signal transducer and activator of transcription-3 (STAT-3) [76]. Furthermore, we also showed that targeting autophagy *in vivo* enhances the anti-tumor effect of tumor vaccine. These findings extend the notion that simultaneously boosting the immune system and targeting autophagy could enhance the therapeutic efficacy of cancer vaccine and may prove beneficial in cancer immunotherapy [75].

Since autophagy can also promote the survival of tumor cells through nutrients recovered from degrading and recycling damaged organelles, it has been recently proposed that chemotherapy-induced autophagy causes the release of ATP from tumor cells, thereby stimulating antitumor immune response. Targeting autophagy blunted the release of ATP by tumor cells in response to chemotherapy without affecting that of other damaged signals. Autophagy-dependent extracellular ATP recruits DCs into tumors and activates a T cell response to tumor cells [77]. Based on this study, it seems that the activation of autophagy in the context of DNA damage-induced apoptosis, causes ATP release which subsequently recruits immune cells.

It is now well established that immune effector cells integrate signals that define the nature and magnitude of the subsequent response. In this context, it has been shown that at high effector-to-target ratios, autophagy was induced in several human tumors by natural killer (NK) cells. Importantly, cell-mediated autophagy promoted resistance from treatment modalities designed to eradicate tumor. Thus, the lymphocyte-induced cell-mediated autophagy promotes cancer cell survival and may represent an important target for development of novel therapies [78].

The complexity of cancer immune response is related to the fact that different immune subsets cooperatively and coordinately act through the secretion of cytokines and other soluble factors. Thus, it stands to reason that antitumor immune responses are not entirely dependent on the presence or absence of any particular subset, but rather on the stoichiometry of immune effectors versus immune suppressors. As a result, any anti-cancer therapies that skew the immune effector to suppressor ratio by impacting autophagy may exert a large effect on overall patient survival [79]. While mounting evidence suggest that autophagy induction enhances immune cell function, autophagy seems to operate as a tumor cells resistance mechanism against immune response. In spite of this, inhibition of autophagy in the clinic can behave as a double-edged sword because it can enhance or suppress cancer immune response. Thus, therapeutic strategies targeting autophagy in tumor cells must consider the potential negative impact on antitumor immunity. The key question that emerged is: what is the net outcome of the autophagy inhibitor in clinic? There are numerous studies supporting that immunotherapy of cancer should focus on inducing and reprogramming cells of the innate and adaptive immune system. Therefore, it is tempting to speculate that combined therapy based on autophagy inhibitor and reprograming immune cells could significantly improve cancer immunotherapy.

1.1.5. Autophagy inhibits metastasis

Metastases are responsible for most cancer-related deaths. Metastatic cascade involves several steps, including: i) invasion from the primary tumor site, ii) intravasation and survival in the systemic circulation, iii) extravasation at the secondary tissue site and, iv) colonization of this target tissue [80]. Autophagy has been found to either promote or prevent the metastatic progression, depending on the step in which it is activated (Figure 2 adapted from [81]). In this section, we will focus on the anti-metastatic activity of autophagy, while its pro-metastatic properties will be overview in the section 1.2.2.

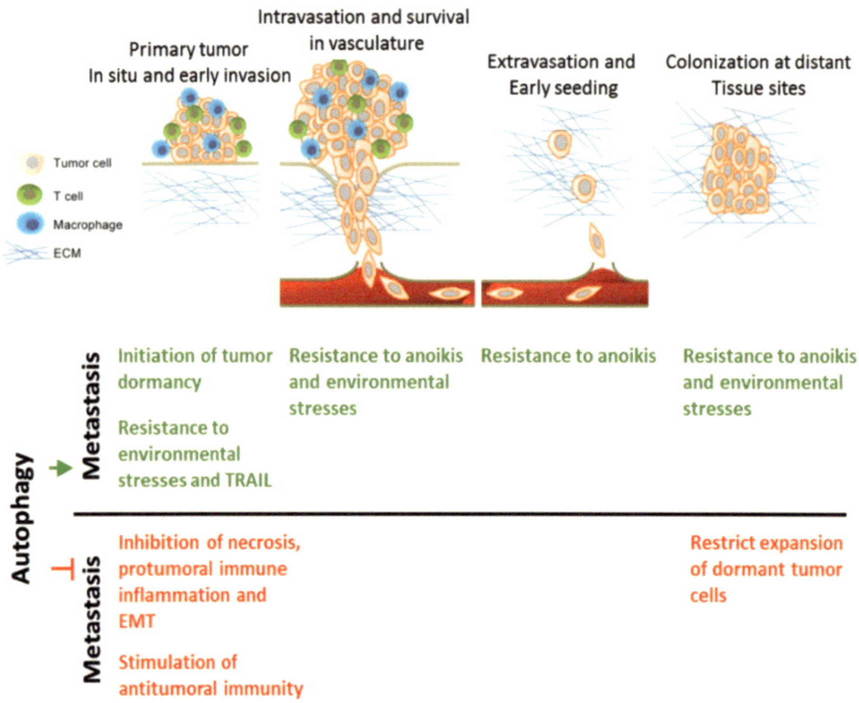

Figure 2. Dual role of autophagy in metastasis (adapted from [80])

1.1.5.1. Autophagy modulates the inflammatory response

At early steps, autophagy is able to limit the metastatic progression from the primary tumor site by restricting inflammatory response. Indeed, infiltrated immune cells can supply some signals within the tumor microenvironment that influence tissue remodeling, angiogenesis, tumor cell survival and spreading. Clinical and experimental data have confirmed the dual role of the immune system in tumor metastasis. As example, Lin *et al.* have demonstrated that macrophages depletion into the primary tumor site in transgenic CFS-1 null mice (colony-stimulating factor-1 is a cytokine involved in the proliferation, differentiation and survival of macrophages) delayed the development of metastasis [82]. While certain immune cells (*e.g.* macrophages, B cells, granulocytes, mast cells) may favor cancer development, others (*e.g.* Natural killer cells and T lymphocytes) may preferentially inhibit it [83] [84]. However, the tumor-promoting or tumor-suppressive properties of immune cells are not clearly defined and seem to be closely dependent on the tissue context and the cellular stimuli [85].

Autophagy can modulate inflammation during metastasis by different ways. On one hand, autophagy may lead to a direct activation of antitumor immunity through the release of high-mobility group box protein 1 (HMGB1) from tumor cells that are destined to die [86]. When released, HMGB1 stimulates the Toll-Like Receptor 4 on dendritic cells and, subsequently, promotes the tumor cell killing by inducing T-cell immunity [87]. On the other hand, autophagy can indirectly attenuate the macrophage infiltration by inhibiting tumor cell necrosis (see section 1.1.1.). Indeed, tumor-associated macrophages (TAMs) are important components of the leukocyte infiltrate and their involvement in metastasis progression have been extensively studied. TAMs positively influence tissue remodeling, angiogenesis, tumor invasion and intravasation through the production of growth factors, cytokines and matrix metalloproteases [88] [85] [89].

1.1.5.2. Autophagy alters the epithelial to mesenchymal transition (EMT)

Many studies have shown that the acquisition of mesenchymal feature by carcinoma cells promotes cancer invasion and metastasis. Epithelial to Mesenchymal Transition (EMT) is a process that leads to the complete loss of epithelial characteristics to achieve a mesenchymal cell phenotype. Initiation and completion of EMT requires the expression of specific transcription factors, microRNAs, cell surface proteins and matrix-degrading proteases [90]. Once undergoing EMT, cancer cells acquire invasive properties that enhance their ability to detach from the primary tumor site and to colonize distant tissues. Recently, two studies have pointed out that autophagy may modulate EMT. Lv *et al.* have shown that the expression of the Death-effector domain-containing DNA-binding protein (DEDD) is inversely correlated with the metastatic phenotype of breast cancer cells. Ectopic expression of DEDD in metastatic MDA-MB-231 cells leads to the autophagy-mediated degradation of the two major EMT inducers Snail and Twist, and subsequently to the loss of the metastatic phenotype. Conversely, knock-down of DEDD in non-metastatic MCF-7 cells reduces autophagy and leads to EMT promotion [91]. Earlier, Sun *et al.* have suggested the implication of the Bcl-2 anti-apoptotic protein, which is also known as an inhibitor of Beclin-1-dependent autophagy, in EMT induction. Under hypoxia condition, Bcl-2 and Twist are coexpressed in hepatocellular carcinoma and physically interact to form a complex that promotes EMT [92].

1.1.5.3. Autophagy restricts expansion of dormant tumor cells

Cancer recurrence is a determinant element for patient life expectancy because this disease presents a high risk of relapse after therapy or a long period of remission. Presence of residual dormant cells in the primary tumor site or in distant organs is one of the major causes of cancer relapse. Tumor dormancy is characterized by a prolonged, but reversible, growth arrest in G0-G1, by which tumor cells survive in a quiescent state. However, dormant tumor cells have to re-activate their proliferative activity to allow the development of micro- or macro-metastasis. Lu *et al.* have reported that induction of aplasia Ras homologue member I (ARHI) gene in ovarian cancer xenografts in mice induced autophagy, led to tumor dormancy and significantly inhibited xenograft growth. Interestingly, a proliferative recovery was obtained when the ARHI-induced autophagy was not maintained supporting the fact that autophagy is required for the establishment of the dormant state [93]. These results illustrate how autophagy can maintain the dormant phenotype of tumor cells and thus inhibit the entry into an active dividing state. By this way, autophagy either prevents the expansion of isolated dormant tumor cells and the development of macrometastases.

1.2. Role of autophagy in tumor progression and metastasis

1.2.1. Autophagy induces survival of tumor cells under a variety of stresses

It has been well documented that tumor cells activate autophagy in response to stress, which enables long-term survival when apoptosis is defective [94]. Autophagy must be a highly selective process to allow extensive cellular degradation while retaining functional integrity. This section will address how autophagy confers tumor cells with superior stress tolerance, which limits damage, maintains viability, sustains dormancy and facilitates recovery.

Cancer cells need to adapt their metabolism to ensure the demands of proliferation enhanced in the microenvironment. The oncogenes affect signaling pathways important in regulation of metabolism, which support cancer growth and proliferation [95]. Autophagy is activated in response to multiple stresses, such as hypoxia, nutrient starvation, and the endoplasmic reticulum (ER) stress [96], during cancer progression. Under metabolic stress, inhibition of autophagy could lead to accelerated apoptosis, thus limiting further tumor progression. In this section, we discuss the role of autophagy regulation in tumor microenvironment and tumor growth [97].

1.2.1.1. Autophagy as adaptive metabolic response to hypoxia

Tumor cells are subjected to elevated metabolic stresses (*i.e.* lack of nutrition, oxygen deprivation) due to a defect in angiogenesis and inadequate blood supply. High levels of hypoxia resulted in an alteration of metabolism, enhanced invasiveness and resistance to therapy. Metabolic stress in tumors is mainly caused by the high metabolic demand required for cell proliferation and the impairment of ATP production [98]. Autophagy acts as an alternative source of energy and promotes tumor cell survival in hypoxic microenvironment. White *et al.* first showed that induction of autophagy in the hypoxic core of tumors promotes cancer cell survival. Hypoxia within tumors can be generated by hypoxia-inducible factor-1 alpha (HIF-1α)-dependent or -independent manner [99]. On one hand, autophagy in hypoxia can be induced through HIF-1α-dependent expression of the BH3-only protein Bcl-2/adenovirus E1B 19 kDa-interacting protein 3 (BNIP3) and the related protein, BNIP3L [100]. These proteins are downstream targets of HIF-1α and are able to induce mitophagy in hypoxia to manage ROS production.

Mechanistically, Bellot *et al.* showed that induction of BNIP3 and BNIP3L in hypoxic cells disrupts the Beclin1-Bcl-2 complex leading to Beclin1 release and subsequently autophagy induction as an adaptive survival response during prolonged hypoxia [101]. We have shown that the acquisition of TNF-resistance in breast cancer cells is correlated with constitutive activation of HIF-1α, even under normoxia, and with the induction autophagy by several signaling pathways highlighting the important role of autophagy in tumor cell adaptation to hypoxia [102] (Figure 3).

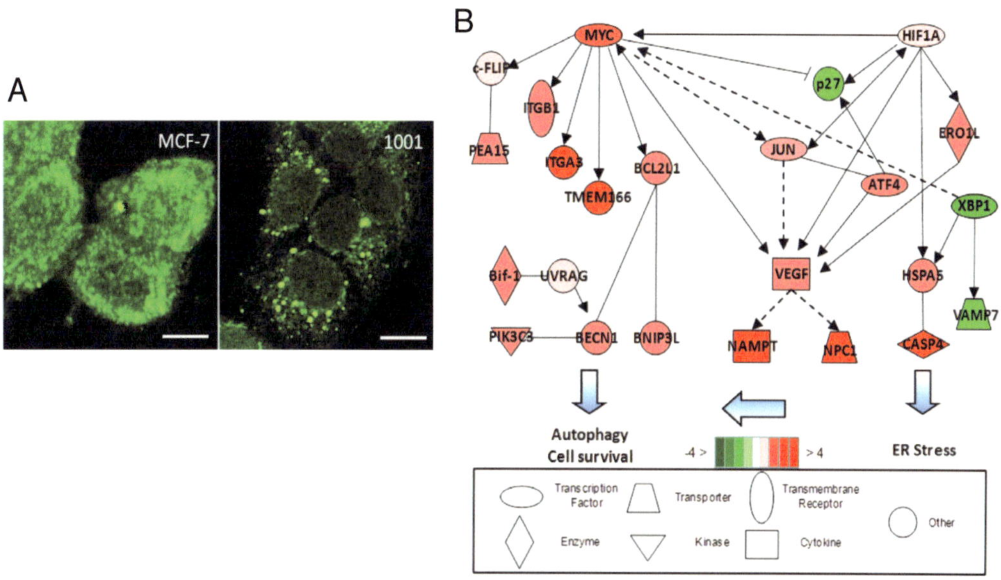

Figure 3. Induction of autophagy in TNF-resistant breast cancer cells. (A) Immunofluorescence analysis of autophagosomes formation in TNF-resistant MCF-7 clone (1001). MCF-7 and 1001 cells were labeled with anti-LC3 primary antibody and Alexa-Fluor 488-conjugated secondary antibody. No autophagosomes were observed in MCF-7 (diffuse green staining) and several autophagosomes (green dot-like structures) were observed in 1001 cells. (B) Data mining of autophagy microarray results performed by Ingenuity software highlights the involvement of MYC and HIF1-α downstream pathways in the activation of autophagy in TNF-resistant cells. Solid lines indicate a direct interaction and dotted lines indicate an indirect interaction; arrows indicate that protein A acts directly (solid line) or indirectly (dotted line) on protein B. Green represents downregulation while red depicts upregulation in 1001 compared to MCF-7 cells. The intensity of color represents the average of log2 fold change from three independent experiments. Symbols affected to each gene reflect cell functions.

Furthermore, Denko *et al.*, showed that autophagy in hypoxia can be induced independently of nutrient deprivation, HIF-1α activity, and expression of BNIP3. The HIF-1 independent hypoxia-induced autophagy involves the activity of the 5′-AMP-activated protein kinase (AMPK) [103]. Recent reports suggest that in addition to its role in the regulation of normal metabolism, AMPK can also regulate cellular energy homeostasis through autophagic degradation of intracellular components. Decrease in the ATP/AMP ratio and the activation of AMPK promotes catabolic pathways instead of anabolic processes. The major downstream pathway activated in HIF-1-independent hypoxia depends on tuberous sclerosis complex (TSC) and mammalian target of rapamycin (mTOR) – a master regulator of cell growth, cellular metabolism, and autophagy [104]. Recently, it has been shown that this pathway can be triggered before any detectable decreases in intracellular ATP levels [105]. Another signaling pathway induced by hypoxic stress and involved in the activation of HIF-1 independent-autophagy is the unfolded protein response (UPR), an evolutionarily conserved pathway activated in response to ER stress. The UPR is activated by three distinct ER stress sensors on the ER membrane: PKR-like ER kinase (PERK), IRE-1, and activating transcription factor 6 (ATF6). PERK kinase phosphorylates the eukaryotic initiation factor 2α (eIF2α) leading to inhibition of the initiation step of mRNA translation [106]. In addition to the inhibition of protein synthesis, PERK phosphorylation of eIF2α enables a selective translation of the ATF4 transcription factor which has a key role in autophagy regulation in response to ER stress. In response to activation of UPR, PERK stimulates autophagy to clear protein aggregates generated by ER stress [107]. Various ER stress–inducing agents have been shown to induce autophagy in yeast and mammalian cells. Additionally, both the PERK/eIF2α and IRE-1 from UPR system have been implicated in regulation of autophagy [108]. Autophagic degradation of cellular components provides the energetic balance governed by AMPK, and suppression of autophagy in cancer cells can increase both resistance to hypoxic stress and tumorigenicity.

1.2.1.2. Autophagy in nutrient starvation

It is now well known that the metabolic stress induced by starvation in tumor microenvironment activates autophagy. Moreover, this metabolic stress is also dependent on autophagy as it allows organelles and proteins recycling in order to provide energy for cell survival. It has been shown that cancer cell lines with Ras activation display elevated levels of basal autophagy essential for survival through starvation and tumor growth [109]. Autophagy induced by starvation (*i.e.* glucose, L-glutamine, pyruvate, serum or amino acids) is mediated by ROS [110]. Indeed, autophagy can be regulated by both superoxide radical and hydrogen peroxide [111]. Another pathway implicated in regulation of autophagy in starvation conditions is dependent on AMP Kinase (AMPK). Specifically, under starvation conditions, activated AMPK inhibits the mTOR pathway and leads to induction of autophagy flux [112]. A new mechanism of regulation of starvation-induced autophagy by microRNAs (miRNAs) has been recently proposed by Gouzacik *et al.*.They have shown that miR-376b was able to attenuate starvation- and rapamycin-induced autophagy in MCF7 and Huh-7 cells. Two direct functional targets of this miRNA were characterized; ATG4C and Beclin1. This finding underlines the importance of miRNAs in tumor microenvironment and as a new regulator of the autophagy pathway [113].

1.2.2. Autophagy promotes tumor cell metastasis

As mentioned in the section 1.1.5., autophagy may also promote different steps of metastatic cascade, mainly by favoring the survival of cancer cells in inhospitable environments (*e.g.* systemic circulation, target organs) (Figure 2). Induction of autophagy allows cancer cells to survive under a variety of stresses (*e.g.* hypoxic, metabolic), that subsequently favors tumor progression (for details, see the section 1.2.1). Several lines of evidence also suggest that autophagy-induced resistance to apoptosis plays a crucial role in cancer progression. It is well documented that metastatic cancer cells are more resistant to apoptosis than their poorly-metastatic counterparts [114]. As example, autophagy is upregulated in tumor cells that are resistant to TRAIL-induced apoptosis [115]. Moreover, blocking specific autophagy actors (*e.g.* Beclin1, ATG7) efficiently restores TRAIL-induced apoptotic cell death, highlighting the cytoprotective role of autophagy in TRAIL-resistance [115] [116].

During the metastatic progression, cancer cells activate mechanisms to resist to anoikis. Anoikis is a form of apoptotic cell death induced by the detachment from the surrounding extracellular matrix (ECM) [117]. Activation of autophagy during anoikis may be a survival strategy developed by the cells to overcome the stress of ECM detachment. Fung *et al.* have pointed out that autophagy is induced in both non-tumorigenic epithelial cell lines and in primary epithelial cells following the ECM detachment [118]. They demonstrated that autophagy protects cells from anoikis as RNAi-depletion of autophagy regulators (ATGs) promotes apoptosis and reduced clonogenic viability upon reattachment. ECM-detached cells activate the PERK-eIF2α-ATF4-CHOP pathway, which is responsible for both autophagy induction and oxidative stress limitation [119]. However, only few studies have confirmed the ability of autophagy to inhibit anoikis in cancer cells. As example, Chen *et al.* demonstrated that loss of autophagy in oncogenic-transformed mammary epithelial cells (PI3K-H1047R) promotes survival and proliferation in 3D organotypic culture [120].

Although autophagy prevents cancer progression by maintaining tumor cells in a dormant state, initiation of dormancy may also promote tumor progression by favoring survival of cancer cells. In this regard, it has been shown that breast cancer cells that lack β1 integrin are in a dormant state, suggesting that dormancy may help cancer cells to overcome the stress of ECM detachment, and subsequently resist to anoikis [121].

1.2.3. Upregulation of autophagy promotes resistance to cancer therapy

Autophagy may function to remove proteins or organelles that are damaged by cancer treatments or, through the degradation of cellular components, may provide nutrients for the rapidly growing cells. Indeed, inhibitors of autophagy can produce different outcomes: cell survival or cell death. Obviously, autophagic cell survival confers tumor cells with superior stress tolerance, which limits damage, maintains viability, sustains dormancy, and facilitates recovery. The dual role of autophagy highlights the need to carefully define its role in tumor cells before applying autophagy-based therapy. It will be important for clinical oncologists and cancer researchers to determine which cancer cell types most commonly undergo autophagy in response to therapy, and whether increased autophagy is a sign of responsiveness or resistance.

Nevertheless, several studies have shown that tumor cells can survive anti-cancer treatment by activating autophagy. This statement was validated using genetic or pharmacological inhibitors of autophagy which led to sensitize tumor cells to cancer therapies. In this context, it has been reported that inhibition of autophagy sensitizes cancer cells to DNA damaging anticancer agents. Evidence has been provided that inhibition of autophagy by 3-methyladenine (3-MA) or by targeting Atg7 enhances the cytotoxicity of 5-fluorouracil in human colorectal cancer cells [122]. Autophagy inhibition also enhances the therapeutic efficacy of cisplatin and 5-fluorouracil in esophageal and colon cancer cells, respectively [122, 123]. Targeting autophagy by genetic approaches using Beclin1, Atg3, and Atg4b siRNA sensitizes resistant cancer cells to ionizing radiation [124]. These studies strongly argue that autophagy operates as a mechanism through which cancer cells acquire resistance to radiotherapy and chemotherapy. There are numerous studies supporting the involvement of autophagy in cancer stem cells resistance to ionizing radiation and other anti- cancer treatments [125]. Thus, in malignant gliomas, the CD133+ cancer stem cells express higher levels of the autophagic proteins LC3, Atg5, and Atg12. In addition, ionizing radiation seems to induce autophagy only in CD133+ cancer stem cells compared to CD133- counterpart [126]. Furthermore, glioma cells treated with autophagy inhibitors exhibit more extensive DNA double-strand breaks than cells treated with radiation alone [127]. We have recently demonstrated that autophagy induction in tumor cells under hypoxia decrease the tumor cell killing by cytotoxic T lymphocytes. Furthermore, we provided evidence that simultaneously boosting the immune system by vaccination and inhibiting autophagy may improve cancer immunotherapy [75, 76].

While the general consensus is that autophagy inhibition is an effective strategy for cancer therapy, some drugs that are being used in the clinic induce autophagy. In most cases, however, it has not been proven that these drugs induce death *via* the autophagy pathway. Indeed, for many of these drugs it is hypothesized that combining them with autophagy inhibitors may improve their efficacy.

2. Autophagy as a target for anti-cancer therapies

Evidence indicated that the modulation of autophagy is an important component of tumorigenesis, making it a possible therapeutic target. Pharmacological inhibitors of autophagy can be broadly classified as early- or late-stage inhibitors of the pathway. Early-stage inhibitors include 3-methyadenine, wortmannin, and LY294002, which target the class III PI3K (Vps34) and interfere with its recruitment to the membranes. Late-stage inhibitors include the antimalarial drugs chloroquine (CQ), hydroxychloroquine (HCQ), bafilomycin A1, and monensin. Bafilomycin A1 is a specific inhibitor of vacuolar-ATPase [128], and monensin and CQ/HCQ are lysosomotropic drugs that prevent the acidification of lysosomes, whose digestive hydrolases depend on low pH. Since autophagosomes and lysosomes move along microtubules, microtubule-disrupting agents (taxanes, nocodazole, colchicine, and vinca alkaloids) can also inhibit the fusion of autophagosomes with lysosomes. Other inhibitors of autophagy that block autophagosome degradation include the tricyclic antidepressant drug clomipramine and the anti-schistome agent lucanthone [129, 130]. Of the known autophagy inhibitors outlined above, only CQ and HCQ have been evaluated in humans, because they are commonly used as antimalarial drugs and in autoimmune disorders. These drugs cross the blood-brain barrier, and HCQ is preferred to CQ in humans because of its more favorable side-effects profile [131]. Quinacrine, which also has been used in patients as an anti-malarial, has been shown to inhibit autophagy similarly to CQ. In fact, quinacrine showed greater cytotoxicity in gastrointestinal stromal tumor (GIST) cell lines treated with imatinib than CQ [132], and therefore this may be a promising anti-autophagy agent for future clinical trials.

Currently, there are nearly 20 clinical trials registered in the National Cancer Institute (www.cancer.gov/clinicaltrials) exploring anti-autophagy strategies in a variety of human cancers. Most of these trials are ongoing, with minimal published results available, and nearly all use HCQ. It is worthy to note that CQ or HCQ are lysosomotropic agents that act at the level of the lysosome by inhibiting acidification, thereby impairing autophagosome degradation. These clinical trials were initiated based on the fact that autophagy is induced in a variety of tumor cells and preclinical models by several types of chemotherapeutic agents as a survival mechanism. Because only a subpopulation of tumor cells undergo autophagy, it is unlikely that autophagy inhibitors are used in cancer therapy as single agent. Indeed, most of these clinical trials used HCQ in combination with other anti-cancer therapies. While these preclinical data are generally supportive of incorporating anti-autophagy therapies in cancer treatment trials, it has been observed, in some circumstances, that inhibition of autophagy decreases therapeutic efficacy. Understanding the circumstances in which autophagy inhibition impairs the therapeutic effect will be of great importance. Importantly, while CQ and HCQ are effective inhibitors of autophagy *in vitro*, whether they will do so at doses used in current clinical trials is still uncertain. An important issue related to the use of these autophagy inhibitors concerns the micromolar concentration that is required to inhibit autophagy and show anti-tumor efficacy in preclinical models. While this is theoretically achievable at tolerated doses after prolonged dosing, it should be better optimized in clinic [133, 134]. Trials combining HCQ as neoadjuvant treatment will provide tumor tissues available for analysis both before and after HCQ treatment. However, the effectiveness of HCQ in the inhibition of autophagy still proves difficult, as HCQ is often combined with other therapies (chemotherapy and radiotherapy) that are also known to modulate autophagy. Alternative biomarkers to predict for autophagy activation as well as autophagy dependence are currently an area of intense investigation [135]. A recently reported phase I trial of HCQ in combination with adjuvant temozolomide and radiation in patients with glioblastoma found that the maximum tolerated dose of HCQ was 600 mg per day, and this dose achieved concentrations of HCQ required for autophagy inhibition in preclinical studies. In this trial, investigators observed a dose-dependent inhibition of autophagy, as indicated by increases in autophagic vesicles (revealed by electron microscopy), and detected elevations in LC3-II in peripheral blood mononuclear cells. In addition, in a phase I trial of 2-deoxyglucose, an agent that blocks glucose metabolism, autophagy occurred in association with a reduction in p62/SQSTM1 in peripheral blood mononuclear cells [136]. These data suggest the potential interest of such biomarkers in the evaluation of autophagy modulation during therapy and in the correlation with treatment outcome [137].

CQ inhibits the last step of autophagy at the level of the lysosome, thereby impacting lysosomal function. Therefore, its effects are not entirely specific to autophagy. Currently, there is a great deal of interest in developing new inhibitors of autophagy. In this regards, and given the complexity of the autophagic process, multiple proteins involved in this process could be good candidates for developing others autophagy inhibitors. It is likely that kinases would be prime candidates for inhibition such as Vps34, a class III PI3K, which has a critical early role in autophagosome development. This is particularly attractive, as there has been significant success in designing effective class I PI3K inhibitors [138]. However, one potential issue which needs to be considered is that Vps34 has roles in other aspects of endosome trafficking, and this may lead to unwanted effects and toxicity [139]. The mammalian orthologs of yeast ATG1, ULK1/2, which acts downstream from AMPK and the TOR complex, have been recently shown as critical proteins for autophagy activation [140-142]. Others potential targets for autophagy inhibitors would be LC3 proteases such as ATG4b, which are necessary for LC3 processing. However, whichever approach is taken, the delicate balance between potency and toxicity must be determined to achieve a clinical success. While there are still uncertainties of how autophagy inhibition will fare as an anti-cancer therapy, the preclinical data generally support this approach. The current clinical trials will hopefully provide insight into whether this will be a viable therapeutic paradigm [135].

Acknowledgements

A part of research results presented in this chapter was generated in the laboratory of BJ and initiated in close collaboration with the team of Salem Chouaib (INSERM U753) at the "Institut de Cancérologie Gustave Roussy". Research projects related to these results were funded by the Luxembourg Ministry of Culture, Higher Education and Research (Grant 2009 0201), Fonds National de la Recherche (AFR Grant 2009 1201) and Fonds National de la Recherche Scientifique "FNRS" (Televie Grant 7.4628.12F).

Author details

Bassam Janji*, Elodie Viry, Joanna Baginska, Kris Van Moer and Guy Berchem

*Address all correspondence to: bassam.janji@crp-sante.lu

Laboratory of Experimental Hemato-Oncology, Department of Oncology, Public Research Center for Health, Luxembourg City, Luxembourg

References

[1] Mathew, R., V. Karantza-Wadsworth, and E. White, *Role of autophagy in cancer.* Nat Rev Cancer, 2007. 7(12): p. 961-7.

[2] Guertin, D.A. and D.M. Sabatini, *Defining the role of mTOR in cancer.* Cancer Cell, 2007. 12(1): p. 9-22.

[3] Maiuri, M.C., et al., *BH3-only proteins and BH3 mimetics induce autophagy by competitively disrupting the interaction between Beclin 1 and Bcl-2/Bcl-X(L).* Autophagy, 2007. 3(4): p. 374-6.

[4] Sinha, S. and B. Levine, *The autophagy effector Beclin 1: a novel BH3-only protein.* Oncogene, 2008. 27 Suppl 1: p. S137-48.

[5] Balaburski, G.M., R.D. Hontz, and M.E. Murphy, *p53 and ARF: unexpected players in autophagy.* Trends Cell Biol, 2010. 20(6): p. 363-9.

[6] Feng, Z., et al., *The coordinate regulation of the p53 and mTOR pathways in cells.* Proc Natl Acad Sci U S A, 2005. 102(23): p. 8204-9.

[7] Crighton, D., et al., *DRAM, a p53-induced modulator of autophagy, is critical for apoptosis.* Cell, 2006. 126(1): p. 121-34.

[8] Tasdemir, E., et al., *Regulation of autophagy by cytoplasmic p53.* Nat Cell Biol, 2008. 10(6): p. 676-87.

[9] Aita, V.M., et al., *Cloning and genomic organization of beclin 1, a candidate tumor suppressor gene on chromosome 17q21.* Genomics, 1999. 59(1): p. 59-65.

[10] Qu, X., et al., *Promotion of tumorigenesis by heterozygous disruption of the beclin 1 autophagy gene.* J Clin Invest, 2003. 112(12): p. 1809-20.

[11] Liang, X.H., et al., *Induction of autophagy and inhibition of tumorigenesis by beclin 1.* Nature, 1999. 402(6762): p. 672-6.

[12] Takamura, A., et al., *Autophagy-deficient mice develop multiple liver tumors.* Genes Dev, 2011. 25(8): p. 795-800.

[13] Kono, H. and K.L. Rock, *How dying cells alert the immune system to danger.* Nat Rev Immunol, 2008. 8(4): p. 279-89.

[14] Degenhardt, K., et al., *Autophagy promotes tumor cell survival and restricts necrosis, inflammation, and tumorigenesis.* Cancer Cell, 2006. 10(1): p. 51-64.

[15] Degterev, A. and J. Yuan, *Expansion and evolution of cell death programmes.* Nat Rev Mol Cell Biol, 2008. 9(5): p. 378-90.

[16] Shen, H.M. and P. Codogno, *Autophagy is a survival force via suppression of necrotic cell death.* Exp Cell Res, 2012. 318(11): p. 1304-8.

[17] Wu, Y.T., et al., *Activation of the PI3K-Akt-mTOR signaling pathway promotes necrotic cell death via suppression of autophagy.* Autophagy, 2009. 5(6): p. 824-34.

[18] Munoz-Gamez, J.A., et al., *PARP-1 is involved in autophagy induced by DNA damage.* Autophagy, 2009. 5(1): p. 61-74.

[19] Pierdominici, M., et al., *Role of autophagy in immunity and autoimmunity, with a special focus on systemic lupus erythematosus.* FASEB J, 2012. 26(4): p. 1400-12.

[20] Levine, B., N. Mizushima, and H.W. Virgin, *Autophagy in immunity and inflammation.* Nature, 2011. 469(7330): p. 323-35.

[21] Smale, S.T., *Hierarchies of NF-kappaB target-gene regulation.* Nat Immunol, 2011. 12(8): p. 689-94.

[22] Xu, C., et al., *Functional interaction of heat shock protein 90 and Beclin 1 modulates Toll-like receptor-mediated autophagy.* FASEB J, 2011. 25(8): p. 2700-10.

[23] Fan, W., et al., *Keap1 facilitates p62-mediated ubiquitin aggregate clearance via autophagy.* Autophagy, 2010. 6(5): p. 614-21.

[24] Trocoli, A. and M. Djavaheri-Mergny, *The complex interplay between autophagy and NF-kappaB signaling pathways in cancer cells.* Am J Cancer Res, 2011. 1(5): p. 629-49.

[25] Strowig, T., et al., *Inflammasomes in health and disease.* Nature, 2012. 481(7381): p. 278-86.

[26] White, E., et al., *Role of autophagy in suppression of inflammation and cancer.* Curr Opin Cell Biol, 2010. 22(2): p. 212-7.

[27] Yue, Z., et al., *Beclin 1, an autophagy gene essential for early embryonic development, is a haploinsufficient tumor suppressor.* Proc Natl Acad Sci U S A, 2003. 100(25): p. 15077-82.

[28] Mathew, R., et al., *Autophagy suppresses tumor progression by limiting chromosomal instability.* Genes Dev, 2007. 21(11): p. 1367-81.

[29] Komatsu, M., et al., *Homeostatic levels of p62 control cytoplasmic inclusion body formation in autophagy-deficient mice.* Cell, 2007. 131(6): p. 1149-63.

[30] Mathew, R., et al., *Autophagy suppresses tumorigenesis through elimination of p62.* Cell, 2009. 137(6): p. 1062-75.

[31] Komatsu, M., et al., *The selective autophagy substrate p62 activates the stress responsive transcription factor Nrf2 through inactivation of Keap1.* Nat Cell Biol, 2010. 12(3): p. 213-23.

[32] Lau, A., et al., *A noncanonical mechanism of Nrf2 activation by autophagy deficiency: direct interaction between Keap1 and p62.* Mol Cell Biol, 2010. 30(13): p. 3275-85.

[33] Duran, A., et al., *The signaling adaptor p62 is an important NF-kappaB mediator in tumorigenesis.* Cancer Cell, 2008. 13(4): p. 343-54.

[34] Jin, S., *Autophagy, mitochondrial quality control, and oncogenesis.* Autophagy, 2006. 2(2): p. 80-4.

[35] Jin, S. and E. White, *Tumor suppression by autophagy through the management of metabolic stress.* Autophagy, 2008. 4(5): p. 563-6.

[36] Rosenfeldt, M.T. and K.M. Ryan, *The multiple roles of autophagy in cancer.* Carcinogenesis, 2011. 32(7): p. 955-63.

[37] Young, A.R., et al., *Autophagy mediates the mitotic senescence transition.* Genes Dev, 2009. 23(7): p. 798-803.

[38] Kroemer, G. and B. Levine, *Autophagic cell death: the story of a misnomer.* Nat Rev Mol Cell Biol, 2008. 9(12): p. 1004-10.

[39] Schweichel, J.U. and H.J. Merker, *The morphology of various types of cell death in prenatal tissues.* Teratology, 1973. 7(3): p. 253-66.

[40] Kosta, A., et al., *Autophagy gene disruption reveals a non-vacuolar cell death pathway in Dictyostelium.* J Biol Chem, 2004. 279(46): p. 48404-9.

[41] Denton, D., et al., *Autophagy, not apoptosis, is essential for midgut cell death in Drosophila.* Curr Biol, 2009. 19(20): p. 1741-6.

[42] Mizushima, N., et al., *Autophagy fights disease through cellular self-digestion.* Nature, 2008. 451(7182): p. 1069-75.

[43] Boya, P., et al., *Inhibition of macroautophagy triggers apoptosis.* Mol Cell Biol, 2005. 25(3): p. 1025-40.

[44] Gonzalez-Polo, R.A., et al., *The apoptosis/autophagy paradox: autophagic vacuolization before apoptotic death.* J Cell Sci, 2005. 118(Pt 14): p. 3091-102.

[45] Yu, S.W., et al., *Autophagic death of adult hippocampal neural stem cells following insulin withdrawal.* Stem Cells, 2008. 26(10): p. 2602-10.

[46] Zhao, Y., et al., *Cytosolic FoxO1 is essential for the induction of autophagy and tumour suppressor activity.* Nat Cell Biol, 2010. 12(7): p. 665-75.

[47] Gozuacik, D., et al., *DAP-kinase is a mediator of endoplasmic reticulum stress-induced caspase activation and autophagic cell death.* Cell Death Differ, 2008. 15(12): p. 1875-86.

[48] Shen, S., O. Kepp, and G. Kroemer, *The end of autophagic cell death?* Autophagy, 2012. 8(1): p. 1-3.

[49] Hanahan, D. and R.A. Weinberg, *Hallmarks of cancer: the next generation.* Cell, 2011. 144(5): p. 646-74.

[50] Noman, M.Z., et al., *Microenvironmental hypoxia orchestrating the cell stroma cross talk, tumor progression and antitumor response.* Crit Rev Immunol, 2011. 31(5): p. 357-77.

[51] Schlie, K., et al., *When Cells Suffocate: Autophagy in Cancer and Immune Cells under Low Oxygen.* Int J Cell Biol, 2011. 2011: p. 470597.

[52] Walmsley, S.R., et al., *Hypoxia-induced neutrophil survival is mediated by HIF-1alpha-dependent NF-kappaB activity.* J Exp Med, 2005. 201(1): p. 105-15.

[53] Mihalache, C.C., et al., *Inflammation-associated autophagy-related programmed necrotic death of human neutrophils characterized by organelle fusion events.* J Immunol, 2011. 186(11): p. 6532-42.

[54] Semenza, G.L., *Hypoxia-inducible factor 1: master regulator of O2 homeostasis.* Curr Opin Genet Dev, 1998. 8(5): p. 588-94.

[55] Naldini, A., et al., *Hypoxia affects dendritic cell survival: role of the hypoxia-inducible factor-1alpha and lipopolysaccharide.* J Cell Physiol, 2012. 227(2): p. 587-95.

[56] Shi, C.S. and J.H. Kehrl, *MyD88 and Trif target Beclin 1 to trigger autophagy in macrophages.* J Biol Chem, 2008. 283(48): p. 33175-82.

[57] Xu, Y., et al., *Toll-like receptor 4 is a sensor for autophagy associated with innate immunity.* Immunity, 2007. 27(1): p. 135-44.

[58] Sanjuan, M.A., et al., *Toll-like receptor signalling in macrophages links the autophagy pathway to phagocytosis.* Nature, 2007. 450(7173): p. 1253-7.

[59] Lee, H.K., et al., *In vivo requirement for Atg5 in antigen presentation by dendritic cells.* Immunity, 2010. 32(2): p. 227-39.

[60] Chemali, M., et al., *Alternative pathways for MHC class I presentation: a new function for autophagy.* Cell Mol Life Sci, 2011. 68(9): p. 1533-41.

[61] Schmid, D., M. Pypaert, and C. Munz, *Antigen-loading compartments for major histocompatibility complex class II molecules continuously receive input from autophagosomes.* Immunity, 2007. 26(1): p. 79-92.

[62] Bell, B.D., et al., *FADD and caspase-8 control the outcome of autophagic signaling in proliferating T cells.* Proc Natl Acad Sci U S A, 2008. 105(43): p. 16677-82.

[63] Pua, H.H., et al., *A critical role for the autophagy gene Atg5 in T cell survival and proliferation.* J Exp Med, 2007. 204(1): p. 25-31.

[64] Hubbard, V.M., et al., *Macroautophagy regulates energy metabolism during effector T cell activation.* J Immunol, 2010. 185(12): p. 7349-57.

[65] Wildenberg, M.E., et al., *Autophagy attenuates the adaptive immune response by destabilizing the immunologic synapse.* Gastroenterology, 2012. 142(7): p. 1493-503 e6.

[66] Li, C., et al., *Autophagy is induced in CD4+ T cells and important for the growth factor-withdrawal cell death.* J Immunol, 2006. 177(8): p. 5163-8.

[67] Kovacs, J.R., et al., *Autophagy promotes T-cell survival through degradation of proteins of the cell death machinery.* Cell Death Differ, 2012. 19(1): p. 144-52.

[68] Jia, W. and Y.W. He, *Temporal regulation of intracellular organelle homeostasis in T lymphocytes by autophagy.* J Immunol, 2011. 186(9): p. 5313-22.

[69] Pua, H.H., et al., *Autophagy is essential for mitochondrial clearance in mature T lymphocytes.* J Immunol, 2009. 182(7): p. 4046-55.

[70] Stephenson, L.M., et al., *Identification of Atg5-dependent transcriptional changes and increases in mitochondrial mass in Atg5-deficient T lymphocytes.* Autophagy, 2009. 5(5): p. 625-35.

[71] Lum, J.J., et al., *Growth factor regulation of autophagy and cell survival in the absence of apoptosis.* Cell, 2005. 120(2): p. 237-48.

[72] Demachi-Okamura, A., et al., *Autophagy Creates a CTL Epitope That Mimics Tumor-Associated Antigens.* PLoS One, 2012. 7(10): p. e47126.

[73] Eskelinen, E.L., *The dual role of autophagy in cancer.* Curr Opin Pharmacol, 2011. 11(4): p. 294-300.

[74] Notte, A., L. Leclere, and C. Michiels, *Autophagy as a mediator of chemotherapy-induced cell death in cancer.* Biochem Pharmacol, 2011. 82(5): p. 427-34.

[75] Noman, M.Z., et al., *Blocking hypoxia-induced autophagy in tumors restores cytotoxic T-cell activity and promotes regression.* Cancer Res, 2011. 71(18): p. 5976-86.

[76] Noman, M.Z., et al., *Hypoxia-induced autophagy: a new player in cancer immunotherapy?* Autophagy, 2012. 8(4): p. 704-6.

[77] Michaud, M., et al., *Autophagy-dependent anticancer immune responses induced by chemotherapeutic agents in mice.* Science, 2011. 334(6062): p. 1573-7.

[78] Buchser, W.J., et al., *Cell-mediated autophagy promotes cancer cell survival.* Cancer Res, 2012. 72(12): p. 2970-9.

[79] Townsend, K.N., et al., *Autophagy inhibition in cancer therapy: metabolic considerations for antitumor immunity.* Immunol Rev, 2012. 249(1): p. 176-94.

[80] Chambers, A.F., A.C. Groom, and I.C. MacDonald, *Dissemination and growth of cancer cells in metastatic sites.* Nat Rev Cancer, 2002. 2(8): p. 563-72.

[81] Kenific, C.M., A. Thorburn, and J. Debnath, *Autophagy and metastasis: another double-edged sword.* Curr Opin Cell Biol, 2010. 22(2): p. 241-5.

[82] Lin, E.Y., et al., *Colony-stimulating factor 1 promotes progression of mammary tumors to malignancy.* J Exp Med, 2001. 193(6): p. 727-40.

[83] Tlsty, T.D. and L.M. Coussens, *Tumor stroma and regulation of cancer development.* Annu Rev Pathol, 2006. 1: p. 119-50.

[84] Mantovani, A., et al., *Cancer-related inflammation.* Nature, 2008. 454(7203): p. 436-44.

[85] Joyce, J.A. and J.W. Pollard, *Microenvironmental regulation of metastasis.* Nat Rev Cancer, 2009. 9(4): p. 239-52.

[86] Thorburn, J., A.E. Frankel, and A. Thorburn, *Regulation of HMGB1 release by autophagy.* Autophagy, 2009. 5(2): p. 247-9.

[87] Apetoh, L., et al., *The interaction between HMGB1 and TLR4 dictates the outcome of anticancer chemotherapy and radiotherapy.* Immunol Rev, 2007. 220: p. 47-59.

[88] DeNardo, D.G., et al., *CD4(+) T cells regulate pulmonary metastasis of mammary carcinomas by enhancing protumor properties of macrophages.* Cancer Cell, 2009. 16(2): p. 91-102.

[89] Qian, B.Z. and J.W. Pollard, *Macrophage diversity enhances tumor progression and metastasis.* Cell, 2010. 141(1): p. 39-51.

[90] Thiery, J.P., et al., *Epithelial-mesenchymal transitions in development and disease.* Cell, 2009. 139(5): p. 871-90.

[91] Lv, Q., et al., *DEDD interacts with PI3KC3 to activate autophagy and attenuate epithelial-mesenchymal transition in human breast cancer.* Cancer Res, 2012. 72(13): p. 3238-50.

[92] Sun, T., et al., *Promotion of tumor cell metastasis and vasculogenic mimicry by way of transcription coactivation by Bcl-2 and Twist1: a study of hepatocellular carcinoma.* Hepatology, 2011. 54(5): p. 1690-706.

[93] Lu, Z., et al., *The tumor suppressor gene ARHI regulates autophagy and tumor dormancy in human ovarian cancer cells.* J Clin Invest, 2008. 118(12): p. 3917-29.

[94] Kroemer, G., G. Marino, and B. Levine, *Autophagy and the integrated stress response.* Mol Cell, 2010. 40(2): p. 280-93.

[95] Jiang, L. and R.J. Deberardinis, *Cancer metabolism: When more is less.* Nature, 2012. 489(7417): p. 511-2.

[96] Mathew, R. and E. White, *Autophagy, stress, and cancer metabolism: what doesn't kill you makes you stronger.* Cold Spring Harb Symp Quant Biol, 2011. 76: p. 389-96.

[97] Janku, F., et al., *Autophagy as a target for anticancer therapy.* Nat Rev Clin Oncol, 2011. 8(9): p. 528-39.

[98] Vander Heiden, M.G., *Targeting cancer metabolism: a therapeutic window opens.* Nat Rev Drug Discov, 2011. 10(9): p. 671-84.

[99] Nelson, D.A., et al., *Hypoxia and defective apoptosis drive genomic instability and tumorigenesis.* Genes Dev, 2004. 18(17): p. 2095-107.

[100] Mazure, N.M. and J. Pouyssegur, *Atypical BH3-domains of BNIP3 and BNIP3L lead to autophagy in hypoxia.* Autophagy, 2009. 5(6): p. 868-9.

[101] Bellot, G., et al., *Hypoxia-induced autophagy is mediated through hypoxia-inducible factor induction of BNIP3 and BNIP3L via their BH3 domains.* Mol Cell Biol, 2009. 29(10): p. 2570-81.

[102] Moussay, E., et al., *The acquisition of resistance to TNFalpha in breast cancer cells is associated with constitutive activation of autophagy as revealed by a transcriptome analysis using a custom microarray.* Autophagy, 2011. 7(7): p. 760-70.

[103] Papandreou, I., et al., *Hypoxia signals autophagy in tumor cells via AMPK activity, independent of HIF-1, BNIP3, and BNIP3L.* Cell Death Differ, 2008. 15(10): p. 1572-81.

[104] Yu, J., A. Parkhitko, and E.P. Henske, *Autophagy: an 'Achilles' heel of tumorigenesis in TSC and LAM.* Autophagy, 2011. 7(11): p. 1400-1.

[105] Liu, W. and J.M. Phang, *Proline dehydrogenase (oxidase), a mitochondrial tumor suppressor, and autophagy under the hypoxia microenvironment.* Autophagy, 2012. 8(9): p. 1407-9.

[106] Rouschop, K.M., et al., *The unfolded protein response protects human tumor cells during hypoxia through regulation of the autophagy genes MAP1LC3B and ATG5.* J Clin Invest, 2010. 120(1): p. 127-41.

[107] Rzymski, T., et al., *Regulation of autophagy by ATF4 in response to severe hypoxia.* Oncogene, 2010. 29(31): p. 4424-35.

[108] Michallet, A.S., et al., *Compromising the unfolded protein response induces autophagy-mediated cell death in multiple myeloma cells.* PLoS One, 2011. 6(10): p. e25820.

[109] Murrow, L. and J. Debnath, *Autophagy as a Stress-Response and Quality-Control Mechanism: Implications for Cell Injury and Human Disease.* Annu Rev Pathol, 2012.

[110] Li, L., Y. Chen, and S.B. Gibson, *Starvation-induced autophagy is regulated by mitochondrial reactive oxygen species leading to AMPK activation.* Cell Signal, 2012.

[111] Azad, M.B., Y. Chen, and S.B. Gibson, *Regulation of autophagy by reactive oxygen species (ROS): implications for cancer progression and treatment.* Antioxid Redox Signal, 2009. 11(4): p. 777-90.

[112] Zoncu, R., A. Efeyan, and D.M. Sabatini, *mTOR: from growth signal integration to cancer, diabetes and ageing.* Nat Rev Mol Cell Biol, 2011. 12(1): p. 21-35.

[113] Korkmaz, G., et al., *miR-376b controls starvation and mTOR inhibition-related autophagy by targeting ATG4C and BECN1.* Autophagy, 2012. 8(2): p. 165-76.

[114] Glinsky, G.V. and V.V. Glinsky, *Apoptosis amd metastasis: a superior resistance of metastatic cancer cells to programmed cell death.* Cancer Lett, 1996. 101(1): p. 43-51.

[115] Han, J., et al., *Involvement of protective autophagy in TRAIL resistance of apoptosis-defective tumor cells.* J Biol Chem, 2008. 283(28): p. 19665-77.

[116] He, W., et al., *Attenuation of TNFSF10/TRAIL-induced apoptosis by an autophagic survival pathway involving TRAF2- and RIPK1/RIP1-mediated MAPK8/JNK activation.* Autophagy, 2012. 8(12).

[117] Frisch, S.M. and H. Francis, *Disruption of epithelial cell-matrix interactions induces apoptosis.* J Cell Biol, 1994. 124(4): p. 619-26.

[118] Fung, C., et al., *Induction of autophagy during extracellular matrix detachment promotes cell survival.* Mol Biol Cell, 2008. 19(3): p. 797-806.

[119] Avivar-Valderas, A., et al., *PERK integrates autophagy and oxidative stress responses to promote survival during extracellular matrix detachment.* Mol Cell Biol, 2011. 31(17): p. 3616-29.

[120] Chen, N., et al., *Autophagy restricts proliferation driven by oncogenic phosphatidylinositol 3-kinase in three-dimensional culture.* Oncogene, 2012.

[121] White, D.E., et al., *Targeted disruption of beta1-integrin in a transgenic mouse model of human breast cancer reveals an essential role in mammary tumor induction.* Cancer Cell, 2004. 6(2): p. 159-70.

[122] Li, J., et al., *Inhibition of autophagy augments 5-fluorouracil chemotherapy in human colon cancer in vitro and in vivo model.* Eur J Cancer, 2010. 46(10): p. 1900-9.

[123] Liu, D., et al., *Inhibition of autophagy by 3-MA potentiates cisplatin-induced apoptosis in esophageal squamous cell carcinoma cells.* Med Oncol, 2011. 28(1): p. 105-11.

[124] Apel, A., et al., *Blocked autophagy sensitizes resistant carcinoma cells to radiation therapy.* Cancer Res, 2008. 68(5): p. 1485-94.

[125] Pajonk, F., E. Vlashi, and W.H. McBride, *Radiation resistance of cancer stem cells: the 4 R's of radiobiology revisited.* Stem Cells, 2010. 28(4): p. 639-48.

[126] Lomonaco, S.L., et al., *The induction of autophagy by gamma-radiation contributes to the radioresistance of glioma stem cells.* Int J Cancer, 2009. 125(3): p. 717-22.

[127] Ito, H., et al., *Radiation-induced autophagy is associated with LC3 and its inhibition sensitizes malignant glioma cells.* Int J Oncol, 2005. 26(5): p. 1401-10.

[128] Shacka, J.J., B.J. Klocke, and K.A. Roth, *Autophagy, bafilomycin and cell death: the "a-B-cs" of plecomacrolide-induced neuroprotection.* Autophagy, 2006. 2(3): p. 228-30.

[129] Carew, J.S., et al., *Lucanthone is a novel inhibitor of autophagy that induces cathepsin D-mediated apoptosis.* J Biol Chem, 2011. 286(8): p. 6602-13.

[130] Rossi, M., et al., *Desmethylclomipramine induces the accumulation of autophagy markers by blocking autophagic flux.* J Cell Sci, 2009. 122(Pt 18): p. 3330-9.

[131] Ruiz-Irastorza, G., et al., *Clinical efficacy and side effects of antimalarials in systemic lupus erythematosus: a systematic review.* Ann Rheum Dis, 2010. 69(1): p. 20-8.

[132] Gupta, A., et al., *Autophagy inhibition and antimalarials promote cell death in gastrointestinal stromal tumor (GIST).* Proc Natl Acad Sci U S A, 2010. 107(32): p. 14333-8.

[133] Tett, S.E., R.O. Day, and D.J. Cutler, *Concentration-effect relationship of hydroxychloroquine in rheumatoid arthritis--a cross sectional study.* J Rheumatol, 1993. 20(11): p. 1874-9.

[134] Munster, T., et al., *Hydroxychloroquine concentration-response relationships in patients with rheumatoid arthritis.* Arthritis Rheum, 2002. 46(6): p. 1460-9.

[135] Kimmelman, A.C., *The dynamic nature of autophagy in cancer.* Genes Dev, 2011. 25(19): p. 1999-2010.

[136] Stein, M., et al., *Targeting tumor metabolism with 2-deoxyglucose in patients with castrate-resistant prostate cancer and advanced malignancies.* Prostate, 2010. 70(13): p. 1388-94.

[137] Yang, Z.J., et al., *The role of autophagy in cancer: therapeutic implications.* Mol Cancer Ther, 2011. 10(9): p. 1533-41.

[138] Wong, K.K., J.A. Engelman, and L.C. Cantley, *Targeting the PI3K signaling pathway in cancer.* Curr Opin Genet Dev, 2010. 20(1): p. 87-90.

[139] Backer, J.M., *The regulation and function of Class III PI3Ks: novel roles for Vps34.* Biochem J, 2008. 410(1): p. 1-17.

[140] Hara, T., et al., *FIP200, a ULK-interacting protein, is required for autophagosome formation in mammalian cells.* J Cell Biol, 2008. 181(3): p. 497-510.

[141] Egan, D.F., et al., *Phosphorylation of ULK1 (hATG1) by AMP-activated protein kinase connects energy sensing to mitophagy.* Science, 2011. 331(6016): p. 456-61.

[142] Kim, J., et al., *AMPK and mTOR regulate autophagy through direct phosphorylation of Ulk1.* Nat Cell Biol, 2011. 13(2): p. 132-41.

Chapter 10

Role of Autophagy in Cancer

Michiko Shintani and Kayo Osawa

Additional information is available at the end of the chapter

http://dx.doi.org/10.5772/55315

1. Introduction

Autophagy is a cellular stress-adaptive process in which double-membrane structures called autophagosomes engage in protein degradation, cellular differentiation, apoptosis and antigen processing, and are recycled to sustain cellular metabolism [1-11]. It is a self-digesting mechanism responsible for removal of long-lived proteins and damaged organelles by lysosomes, and opposing roles in cell death and survival have been described for autophagy.

Autophagy is a multifaceted process, and alterations in autophagic signaling pathways are frequently observed in cancer. Cancer is a disease generated by mutation, selection and genome instability in the resulting tumor tissue, and is considered to be the second leading cause of death in western countries after heart disease [12, 13]. Autophagy can be activated by various stimuli including hypoxia during the tumor formation [14]. One hypothetical mechanism is that autophagy promotes tumor cell survival in response to diverse stresses [15]. Furthermore, autophagy spatially and temporally regulates tumor development by suppressing tumor growth through regulating cell proliferation in the early stages of tumorigenesis [16]. Conversely, when autophagy is reduced, it contributes to tumor formation and growth by the breakdown of tumor cells following autophagy-related cell death, leading to tumor cell survival [17]. There is a controversy about the roles of autophagy in cancer [1, 3, 18]. In this review, we outline the multiple roles of autophagy in cancer, including gene expression, gene mutation, and chemotherapy.

2. Autophagy-related genes in cancer

2.1. ATG genes

Most recently, molecular genetic analyses have focused on the function of autophagy-related gene (*ATG*) products. *ATG* products are implicated in autophagosome formation and associated pathways. In humans, there are more than 30 known *ATG* genes, some of which have mononucleotide repeats with seven or more nucleotides. Of the many genes associated with autophagy, *ATG* genes are the main regulators and implementers of the autophagy process [19].

Beclin-1 (encoded by *BECN1* gene, a mammalian orthologue of yeast Atg6) protein, a component of PI3-kinase complexes, is a key regulator in the vesicle nucleation process of autophagic programmed cell death [20-22].The role of autophagy in tumor suppression is known to be as a result of allelic loss of the essential autophagy genes. Beclin-1 and Beclin-1$^{+/-}$ mice were shown to be tumor prone, indicating that *BECN1* is a haploinsufficient tumor suppressor gene [20, 21], and allelic deletion and point mutations of *BECN1* gene and loss of Beclin-1 expression is found with high frequency in human breast, ovarian and prostate cancers [22, 23]. Lee et al. detected 11 somatic mutations of the *BECN1* gene, including three missense mutations (N8K, P350R and R389C) in coding sequences and eight mutations in introns [24]. These mutations were observed in five gastric, three colorectal, one lung and one breast carcinoma. However, the expression of Beclin-1 is known to be upregulated in colon and gastric cancers [25]. It also reported that *Atg4C*-deficient mice are prone to tumors [26].

Frameshift mutations of genes with mononucleotide repeats are features of cancers with microsatellite instability (MSI). Mononucleotide repeat frameshift mutations in *ATG* genes are common in gastric and colorectal carcinomas with high MSI, and possibly contribute to cancer development by deregulating the autophagy process. Kang et al. detected truncation mutations of three genes (*ATG2B*; c.3120delA, *ATG5*; c.704delA and *ATG9B*; c.293delC) in high MSI cancers (gastric and colorectal) by single-strand conformation polymorphism analysis [27]. In particular, ATG5 is a protein involved in the early stage of autophagosome formation [18, 28]. ATG5 high expression was altered in prostate cancers and other data showed a low incidence of *ATG5* mutations in gastric hepatocellular, and colorectal cancers with MSI [29, 30]. It is important to identify the expression and mutation status of a gene in cancers to understand its role in cancer development. These frameshift mutations or SNPs in *ATG* genes may alter the autophagic cell death in cancers and might contribute to the pathogenesis of human cancers.

2.2. UVRAG

As an *ATG*-related gene, the ultraviolet (UV) radiation resistance-associated gene (*UVRAG*) was initially identified as a gene that is responsible for the partial complementation of UV sensitivity in xeroderma pigmentosum cells, and binds with Beclin-1/PI3-kinase and Bif-1, a Bax activator to induce autophagy formation and suppress the tumorigenic activity of cancer cells [31, 32]. It has been reported that *UVRAG* exon 8 frameshift mutations containing c.709delA or c.708_709delAA mutations were found in gastric and colorectal cancers with MSI [33, 34].

2.3. IRGM

In the autophagy pathway, the immunity-related guanosine triphosphatase (GTPase) family, M (IRGM), plays a central function and appears to have an important role in the activation of the pathway. *IRGM* is located on chromosome 5q33.1, and its mRNA transcripts can be found in five different 3′-splicing isoforms [35, 36]. Recent evidence indicates that variants of the *IRGM* locus, especially those in the promoter region, may be correlated with differential expression, and consequently the efficacy of autophagy is affected by alterations in IRGM regulation [36-38]. *IRGM* has two major SNPs (rs13361189 and rs4958847) associated with chronic inflammatory digestive diseases. It is not known exactly why *IRGM* rs4958847 but not rs13361189 polymorphism has reported to influence susceptibility to gastric cancer [39].

2.4. RASSF1

The RAS association domain family 1A (RASSF1A) is one of the most epigenetically silenced elements in human cancers. The tumor suppressor gene, *RASSF1A*, has been reported to play a role in diverse activities including cell cycle regulation, apoptosis and modulation of autophagy or genomic instability [40]. It is also associated with epigenetic silencing of other proteins including that of death-associated protein kinase (DAPK) [41-44]. DAPK is a unique calcium/calmodulin-activated serine/threonine kinase involved in autophagy-related signaling pathways [45-48]. RASSF1A can also promote cell death utilizing the association with the anaphase promoting complex protein cdc20 and the autophagic protein, C19ORF5/MAP1S [49]. Expression of the longer isoform of RASSF1A (39 kDa predicted peptide) is lost or downregulated in many lung tumor lines [50, 51]. Agatheanggelou et al. also reported that *RASSF1A* inactivation by methylation and loss is a critical step in lung cancer [52]. Epidemiological studies have identified an association between the *RASSF1A* A133S polymorphism and cancer risk including breast cancer, lung cancer, and hepatocellular carcinoma [53-57]. Moreover, several studies have shown that expression loss by promoter-specific hypermethylation of *RASSF1A* is one of the most common early events in hepatocellular carcinoma that play important roles in tumorigenesis and metastasis of hepatocellular carcinoma [58, 59]. A133S and S131F polymorphisms resulted in the lost ability of RASSF1A to inhibit growth and cyclin D1 expression, suggesting an important role in tumor suppression [60, 61]. Moreover, Gordon et al. reported that E246K, C65R, R257Q *RASSF1A* polymorphisms were related to tumor suppressor function [62]. Additional evidence suggests that *RASSF1C* may be a tumor suppressor gene in prostate and renal carcinoma cells but not in lung cancer cells [63]. It has reported that the loss of RASSF1C results in the downregulation of proliferation of lung and breast cancer cells, suggesting a prosurvival role for RASSF1C [64-66]. Recently, it has been suggested that a possible pathogenic role for RASSF1C in cancer may exist, as its expression was more than 11-fold greater in pancreatic endocrine tumors than in normal tissue [67].

2.5. NOD2

The nucleotide-binding oligomerization domain-containing protein 2 (NOD2) is a member of the Nod-like receptor family and associates with the cell surface membrane. NOD2 activation controls the induction of autophagy, or apoptosis [68-70]. Four major *NOD2* single nucleotide polymorphisms are correlated with increased risk of colorectal cancer, and a possible association of the *NOD2* P268S polymorphism with rectal and gastric cancers has been identified [71-78]. A recent meta-analysis also provided good evidence that *NOD2* R702W, G908R, and most significantly, 3020insC, polymorphisms were associated with increased risk of colorectal cancer [79]. Other studies also found significant associations with laryngeal, lung, and ovarian cancers [80, 81]. In contrast, Suchy et al. found the association of the TNFα-1,031 T/T genotype and *NOD2* 3020insC polymorphism may act as a modifier to reduce colorectal cancer risk [82]. Further research of *NOD2* polymorphisms and gene–gene interactions will provide a more comprehensive insight into the associations described here.

3. Analysis of autophagy by immunohistochemistry

Recently, the role of autophagy in cancer development and progression has been investigated using immunohistochemistry. Immunohistochemical methods have been developed that supplement the detection of autophagy via genetic analyses. Many antibodies for autophagy detection are routinely used for immunohistochemistry against proteins involved in autophagy pathways [83-86] (Table 1).

Antibody		***ref. No***
	LC3 (rabbit polyclonal antibody)	[86]
Source; (1: x, dilution rate)	Medical & Biological Laboratories, Japan	
Antigen retrieval method	Pressure cooker (110C-120C) for 10 min; 10 mM citrate buffer, pH 6.0	
Sample type	Formalin-fixed, paraffin-embedded specimens	
Staining pattern	Invariably granular cytoplasmic staining	
	LC3	[100]
Source; (1: x, dilution rate)	Novus Biologicals, USA; (1:400)	
Antigen retrieval method	High temperature and pressure, citrate buffer	
Sample type	Formalin-fixed, paraffin-embedded specimens	
Staining pattern	Cytoplasmic staining	
	Beclin-1 (rabbit monoclonal antibody)	[95]
Source; (1: x, dilution rate)	Abcam, UK; (1:100)	
Antigen retrieval method	Microwave oven for 15 min, 10 mM citrate buffer, pH 6	
Sample type	Formalin-fixed, paraffin-embedded specimens	
Staining pattern	Cytoplasmic staining	
	Beclin-1 (rabbit polyclonal antibody)	[97]
Source; (1: x, dilution rate)	Abcam, UK; (1:100)	
Antigen retrieval method	Microwave oven, 10 mM citrate buffer, pH 6	

Antibody		ref. No
Sample type	Formalin-fixed, paraffin-embedded specimens	
Staining pattern	Membrane-plasma, cytoplasm and nucleus in the cancer cells and no or modest staining in the adjacent noncancerous tissue	
Beclin-1 (rabbit polyclonal antibody)		[25]
Source; (1: x, dilution rate)	Novus Biologicals, USA	
Antigen retrieval method	Pressure cooker inside a microwave oven at 700 W for 30 min, 10 mM citrate buffer, pH 6.0	
Sample type	Microarray recipient block was constructed containing paraffin-embedded colorectal adenocarcinoma tissue samples from 103 archival patient specimens	
Staining pattern	Cytoplasmic staining	
Beclin-1		[100]
Source; (1: x, dilution rate)	Cell Signaling, USA; (1:100)	
Antigen retrieval method	High temperature and pressure, citrate buffer	
Sample type	Formalin-fixed, paraffin-embedded specimens	
Staining pattern	Cytoplasmic staining	
BIF-1 (mouse monoclonal antibody)		[98]
Source; (1: x, dilution rate)	Imgenex, USA; (1:2500)	
Antigen retrieval method	standard cell conditioning (Ventana Medical Systems, USA)	
Sample type	Formalin-fixed, paraffin-embedded core sections on a tissue array	
Staining pattern	Cytoplasmic staining	
ATG5 (rabbit polyclonal antibody)		[30]
Source; (1: x, dilution rate)	Abcam, UK; (1:800)	
Antigen retrieval method	Pressure cooker inside a microwave oven at 700 W for 30 min, 10 mM citrate buffer, pH 6.0	
Sample type	Formalin-fixed, paraffin-embedded specimens	
Staining pattern	Cytoplasmic and/or nuclear	

Table 1. Immunohistochemical analysis of autophagy-related proteins.

3.1. Proteins involved in autophagy

3.1.1. LC3

Microtubule-associated protein 1 light chain 3 (LC3) is an autophagosomal orthologue of yeast ATG8, with approximately 30% amino acid homology [87, 88]. LC3 is a specific marker of autophagosome formation. LC3-I is localized to the cytoplasm, whereas LC3-II binds to autophagosomes [89].

3.1.2. Beclin-1 (ATG6)

Beclin-1 is a mammalian homolog of the yeast ATG6 protein. The expression of Beclin-1 protein has been reported in tumor tissues such as breast, ovarian, prostate, lung, brain, stomach and colorectum [25, 90]. Beclin-1 was found to be deregulated in human cancers and may play a role in the tumorigenesis and/ or progression of human cancers [21, 91]. It is required for autophagic induction and is a haploinsufficient tumor suppressor.

3.1.3. ATG5

ATG5 is a key regulator of autophagic and apoptotic cell death, and is involved in the early stages of autophagosome formation [18, 28]; binding of ATG5 with ATG12 contributes to autophagosome formation, which sequesters cytoplasmic materials before lysosomal delivery [18]. It is suggested that ATG5 is involved in both apoptotic and autophagic cell death [92].

3.1.4. Bax-interacting factor -1

Bax-interacting factor-1 (Bif-1) protein is a member of the endophilin B family, which plays a critical role in cell death, including autophagy and apoptosis. Loss of Bif-1 suppresses programmed cell death and promotes tumorigenesis [93, 94].

3.1.5. GABARAP

Gamma-aminobutyric acid type A receptor-associated protein (GABARAP) is one of the mammalian homologue of yeast ATG8. It is involved in autophagosome formation during autophagy and was first identified in the brain, but is widely expressed in a variety of normal tissues. Recent reports have suggested that GABARAP is an essential component of autophagic vacuoles in addition to its role as an intracellular trafficking molecule [87,88].

3.2. Expression of autophagy-related proteins in gastrointestinal cancers

Resent reports have demonstrated the expression of autophagy-related proteins in gastrointestinal carcinomas. Chen et al. examined the expression levels of Beclin1 in gastric carcinomas and adjacent normal gastric mucosal tissues by immunohistochemistry. According to their results, high levels of Beclin-1 expression were observed in 90/155 (58.1%) of gastric carcinomas, in 24/60 (40.0%) of adjacent mucosal tissues and in 13/30 (43.3%) of normal gastric mucosa tissues (P=0.036). Decreased expression of Beclin-1 in cancer cells was significantly correlated with poor differentiation, nodal and distant metastasis, advanced TNM stage, and tumor relapse. More importantly, decreased expression of Beclin-1 was associated with shorter survival as evidenced by univariate and multivariate analysis. Chen et al. concluded that decreased expression of Beclin-1 in gastric carcinoma may be important in the acquisition of a metastatic phenotype, suggesting that decreased Beclin-1 expression, as examined by immunohistochemistry, is an independent biomarker for poor prognosis of patients with gastric carcinoma [95].

In contrast, using a tissue microarray approach, Ahn et al. investigated Beclin-1 protein expression in 103 colorectal and 60 gastric carcinoma tissues by immunohistochemistry. The expression of Beclin-1 was detected in 50/60 (83%) of gastric carcinomas and 98/103 (95%) of colorectal carcinomas. Conversely, the normal mucosal cells of both the stomach and colon showed no or very weak expression of Beclin-1. There was no significant association of Beclin-1 expression with clinicopathological characteristics, including invasion, metastasis and stage. Their data indicate that Beclin-1 inactivation by loss of expression may not occur in colorectal and gastric cancers. Rather, increased expression of Beclin-1 in the malignant colorectal and gastric epithelial cells compared with their normal mucosal epithelial cells suggests that neo-expression of Beclin-1 may play a role in both colorectal and gastric tumorigenesis [25].

An et al. analyzed ATG5 protein expression by immunohistochemistry and *ATG5* somatic mutations by single-strand conformation polymorphism in cancer cells and the normal mucosal cells of gastrointestinal tissues. Their results showed that ATG5 protein was well expressed in normal stomach, colon, and liver epithelial cells, while it was lost in 21/100 (21%) of gastric carcinomas, 22/95 (23%) of colorectal carcinomas, and 5/50 (10%) of hepatocellular carcinomas. Furthermore, such loss of ATG5 expression was observed in the cancers irrespective of the histological subtypes and TNM stages. Also, they found that only 1.5% (2/135) of these cancers harbored *ATG5* mutations. They suggested that loss of ATG5 expression may play a role in the pathogenesis of some gastric and colorectal cancers [30].

Colorectal carcinoma is one of the most common cancers in the world and the incidence rate is rising. Miao et al. performed experiments to investigate a possible correlation between GABARAP expression in colorectal carcinoma and clinicopathological parameters, including patient survival times. Their results showed that the expression of GABARAP protein was significantly higher in colorectal cancers (51.5%) than the adjacent matched non-tumor tissues (33.0%), and overexpression of GABARAP was significantly correlated with a low grade of differentiation and shortened overall survival. They described GABARAP protein expression as a new prognosis marker in colorectal carcinoma [96].

Li et al. analyzed the expression of Beclin-1 protein in stage IIIB colon carcinoma by immunohistochemistry and correlated it with survival. Their results showed Beclin-1 immunostaining was distributed in the plasma membrane, cytoplasm and nuclei of tumor cells in 98/115 cases (85.2%). Modest or no Beclin-1 expression was observed in adjacent non-cancerous tissues. Higher levels of Beclin-1 expression were strongly associated with longer survival. Both univariate analysis and multivariate analysis showed that Beclin-1 expression levels and invasive depth of primary mass (T stage) were independent prognostic factors. They suggested that Beclin-1 is a favorable prognostic biomarker in locally advanced colon carcinomas [97].

Bif-1 protein plays a critical role in cell death, including autophagy and apoptosis. Coppola et al. examined Bif-1 expression level in colorectal carcinoma using semiquantitative immunohistochemistry and microarray analysis of archival specimens. Bif-1 expression was negative in 23/102 (22.5%) of colorectal carcinomas. Moderate to strong Bif-1 staining was identified in 37/102 (36.3%) of the tumors, and weak staining was noted in 42/102 (41.2%). Moderate to strong Bif-1 immunoreactivity was shown in 26/38 (68.4%) normal colorectal mucosa, and none were negative. In 12/38 (31.6%) cases, the normal colorectal mucosa demonstrated weak Bif-1 stain. The mean staining scores (intensity and percentage of positively stained cells) for colorectal carcinomas and normal colorectal mucosa differed significantly (P=0.0003). The percentage of cases with negative expression also differed significantly between normal colorectal mucosa and colorectal carcinoma (P=0.002). Decreased Bif-1 expression in colorectal carcinomas was confirmed at the mRNA level by microarray analysis. They concluded Bif-1 was downregulated during the transition from normal colorectal mucosa to colorectal adenocarcinoma, a novel finding in agreement with the tumor suppressor function of Bif-1 [98].

LC3 is a one of the most useful markers of autophagy. Yoshioka et al. evaluated LC3 expression in gastrointestinal cancers by immunohistochemistry to elucidate the role of autophagy in human cancer development. LC3 expression was compared with Ki-67 staining and expression of carbonic anhydrase IX, a hypoxic marker. LC3 was expressed in the cytoplasm of cancer cells, but not in non-cancerous epithelial cells. Furthermore, high expression of LC3 was observed in 56/106 (53%) of esophageal, 22/38 (58%) of gastric and 12/19 (63%) of colorectal cancers. The immunoreactive score (intensity and percentage of positively stained cells) of LC3 gradually increased during the early stages of esophageal carcinogenesis in low- and high-grade intraepithelial neoplasia and T1 carcinoma, but did not change in later cancer progression (T2–T4 carcinomas). In early esophageal carcinogenesis, LC3 expression correlated with the Ki-67 labeling index (P=0.0001), but showed no significant association with carbonic anhydrase IX expression. In esophageal cancers, LC3 expression did not correlate with various clinicopathological factors, including survival. LC3 is also upregulated in various gastrointestinal cancers and is partly associated with Ki-67 index. Their results suggest that LC3 expression is advantageous to cancer development, especially in early-phase carcinogenesis. Taken together, these findings suggest that LC3 expression is advantageous to cancer development in early phase of carcinogenesis [99].

Ahn et al. reported that Beclin-1 expression was detected in 95% of colorectal carcinomas examined. In contrast, normal mucosal cells of colon showed no or very weak expression of Beclin-1. There was no significant association of Beclin-1 expression with clinicopathological characteristics, including invasion, metastasis and stage [25].

Guo et al. performed experiments to investigate the utility of Beclin-1 and LC3, in predicting the efficiency of cetuximab in the treatment of advanced colorectal cancer. Their results showed that Beclin-1 and LC3 expression was significantly correlated (r=0.44, P<0.01), and patients with low Beclin-1 expression had longer progression-free survival than those with high Beclin-1 expression [100].

4. Autophagy in cancer chemotherapy

One of the standard modalities for treatment of patients with cancer is chemotherapy. Cytotoxic drug treatment often triggers autophagy, particularly in apoptosis-defective cells, and this excessive cellular damage combined with attempts to remediate that damage through progressive autophagy can promote autophagic cell death [101]. Platinum-containing cisplatin is one of the most extensively used chemotherapeutic agents, and remains the first-line treatment in various types of cancer [102]. Cisplatin-based chemotherapy frequently resulted in acquired resistance of cancer cells. Sirichanchuen et al. indicated that the levels of LC3-related autophagy were significantly lower in cisplatin resistant cells, and autophagosome formation was dramatically reduced in the resistant cells [103]. Patients with low LC3 expression had a higher objective response rate amongst advanced colorectal cancer patients treated with cetuximab-containing chemotherapy [100]. Expression of *ATG5* sensitizes tumor cells to chemotherapy, but its silencing results in resistance to cisplatin therapy combined with AKT inhibitor treatment, thus revealing a key role for autophagy in chemoresistance [92]. Autophagic cell death is activated in cancer cells that are derived from different tissues in response to anticancer therapies [101, 104]. Combination therapy with erlotinib and cisplatin is an effective treatment against erlotinib-resistant cancer by targeting (downregulating) ATG3-mediated autophagy and induction of apoptotic cell death. Autophagy may delay apoptotic cell death caused by DNA-damaging agents and hormonal therapies such as tamoxifen. On the contrary, autophagy has a role as a cell survival pathway. Therefore, autophagy is also induced as a protective and survival mechanism. A major regulator of autophagy is the mammalian target of rapamycin (mTOR) pathway, which consists of two distinct signaling complexes known as mTORC1 and mTORC2 [105]. Thus, results all suggest the role of autophagy in attenuation of chemotherapy-induced cell death or survival.

5. Conclusion

Autophagy is involved in metabolism, cell-death, stress response and carcinogenesis. Several key autophagic mediators containing ATG-related proteins, LC3, Bif-1, GABARAP, UVRAG, IRGM, RASSF1, or NOD2, play pivotal roles in autophagic signaling networks in cancer. By these tumor-suppressive mechanisms in early-stage carcinogenesis, autophagy promotes genomic stability in carcinomas, and possibly contributes to cancer development.

Furthermore, immunohistochemical methods have been developed that supplement the detection of autophagy via genetic analyses. These are especially important since diagnosis of autophagic vacuoles using the classical method of electron microscopy is time-consuming, labor-intensive and costly. Many antibodies for autophagy detection are routinely used for immunohistochemistry. These autophagosomes then fuse with lysosomes to generate autolysosomes. Therefore, LC3 is an efficient and reliable marker for the detection of autophagosome formation.

Autophagy or 'self-eating' is frequently activated in tumor cells treated with chemotherapy. In cancer therapy, adaptive autophagy in cancer cells sustains tumor growth and survival in the face of the toxicity of cancer therapy. However, in certain circumstances, autophagy mediates the therapeutic effects of some anticancer agents. During tumor development and in cancer therapy, autophagy has been reported to have paradoxical roles in promoting both cell survival and cell death.

Autophagy may play a variety of physiological roles in cancer progression at each stage in various cancers. Further investigations are required to clarify the biological role of autophagy-related proteins so as to estimate their potential value in the diagnosis and treatment of cancer.

Author details

Michiko Shintani[1*] and Kayo Osawa[2]

1 Department of Biophysics, Kobe University Graduate School of Health Sciences, Kobe, Japan

2 Department of International health, Kobe University Graduate School of Health Sciences, Kobe, Japan

References

[1] Edinger AL, Thompson CB. Death by design: apoptosis, necrosis and autophagy. Curr Opin Cell Biol 2004;16(6) 663–669.

[2] Levine B, Klionsky, DJ. Development by self-digestion: molecular mechanisms and biological functions of autophagy. Dev. Cell 2004; 6(4) 463–477.

[3] Baehrecke EH. Autophagy: dual roles in life and death? Nat Rev Mol Cell Biol 2005;6(6) 505–510.

[4] Deretic V. Autophagy in innate and adaptive immunity. Trends Immunol 2005; 26(10) 523–528.

[5] Degenhardt K, Mathew R, Beaudoin B, Bray K, Anderson D, Chen G, Mukherjee C, Shi Y, Gélinas C, Fan Y, Nelson DA, Jin S, White E. Autophagy promotes tumor cell survival and restricts necrosis, inflammation, and tumorigenesis. Cancer Cell 2006; 10(1) 51–64.

[6] Berry DL, Baehrecke EH. Growth arrest and autophagy are required for salivary gland cell degradation in Drosophila. Cell 2007; 131(6) 1137–1148.

[7] Levine B, Deretic V. Unveiling the roles of autophagy in innate and adaptive immunity. Nat Rev Immunol 2007; 7(10) 767–777.

[8] Maiuri MC, Zalckvar E, Kimchi A, Kroemer G. Self-eating and self-killing: crosstalk between autophagy and apoptosis. Nat Rev Mol Cell Biol 2007; 8(9) 741–752.

[9] Mizushima N. Autophagy: process and function. Genes Dev. 2007;21(22) 2861–2873.

[10] Schmid D, Munz C. Innate and adaptive immunity through autophagy. Immunity 2007; 27(1) 11–21.

[11] Maiuri MC, Zalckvar E, Kimchi A, Kroemer G. Self-eating and self-killing: crosstalk between autophagy and apoptosis. Nat Rev Mol Cell Biol. 2007; 8(9) 741–752.

[12] Minino AM. Tech. Rep. 64. Hyattsville, Md, USA: NCHS Data Brief; 2009. Death in the United States.

[13] Canada S. Leading Causes of Death in Canada. Government of Canada Publications; 2011.

[14] Liang C, Lee JS, Inn KS, Gack MU, Li Q, Roberts EA, Vergne I, Deretic V, Feng P, Akazawa C, Jung JU. Beclin1-binding UVRAG targets the class C Vps complex to coordinate autophagosome maturation and endocytic trafficking. Nat Cell Biol. 2008; 10(7) 776–787.

[15] Mizushima N, Levine B, Cuervo AM, Klionsky DJ. Autophagy fights disease through cellular self-digestion. Nature 2008; 451(7182) 1069–1075.

[16] Levine B. Cell biology: autophagy and cancer. Nature. 2007;446(7137) 745-747.

[17] Tsuchihara K, Fujii S, Esumi H. Autophagy and cancer: dynamism of the metabolism of tumor cells and tissues. Cancer Lett 2009; 278(2) 130–138.

[18] Klionsky DJ. Autophagy: from phenomenology to molecular understanding in less than a decade. Nat Rev Mol Cell Biol 2007; 8(11) 931–937.

[19] Mizushima N, Yoshimori T, Ohsumi Y. The role of Atg proteins in autophagosome formation. Annu Rev Cell Dev Biol. 2011; 27:107-32.

[20] Yue Z, Jin S, Yang C, Levine AJ, Heintz N. Beclin 1, an autophagy gene essential for early embryonic development, is a haploinsufficient tumor suppressor. Proc Natl Acad Sci U S A 2003;100(25) 15077–15082.

[21] Qu X, Yu J, Bhagat G, Furuya N, Hibshoosh H, Troxel A, Rosen J, Eskelinen EL, Mizushima N, Ohsumi Y, Cattoretti G, Levine B. Promotion of tumorigenesis by heterozygous disruption of the beclin 1 autophagy gene. J Clin Invest 2003;112(12) 1809–1820.

[22] Liang XH, Jackson S, Seaman M, Brown K, Kempkes B, Hibshoosh H, Levine B. Induction of autophagy and inhibition of tumorigenesis by beclin 1. Nature 1999; 402(6762) 672–676.

[23] Aita VM, Liang XH, Murty VV, Pincus DL, Yu W, Cayanis E, Kalachikov S, Gilliam TC, Levine B. Cloning and genomic organization of beclin 1, a candidate tumor suppressor gene on chromosome 17q21. Genomics 1999; 59(1) 59–65.

[24] Lee JW, Jeong EG, Lee SH, Yoo NJ, Lee SH. Somatic mutations of BECN1, an autophagy-related gene, in human cancers. APMIS 2007; 115(6) 750–756.

[25] Ahn CH, Jeong EG, Lee JW, Kim MS, Kim SH, Kim SS, Yoo NJ, Lee SH. Expression of beclin-1, an autophagy-related protein, in gastric and colorectal cancers. APMIS 2007; 115(12) 1344–1349.

[26] Mariño G, Salvador-Montoliu N, Fueyo A, Knecht E, Mizushima N, López-Otín C. Tissue-specific autophagy alterations and increased tumorigenesis in mice deficient in Atg4C/autophagin-3. J Biol Chem. 2007; 282(25) 18573–18583.

[27] Kang MR, Kim MS, Oh JE, Kim YR, Song SY, Kim SS, Ahn CH, Yoo NJ, Lee SH. Frameshift mutations of autophagy-related genes ATG2B, ATG5, ATG9B and ATG12 in gastric and colorectal cancers with microsatellite instability. J Pathol. 2009; 217(5) 702–706.

[28] Hammond EM, Brunet CL, Johnson GD, Parkhill J, Milner AE, Brady G, Gregory CD, Grand RJ. Homology between a human apoptosis specific protein and the product of APG5, a gene involved in autophagy in yeast. FEBS Lett. 1998; 425 (3) 391–395.

[29] Kim MS, Song SY, Lee JY, Yoo NJ, Lee SH. Expressional and mutational analyses of ATG5 gene in prostate cancers. APMIS. 2011;119(11) 802–807.

[30] An CH, Kim MS, Yoo NJ, Park SW, Lee SH. Mutational and expressional analyses of ATG5, an autophagy-related gene, in gastrointestinal cancers. Pathol Res Pract. 2011; 207(7): 433–437.

[31] Liang C, Feng P, Ku B, Dotan I, Canaani D, Oh BH, Jung JU. Autophagic and tumour suppressor activity of a novel Beclin1-binding protein UVRAG. Nat Cell Biol. 2006; 8(7) 688–699.

[32] Takahashi Y, Coppola D, Matsushita N, Cualing HD, Sun M, Sato Y, Liang C, Jung JU, Cheng JQ, Mulé JJ, Pledger WJ, Wang HG. Bif-1 interacts with Beclin 1 through UVRAG and regulates autophagy and tumorigenesis. Nat Cell Biol. 2007; 9(10) 1142–1151.

[33] Ionov Y, Nowak N, Perucho M, Markowitz S, Cowell JK. Manipulation of nonsense mediated decay identifies gene mutations in colon cancer cells with microsatellite instability. Oncogene 2004; 23: 639–645.

[34] Kim MS, Jeong EG, Ahn CH, Kim SS, Lee SH, Yoo NJ. Frameshift mutation of UVRAG, an autophagy-related gene, in gastric carcinomas with microsatellite instability. Hum Pathol 2008; 39(7) 1059–1063.

[35] Bekpen C, Xavier RJ, Eichler EE. Human IRGM gene "to be or not to be". Semin Immunopathol 2010; 32(4) 437–444.

[36] Parkes M, Barrett JC, Prescott NJ, Tremelling M, Anderson CA, Fisher SA, Roberts RG, Nimmo ER, Cummings FR, Soars D, Drummond H, Lees CW, Khawaja SA, Bagnall R, Burke DA, Todhunter CE, Ahmad T, Onnie CM, McArdle W, Strachan D, Be-

thel G, Bryan C, Lewis CM, Deloukas P, Forbes A, Sanderson J, Jewell DP, Satsangi J, Mansfield JC; Wellcome Trust Case Control Consortium, Cardon L, Mathew CG. Sequence variants in the autophagy gene IRGM and multiple other replicating loci contribute to Crohn's disease susceptibility. Nat Genet 2007; 39(7) 830–832.

[37] Intemann CD, Thye T, Niemann S, Browne EN, Amanua Chinbuah M, Enimil A, Gyapong J, Osei I, Owusu-Dabo E, Helm S, Rüsch-Gerdes S, Horstmann RD, Meyer CG. Autophagy gene variant IRGM -261T contributes to protection from tuberculosis caused by Mycobacterium tuberculosis but not by M. africanum strains. PLoS Pathog. 2009; 5(9) e1000577.

[38] Prescott NJ, Dominy KM, Kubo M, Lewis CM, Fisher SA, Redon R, Huang N, Stranger BE, Blaszczyk K, Hudspith B, Parkes G, Hosono N, Yamazaki K, Onnie CM, Forbes A, Dermitzakis ET, Nakamura Y, Mansfield JC, Sanderson J, Hurles ME, Roberts RG, Mathew CG. Independent and population-specific association of risk variants at the IRGM locus with Crohn's disease. Hum Mol Genet 2010; 19(9) 1828–1839.

[39] Burada F, Plantinga TS, Ioana M, Rosentul D, Angelescu C, Joosten LA, Netea MG, Saftoiu A.IRGM gene polymorphisms and risk of gastric cancer. J Dig Dis. 2012;13(7) 360–365.

[40] Lerman MI, Minna JD The 630-kb lung cancer homozygous deletion region on human chromosome 3p21.3: identification and evaluation of the resident candidate tumor suppressor genes. The International Lung Cancer Chromosome 3p21.3 Tumor Suppressor Gene Consortium. Cancer Res. 2000; 60(21) 6116 –6133.

[41] Yu J, Ni M, Xu J, Zhang H, Gao B, Gu J, Chen J, Zhang L, Wu M, Zhen S, Zhu J. Methylation profiling of twenty promoter-CpG islands of genes which may contribute to hepatocellular carcinogenesis. BMC Cancer. 2002;2: 29

[42] Schagdarsurengin U, Wilkens L, Steinemann D, Flemming P, Kreipe HH, Pfeifer GP, Schlegelberger B, Dammann R. Frequent epigenetic inactivation of the RASSF1A gene in hepatocellular carcinoma. Oncogene. 2003; 22(12)1866–1871.

[43] Fischer JR, Ohnmacht U, Rieger N, Zemaitis M, Stoffregen C, Kostrzewa M, Buchholz E, Manegold C, Lahm H. Promoter methylation of RASSF1A, RAR beta and DAPK predict poor prognosis of patients with malignant mesothelioma. Lung Cancer. 2006; 54(1) 109–116.

[44] Jing F, Yuping W, Yong C, Jie L, Jun L, Xuanbing T, Lihua H. CpG island methylator phenotype of multigene in serum of sporadic breast carcinoma. Tumor Biology. 2010; 31(4):321–331.

[45] Dallol A, Agathanggelou A, Fenton SL, Ahmed-Choudhury J, Hesson L, Vos MD, Clark GJ, Downward J, Maher ER, Latif F. RASSF1A interacts with microtubule-associated proteins and modulates microtubule dynamics. Cancer Research. 2004; 64(12) 4112–4116.

[46] Min SS, Jin SC, Su JS, Yang TH, Lee H, Lim DS. The centrosomal protein RAS association domain family protein 1A (RASSF1A)-binding protein 1 regulates mitotic progression by recruiting RASSF1A to spindle poles. Journal of Biological Chemistry. 2005; 280(5) 3920–3927.

[47] Bialik S, Kimchi A. The death-associated protein kinases: structure, function, and beyond. Annual Review of Biochemistry. 2006; 75:189–210.

[48] Gozuacik D, Kimchi A. DAPk protein family and cancer. Autophagy. 2006; 2(2)74–79.

[49] Liu L, Xie R, Yang C, McKeehan WL. Dual function microtubule- and mitochondria-associated proteins mediate mitotic cell death. Cellular Oncology. 2009; 31(5) 393–405.

[50] Whang YM, Kim YH, Kim JS, Yoo YD. RASSF1A suppresses the c-Jun-NH2-kinase pathway and inhibits cell cycle progression.Cancer Res. 2005; 65(9) 3682–3690.

[51] Whang YM, Park KH, Jung HY, Jo UH, Kim YH. Microtubule-damaging agents enhance RASSF1A-induced cell death in lung cancer cell lines. Cancer. 2009; 115(6)1253–1266.

[52] Agathanggelou A, Bièche I, Ahmed-Choudhury J, Nicke B, Dammann R, Baksh S, Gao B, Minna JD, Downward J, Maher ER, Latif F. Identification of novel gene expression targets for the Ras association domain family 1 (RASSF1A) tumor suppressor gene in non-small cell lung cancer and neuroblastoma. Cancer Res. 2003; 63(17) 5344–5351.

[53] Schagdarsurengin U, Seidel C, Ulbrich EJ, Kölbl H, Dittmer J, Dammann R. A polymorphism at codon 133 of the tumor suppressor RASSF1A is associated with tumorous alteration of the breast. Int J Oncol. 2005; 27 (1) 185–191.

[54] Gao B, Xie XJ, Huang C, Shames DS, Chen TT, Lewis CM, Bian A, Zhang B, Olopade OI, Garber JE, Euhus DM, Tomlinson GE, Minna JD. RASSF1A polymorphism A133S is associated with early onset breast cancer in BRCA1/2 mutation carriers. Cancer Res. 2008; 68(1) 22–25.

[55] Bergqvist J, Latif A, Roberts SA, Hadfield KD, Lalloo F, Howell A, Evans DG, Newman WG. RASSF1A polymorphism in familial breast cancer. Fam Cancer. 2010; 9(3) 263–265.

[56] Kanzaki H, Hanafusa H, Yamamoto H, Yasuda Y, Imai K, Yano M, Aoe M, Shimizu N, Nakachi K, Ouchida M, Shimizu K. Single nucleotide polymorphism at codon 133 of the RASSF1 gene is preferentially associated with human lung adenocarcinoma risk. Cancer Lett. 2006; 238(1) 128–134.

[57] Bayram S.RASSF1A Ala133Ser polymorphism is associated with increased susceptibility to RASSF1A Ala133Ser polymorphism is associated with increased susceptibility to hepatocellular carcinoma in a Turkish population. Gene. 2012; 498(2) 264–269.

[58] Schagdarsurengin U, Wilkens L, Steinemann D, Flemming P, Kreipe HH, Pfeifer GP, Schlegelberger B, Dammann R. Frequent epigenetic inactivation of the RASSF1A gene in hepatocellular carcinoma. Oncogene. 2003; 22 (12) 1866–1871.

[59] Hu L, Chen G, Yu H, Qiu X. Clinicopathological significance of RASSF1A reduced expression and hypermethylation in hepatocellular carcinoma. Hepatol Int. 2010; 4 (1) 423–432.

[60] Hamilton G, Yee KS, Scrace S, O'Neill E. ATM regulates a RASSF1A-dependent DNA damage response. Curr Biol. 2009;19(23) 2020–2025.

[61] Bayram S. RASSF1A Ala133Ser polymorphism is associated with increased susceptibility to hepatocellular carcinoma in a Turkish population.Gene. 2012; 498(2) 264–269.

[62] Gordon M, El-Kalla M, Baksh S. RASSF1 Polymorphisms in Cancer. Mol Biol Int. 2012;2012: 365213.

[63] Li J, Wang F, Protopopov A, Malyukova A, Kashuba V, Minna JD, Lerman MI, Klein G, Zabarovsky E.Inactivation of RASSF1C during in vivo tumor growth identifies it as a tumor suppressor gene. Oncogene. 2004; 23(35):5941–5949.

[64] Amaar YG, Minera MG, Hatran LK, Strong DD, Mohan S, Reeves ME. Ras association domain family 1C protein stimulates human lung cancer cell proliferation. Am J Physiol Lung Cell Mol Physiol. 2006; 291(6) L1185–L1190.

[65] Reeves ME, Baldwin SW, Baldwin ML, Chen ST, Moretz JM, Aragon RJ, Li X, Strong DD, Mohan S, Amaar YG. Ras-association domain family 1C protein promotes breast cancer cell migration and attenuates apoptosis. BMC Cancer. 2010;10: 562.

[66] Malpeli G, Amato E, Dandrea M, Fumagalli C, Debattisti V, Boninsegna L, Pelosi G, Falconi M, Scarpa A. Methylation-associated down-regulation of ,RASSF1A and up-regulation of RASSF1C in pancreatic endocrine tumors. BMC Cancer. 2011;11: 351

[67] Barnich N, Aguirre JE, Reinecker HC, Xavier R, Podolsky DK. Membrane recruitment of NOD2 in intestinal epithelial cells is essential for nuclear factor-{kappa}B activation in muramyl dipeptide recognition. J Cell Biol, 2005; 170 (1) 21–26.

[68] Inohara N, Koseki T, del Peso L, Hu Y, Yee C, Chen S, Carrio R, Merino J, Liu D, Ni J, Núñez G. Nod1, an Apaf-1-like activator of caspase-9 and nuclear factor-kappaB. J Biol Chem. 1999; 274 (21) 14560–14567.

[69] Ogura Y, Inohara N, Benito A, Chen FF, Yamaoka S, Nunez G. Nod2, a Nod1/Apaf-1 family member that is restricted to monocytes and activates NF-kappaB. J Biol Chem. 2001; 276 (7) 4812–4818.

[70] Travassos LH, Carneiro LA, Ramjeet M, Hussey S, Kim YG, Magalhães JG, Yuan L, Soares F, Chea E, Le Bourhis L, Boneca IG, Allaoui A, Jones NL, Nuñez G, Girardin

SE, Philpott DJ. Nod1 and Nod2 direct autophagy by recruiting ATG16L1 to the plasma membrane at the site of bacterial entry. Nat Immunol, 2010; 11 (1) 55–62.

[71] Alhopuro P, Ahvenainen T, Mecklin JP, Juhola M, Järvinen HJ, Karhu A, Aaltonen LA. NOD2 3020insC alone is not sufficient for colorectal cancer predisposition. Cancer Res. 2004; 64(20) 7245–7247.

[72] Kurzawski G, Suchy J, Kładny J, Grabowska E, Mierzejewski M, Jakubowska A, Debniak T, Cybulski C, Kowalska E, Szych Z, Domagała W, Scott RJ, Lubiński J. The NOD2 3020insC mutation and the risk of colorectal cancer. Cancer Res. 2004; 64(5)1604–1606.

[73] Papaconstantinou I, Theodoropoulos G, Gazouli M, Panoussopoulos D, Mantzaris GJ, Felekouras E, Bramis J. Association between mutations in the CARD15/NOD2 gene and colorectal cancer in a Greek population. Int J Cancer. 2005; 114(3) 433–435

[74] Roberts RL, Gearry RB, Allington MD, Morrin HR, Robinson BA, Frizelle FA. Caspase recruitment domain-containing protein 15 mutations in patients with colorectal cancer. Cancer Res. 2006; 66(5) 2532–2535

[75] Lakatos PL, Hitre E, Szalay F, Zinober K, Fuszek P, Lakatos L, Fischer S, Osztovits J, Gemela O, Veres G, Papp J, Ferenci P. Common NOD2/CARD15 variants are not associated with susceptibility or the clinicopathologic characteristics of sporadic colorectal cancer in Hungarian patients. BMC Cancer. 2007; 7: 54.

[76] Tuupanen S, Alhopuro P, Mecklin JP, Jarvinen H, Aaltonen LA.No evidence for association of NOD2 R702W and G908R with colorectal cancer. Int J Cancer. 2007; 121(1) 76–79.

[77] Rigoli, L, Di Bella, C, Fedele, F, Procopio, V, Amorini M, Lo Giudice,G, Romeo, P, Pugliatti, F, Finocchiaro, G, Lucianò, R, & Caruso, RA. TLR4 and NOD2/CARD15 genetic polymorphisms and their possible role in gastric carcinogenesis. Anticancer Res. (2010). , 30(2), 513-517.

[78] Szeliga J, Sondka Z, Jackowski M, Jarkiewicz-Tretyn J, Tretyn A, Malenczyk M. NOD2/CARD15 polymorphism in patients with rectal cancer. Med Sci Monit. 2008; 14(9) CR480–CR484.

[79] Tian Y, Li Y, Hu Z, Wang D, Sun X, Ren C. Differential effects of NOD2 polymorphisms on colorectal cancer risk: a meta-analysis. Int J Colorectal Dis. 2010 Feb;25(2): 161–168.

[80] Lubiński J, Huzarski T, Kurzawski G, Suchy J, Masojć B, Mierzejewski M, Lener M, Domagała W, Chosia M, Teodorczyk U, Medrek K, Debniak T, Złowocka E, Gronwald J, Byrski T, Grabowska E, Nej K, Szymańska A, Szymańska J, Matyjasik J, Cybulski C, Jakubowska A, Górski B, Narod SA. The 3020insC Allele of NOD2 Predisposes to Cancers of Multiple Organs. Hered Cancer Clin Pract. 2005; 3(2) 59–63.

[81] Kutikhin AG. Role of NOD1/CARD4 and NOD2/CARD15 gene polymorphisms in cancer etiology.Hum Immunol. 2011;72(10) 955–968.

[82] Suchy J, Kłujszo-Grabowska E, Kładny J, Cybulski C, Wokołorczyk D, Szymańska-Pasternak J, Kurzawski G, Scott RJ, Lubiński J. Inflammatory response gene polymorphisms and their relationship with colorectal cancer risk. BMC Cancer. 2008; 8:112.

[83] Rosenfeldt MT, Nixon C, Liu E, Mah LY, Ryan KM. Analysis of macroautophagy by immunohistochemistry. Autophagy. 2012; 8(6) 963-969.

[84] Martinet W, De Meyer GR, Andries L, Herman AG, Kockx MM. Detection of autophagy in tissue by standard immunohistochemistry: possibilities and limitations. Autophagy. 2006; 2(1) 55-57.

[85] Yang ZJ. Measurement of autophagy-related proteins by immunohistochemistry/tissue microarray to characterize autophagy: problems and considerations. Am J Physiol Gastrointest Liver Physiol. 2012; 302(11) G1356-G1357.

[86] Shintani M, Sangawa A, Yamao N, Miyake T, Kamoshida S. Immunohistochemical analysis of cell death pathways in gastrointestinal adenocarcinoma. Biomed Res. 2011; 32(6) 379-386.

[87] Tanida I, Ueno T, Kominami E. LC3 conjugation system in mammalian autophagy. Int J Biochem Cell Biol. 2004; 36(12) 2503-2518.

[88] Hemelaar J, Lelyveld VS, Kessler BM, Ploegh HL. A single protease, Apg4B, is specific for the autophagy-related ubiquitin-like proteins GATE-16, MAP1-LC3, GABARAP, and Apg8L. J Biol Chem. 2003; 278(51) 51841-51850.

[89] Kabeya Y, Mizushima N, Ueno T, Yamamoto A, Kirisako T, Noda T, Kominami E, Ohsumi Y, Yoshimori T. LC3, a mammalian homologue of yeast Apg8p, is localized in autophagosome membranes after processing. EMBO J. 2000; 19 (21) 5720-5728.

[90] Karantza-Wadsworth V, White E. Role of autophagy in breast cancer. Autophagy 2007; 3(6) 610-613.

[91] Cai MY, Zhang B, He WP, Yang GF, Rao HL, Rao ZY, Wu QL, Guan XY, Kung HF, Zeng YX, Xie D. Decreased expression of PinX1 protein is correlated with tumor development and is a new independent poor prognostic factor in ovarian carcinoma. Cancer Sci. 2010; 101(6) 1543–1549.

[92] Yousefi S, Perozzo R, Schmid I, Ziemiecki A, Schaffner T, Scapozza L, Brunner T, Simon HU. Calpain-mediated cleavage of Atg5 switches autophagy to apoptosis. Nat Cell Biol. 2006; 8(10) 1124–1132.

[93] Cuddeback SM, Yamaguchi H, Komatsu K, Miyashita T, Yamada M, Wu C, Singh S, Wang HG. Molecular cloning and characterization of Bif-1: a novel SH3 domaincontaining protein that associates with Bax. J Biol Chem. 2001; 276 (23) 20559-20565.

[94] Pierrat B, Simonen M, Cueto M, Mestan J, Ferrigno P, Heim J. SH3GLB, a new endophilin-related protein family featuring an SH3 domain. Genomics. 2001; 71(2) 222-234.

[95] Chen YB, Hou JH, Feng XY, Chen S, Zhou ZW, Zhang XS, Cai MY. Decreased expression of Beclin 1 correlates with a metastatic phenotypic feature and adverse prognosis of gastric carcinomas. J Surg Oncol. 2012; 105(6) 542-547.

[96] Miao Y, Zhang Y, Chen Y, Chen L, Wang F. GABARAP is overexpressed in colorectal carcinoma and correlates with shortened patient survival.Hepatogastroenterology. 2010; 57(98) 257-261.

[97] Li BX, Li CY, Peng RQ, Wu XJ, Wang HY, Wan DS, Zhu XF, Zhang XS. The expression of beclin 1 is associated with favorable prognosis in stage IIIB colon cancers. Autophagy. 2009; 5(3) 303-306.

[98] Coppola D, Khalil F, Eschrich SA, Boulware D, Yeatman T, Wang HG. Down-regulation of Bax-interacting factor-1 in colorectal adenocarcinoma. Cancer. 2008; 113(10) 2665-2670.

[99] Yoshioka A, Miyata H, Doki Y, Yamasaki M, Sohma I, Gotoh K, Takiguchi S, Fujiwara Y, Uchiyama Y, Monden M. LC3, an autophagosome marker, is highly expressed in gastrointestinal cancers. Int J Oncol. 2008; 33(3) 461-468.

[100] Guo GF, Jiang WQ, Zhang B, Cai YC, Xu RH, Chen XX, Wang F, Xia LP. Autophagy-related proteins Beclin-1 and LC3 predict cetuximab efficacy in advanced colorectal cancer. World J Gastroenterol. 2011; 17(43) 4779-4786.

[101] Kondo Y, Kanzawa T, Sawaya R, Kondo S. The role of autophagy in cancer development and response to therapy. Nat Rev Cancer 2005; 5(9) 726–734.

[102] Yin M, Yan J, Voutsina A, Tibaldi C, Christiani DC, Heist RS, Rosell R, Booton R, Wei Q. No evidence of an association of ERCC1 and ERCC2 polymorphisms with clinical outcomes of platinum-based chemotherapies in non-small cell lung cancer: a meta-analysis. Lung Cancer. 2011; 72(3) 370–377.

[103] Sirichanchuen B, Pengsuparp T, Chanvorachote P. Long-term cisplatin exposure impairs autophagy and causes cisplatin resistance in human lung cancer cells. Mol Cell Biochem. 2012; 364(1-2)11–18.

[104] Apel A, Herr I, Schwarz H, Rodemann HP, Mayer A. Blocked autophagy sensitizes resistant carcinoma cells to radiation therapy. Cancer Res. 2008; 68(5) 1485–1494.

[105] Kim KW, Hwang M, Moretti L, Jaboin JJ, Cha YI, Lu B. Autophagy upregulation by inhibitors of caspase-3 and mTOR enhances radiotherapy in a mouse model of lung cancer. Autophagy. 2008; 4(5) 659–668.

Chapter 11

Regulation of Autophagy by Short Chain Fatty Acids in Colon Cancer Cells

Djamilatou Adom and Daotai Nie

Additional information is available at the end of the chapter

http://dx.doi.org/10.5772/54999

1. Introduction

Short chains fatty (SCFAs) acids are organic fatty acids that are the major products of bacterial fermentation of undigested dietary fiber and resistant starch in the colon. Propionate, acetate and butyrate are the main SCFAs produced from fermentation and serve as fuel for colonocytes. SCFAs serve as regulators of intracellular pH, cell volume, and other functions associated with ion transport. Moreover, SCFAs act as regulators of proliferation, differentiation and gene expression. Our recent studies reported that SCFAs promote autophagy in colon cancer cells. In this chapter, the regulation of autophagy by short chain fatty acids in colon cancer cells will be discussed in details including the mechanism of action.

2. Short chain fatty acids and the colon

Short chain fatty acids are organic fatty acids with 1 to 6 carbon atoms and the major products of bacterial fermentation in the human large intestine. Those are mostly derived from polysaccharides, oligosaccharides, proteins, peptides and glycoproteins precursors by anaerobic microorganisms [1]. Diets high in fiber, resistant starches and complex carbohydrates lead to an increase in the levels of SCFAs. The principal SCFA involved in mammalian physiology are acetate c2, propionate c3 and butyrate c4. Formate, valerate, caproate, lactate and succinate are other fermentation products which are produced but to a lesser extent [2]. The scope of this chapter will be limited to propionate, butyrate and acetate.

In the colon, short chain fatty acids are absorbed at the same time as sodium and water absorption. Two mechanisms of absorption have been proposed. During *diffusion of protonated* SCFAs, luminal protons (Na^+/H^+ exchange, K^+ H^+-ATPase or bacterial metabolic activity) acidify the colonic lumen. This creates a pH between the colonic lumen compared to the systemic circulation. This pH gradient can promote the diffusive movement of SCFAs. With *anion exchange,* a family of anion exchangers mediates SCFAs and HCO3 exchange and entry across the membrane[3]. Once absorbed, SCFAs are used preferentially as fuel for colonic epithelial cells [4]. Butyrate is used preferentially over propionate and acetate. Those SCFAs are later transported to the liver. There, propionate acts as a substrate for gluconeogenesis and inhibits cholesterol synthesis in hepatic tissue. Acetate is utilized in the synthesis of long chain fatty acids, glutamine, glutamate and beta-hydroxybutyrate [2, 4]. Over the last a few decades, understanding the role of short chain fatty acids in the gastrointestinal physiology has grown considerably. They are attractive because of their role as potential therapeutic agents in diversion colitis, ulcerative colitis, radiation proctitis, pouchitis and antibiotic- associated diarrhea [4]. Furthermore, SCFAs have anti-tumor activities in colon cancer by promoting autophagy and apoptosis. In the next section, the role of SCFAs in cancer, mostly in the colon will be discussed.

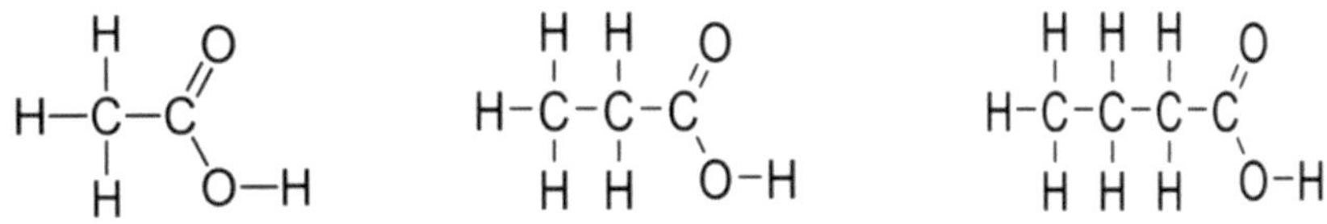

Acetic acid (acetate) Propionic acid (propionate) Butyric acid (butyrate)

Figure 1. Nomenclature of short chain fatty acids (SCFAs). Different SCFAs (acetate, propionate and butyrate) are illustrated in this figure.

3. The role of short chain fatty acids in colon cancer

Diet has a considerable influence on the risk of colon cancer. A diet high in fat has been considered to promote colon cancer while increased fiber and complex carbohydrates in the diet may protect against colon cancer. Butyrate is believed to be mostly responsible for the tumor inhibitory effects of dietary fiber. Sodium butyrate is known to be an effective inducer of cell differentiation. Colorectal cancer cells treated with sodium butyrate showed a more differentiated cell state [5, 6]. It is believed that the protective effect of dietary fibers is associated with butyrate production in the colon which possibly decreases the occurrence of neoplasia in colonocytes. For example, in reference [7], fibers associated with high butyrate were protective against colon cancer. Patients with familial polyposis (FAP) syndrome develop hundreds to thousands of begin tumors of the colon, some of which will progress to colon cancer if not removed. These FAP patients produced less butyrate than healthy controls and patients without polyps. All the studies suggest that butyrate has a protective effect against colon cancer. In the next section, different mechanisms of action of short chain fatty acids on colon cancer will be discussed.

4. Mechanisms of action of short chain fatty acids on colon cancer cells

Different mechanisms of action has been proposed through which sodium butyrate regulate different genes to exert an inhibitory effect on colon cancer development. Sodium butyrate induces growth inhibition in colon cancer cells by promoting histone hyperacetylation and induction of the cell cycle inhibitor p21 [8]. Other short chain fatty acids such as propionate and valerate were also shown to inhibit cell tumor cells but to a lesser extent than butyrate. However for acetate, no effects were observed on cell proliferation when it was used at a concentration of 20 mM [9]. Cells other than colonocytes, mainly smooth muscle cells are also present in the colon. The effect of SCFAs on colon smooth muscle cells was quiet different. Sodium butyrate promotes the proliferation of smooth muscle cells. Propionate also promotes colon smooth muscle cells proliferation to a lesser extent than butyrate while acetate had no effects [9]. The cell growth inhibitory effect of butyrate on colon cancer cells is attributed to its ability to induce histone hyperacetylation through inhibition of histone deacetylase (HDAC). Histone hyperacetylation usually results in relaxation of chromatin, thus making DNA more accessible to transcription factors. For example in reference [10], the cell cycle inhibitor p21 gene was increased and involved in the butyrate effect on colon cancer cell proliferation. The promoter region of p21 was shown to harbor butyrate-responsive elements. Upon butyrate treatment, p21 was induced due to HDAC inhibition resulting in G1 phase arrest [9, 11-13].

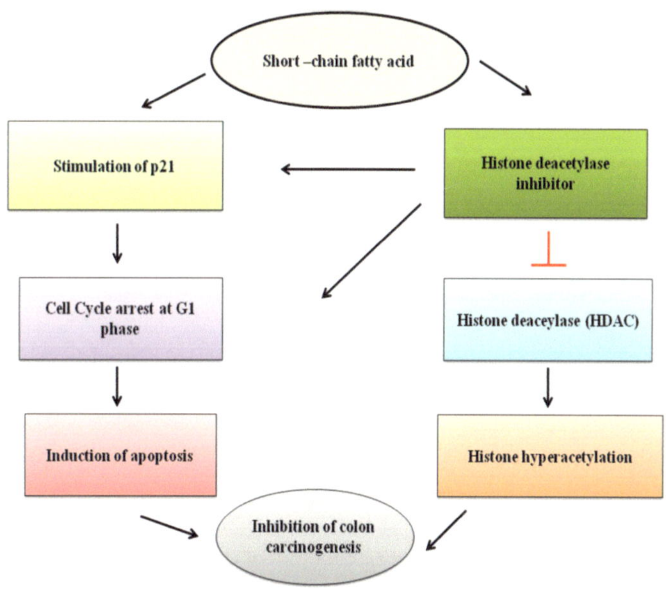

Figure 2. Mechanism of inhibition of colon cancer cells by short chain fatty acids. Short chain fatty acids, particularly butyrate can inhibit colon cancer by different mechanisms. In this figure, treatment of colon cancer cells with butyrate leads to the activation of the cycle inhibitor p21, followed by cell cycle arrest at G1 phase and induction of apoptosis. Moreover, SCFAs treatment can promote the inhibition of histone deacetylase, leading to histone hyperacetylation and availability of chromatin structure for binding by different transcription factors. Both histone hyperacetylation and cell cycle protein p21 inhibit colon cancer growth.

Sodium butyrate also inhibits colon cancer cell invasion by activating tissue inhibitor matrix metalloproteinase (TIM) 1- and 2, thus inhibiting the activity of metalloproteinases (MMPs). Furthermore, sodium butyrate reduces the adherence of colon cancer cells to the basement membrane protein laminin substrate via fibronection or type IV collagen resulting in the inhibition of cancer growth. Inflammatory cytokines such as IL-4 and TNF-α also play a role in the inhibitory role of butyrate in colon cancer [14]. G-protein coupled receptors are also found to play a role on colon cancer and their role will be discussed in the next section.

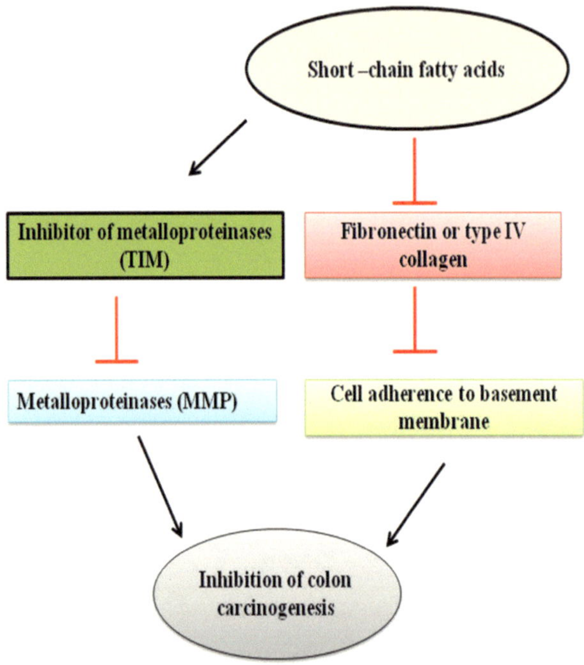

Figure 3. Other mechanisms used by short chain fatty acids to inhibit colon cancer carcinogenesis. Treatment with SCFAs can inhibit the destruction of the basement membrane through the stimulation of the inhibitor of metalloproteinases (TIM). TIM will inhibit metalloproteinases and prevent them from destroying the basement membrane. SCFAs also inhibit the adherence of colon cancer cells to the basement membrane by reducing fibronection of type IV collagen levels.

5. The roles of short chain fatty acid receptors GPR41 and GPR43 on colonic functions

G-protein coupled receptors (GPCRs) consist of a large and diverse family of proteins that mainly transduce extracellular stimuli to intracellular signals. GPRC family is among the largest and more diverse family of proteins in the mammalian genome and contain 7 spanning membrane helixes, an extracellular N-terminus and an intracellular C-terminus. GPCRs can be coupled by at least 18 Gα which forms a heterodimer with Gβ subunits, which have at least 5 types, and Gγ subunit, of which there are at least 11 types. Over 800 GPCRs have been identified in the human genome. The family of GPCR protein is activated upon binding of a ligand or agonist on the extracellular N-terminus that leads to a conformational change and activation of the G-protein heterodimer. At least 50 GPCRs have unknown ligands and are referred as orphans. Depending on the type of GPCRs that is being activated, diverse downstream signaling will be activated.GPRCs respond to different stimuli such as light, neurotransmitters, amino acids, hormones and activate different signaling pathways [15, 16].

Recently, short chain fatty acids (acetate, propionate and butyrate) were reported as ligands for two orphan GPRCs, GPR41 and GPR43. GPR43 is expressed in immune cells whereas GPR41 is present in blood vessel endothelial cells, particularly in adipose tissue with significant expression also in immune cells and endothelial cells of other tissues [17]. Both GPR41 and GPR43 are expressed in colonic mucosa suggesting their role in the normal development or functions of the colon tissue [16]. A study by Tang and al. revealed more information on the function of GPR43 in colon cancer. Immunohistochemistry showed a reduction of GPR43 in human colon cancers compared to normal human colon tissues. No epigenetic changes such as promoter hypermethylation or chromatin compaction due to histone deacetylation (HDAC) were found responsible for the repression or silencing of GPR43 in colon cancer. In order to determine the function of GPR43 in colon cancer, GPR43 was restored in colon cancer cells. Treatment of those GPR43 expressing cells with short chain fatty acids propionate and butyrate rendered the cells more sensitive, caused cell death and promoted cell cycle arrest at Go/G1 phase. The study suggested that loss of GPR43 expression may contribute to colon cancer development and progression [18]. In the next section, the role of short chain fatty acids in the fate of colon cancer cells will be discussed.

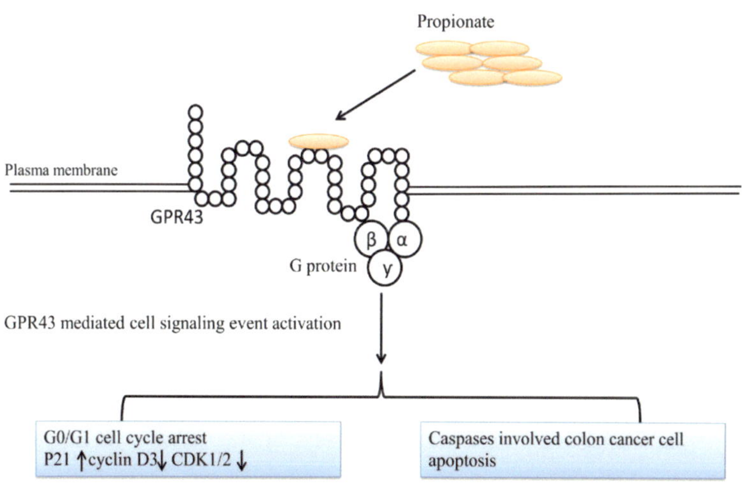

Figure 4. Mechanism of GPR43 role as a tumor suppressor in colon cancer. Treatment of colon cancer cells with SCFAs (propionate) results in binding of the GPR43 on the extracellular surface. This leads to the activation of intracellular downstream signaling such as increase in cell cycle protein p21 and CDK1/2 while cyclin D3 is increased resulting in cell cycle arrest at Go/G1 phase. Propionate treatment also stimulates different caspases which activates apoptosis and death of colon cancer cells.

6. Regulation of autophagy by short chain fatty acids

Short chain fatty acids were initially reported to induce apoptosis (type I programmed cell death) in colon cancer. Treatment of colon cancer cells with butyrate inhibited cell proliferation, promoted apoptosis in 79% of cells through the activation of caspase-3 and the degradation of PARP [19] [20]. Recently, SCFAs, particularly propionate was reported to induce autophagy as evidenced by an increased LC3 punctuates formation and upregulation of LAMP-2 [21]. In this section, the regulation of autophagy by short chain fatty acids will be discussed. The mechanisms involved will be discussed in details, including the signaling pathways involved.

6.1. Mechanism of autophagy regulation by short chain fatty acids in colon cancer

In autophagy studies, it is very well established that the mammalian target of rapamycin (mTOR) negatively regulates autophagy. The autophosphorylation at Ser2481 is regarded as an indicator of its catalytic activity. Post propionate treatment of colon cancer cells HCT116, a strong time dependent reduction in the phosphorylation state at Ser2481 was observed, while there were no changes in the total mTORC levels. Another key downstream effector of mTOR, p70S6K, whose phosphorylation status at Thr389 reflects mTOR activity, whereas phosphorylation at Thr421/Ser424 is thought to activate p70S6K was also performed. Following propionate treatment, a reduced phosphorylation of p70S6K at Thr389 was observed by 7 h confirming that downregulation of the mTOR signaling pathway is a mechanism for propionate to induce autophagy. The group then hypothesized that propionate must induce mTOR signaling from the inhibition of PI3/Akt pathway, which was shown to activate mTOR in response to the introduction of nutrient and growth factors. However, no changes in the phosphorylation state of Akt at S473 or Thr308 or the total Akt were observed post propionate treatment. Another pathway which acts upstream of the mTOR pathway, the AMP-activated protein kinase (AMPK) [22], an inhibitor of the mTOR protein and a sensor of cellular bioenergetics, was significantly activated.

Interestingly, propionate mediated AMPK activation caused a decrease in ATP levels in colon cancer cells due to a breach of the mitochondrial membrane potential. In more details, propionate depolarized the mitochondrial membrane, which was shown using mitotracker deep red, a dye that stains the mitochondria in live cells and accumulates in proportion to the membrane potential. The proportion of mitochondria with lower fluorescence intensity, which represents the depolarized mitochondria, was increased post propionate treatment in a dose- and time dependent manner in the colon cancer cells treated. The study in references [21, 23] demonstrated that propionate causes mitochondrial defect leading to ATP depletion and release of reactive oxygen species (ROS). Excessive ROS levels have been attributed to induction of autophagy. The defective mitochondria post treatment is removed by a selective autophagy process known as mitophagy. Mitophagy will be discussed in the next section and how it is regulated in colon cancer cells post propionate treatment.

6.2. Propionate treatment causes mitophagy in colon cancer cells

Mitochondria are cells organelles that primarily produce ATP via oxidative phosphorylation in the inner membrane of the mitochondria. During changes in the environment, ATP synthesis can be disrupted leading to the production of reactive oxygen species (ROS) and release of proteins to promote cell death. Several pathologies have impaired mitochondria, oxidative stress, accumulation of protein aggregates and autophagic stress. Oxidative stress can lead to the nonspecific modification of proteins and contributes to protein aggregation. Interestingly, the cell can adopt its own defense mechanism against aberrant mitochondria, which can be harmful to the cell [24]. This mechanism termed mitophagy was first observed in mammalian cells by early electron microscopy studies, where increased mitochondrial sequestration was identified in lysosomes following stimulation of hepatocytes catabolism with glucagon [25]. This selective autophagy process is characterized by the removal of excess or damaged mitochondria in order to prevent activation of apoptotic cell death [26]. Mitophagy has been shown to play a role during cellular quality control. For instance, in yeast and in mammalian cells, mitophagy is preceded by mitochondrial fission, which divides elongated mitochondria into pieces of manageable size for encapsulation and also quality control of segregation of damaged mitochondrial material for selective removal by mitophagy. Another process, mitochondrial fusion, occurs every 5 to 20 minutes and was shown to reduce mitochondrial depolarization in two cell lines (COS7 and INS1). Mitochondrial fission, fusion or mitophagy are all important for mitochondrial homeostasis [27]. Mitophagy has also been shown to be required for steady-state turnover of mitochondria, for the adjustment of mitochondrion numbers to changing metabolic requirements and during specialized developmental stages in mammalians cells such as during red blood cell differentiation [25]. Some important proteins such as ULK1 and ULK2 (two Atg1 homologues), Parkinson's disease genes α-synuclein, parkin, PINK1 and DJ-1 are all involved in mitophagy. During the selection of mitochondria for mitophagy, mitochondrial components of the cell are identified by partner cytosolic proteins such as Parkin or Nix that bind to the surface and tag it for degradation [24-26].

Treatment of colon cancers cells HCT116 with propionate resulted in reduced staining intensity of mitochondria and an increased colocalization between mitochondria and punctuates GFP-LC3. COXIV, a mitochondrial marker was also reduced and was localized as defective mitochondria by autolysosomes. An ubiquitin-binding protein-p62, a protein that interacts with LC3 and regulates autophagosome formation, significantly colocalized with mitochondrial COXIV. Flow cytometry analysis showed that most of colon cancer cells treated with propionate showed a reduced Mito Tracker Deep Red staining and enhanced GFP-LC3 fluorescence. Addition of chloroquine, an inhibitor of autophagic degradation, dramatically increased the accumulation of defective mitochondria. All the experiments performed in HCT116 colon cancer cells post treatment suggest that propionate triggers mitophagy. This mitophagy selectively targets mitochondria with a depolorized membrane potential [21, 23].

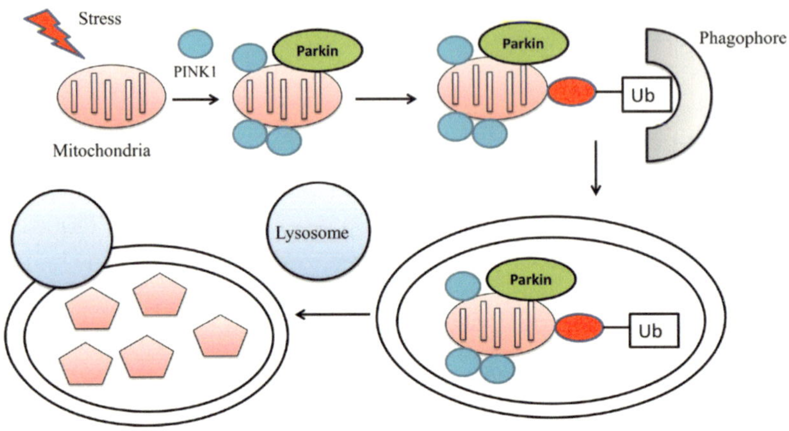

Figure 5. Mechanisms of mitophagy and representation of proposed role of short chain fatty acids on mitochondria. Treatment of colon cancer cells with SCFAs depolarizes and damages the mitochondria. Reduced mitochondrial membrane potential leads to the accumulation of phosphatase and tensin homolog-induced putative kinase 1 (PINK1) and recruitment of the E3 ubiquitin ligase Parkin to mitochondria. Parkin, then promotes the ubiquitination of proteins in the mitochondrial membrane, which targets the damaged mitochondrion for removal by an autophagasome.

6.3. Short chain fatty acids and lipid metabolism in colon cancer cells

Interestingly, treatment of HCT116 cells with propionate altered the lipid metabolism. Lipids play an important role in cell structure and metabolism. Post treatment with propionate, the expression of fatty acid synthase, the enzyme that catalyzes the synthesis of long chain-fatty acids (LCFAs) from acetyl-CoA (ACC) and malonyl-CoA, was reduced. Furthermore, GSK-3β, which inhibits endergonic glycogen synthesis by phosphorylation and activation of glycogen synthase, was also downregulated. AMPK kinase, which was increased post propionate treatment further phosphorylates and inhibit acetyl-CoA, thus de novo lipid synthesis. The inactivation of ACC by AMPK mediates the increase in mitochondrial import and oxidation of LCFAs, resulting in the generation of ATP.

Energy deprivation can stimulate mitochondrial biogenesis in skeletal muscle in an AMPK-dependent manner. When the effects of propionate treatment in mitochondrial biogenesis were investigated, the mRNA levels of nuclear gene mitochondrial transcription factor A (Tfam) and mitochondrial transcription factor B (mtTFB) were stimulated, peaking at 8.5 and 11.5 h respectively, before returning to the initial pre-stimulatory levels. Other transcription factors related to heme biosynthesis and mitochondria biogenesis named nuclear respiratory factors-1 and 2 (NRF-1 and NRF-2) and polymerase gamma (pol-γ) expression were also stimulated at the mRNA level in colon cancer cells after propionate treatment. On the other hand, peroxisome proliferator activated receptor-γ was reduced.

Moreover, an increase in mitochondria complex subunits and increase in Mitotracker greem FM fluorescence also indicated and confirmed that HCT116 cells adapt to propionate-induced ATP depletion by downregulating anabolic processes such as glycogen and lipid synthesis, while stimulating mitochondrial biogenesis in an attempt to resume cellular energy homeostatis [21, 23].

6.4. Propionate, autophagy and apoptosis

Autophagy and apoptosis are two programmed-cell deaths that may be interconnected and even simultaneously regulated by the same trigger in tumor cells. During apoptosis, cells are destroyed as an end result of caspase mediated destruction of the cellular structure. There exist two core pathways inducing apoptosis: the extrinsic and intrinsic pathways. The extrinsinc pathway is triggered by the Fas death receptor (DR), which depends on the combination of FasL and Fas. The other process, the intrinsic pathway leads to apoptosis upon sensing of an extracellular stimuli or intracellular signal that renders the mitochondrial membrane permeable and releases cytochrome c [28-31].

Molecular pathways leading to cancer can cross-talk. Autophagy and apoptosis can particularly act as partners to induce cell death in a coordinated or cooperative fashion. Although both can be triggered by common upstream signals, this will have different effects on the cell fate. For instance, autophagy can function as a double edged sword to either promote or inhibit cell death. In most cases, inhibition of autophagy leads to an increase susceptibility to apoptotic stimuli [29]. For instance when colon cancer cells were treated with propionate, [21] found that the induction of autophagy has a protective effect on HCT116 cells. SCFAs were previously shown to induce caspase-3-mediated apoptosis. Cotreatment of colon cancer cells (HCT116 and SW480) with 3-methyladenine (3-MA), an inhibitor of autophagy, significantly reduced the percentage of GFP-LC3 formation. However, 12 h after the initiation of treatment of propionate/3MA, the number of apoptotic cells increased as indicated by the high annexin-V staining. Western blot analysis also revealed increased cleavages of the pro-apoptotic cascapase-7 and executioner caspase-3, which are all critical mediators of the mitochondrial events of apoptosis in cells treated with propionate/3MA compared to nontreated group. Addition of another inhibitor of autophagy, chloroquine, enhanced apoptosis in HCT116 cells, especially at the later stages of treatment. Since depletion of AMPKα using shRNA was also shown to mimic the effects of autophagy silencing,colon cancer cells depleted of the AMPKα were also treated with propionate. AMPKα depleted cells showed a more significant cytotoxicity post propionate treatment. Further depletion of autophagy by knocking down ATG5 expression, an important protein required for autophagy, reduced the ability of propionate to induce GFP-LC3 punctae formation, indicating successful depletion. This inhibition of autophagy also confirmed the protective role of autophagy in colon cancer cells post treatment with the short chain fatty acid, propionate. All the findings by [21, 23] suggested that autophagy confers a protective role in propionate-mediated cell death in colon cancer cell.

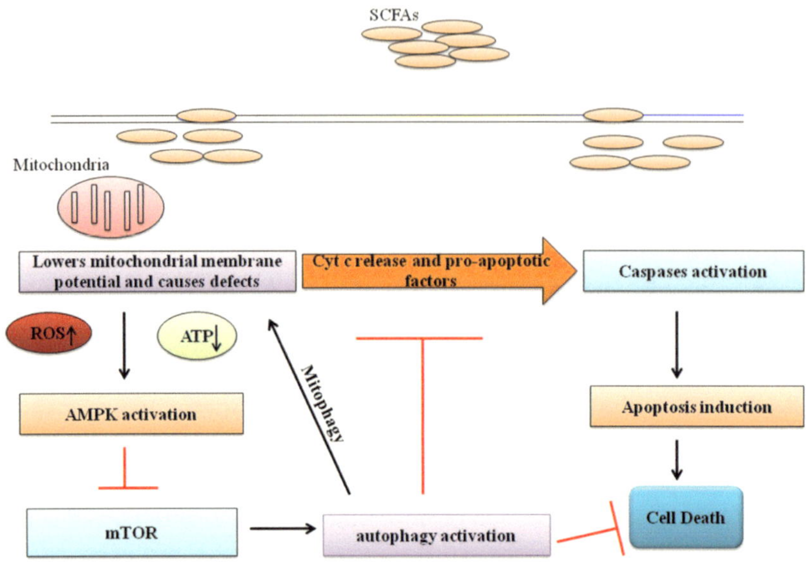

Figure 6. Proposed model for regulation of autophagy by short chain fatty acids in colon cancer cells. Treatment of colon cancer cells with short chain fatty acids increases ROS levels, lowers the mitochondrial membrane potential and ATP levels. This results in a reduction of cellular energy and defects in mitochondria. As a result, downstream signaling pathway AMPK is activated while mTOR is inhibited and resulting in the activation of autophagy. At the same time, defective mitochondria is removed by a process known as mitophagy in order to rescue the cells from apoptosis and cell death. The activation of autophagy has a protective role on the cell as it prevents cell death.

7. Conclusion

Short chain fatty acids are the major by-products of bacterial fermentation of undigested fibers in the colon. Short chain fatty acids, particularly propionate has been shown to promote differentiation, growth arrest and apoptosis in colon cancer cells. Furthermore, SCFAs can promote autophagy to retard mitochondrial defect stimulated apoptosis. Although the in vitro studies demonstrated a role of short chain fatty acids in promoting autophagy and protecting the cells from apoptotic death, the relevance of this finding has yet to be investigated in vivo. Moreover, cotreatment of colon cancer cells with SCFAs and chloloroquine, an inhibitor of autophagy may be an effective therapy approach to be investigated.

Author details

Djamilatou Adom and Daotai Nie*

Department of Medical Microbiology, Immunology, and Cell Biology, Southern Illinois University School of Medicine and Simmons Cancer Institute, Springfield, IL, USA

References

[1] Macfarlane S, Macfarlane GT. Regulation of short-chain fatty acid production; 2003.

[2] Cummings JH, Macfarlane GT. The control and consequences of bacterial fermentation in the human colon. The Journal of applied bacteriology. 1991 Jun;70(6):443-59.

[3] Augenlicht LH, Mariadason JM, Wilson A, Arango D, Yang W, Heerdt BG, et al. Short chain fatty acids and colon cancer. The Journal of nutrition. 2002 Dec;132(12): 3804S-8S.

[4] Cook SI, Sellin JH. Review article: short chain fatty acids in health and disease. Alimentary pharmacology & therapeutics. 1998 Jun;12(6):499-507.

[5] Fu H, Shi YQ, Mo SJ. Effect of short-chain fatty acids on the proliferation and differentiation of the human colonic adenocarcinoma cell line Caco-2. Chinese journal of digestive diseases. 2004;5(3):115-7.

[6] D'Argenio G, Mazzacca G. Short-chain fatty acid in the human colon. Relation to inflammatory bowel diseases and colon cancer. Adv Exp Med Biol. 1999;472:149-58.

[7] McIntyre A, Young GP, Taranto T, Gibson PR, Ward PB. Different fibers have different regional effects on luminal contents of rat colon. Gastroenterology. 1991 Nov; 101(5):1274-81.

[8] Hinnebusch BF, Meng S, Wu JT, Archer SY, Hodin RA. The effects of short-chain fatty acids on human colon cancer cell phenotype are associated with histone hyperacetylation. The Journal of nutrition. 2002 May;132(5):1012-7.

[9] Blottiere HM, Buecher B, Galmiche JP, Cherbut C. Molecular analysis of the effect of short-chain fatty acids on intestinal cell proliferation. The Proceedings of the Nutrition Society. 2003 Feb;62(1):101-6.

[10] Nakano K, Mizuno T, Sowa Y, Orita T, Yoshino T, Okuyama Y, et al. Butyrate activates the WAF1/Cip1 gene promoter through Sp1 sites in a p53-negative human colon cancer cell line. J Biol Chem. 1997 Aug 29;272(35):22199-206.

[11] Archer S, Meng S, Wu J, Johnson J, Tang R, Hodin R. Butyrate inhibits colon carcinoma cell growth through two distinct pathways. Surgery. 1998 Aug;124(2):248-53.

[12] Archer SY, Meng S, Shei A, Hodin RA. p21(WAF1) is required for butyrate-mediated growth inhibition of human colon cancer cells. Proceedings of the National Academy of Sciences of the United States of America. 1998 Jun 9;95(12):6791-6.

[13] Siavoshian S, Segain JP, Kornprobst M, Bonnet C, Cherbut C, Galmiche JP, et al. Butyrate and trichostatin A effects on the proliferation/differentiation of human intestinal epithelial cells: induction of cyclin D3 and p21 expression. Gut. 2000 Apr;46(4): 507-14.

[14] Andoh A, Tsujikawa T, Fujiyama Y. Role of dietary fiber and short-chain fatty acids in the colon. Current pharmaceutical design. 2003;9(4):347-58.

[15] Kroeze WK, Sheffler DJ, Roth BL. G-protein-coupled receptors at a glance. J Cell Sci. 2003 Dec 15;116(Pt 24):4867-9.

[16] Tazoe H, Otomo Y, Kaji I, Tanaka R, Karaki SI, Kuwahara A. Roles of short-chain fatty acids receptors, GPR41 and GPR43 on colonic functions. J Physiol Pharmacol. 2008 Aug;59 Suppl 2:251-62.

[17] Brown AJ, Goldsworthy SM, Barnes AA, Eilert MM, Tcheang L, Daniels D, et al. The Orphan G protein-coupled receptors GPR41 and GPR43 are activated by propionate and other short chain carboxylic acids. J Biol Chem. 2003 Mar 28;278(13):11312-9.

[18] Tang Y, Chen Y, Jiang H, Robbins GT, Nie D. G-protein-coupled receptor for short-chain fatty acids suppresses colon cancer. Int J Cancer. Feb 15;128(4):847-56.

[19] Ruemmele FM, Schwartz S, Seidman EG, Dionne S, Levy E, Lentze MJ. Butyrate induced Caco-2 cell apoptosis is mediated via the mitochondrial pathway. Gut. 2003 Jan;52(1):94-100.

[20] Ruemmele FM, Dionne S, Qureshi I, Sarma DS, Levy E, Seidman EG. Butyrate mediates Caco-2 cell apoptosis via up-regulation of pro-apoptotic BAK and inducing caspase-3 mediated cleavage of poly-(ADP-ribose) polymerase (PARP). Cell Death Differ. 1999 Aug;6(8):729-35.

[21] Tang Y, Chen Y, Jiang H, Nie D. Short-chain fatty acids induced autophagy serves as an adaptive strategy for retarding mitochondria-mediated apoptotic cell death. Cell Death Differ. Apr;18(4):602-18.

[22] Hardie DG. AMP-activated protein kinase: an energy sensor that regulates all aspects of cell function. Genes Dev. Sep 15;25(18):1895-908.

[23] Tang Y, Chen Y, Jiang H, Nie D. The role of short-chain fatty acids in orchestrating two types of programmed cell death in colon cancer. Autophagy. Feb;7(2):235-7.

[24] Kubli DA, Gustafsson AB. Mitochondria and mitophagy: the yin and yang of cell death control. Circ Res. Oct 12;111(9):1208-21.

[25] Youle RJ, Narendra DP. Mechanisms of mitophagy. Nat Rev Mol Cell Biol. Jan;12(1): 9-14.

[26] Kim I, Rodriguez-Enriquez S, Lemasters JJ. Selective degradation of mitochondria by mitophagy. Arch Biochem Biophys. 2007 Jun 15;462(2):245-53.

[27] Lee J, Giordano S, Zhang J. Autophagy, mitochondria and oxidative stress: cross-talk and redox signalling. Biochem J. Jan 15;441(2):523-40.

[28] Maiuri MC, Zalckvar E, Kimchi A, Kroemer G. Self-eating and self-killing: crosstalk between autophagy and apoptosis. Nature reviews. 2007 Sep;8(9):741-52.

[29] Rikiishi H. Novel Insights into the Interplay between Apoptosis and Autophagy. International journal of cell biology.2012:317645.

[30] Eisenberg-Lerner A, Bialik S, Simon HU, Kimchi A. Life and death partners: apoptosis, autophagy and the cross-talk between them. Cell death and differentiation. 2009 Jul;16(7):966-75.

[31] Galluzzi L, Vicencio JM, Kepp O, Tasdemir E, Maiuri MC, Kroemer G. To die or not to die: that is the autophagic question. Current molecular medicine. 2008 Mar;8(2): 78-91.

Chapter 12

Natural Compounds and Their Role in Autophagic Cell Signaling Pathways

Azhar Rasul and Tonghui Ma

Additional information is available at the end of the chapter

http://dx.doi.org/10.5772/55447

1. Introduction

The term autophagy (from the Greek, *auto* – oneself, *phagy* - to eat or autophagy-self eating) was first coined for structures that were observed under an electron microscope and that were consisted of single-or double-membrane lysosomal-derived vesicles containing cytoplasmic particles including organelles during various stages of disintegration [1,2]. Autophagic cell death or autophagy is a type of cell death that occurs in the absence of chromatin condensation but is associated with the massive autophagic vacuolization of the cytoplasm [3]. Autophagy is the process by which cells recycle their own nonessential, redundant, and damaged organelles and macromolecular components. Autophagy also plays its role in the suppression of tumor growth, deletion of toxic misfolded proteins, elimination of intracellular microorganisms, and pathogenesis of several diseases such as cancer and muscular disorders [4-6].

Like apoptosis, autophagy is an evolutionary conserved process that occurs in all eukaryotic cells [7]. Autophagy can be triggered as a result of nutrient deprivation, differentiation, and developmental factors. In contrast to the apoptosis, cells that die with an autophagic morphology have little or no association with phagocytes. Excessive autophagy may be attributed to crumple the cellular functions and induce cell death directly. On the other hand, autophagy can lead to the execution of apoptotic or necrotic cell death programs. It has also been suggested that autophagy may occur as a process alongside apoptosis or it may play a supportive role in apoptosis [8]. Autophagy and apoptosis differ in morphological characteristics. The causative relationship between these two has not been elucidated yet. The increase in autophagosomes indicates an increase in autophagic activity or decreased autophagosome-lysosome fusion. The most important characteristic of autophagic cell death is the appearance of double- or multiple-membrane vesicles (autophagosomes) in cytoplasm, which sequesters cytoplasmic components and organelles such as mitochondria and endoplasmic reticulum [9].

1.1. Role of autophagy in cancer

Cancer is an umbrella term covering a plethora of conditions characterized by unscheduled and uncontrolled cellular proliferation [10]. Cancer is the second leading cause of mortality with an incident rate of about 2.6 million cases reported annually across Europe and USA [11]. Autophagy has a multifaceted role in cancer [12,13]. Even though, present studies are only associative, autophagy most probably functions to curtail neoplasia [14]. There are several oncogenes including PI3K and Akt family members, MTOR, and Bcl2 restrain autophagy, while tumor suppressors such as PTEN, HIF1A, and TSC2 endorse autophagy [15]. Paradoxically, autophagy is double-edged sword having a role in promoting both cell survival and cell death [16]. The role of autophagy in the demise of a cell is contentious [17]. Despite the fact that number of autophagosomes increases in some dying cells, it is still ambiguous whether these structures are involved or just facilitate cell death [18]. The genetic deletion of key autophagic genes pick up the pace rather than to inhibit cell death, which accentuate the predominant survival role of autophagy [19].

Although, apoptosis and autophagy are markedly different processes but several lines of evidence have portrayed interplay between these two processes; such as the proteins from the Bcl2 regulate both autophagic and apoptotic machinery [17]. Furthermore, three types of interplay exist between autophagic and apoptotic pathways. Both apoptosis and autophagy function as a collaborator to induce cell death; autophagy act as anagonistic to hamper apoptotic cell death by promoting cell survival, autophagy act as enabler of apoptosis, and contributing in certain morphological and cellular events that occur during apoptotic cell death without leading to death in itself [8].

1.1.1. Autophagy in tumor suppression

Cancer is considered as a complex group of genetic disorder with multiple causes. It is thought to be involved in perturbation of several different pathways that control and regulate the cell differentiation, cell proliferation, and cell survival. Another enigma has been the role of autophagy in tumor suppression; cancer may be protected by macroautophagy by sequestering damaged organelles, permitting cellular differentiation, increasing protein catabolism, and promoting autophagic cell death [20]. There are some experimental evidences which support the possibility that autophagy promotes the survival of nutrient-starved tumor cells and in turn contribute to cancer. Recent advances give deep insight into the molecular mechanism of autophagy. These findings more likely favor the concept that autophagy and defects in autophagy contribute to tumor suppression and oncogenesis respectively. Biochemical studies and genetic evidences designed in mammalian cells and in C. elegans respectively suggest that autophagy is positively regulated by the PTEN tumor suppressor gene and negatively regulated by the oncogenic Class I phosphatidylinositol 3-kinase signaling pathway. Furthermore Beclin 1, the mammalian APG gene has tumor suppressor activity and maps a tumor susceptibility locus, which is commonly deleted in human breast and ovarian cancers. The molecular mechanism of oncogenesis in human cancer can be fairly understood through genetic disruption of autophagy control. Such insights may foster the development of novel approaches to restore autophagy in the chemoprevention or treatment of human malignancies [21].

Autophagy is a kind of homeostatic mechanisms which accelerates and induces tumorigenesis when it disrupts. It removes the damaged organelles/proteins, limiting cell growth and causes genomic instability which are involved in the tumor suppression mechanism [22]. The experimental studies show that Beclin 1 is a haploin sufficient tumor suppressor gene. As this protein is used for autophagy induction and Beclin $1^{+/-}$ mice were shown to be tumor prone [23]. The excessive stimulation of autophagy due to Beclin 1 protein overexpression can inhibit tumor development [24]. The accumulation of p62/SQSTM 1 protein aggregates, damaged mitochondria, and misfolded proteins due to the formation of molecular link between defective autophagy and tumorigenesis generate the reactive oxygen species (ROS) and genomic instability is observed due to the damage of DNA. ROS and the DNA damage can be prevented by knockdown of p62/SQSTM 1 in autophagy-defective cells [22]. The relationship between defective autophagy and p62/SQSTM 1 accumulation with tumorigenesis is further evidenced from a study involving p62/SQSTM $1^{-/-}$ mice protected from *Ras*-induced lung carcinoma compared with wild-type animals [25]. It is further concluded that autophagy may also provide protection against tumorigenesis by limiting necrosis and chronic inflammation, which are associated with the release of pro-inflammatory HMGB1 [26]. All above findings give a concentric remark about the role of the autophagy as a mechanism of tumor suppression.

1.1.2. Autophagy in tumor cell survival

The autophagy plays a predominant role in cancer cells to confer stress tolerance, which serves to maintain tumor cell survival [20]. The induction of cell death mainly relates to the knock-down of essential autophagy genes in tumor cells [27]. Cancer cells have high metabolic demands. The exposure of increased cellular proliferation and *in vivo* models to metabolic stress was shown to impair the survival of autophagy-deficient cells with compared to autophagy-proficient cells [22]. Moreover, cytotoxic and metabolic stresses, including hypoxia and nutrient deprivation, can activate autophagy for recycling of ATP and in maintaining the cellular biosynthesis and survival. Autophagy is mainly considered to be induced in hypoxic tumor cells from regions that are distal to blood vessels and HIF-1α-dependent and -independent activation have been described [28]. The expression of angiogenic factors, such as vascular endothelial growth factor, platelet-derived growth factor, and nitric oxide synthtase are HIF-1α [28]. Human pancreatic cancer cell lines have increased basal levels of autophagy. These enable tumor cell growth by maintaining cellular energy production. Autophagic inhibition may lead to tumor regression and extended survival in pancreatic cancer xenografts [29]. In the survived cancer cells autophagy generates a state of dormancy in residual cancer cells that may further contribute to tumor recurrence and progression [30]. The increased efficacy of anticancer drugs, in response to inhibition of autophagy, supports cytoprotective role of autophagy in cancer cells. The research data indicate that *H-ras* or *K-ras* bearing activating mutations show high basal levels of autophagy in human cell lines irrespective of abundant nutrients [31]. The cell growth in these cell lines was associated with autophagic proteins. In conclusion, it is the autophagy that maintains tumor cell survival. Moreover it suggests that by blocking autophagy in tumors is an effective treatment approach [21].

1.2. Cross-talk between different cell-death modes

Under the physiological conditions cells show compliance with respect to how they die responding various stimuli. There are certain factors such as the type of cell, type and intensity of noxious signals, and ATP concentration that determine how cells die [19]. Acute myocardial ischemia (which is involved in the sudden fall in ATP level) induces necrosis, whereas chronic congestive heart failure (with more modest yet chronic decrease in ATP) induces apoptosis [32]. Although, a particular cell death program may preferentially be triggered in different circumstances and multiple pathways may be activated concomitantly or successively in individual dying cells [33]. Furthermore, there seems to exist an interplay among different cell death pathways. Even though apoptosis and autophagy both bear distinct morphological characteristics and physiological processes, still there exist some intricate interrelationships between them. Apoptosis and autophagy, under some conditions, play synergistic effects; while other times autophagy onsets only when apoptotic suppression occurs. Moreover, recent studies have markedly pointed out the existence of strong interconnection between apoptosis and autophagy and also strengthened the concept of simultaneous regulation of both A's that trigger cell death in cancer. The obstruction of a particular pathway of cell death may not avert the annihilation of the cell but instead may recruit an alternative path such as the broad-spectrum anti-apoptotic caspase inhibitors, zVAD-fmk, modulates the three major types of cell death. Addition of zVAD-fmk blocks apoptotic cell death, sensitizes cells to necrotic cell death, and induces autophagic cell death [34].

The overexpression of anti-apoptotic proteins may lead to the survival of injured cells where critical metabolites are provided by autophagy [35]. Nevertheless, if death stimuli persist, anti-apoptotic pathways and autophagy are unlikely to prolong and necrosis ensues [36]. Most likely, NF-κB, ATG5, ATP, and PARP function as molecular switches that determine whether a cell undergoes apoptosis, necrosis, or autophagy [37-39]. Protein p53 also modulates autophagy and other responses to cell stress. Recent studies reveal that basal p53 activity suppresses autophagy, whereas the activation of p53 by certain stimuli induces autophagy and the activation of p53 by different stimuli results in the PUMA- and NOXA-mediated apoptosis [40-42]. In addition, low and moderate concentrations of some agents have been revealed to induce apoptosis, but increasing the concentration of the same agent triggers necrosis. The challenge is, therefore, not only to understand the mechanisms leading to cell death but also to categorize the connection at the molecular level between different modes of the cell death.

In this chapter, we discussed the natural compounds and their mechanisms by which they induce apoptotic and autophagic cell death in cancer cells and their potential as a novel strategy for the treatment of cancer. We also presented the results of our previously published natural compounds screened against gastric cancer [43]. The screen was used to identify new targets to combat cancer or to identify selective natural compounds those target to apoptosis or autophagy signaling pathways. In this chapter, we reviewed the main effects of natural compounds on the different autophagic cell death signaling pathways. In addition, we focused on highlighting several representative plant-derived natural compounds such as curcumin, resveratrol, evodiamine, oridonin, and magnolol (structures of these compounds are shown

http://dx.doi.org/10.5772/55447

in Fig. 1) that may lead to cancer cell death - for regulation of some core autophagic pathways, involved in Ras-Raf signaling, Beclin-1 interactome, BCR-ABL, PI3KCI/Akt/mTOR, FOXO1 signaling, the NF-κB-mediated pathway, the PI3K/Akt signaling pathway, p53 and other main pathways. Two of the identified autophagy inducer natural compounds, magnolol and evodiamine, have been discussed in detail, while the other natural compounds, which had shown an essential role in autophagic cell signaling pathways, been reviewed recently by Zhang et al., 2012 [44].

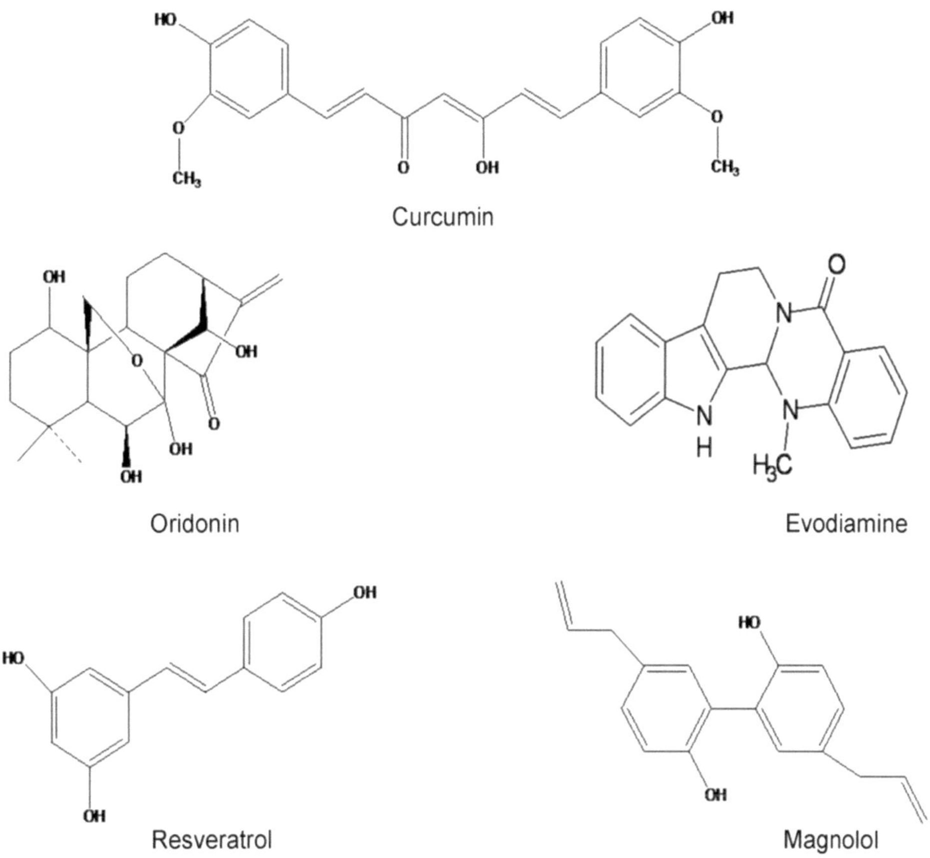

Figure 1. The structure of natural compounds that act on autophagic cell signaling pathways.

2. Methods

The rationale for overall project design was based on assumptions as presented above, which were motivated by the set of issues. The main goal of the study was to explore naturally occurring compounds that exhibit cytotoxic activity toward cancer cells. To search for compounds with significant cytotoxic activity and unprecedented chemical structures from a variety of traditional Chinese medicines, the crude ethanolic extracts of 300 species of herbal plants, traditionally used in China for the treatment of a variety of diseases, and four hundred TCM compounds were screened. Understanding the interplay of different cancer-related signaling pathways is important for the development of efficacious multi-targeted anticancer drugs. Hence, the underlying molecular mechanisms of the above mentioned natural compounds have been elucidated, which induced cell death.

2.1. Screening strategies

The screening strategy has been shown in schematic form in Fig. 2. Several cancer cells, including SGC-7901, U87, PANC-1 and A-375 cells were cultured in DMEM supplemented with 10% fetal bovine serum and were exposed to different Chinese medicinal herbs extracts and TCM compounds.

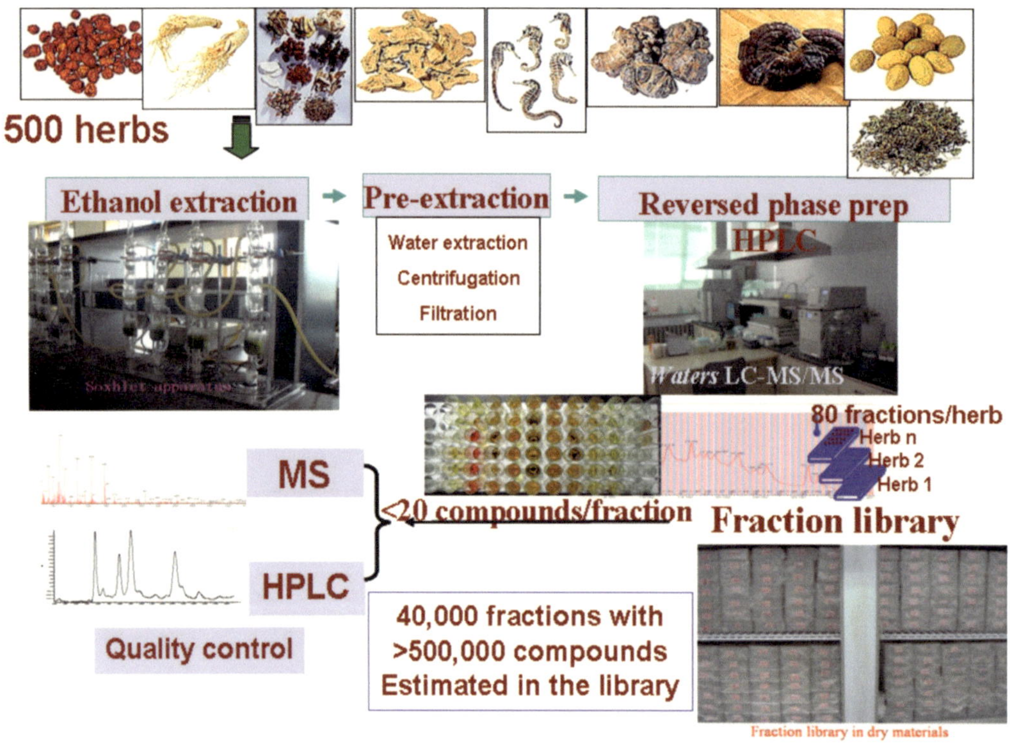

Figure 2. Schematic presentation of strategy of identification, isolation, structure elucidation, and screenings of Traditional Chinese medicines (TCM) against cancer cells.

2.2. Acridine orange staining

Acridine orange staining assay was performed according to published procedure [45]. In brief, cells were incubated without (control) and with respective compounds and with rapamycin (positive control group) for indicated time periods and then acridine orange at a final concentration of 1 mg ml^{-1} was added to cells for a period of 20 minutes in the dark at 37 °C. Then, cells were washed twice with PBS. Images of cells were obtained under fluorescence microscopy

2.4. Flow cytometric quantification of Acidic Vesicular Organelles (AVOs)

AVOs formation (autophagosomes and autolysosomes) is a characteristic feature of Autophagy [17]. Furthermore, for quantification of AVOs, we used flow cytometry after the cells were stained by Acridine Orange (AO) [46]. AO is a weak base that accumulates in acidic spaces and gives bright red fluorescence (punctate staining (dots) in the cytoplasm which is detected by fluorescence microscopy and the formation of AVOs can be quantified by flow cytometry. The intensity of the red fluorescence is proportional to the degree of acidity.

3. Results

The large collections of Traditional Chinese medicinal herbs and natural compounds libraries have been used to identify anticancer TCM herbs and natural compounds associated with various cancer specific cellular processes such as apoptosis or autophagy. A list of natural compounds those expressed cytotoxic activity against gastric cancer cells has been presented in Table 1 [43]. Using Traditional Chinese medicinal herbs and natural compounds libraries screen, we discovered several TCM compounds that showed potential anticancer activities [43]. In a natural compounds screen with glioma brain tumor cells, our results revealed that alantolactone and Pseudolaric acid B have shown selective anti-glioma activity with lesser toxic effect over liver and kidney [47,48]. Furthermore, we reported that Dracorhodin perchlorate regulates PI3K/Akt, p53 and NF-κB pathways that are frequently deregulated in cancer and their simultaneous targeting by Dracorhodin perchlorate could result in efficacious and selective killing of cancer cells [49]. Through the screen of natural compounds for apoptosis and autophagic cell signaling pathways, we identified several compounds including costunolide and xanthoxyletin that induce cell death via apoptotic pathways [50,51]. In addition, we also found that several compounds such as curcumin, resveratrol, evodiamine, oridonin, and magnolol induce autophagy and act on autophagic cell signaling pathways (unpublished data). Furthermore, we examined the role of evodiamine- and magnolol-induced autophagy in cancer cell death [52,53]. The role of each of the natural compound in autophagic cell signaling will be discussed later in this chapter.

4. Discussion

4.1. The role of natural compounds in autophagic cell signaling

Plants have a long recorded history to employ in the treatment of cancer [54] and represent the most important direct antecedent to contemporary anticancer drugs [55]. Recently, some of the most encouraging clinical evidences and promising anticancer natural herbal compounds let us to reconstruct the story of these plants and their ultimate role in chemotherapy [56]. To provide a paradigm of the most contemporary progress in this field, there were number of compounds namely artesunate, homoharringtonine, arsenic trioxide and cantharidin isolated from natural products and have the potential for use in cancer therapy. For many years, apoptosis has taken a center stage as the most important mechanism of programmed cell death in mammalian tissues. Apoptosis is a common mode of action for chemotherapeutic agents including natural product-derived drugs [57,58]. Four categories of dynamic cellular activities, which lead to cell death, have been described: apoptosis, autophagy, necrosis, and mitotic catastrophe [59]. With contemporary development in cancer research, it has also been increasingly noted that conventional chemotherapeutic agents not only elicit apoptosis but also activate other modes of cell death such as necrosis, mitotic catastrophe, senescence, and autophagy [60].

Sr. No	English Name	Chinese Name	M.W	IC_{50} (μM)
1	Artesunate	青蒿琥酯	384.43	44.7±5.3
2	Isoalantolactone	异土木香内酯	232.318	34.9±3.4
3	Cucurbitacin Ⅱa	雪胆素甲（雪胆甲素）	574.702	17.9±3.4
4	Tubeimiside-1	土贝母苷甲	1319.43	20.7±1.3
5	20(S)-Ginsenoside Rh2	20(S)-人参皂苷Rh2	622.6	18.9±1.3
6	Shikonin	左旋紫草素	288.295	19.7±0.9
7	Cepharanthin	千金藤素	606.707	20.4±3.6
8	Evodiamine	吴茱萸碱	303.358	11.7±1.9
9	Chelerythrine	白屈菜红碱	348.36	18.4±2.6
10	Patchouli alcohol	百秋李醇	222.366	29.3±3.8
11	Dracorhodin perchlorate	血竭素高氯酸盐	366.75	54.9±4.3
12	Resveratrol	白藜芦醇	390	16.4±2.7
13	Podophyllotoxin	鬼臼毒素	414.405	19.4±2.7
14	Oridonin	冻凌草甲素	364.43	18.4±0.7
15	Curcumin	姜黄素	368.38	28.7±2.3
16	Magnolol	厚朴酚	266.33	64.9±4.3
17	Costunolide	木香烃内酯	232.32	37.7±3.3
18	Pseudolaric acid	土荆皮乙酸	430.491	8.7±1.9

Table 1. MMT assay results of the cytotoxic activities of various compounds against cancer cells with their IC_{50} values.

Autophagy represents a major route for degradation of aggregated cellular proteins and dysfunctional organelles. Accumulated lines of evidence have recently revealed that targeting autophagic signaling pathways might be a promising avenue for potential therapeutic purposes. Alterations in autophagy are thought to play an important role in the pathogenesis of many diseases—for example, Autophagy is closely associated with tumors and plays an important role in human tumor suppression, so inducing autophagy is a potential therapeutic strategy in adjuvant chemotherapy [61,62]. Many studies have demonstrated that anticancer agents induce autophagy, leading to the implications that autophagic cell death may be a vital mechanism for tumor cell killing by these agents [63] and are beneficial in the context of various models of cancer cells. The autophagy machinery interfaces many cellular stress-response pathways, and recent studies depict that defects in autophagy lead to cancer cell proliferation [64]. The regulation of autophagy in cancer cells can enhance tumor cell survival, yet can also suppress the initiation of tumor growth. Understanding the signaling pathways involved in the regulation of autophagy is crucial to the development of anticancer therapies [21]. In this chapter, we reviewed the natural compounds molecular mechanisms of autophagy and examined ongoing drug discovery strategies for modulating autophagy for therapeutic benefits. The natural antitumor agents have led to enhanced enthusiasm for the development of drugs that target the various aspects of the autophagic pathways. Some of these autophagic cellular approaches by representative natural compounds in autophagic induced cell death have been outlined in Fig. 3. In addition, magnolol and evodiamine have been illustrated in detail.

4.2. Magnolol

In our own studies, performing the screen for natural compounds that induce autophagy, we identified magnolol [52]. Magnolol, a natural compound, has been reported to inhibit growth in a variety of tumor cells [65]. Several researchers reported that magnolol-induced cell death involve apoptosis while Li *et al* [66] reported that magnolol-induced death occurs via autophagy but not apoptosis. We observed that there was no significant formation of AVOs at low concentration while AVOs were formed at a higher concentration of magnolol treated cells. The formation of acidic vesicular organelles (AVOs) is one of the characteristic features of cells, which passes through process of autophagy after their exposure to different autophagy inducer agents [67,68]. Autophagic vacuoles (AV) or autophagosomes are formed as result of sequestering of parts of the cytoplasm or entire organelles respectively during the process of autophagy [62]. Currently autophagic cell death has been studied as a potential method for cancer therapy. To determine the role of magnolol-induced autophagy in killing the SGC-7901 cells, we added the autophagy inhibitor, 3-methyladenine (3-MA), which controlled autophagy pathway at various points [8]. In contrast to the previous report [69], it was found that magnolol-induced cell death was not suppressed when treating the cells in combination with 3-MA. These results showed that magnolol-induced autophagy is not involved in the induction of SGC-7901 cell death. In addition, the findings also demonstrated that magnolol-induced autophagy may has an effect on ATP level in the SGC-7901 cells and supported those observations which showed that autophagy may alter the morphological and cellular events (ATP, cells blebbing and DNA fragmentation) that take place in apoptotic cell death, without leading to cell death in itself [8].

4.3. Evodiamine

Evodiamine is a naturally occurring quinolone alkaloid found in the fruit of *Evodia rutaecarpa*. The data of several studies concerning the cytotoxic activity on cancer cells demonstrated that evodiamine inhibited the growth of several tumor cells [70]. Results from our screen indicate that evodiamine induced apoptosis and autophagy simultaneously in human gastric cancer cells. Evodiamine has been reported as an inducer of autophagy in human cervical carcinoma HeLa cells [71]. Autophagy is closely associated with tumors and plays an important role in human tumor suppression, so inducing the autophagy is a potential therapeutic strategy in adjuvant chemotherapy [61,62]. When the cells are exposed to various autophagy inducer agents, they form acidic vesicular organelles, which is an important characteristic of autophagy [67,68]. Thus, we observed the effect of evodiamine treatment on the formations of AVOs in SGC-7901 cells using fluorescence microscopy after staining with the lysosomotropic agent, acridine orange (AO). These findings indicate that evodiamine, a natural compound, has the potential to activate autophagy in gastric cancer cells. This result of evodiamine is also consistent with the results of the studies reporting that natural compounds can induce autophagy in various cancer cells. We also demonstrated that evodiamine-induced cell death was partially suppressed when the cells were treated in combination with specific autophagy inhibitor, 3-MA. These results showed that evodiamine-induced autophagy was partially involved in the cell death of cancer cells [52]. While our recent studies demonstrated that autophagy inhibition enhanced evodiamine-induced apoptosis in prostate cancer cells, indicating a survival function of autophagy (unpublished data). These results corroborate the line of evidence demonstrating that evodiamine-induced autophagy, implicated in cell survival, contributes to the cytoprotective role of autophagy [71,72]. These facts demonstrated that there is still a great discrepancy between roles of natural compounds-induced autophagy in cancer cells. The role (or more likely roles because as we discuss, distinct functions for autophagy occur at different times) of natural compounds-induced autophagy in cancer, is a topic of intense debate. These responses might vary with cell type and type of stress, and will undoubtedly reflect the nature of the mutational events occurring in the tumor cells, not only that of *BECN1* and the PI3K pathway as described above, but also p53 status [73,74]. Moreover, it is believe that in short term assays, the phenomenal protection caused by autophagy inhibition may be due to delay in cell death instead of true protective effect and this inhibition causes an increase in tumor cell clonogenic growth after drug treatment. In most of the examples cited above, this appears as a starking effect, as the whole debate is about the drug induced autophagy and in response to which the cells die (or cells found dead). This novel approach emerged as an important point because a recent study also supports this myth that rapamycin-induced autophagy can protect various tumor cell lines against apoptosis induced by general apoptotic stimuli [75] and may have a similar effect on the action of anticancer agents. Moreover, similar to etoposide [74], it has been observed that knockdown of *Atg* genes does confer a clonogenic survival advantage to cells after treating with anticancer agents and the used cells have profound defects in their apoptosis machinery [73,74].

4.4. Role of autophagy in cancer: Science or myth?

It is generally believed that the complex two-faced nature of autophagy in tumor cell survival versus death may help in determining cancer therapeutic potential. So inhibiting autophagy may enhance anti-cancer drugs efficacy fairly used in chemo- and radiotherapy-induced activation of autophagic signaling pathways and which may augment anti-tumor activity, and thus efficacy of radiation and/or anti-cancer drugs. We are still at the initial stages of understanding the complex interplay of autophagy and cancer, but it is incontrovertible that autophagy is deeply integrated into metabolism, stress response and cell-death pathways [64]. Preliminary evidences, in addition to some natural compounds that induced autophagic cell

http://dx.doi.org/10.5772/55447

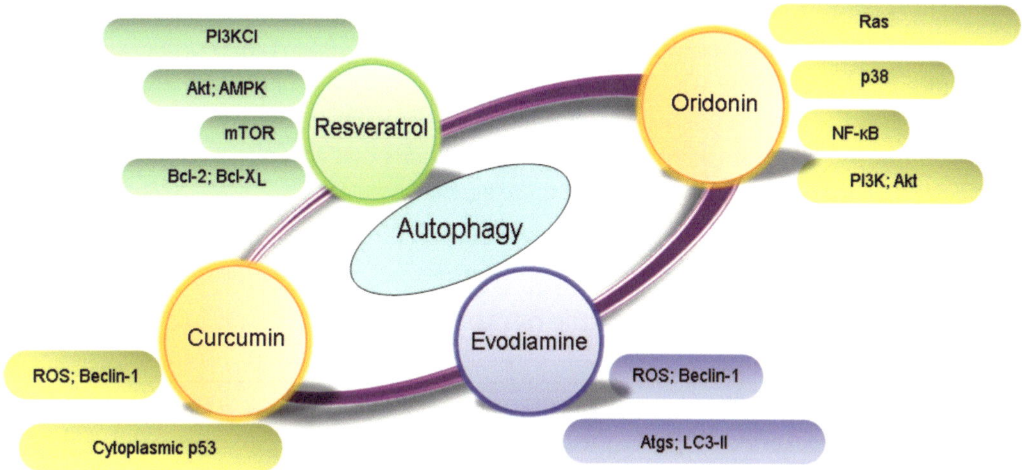

Figure 3. Representative natural compounds (Resveratrol, Oridonin, Curcumin, and Evodiamine) targeting autophagic pathways.

death, support the idea that natural compounds-induced autophagy enhances tumor cell survival. Anticancer agents that can be involved in the induction of autophagy include tamoxifen, arsenic trioxide, rapamycin, histone deacetylase inhibitors, temozolomide, ionizing radiation [63], vitamin D analogues [76], and etoposide [74]. In addition, several natural compounds, (curcumin, resveratrol, evodiamine, oridonin, and magnolol) in our natural compounds libraries screen for autophagy inducer, were found to be involved in autophagy. However, despite the above examples, is autophagy really an important cell death mechanism? is highly controversial.

Controversy remains as whether autophagy limits or promotes tumor malignancy, till genetic inactivation of autophagy, is found to promote tumorigenesis constituting a new category of tumor suppressors including Beclin 1. Some of the oncogenes including PI3K/AKT/mTOR and Bcl-2 inhibit autophagy causing tumor cells proliferation, while the other oncogenes including Ras and myc stimulate autophagy [63]. The significance of autophagy at different stages of tumor progression can be evaluated considering these kaleidoscopic effects. Further investigations on natural compounds into the impact of autophagy inactivation are warranted. All of the above data draw many questions in autophagy mechanistic pool focusing whether autophagy is really an important mechanism of tumor cell killing by anticancer agents in cells having ability to undergo apoptosis. Rigorous examination also manifest the speculation whether bona-fide cancer drugs are actually capable of killing tumor cells via autophagy, is needed. To answer these questions is the need of the hour, as it may determine the route causes and may best streamline the contradictory approaches in developing effective combination therapies by regulating autophagy along with anticancer agents. In conclusion, we now have sound justifications to visualize that manipulation of autophagy may provide a useful way to prevent cancer development, limit tumor progression, and increase the efficacy of cancer treatments. This comprehension seems reasonable due to the fact that drugs induce autophagy, such as rapamycin (as discussed above), and is rapidly gaining a better understanding of how this process works based on the effects of targeted inactivation of autophagy regulators in mouse models and human tumor cells. More contradictory messages come when we consider how autophagy affects the ways by which the tumor cells die when we treat them with anticancer agents. Over the last several decades the therapeutic use of natural compounds that induce autophagy has been leading us to the implications that autophagic cell death may be a vital mechanism of tumor cell killing by these agents.

5. Concluding remarks and future perspectives

Accumulated lines of evidence have recently revealed that targeting autophagic signaling pathways may be a promising avenue for potential therapeutic purposes. Although this chapter has focused on natural compounds and their role in autophagic cell signaling pathways, future studies investigating the mechanisms of natural compounds-induced autophagy and their role in cancer cell death. Progress towards better treatment and understanding by natural compounds may be made by further examining the role of natural compounds and crosstalk between the apoptosis and the autophagy. Despite these obstacles, many compounds bring the hope that with sufficient modification by tools of structural biology and combinatorial chemistry, it might be possible to derive sufficiently potent drugs to target core autophagy pathways, and even autophagic networks in cancer cells, rather than their individual gene or protein components. Indeed, as discussed above, the generally used cancer therapeutics, especially natural compounds abolish tumors by inducing apoptosis and autophagy. On the other hand, a better but growing setting approach is required to make a distinction between the survival-supporting and death-promoting roles of autophagy. Furthermore, for selectivity and specificity, role of autophagy, along with the elucidation of the signaling pathways those confer the autophagic response downstream of different stimuli and activate the specific and therapeutic response, is desired.. In the end we have coherent arguments in favor of principal paradigm that disease-associated autophagy could be selectively targeted for therapeutics.

Author details

Azhar Rasul and Tonghui Ma*

Central Research Laboratory, Jilin University, Bethune Second Hospital, Changchun, China

References

[1] Ashford T PPorter K R. Cytoplasmic components in hepatic cell lysosomes. J Cell Biol 1962;12 198-202.

[2] Watanabe E, Muenzer J T, Hawkins W G, Davis C G, Dixon D J, McDunn J E, Brackett D J, Lerner M R, Swanson P EHotchkiss R S. Sepsis induces extensive autophagic vacuolization in hepatocytes: a clinical and laboratory-based study. Lab Invest 2009;89 549-561.

[3] Schweichel J UMerker H J. The morphology of various types of cell death in prenatal tissues. Teratology 1973;7 253-266.

[4] Klionsky D J. Autophagy: from phenomenology to molecular understanding in less than a decade. Nat Rev Mol Cell Biol 2007;8 931-937.

[5] Kroemer GJaattela M. Lysosomes and autophagy in cell death control. Nat Rev Cancer 2005;5 886-897.

[6] Levine BDeretic V. Unveiling the roles of autophagy in innate and adaptive immunity. Nat Rev Immunol 2007;7 767-777.

[7] Levine BKlionsky D J. Development by self-digestion: molecular mechanisms and biological functions of autophagy. Dev Cell 2004;6 463-477.

[8] Eisenberg-Lerner A, Bialik S, Simon H UKimchi A. Life and death partners: apoptosis, autophagy and the cross-talk between them. Cell Death Differ 2009;16 966-975.

[9] Galluzzi L, Vitale I, Abrams J M, Alnemri E S, Baehrecke E H, Blagosklonny M V, Dawson T M, Dawson V L, El-Deiry W S, Fulda S, Gottlieb E, Green D R, Hengartner M O, Kepp O, Knight R A, Kumar S, Lipton S A, Lu X, Madeo F, Malorni W, Mehlen P, Nunez G, Peter M E, Piacentini M, Rubinsztein D C, Shi Y, Simon H U, Vandenabeele P, White E, Yuan J, Zhivotovsky B, Melino GKroemer G. Molecular definitions of cell death subroutines: recommendations of the Nomenclature Committee on Cell Death 2012. Cell Death Differ 2012;19 107-120.

[10] Evan G IVousden K H. Proliferation, cell cycle and apoptosis in cancer. Nature 2001;411 342-348.

[11] Jemal A, Siegel R, Ward E, Hao Y, Xu J, Murray TThun M J. Cancer statistics, 2008. CA Cancer J Clin 2008;58 71-96.

[12] Amaravadi R KThompson C B. The roles of therapy-induced autophagy and necrosis in cancer treatment. Clin Cancer Res 2007;13 7271-7279.

[13] Marx J. Autophagy: is it cancer's friend or foe? Science 2006;312 1160-1161.

[14] Hotchkiss R S, Strasser A, McDunn J ESwanson P E. Cell death. N Engl J Med 2009;361 1570-1583.

[15] Maiuri M C, Tasdemir E, Criollo A, Morselli E, Vicencio J M, Carnuccio RKroemer G. Control of autophagy by oncogenes and tumor suppressor genes. Cell Death Differ 2009;16 87-93.

[16] Rosenfeldt M TRyan K M. The multiple roles of autophagy in cancer. Carcinogenesis 2011;32 955-963.

[17] Levine BYuan J. Autophagy in cell death: an innocent convict? J Clin Invest 2005;115 2679-2688.

[18] Kroemer G, Galluzzi L, Vandenabeele P, Abrams J, Alnemri E S, Baehrecke E H, Blagosklonny M V, El-Deiry W S, Golstein P, Green D R, Hengartner M, Knight R A, Kumar S, Lipton S A, Malorni W, Nunez G, Peter M E, Tschopp J, Yuan J, Piacentini M, Zhivotovsky BMelino G. Classification of cell death: recommendations of the Nomenclature Committee on Cell Death 2009. Cell Death Differ 2009;16 3-11.

[19] Galluzzi L, Maiuri M C, Vitale I, Zischka H, Castedo M, Zitvogel LKroemer G. Cell death modalities: classification and pathophysiological implications. Cell Death Differ 2007;14 1237-1243.

[20] Degenhardt K, Mathew R, Beaudoin B, Bray K, Anderson D, Chen G, Mukherjee C, Shi Y, Gelinas C, Fan Y, Nelson D A, Jin SWhite E. Autophagy promotes tumor cell survival and restricts necrosis, inflammation, and tumorigenesis. Cancer Cell 2006;10 51-64.

[21] Yang Z J, Chee C E, Huang SSinicrope F A. The role of autophagy in cancer: therapeutic implications. Mol Cancer Ther 2011;10 1533-1541.

[22] Mathew R, Karp C M, Beaudoin B, Vuong N, Chen G, Chen H Y, Bray K, Reddy A, Bhanot G, Gelinas C, Dipaola R S, Karantza-Wadsworth VWhite E. Autophagy suppresses tumorigenesis through elimination of p62. Cell 2009;137 1062-1075.

[23] Qu X, Yu J, Bhagat G, Furuya N, Hibshoosh H, Troxel A, Rosen J, Eskelinen E L, Mizushima N, Ohsumi Y, Cattoretti GLevine B. Promotion of tumorigenesis by heterozygous disruption of the beclin 1 autophagy gene. J Clin Invest 2003;112 1809-1820.

[24] Liang X H, Jackson S, Seaman M, Brown K, Kempkes B, Hibshoosh HLevine B. Induction of autophagy and inhibition of tumorigenesis by beclin 1. Nature 1999;402 672-676.

[25] Duran A, Linares J F, Galvez A S, Wikenheiser K, Flores J M, Diaz-Meco M TMoscat J. The signaling adaptor p62 is an important NF-kappaB mediator in tumorigenesis. Cancer Cell 2008;13 343-354.

[26] Tang D, Kang R, Livesey K M, Cheh C W, Farkas A, Loughran P, Hoppe G, Bianchi M E, Tracey K J, Zeh H J, 3rdLotze M T. Endogenous HMGB1 regulates autophagy. J Cell Biol 2010;190 881-892.

[27] White EDiPaola R S. The double-edged sword of autophagy modulation in cancer. Clin Cancer Res 2009;15 5308-5316.

[28] Semenza G L. HIF-1: upstream and downstream of cancer metabolism. Curr Opin Genet Dev 2010;20 51-56.

[29] Yang S, Wang X, Contino G, Liesa M, Sahin E, Ying H, Bause A, Li Y, Stommel J M, Dell'antonio G, Mautner J, Tonon G, Haigis M, Shirihai O S, Doglioni C, Bardeesy NKimmelman A C. Pancreatic cancers require autophagy for tumor growth. Genes Dev 2011;25 717-729.

[30] Lu Z, Luo R Z, Lu Y, Zhang X, Yu Q, Khare S, Kondo S, Kondo Y, Yu Y, Mills G B, Liao W SBast R C, Jr. The tumor suppressor gene ARHI regulates autophagy and tumor dormancy in human ovarian cancer cells. J Clin Invest 2008;118 3917-3929.

[31] Guo J Y, Chen H Y, Mathew R, Fan J, Strohecker A M, Karsli-Uzunbas G, Kamphorst J J, Chen G, Lemons J M, Karantza V, Coller H A, Dipaola R S, Gelinas C, Rabinowitz J DWhite E. Activated Ras requires autophagy to maintain oxidative metabolism and tumorigenesis. Genes Dev 2011;25 460-470.

[32] Schneider M D. Cyclophilin D: knocking on death's door. Sci STKE 2005;2005 pe26.

[33] Orrenius S, Nicotera PZhivotovsky B. Cell death mechanisms and their implications in toxicology. Toxicol Sci 2011;119 3-19.

[34] Vandenabeele P, Vanden Berghe TFestjens N. Caspase inhibitors promote alternative cell death pathways. Sci STKE 2006;2006 pe44.

[35] Levine BAbrams J. p53: The Janus of autophagy? Nat Cell Biol 2008;10 637-639.

[36] Long J SRyan K M. New frontiers in promoting tumour cell death: targeting apoptosis, necroptosis and autophagy. Oncogene 2012;

[37] Luo J L, Kamata HKarin M. IKK/NF-kappaB signaling: balancing life and death--a new approach to cancer therapy. J Clin Invest 2005;115 2625-2632.

[38] Moubarak R S, Yuste V J, Artus C, Bouharrour A, Greer P A, Menissier-de Murcia JSusin S A. Sequential activation of poly(ADP-ribose) polymerase 1, calpains, and Bax is essential in apoptosis-inducing factor-mediated programmed necrosis. Mol Cell Biol 2007;27 4844-4862.

[39] Yousefi S, Perozzo R, Schmid I, Ziemiecki A, Schaffner T, Scapozza L, Brunner TSimon H U. Calpain-mediated cleavage of Atg5 switches autophagy to apoptosis. Nat Cell Biol 2006;8 1124-1132.

[40] Maclean K H, Dorsey F C, Cleveland J LKastan M B. Targeting lysosomal degradation induces p53-dependent cell death and prevents cancer in mouse models of lymphomagenesis. J Clin Invest 2008;118 79-88.

[41] Murray-Zmijewski F, Slee E ALu X. A complex barcode underlies the heterogeneous response of p53 to stress. Nat Rev Mol Cell Biol 2008;9 702-712.

[42] Tasdemir E, Maiuri M C, Galluzzi L, Vitale I, Djavaheri-Mergny M, D'Amelio M, Criollo A, Morselli E, Zhu C, Harper F, Nannmark U, Samara C, Pinton P, Vicencio J M, Carnuccio R, Moll U M, Madeo F, Paterlini-Brechot P, Rizzuto R, Szabadkai G,

Pierron G, Blomgren K, Tavernarakis N, Codogno P, Cecconi FKroemer G. Regulation of autophagy by cytoplasmic p53. Nat Cell Biol 2008;10 676-687.

[43] Rasul AMa T. In vitro cytotoxic screening of 300 selected Chinese medicinal herbs against human gastric adenocarcinoma SGC-7901 cells. Afr J Pharm Pharmaco 2012;6 592 - 600.

[44] Zhang X, Chen L X, Ouyang L, Cheng YLiu B. Plant natural compounds: targeting pathways of autophagy as anti-cancer therapeutic agents. Cell Prolif 2012;45 466-476.

[45] Kanzawa T, Kondo Y, Ito H, Kondo SGermano I. Induction of autophagic cell death in malignant glioma cells by arsenic trioxide. Cancer Res 2003;63 2103-2108.

[46] Kanzawa T, Zhang L, Xiao L, Germano I M, Kondo YKondo S. Arsenic trioxide induces autophagic cell death in malignant glioma cells by upregulation of mitochondrial cell death protein BNIP3. Oncogene 2005;24 980-991.

[47] Khan M, Yi F, Rasul A, Li T, Wang N, Gao H, Gao RMa T. Alantolactone induces apoptosis in glioblastoma cells via GSH depletion, ROS generation, and mitochondrial dysfunction. IUBMB Life 2012;64 783-794.

[48] Khan M, Zheng B, Yi F, Rasul A, Gu Z, Li T, Gao H, Qazi J I, Yang HMa T. Pseudolaric Acid B induces caspase-dependent and caspase-independent apoptosis in u87 glioblastoma cells. Evid Based Complement Alternat Med 2012;2012 957568.

[49] Rasul A, Ding C, Li X, Khan M, Yi F, Ali MMa T. Dracorhodin perchlorate inhibits PI3K/Akt and NF-kappaB activation, up-regulates the expression of p53, and enhances apoptosis. Apoptosis 2012;17 1104-1119.

[50] Rasul A, Khan M, Yu B, Ma TYang H. Xanthoxyletin, a coumarin induces S phase arrest and apoptosis in human gastric adenocarcinoma SGC-7901 cells. Asian Pac J Cancer Prev 2011;12 1219-1223.

[51] Rasul A, Yu B, Yang L, Arshad M, Khan M, Ma TYang H. Costunolide, a sesquiterpene lactone induces G2/M phase arrest and mitochondria-mediated apoptosis in human gastric adenocarcinoma SGC-7901 cells. J Med Plant Res 2012;6 1191-1200.

[52] Rasul A, Yu B, Khan M, Zhang K, Iqbal F, Ma TYang H. Magnolol, a natural compound, induces apoptosis of SGC-7901 human gastric adenocarcinoma cells via the mitochondrial and PI3K/Akt signaling pathways. Int J Oncol 2012;40 1153-1161.

[53] Rasul A, Yu B, Zhong L, Khan M, Yang HMa T. Cytotoxic effect of evodiamine in SGC-7901 human gastric adenocarcinoma cells via simultaneous induction of apoptosis and autophagy. Oncol Rep 2012;27 1481-1487.

[54] Hartwell J L. Plants used against cacer. Quarterman, Lawrence, MA. 1982;

[55] Cragg G MNewman D J. Plants as a source of anti-cancer agents. J Ethnopharmacol 2005;100 72-79.

[56] Nobili S, Lippi D, Witort E, Donnini M, Bausi L, Mini ECapaccioli S. Natural compounds for cancer treatment and prevention. Pharmacol Res 2009;59 365-378.

[57] Fulda SDebatin K M. Targeting inhibitor of apoptosis proteins (IAPs) for diagnosis and treatment of human diseases. Recent Pat Anticancer Drug Discov 2006;1 81-89.

[58] Kaufmann S HEarnshaw W C. Induction of apoptosis by cancer chemotherapy. Exp Cell Res 2000;256 42-49.

[59] Okada HMak T W. Pathways of apoptotic and non-apoptotic death in tumour cells. Nat Rev Cancer 2004;4 592-603.

[60] Ricci M SZong W X. Chemotherapeutic approaches for targeting cell death pathways. Oncologist 2006;11 342-357.

[61] Scherz-Shouval RElazar Z. ROS, mitochondria and the regulation of autophagy. Trends Cell Biol 2007;17 422-427.

[62] Yoshimori T. Autophagy: a regulated bulk degradation process inside cells. Biochem Biophys Res Commun 2004;313 453-458.

[63] Kondo Y, Kanzawa T, Sawaya RKondo S. The role of autophagy in cancer development and response to therapy. Nat Rev Cancer 2005;5 726-734.

[64] Mathew R, Karantza-Wadsworth VWhite E. Role of autophagy in cancer. Nat Rev Cancer 2007;7 961-967.

[65] Lee Y J, Lee Y M, Lee C K, Jung J K, Han S BHong J T. Therapeutic applications of compounds in the Magnolia family. Pharmacol Ther 2011;130 157-176.

[66] Li H B, Yi X, Gao J M, Ying X X, Guan H QLi J C. Magnolol-induced H460 cells death via autophagy but not apoptosis. Arch Pharm Res 2007;30 1566-1574.

[67] Daido S, Kanzawa T, Yamamoto A, Takeuchi H, Kondo YKondo S. Pivotal role of the cell death factor BNIP3 in ceramide-induced autophagic cell death in malignant glioma cells. Cancer Res 2004;64 4286-4293.

[68] Paglin S, Hollister T, Delohery T, Hackett N, McMahill M, Sphicas E, Domingo DYahalom J. A novel response of cancer cells to radiation involves autophagy and formation of acidic vesicles. Cancer Res 2001;61 439-444.

[69] Li H B, Yi X, Gao J M, Ying X X, Guan H QLi J C. Magnolol-induced H460 cells death via autophagy but not apoptosis. Arch. Pharm. Res. 2007;30 1566-1574.

[70] Jiang JHu C. Evodiamine: a novel anti-cancer alkaloid from Evodia rutaecarpa. Molecules 2009;14 1852-1859.

[71] Yang J, Wu L J, Tashino S, Onodera SIkejima T. Reactive oxygen species and nitric oxide regulate mitochondria-dependent apoptosis and autophagy in evodiamine-treated human cervix carcinoma HeLa cells. Free Radic Res 2008;42 492-504.

[72] Tu Y J, Fan X, Yang X, Zhang CLiang H P. Evodiamine activates autophagy as a cytoprotective response in murine Lewis lung carcinoma cells. Oncol Rep 2012;

[73] Crighton D, Wilkinson S, O'Prey J, Syed N, Smith P, Harrison P R, Gasco M, Garrone O, Crook TRyan K M. DRAM, a p53-induced modulator of autophagy, is critical for apoptosis. Cell 2006;126 121-134.

[74] Shimizu S, Kanaseki T, Mizushima N, Mizuta T, Arakawa-Kobayashi S, Thompson C BTsujimoto Y. Role of Bcl-2 family proteins in a non-apoptotic programmed cell death dependent on autophagy genes. Nat Cell Biol 2004;6 1221-1228.

[75] Ravikumar B, Berger Z, Vacher C, O'Kane C JRubinsztein D C. Rapamycin pre-treatment protects against apoptosis. Hum Mol Genet 2006;15 1209-1216.

[76] Hoyer-Hansen M, Bastholm L, Mathiasen I S, Elling FJaattela M. Vitamin D analog EB1089 triggers dramatic lysosomal changes and Beclin 1-mediated autophagic cell death. Cell Death Differ 2005;12 1297-1309.

Section 6

Autophagy in Infectious Diseases

Chapter 13

Infectious Agents and Autophagy: Sometimes You Win, Sometimes You Lose

Patricia Silvia Romano

Additional information is available at the end of the chapter

http://dx.doi.org/10.5772/55179

1. Introduction

Successful microorganisms are those that can evade the immune responses of the host. To reach this purpose, many pathogens have evolved as intracellular organisms, acquiring the capacity to live and to develop inside cells. This cellular parasitism has many benefits for pathogens such as protection from circulating antibodies, free access to nutrients and to specialized compartments that microorganisms use to establish their replicative niches. According to their particular lifestyles and requirements, many pathogens such as *L.monocytogenes, Shigella* or *T. cruzi*; reproduce in cell cytosol, while others target specific vesicles to generate particular compartments called parasitophorous vacuoles (PVs). This class of parasitism is utilized by *M. tuberculosis, C. burnetii* or *T. gondii*. On the other hand, in response to this level of adaptation, mammalian cells have developed different processes for eliminating intracellular microorganisms or for keeping them under strict control. These mechanisms are part of the innate immune responses, being phagocytosis (and the related processes) the best characterized. Innate cellular immunity also encompasses the autophagic process, a well conserved eukaryotic pathway that interacts with intracellular pathogens under certain circumstances to produce the destruction of the foreign organism. Autophagy comprises the inclusion of pathogens in autophagic-derived compartments and delivers them in lysosomes for digestion. Some pathogens, however, have acquired the capacity to subvert autophagy for their own benefit. This chapter will describe the interaction between intracellular microorganisms and the defense mechanisms of host cells, with special focus on the dual involvement of autophagy against pathogens, and the net outcome of this interaction.

2. The phagocytosis process

Phagocytosis is a form of endocytosis mainly present in specialized types of cells: the professional phagocytes that include the neutrophil, monocytes/macrophages, and dendritic cells. In the initial stage of phagocytosis, the cells change shape by sending out membrane projections (pseudopodia) that contact and surround the particle (bacteria, apoptotic bodies, etc) in a receptor-mediated and actin-dependent process (Figure 1). When the tips of the pseudopodia meet each other, membrane fusion occurs and the particle is enveloped in a vesicular compartment called phagosome. According to the type of internalized particle and the class of phagocyte involved, a different set of processes are activated that end in the destruction of the enclosed material. These mechanisms involve:

2.1. The respiratory or oxidative burst

The respiratory or oxidative burst is produced by the activity of specialized enzymes such as NADPH oxidase, which generates superoxide that recombines with other molecules as NO to form peroxynitrite, a potent oxidant agent against bacteria and parasites. Neutrophils and monocytes also utilize myeloperoxidase to further combine H_2O_2 with Cl^- to produce hypochlorite, a harmful component of phagosomes [1]

2.2. The production of microbicidal substances

Lysozime and defensins attack cell walls and membranes of certain bacteria.

2.3. Phagosome maturation

Phagosome maturation is a process that confers to nascent phagosomes the ability to kill pathogens or to degrade the ingested materials. Phagosomal maturation involves a complex sequence of reactions that result in the drastic remodeling of the phagosomal membrane and contents, produced as a consequence of vesicular fusion and fission events between the nascent vacuole and other cellular compartments mainly belonging to the endocytic pathway [2]. Rab and SNARES proteins are the main molecular components that regulate these events. Rabs are small GTP-binding proteins that control intracellular trafficking and supervise the maintenance of specific organellar identity, whereas SNARES are transmembrane proteins that, associating with their specific partners, form complexes that are the final executioners of the fusion processes between membranes. Interactions between phagosomes and endosomes commence soon after phagosome sealing, in a fashion that recapitulates the endocytic sequence: nascent (early) phagosomes (Eph) seemingly fuse initially with sorting (early) endosomes (EE), followed by late endosomes (LE) and ultimately lysosomes (Ly). Therefore, the membrane of Eph initially acquires components present in early endosomes such as Rab 5, phosphatidylinositol 3-phosphate (PI3P), Early Endosomal Antigen 1 (EEA-1) and Vamp-3. In contrast, late phagosomes (Lph) present Rab 7, Mannose-6-phosphate receptor, Vamp-7 and Lysosomal associated membrane proteins 1 and 2 (LAMPS 1 and 2) [2]. Furthermore, the luminal environment of phagosomes turned progressively more acidic due to the accumula-

http://dx.doi.org/10.5772/55179

tion of H+ ATPase complexes in the phagosome membrane [3]. Phagolysosomes (PLy), the hybrid compartment generated by fusion between Lph and lysosomes, reach a pH of around 4.5, favoring the maturation of acidic hydrolases that will finally digest the materials (Figure 1, red compartments). Lysosomes contain several proteases, including Cathepsin D and Elastase, which are essential for killing various bacteria.

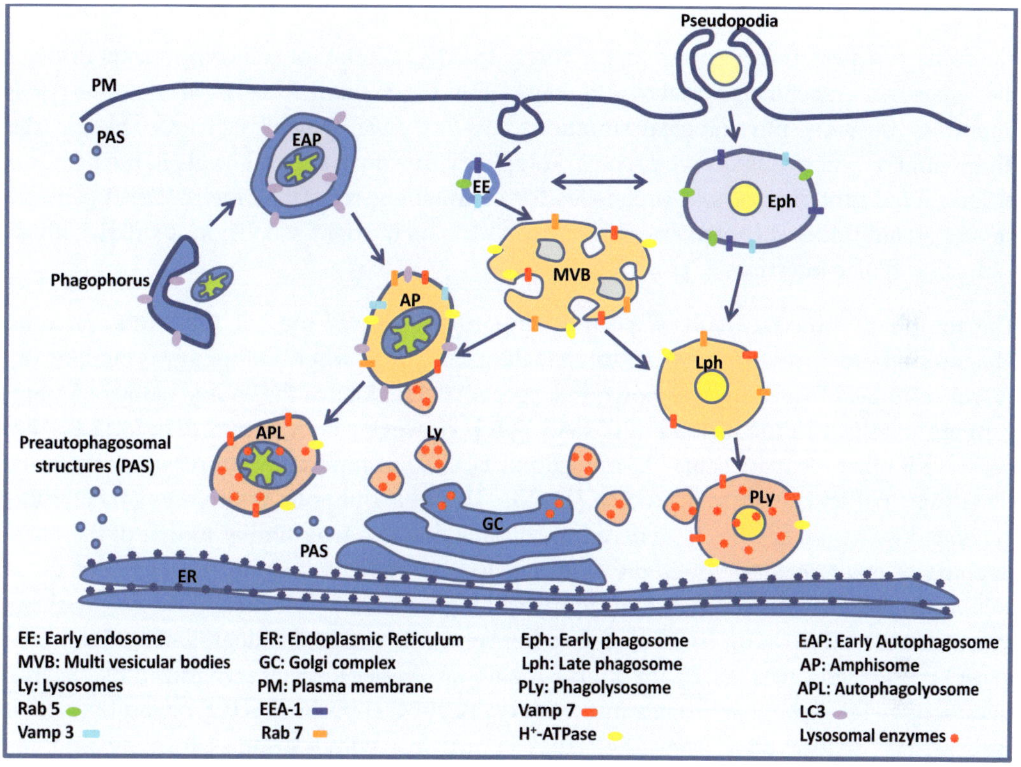

Figure 1. Phagocytosis and autophagy are the main cellular processes involved in the innate immune responses against pathogens. The scheme depicts the vesicular compartments that belong to each process and the main molecules that characterize them. Note that earlier, non-degradative compartments are colored in blue whereas acidic, lysosomal derived compartments are showed in red.

All these mechanisms generate a high level of protection against a wide range of pathogens. Paradoxically, phagocytosis can also have deleterious effects for the host: certain pathogens, exemplified by *Mycobacteria*, take advantage of the phagocytic machinery to gain access to the cell interior where, by subverting the maturation process, become intracellular pathogens [4–6].

Phagocytosis can also be produced in a class of cells different from immune cells. These "non -professional phagocytes" are cells with low phagocytic competence. Pathogens that can colonize these cells avoid the harmful ambient of the phagocyte-derived phagosome because many of the killing processes described for phagocytes are low or absent in non-professional phagocytes, keeping lysosomal degradation as the main defense system. In this way, the autophagic pathway which delivers cytoplasmic materials to lysosomes constitutes an important mechanism for eliminating pathogens, especially in these non-immune cells.

3. The autophagic pathway

Autophagy is a catabolic process that involves the degradation of cell components through the lysosomal machinery. Macroautophagy, the most studied type of autophagy, is important in many physiological situations such as development, cell growth, and cell differentiation. As a constitutive process, autophagy functions at basal levels in the turnover of long lived proteins and old organelles for maintaining cellular homeostasis. It can also be stimulated under different stress situations such as nutrient starvation, oxidative stress and intracellular infections [7].

The autophagic process involves specific compartments inside the cell. The initial preautophagosomal structures (PAS) are recruited to the cellular sites where autophagy is initiated [8]. A large number of studies have shown that specialized regions of the endoplasmic reticulum (ER) are involved in the formation of PAS [9,10]. However, more recent data indicate that besides ER other compartments like mitochondria, Golgi complex (GC) or plasma membrane (PM) may participate in this process [11–13]. The phagophorus, or isolation membrane, generated by fusion of PAS, is a curved membrane that in a way similar to that of the pseudopodia of macrophages, wraps the materials to be trapped which, in this case, consist of soluble or membranous content from cytoplasm (Figure 1, left). This membrane finally closes in a structure called autophagosome that transports the cargo for degradation. Immature (early) autophagosomes (EAP) are double-membrane vesicles easily recognized by electron microscopy [14]. Autophagosomes fuse with endocytic compartments (LE or multivesicular bodies, MVB) to form amphisomes (AP) that, in turn, fuse with lysosomes; forming autophagolysosomes (APL) where the materials are degraded (Figure 1, red compartments).

At the molecular level, a large number of proteins engage in autophagy. The specific autophagy related proteins (Atgs) are a large family of proteins that regulate the nucleation of PAS, and the formation and elongation of phagophorus and autophagosomes. At least 16 genes were found to be important for autophagy in yeasts, especially in the PAS nucleation [15]. Two protein conjugation reactions, both catalyzed by the action of Atg7, (an E3-like ubiquitin ligase activity), are mainly required for autophagosome formation in mammals. The mammalian Atg5-Atg12-Atg16L complex is recruited to the isolation membranes, favoring the elongation of the precursor membrane. The second conjugation system yields LC3-II which inserts into the autophagosomal membrane and contributes to vesicle elongation [16]. Pro-LC3 is initially cleaved by Atg4 to produce LC3-I. This molecule is a soluble protein distributed in the cytoplasm. After autophagic induction, LC3-I is conjugated with phosphatidylethanolamine (PE), allowing the insertion in the membrane of autophagic vesicles [16]. Two key signaling nodes converge to correlate autophagy with cell nutrient or stress conditions. The Tor-Atg1 signaling cascade transduces the response from growth factors, via class I PI3K, Akt/PKB, and so forth, to negatively regulate autophagy [17]. The second system, formed by Beclin1 (Atg6) and hVps34, is a lipid kinase that produces PI3P, which plays a pivotal role in early autophagosome formation, LC3 lipidation and the maturation of autophagosomes into autolysosomes [18].

Proteins that regulate transport and fusion events between vesicles are also important in autophagosome formation and maturation. Rab 7, a protein involved in transport to late endosomes and in the biogenesis of the perinuclear lysosome compartment is required for the normal progression of autophagosomes to autophagolysosomes [19]. The N-ethylmaleimide-sensitive factor attachment protein receptors (SNARES) Vamp3 and Vamp7 are important during the first steps of autophagy [20,21], whereas Vamp7 and Vamp8 also participate in the autophagosome-lysosome fusion [20,22]. Furthermore, it has been recently shown that actin has a role in the very early stages of autophagosome formation, linked to the PI3P generation step [23]. The description above shows that autophagosome formation and maturation engage similar molecular transport components and fusion machinery than that required for progression in endocytic and phagocytic pathways (Figure 1).

The process of autophagy can be monitored intracellularly by utilizing LC3 fused to a fluorescent protein (GFP-LC3 or mCherry-LC3). As the fluorescent LC3 is incorporated into autophagosomes, they can be seen as small puncta within the cell. As autophagy is a highly dynamic process, the number of puncta seen in a cell is a function of initiation as well as clearance (lysosomal fusion and subsequent degradation) [24]. Autophagosome initiation can be inhibited by blocking Class III PI3 kinases (Vps34) with 3-methyladenine or wortmannin, or by knockdown of essential factors such as Atg5 or Beclin-1, a component of Vps34 kinase complex. Autophagosome clearance can be prevented by interfering with lysosomal fusion by Bafilomycin A1, chloroquine, and other agents that tend to alkalinize the lysosome (e.g. NH_4Cl). Rapamycin inhibits the TOR signaling pathway, leading to induction of autophagy. Spermidine and resveratrol have been recently characterized as autophagy inducers. Genetic and functional studies indicate that spermidine inhibits histone acetylases, while resveratrol activates the histone deacetylase Sirtuin 1. Although it remains elusive whether the same histones (or perhaps other nuclear or cytoplasmic proteins) act as the downstream targets of spermidine and resveratrol, these results point to an essential role of protein hypoacetylation in autophagy control [25]. This hipoacetylated protein status leads to upregulation of several *atg* genes, including *atg7*, *atg11* and *atg15* in several organisms such as mammals, yeasts, nematodes and flies [26].

As explained above, host autophagy is a component of the innate responses against intracellular pathogens that generally functions as a second barrier when phagocytic or other defense mechanisms are exceeded. However, some pathogens have the capacity to evade autophagic responses or to subvert the autophagic pathway and to live and replicate inside autophagosomal compartments. The following sections describe the opposite effects of autophagic response against microorganisms.

3.1. Autophagy as a component of the innate immune responses

In order to internalize host cells, many pathogens induce their own ingestion in phagocytic cells. After entry, pathogenic microorganisms manipulate the normal (or "canonical") phagocytic pathway to evade lysosomal degradation and to achieve the maximal benefits (protection, nutrition, survival, and replication) from cells. These actions include inhibition of phagosome maturation, escape from phagosome to cytoplasm and development in a vacuole with particular characteristics. As a component of immune responses, autophagy hampers these mechanisms enclosing viruses, bacteria or parasites in compartments which share characteristics and molecular machinery with canonical autophagosomes, a process usually named as xenophagy [27]. For a better comprehension of the autophagic action, each pathogenic strategy will be described, and the most characteristic pathogen of each group will be exemplified in the following paragraphs.

- *The Mycobacteria case*: A marquee feature of the powerful human pathogen *Mycobacterium tuberculosis* is its macrophage parasitism. The intracellular survival of this microorganism rests upon its ability to arrest phagolysosome biogenesis, avoid direct cidal mechanisms in macrophages, and block efficient antigen processing and presentation. Lipoarabinomannan (LAM) and phoshatidylinositol mannoside (PIM) are two toxins elaborated by *Mycobacterium tuberculosis* that stimulate fusion between phagosomes and early endosomes and prevent Rab conversion on phagosomes by interference with Rab effectors, especially the type III PI3K (hVps34). LAM abolishes the normal recruitment of PI3K to mycobacterium phagosomal membrane decreasing the levels of PI3P. The final result is the reduction of the recruitment of EEA-1 and other effectors and the inhibition of the normal progression from early to late phagosomes [28]. Thus, a critical feature of the *M. tuberculosis* phagosome is its lack of the vacuolar H+ ATPase [29] and mature lysosomal hydrolases, such as Cathepsin D. In stark contrast, the induction of autophagy by physiological, pharmacological or immunological signals, including the major antituberculosis Th1 cytokine IFN-gamma, can overcome mycobacterial phagosome maturation block. Almost ten years ago, Gutierrez and colleagues demonstrated that when infected macrophages were treated in conditions that induce autophagy, mycobacterium-containing phagosomes become more acidic and also acquire markers of maturation, including the vacuolar H+ ATPase, LAMP-1, LBPA and Cathepsin D. Additionally, starvation promotes recruitment to mycobacterial phagosomes of critical autophagy components such as LC3, indicating that these phagosomes are redirected to a compartment with autophagic characteristics that finally fuses with lysosomes [30]. The most remarkable finding of this work was the demonstration that autophagy induction hampered the survival of this intracellular pathogen, recognizing to autophagy as an effector of innate immunity (see Figure 2A).

- *The* group A *Streptococcus case:* The second case belongs to pathogens that escape from phagosomes and turn into cytosolic invaders. The Group A of *Streptococcus* (GAS) is often internalized into nonphagocytic epithelial cells via the endocytic pathway. At early times after infection, GAS secrets its major virulence factor: the cytolysin streptolysin O (SLO) that supports the escape of GAS into the cytoplasm from endosomes [31]. After escaping, GAS became enveloped by autophagosome-like compartments and were killed upon fusion of

http://dx.doi.org/10.5772/55179

these compartments with lysosomes. In contrast, in autophagy-deficient Atg5-/- cells, GAS survived, multiplied, and were released from the cells [32,33]. Additionally a SLO-deficient mutant of GAS was viable for a longer time than the wild-type strain, although it failed to escape the endosomes [31]. Both results highlight the crucial role of autophagy in the suppression of intracellular survival of this pathogen. A similar conclusion was recently obtained with *Staphylococcus aureus*. After invasion of non-phagocytic cells, virulent strains of this gram positive bacterium stimulate autophagy and become entrapped in intracellular PI3P-enriched vesicles and its effector WIPI-1, a protein present in the membrane of both phagophores and autophagosomes. This interaction seems to be deleterious for bacteria, given that these autophagosome-like WIPI-1 positive vesicles that envelope *S. aureus* are finally targeted for lysosomal degradation [34].

- *The Toxoplasma case*: The third strategy used by pathogenic organisms is to create a specialized compartment that remains isolated from the host endocytic or phagocytic networks. *Toxoplasma gondii* relies on this mechanism; the membrane of its PV is nonfusogenic due to its unique composition lacking host proteins. Nonetheless, macrophages infected with *Toxoplasma* can reroute the pathogen-containing compartment to lysosomes. Autophagy

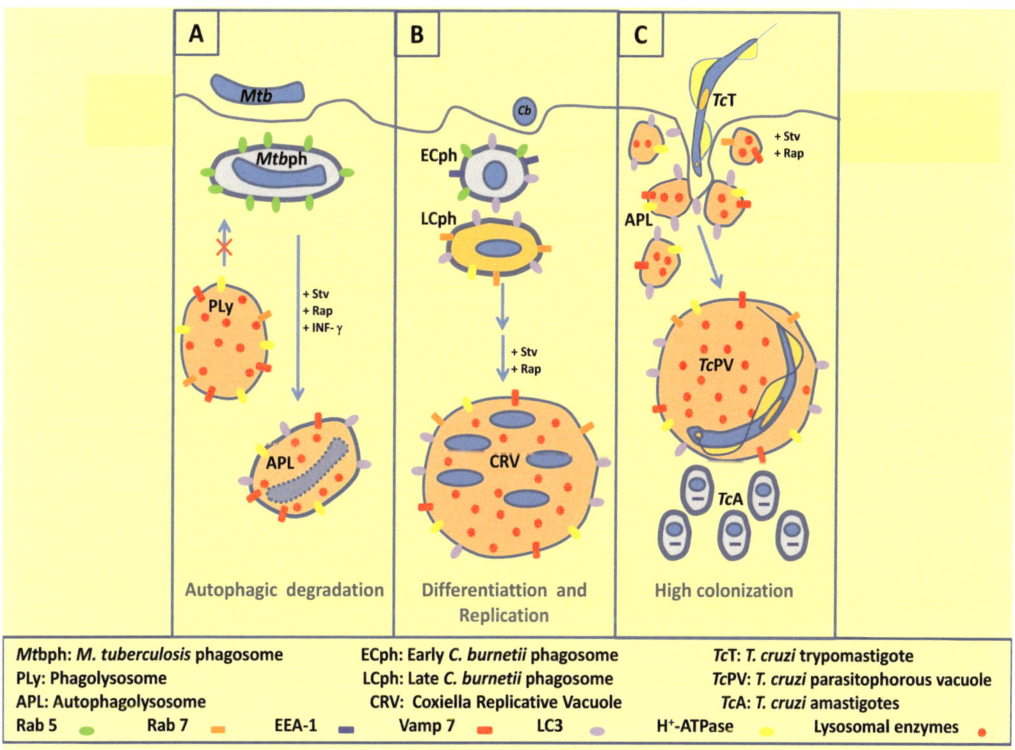

Figure 2. Mechanisms of pathogen-host autophagy interactions. As a component of innate immune responses against intracellular pathogens, autophagy can effectively eliminate some pathogens, re-routing them to lysosomes (autophagolysosomes). That is the case of *Mycobacterium tuberculosis* (A). In contrast, some microorganisms such as *Coxiella burnetii* utilize autophagic compartments to delay lysosomal fusion until differentiating into more resistant forms (B). On the other hand, *Trypanosoma cruzi* exploits the autophagic pathway to efficiently colonize host cells (C).

plays a key role in this process [35,36]. When CD40 of human or mouse macrophages infected with *T. gondii* is stimulated with (CD154+) CD4+ T cells or exposed to anti-CD4 antibodies, the nonfusogenic nature of the PV is reversed and the vacuole fuses with late endosomes and lysosomes. This fusion is dependent on autophagy, as indicated by the inhibition of this mechanism in cells knockdown for Beclin 1 or treated with 3-MA, an inhibitor of phagophore formation. CD40 activation also stimulates expression of LC3 that localizes to the PV [36].

3.2. Evasion of autophagic responses

Despite the potent effect of autophagy in killing intracellular pathogens, some microbial pathogens have the capacity to control cellular autophagy and successfully parasitize eukaryotic cells. These highly evolved microorganisms have developed specific virulence factors to protect themselves from autophagic elimination by producing:

- *Prevention of autophagy induction:* Several viruses direct their products to essential autophagic proteins, causing them to be functionally inhibited. The herpes virus family can produce autophagy blockage through different mechanisms. The HSV-1 ICP34.5 viral protein encoded by herpes simplex virus type 1 blocks Beclin1 function and confers neurovirulence in mice [37]. A similar mechanism was recently shown for human cytomegalovirus; the virus protein TRS1 interacts with Beclin 1 to inhibit autophagy [38]. Gamma-herpesviruses, including important human pathogens such as Epstein Barr virus or Kaposi's sarcoma-associated HIV, displayed a different type of inhibition. They encode homologs of the antiapoptotic, host Bcl-2 protein to promote viral replication and pathogenesis. Cellular Bcl-2 and their viral homologs have the property to bind and inhibit Beclin1, suppressing both apoptotic and autophagic responses [39]. It is not yet clear whether other intracellular pathogens besides viruses also actively suppress initiation of the autophagy pathway. In contrast, many bacteria display the following actions.

- *Suppression of autophagosome maturation into autolysosome*: Similar to mycobacterium phagosome maturation arrest, other pathogens have the ability to suppress autophagosome maturation. They specifically reside in vacuoles with autophagosomal characteristics in order to survive and replicate, but avoid transient or permanent fusion with lysosomes. *Porphyromonas gingivalis*, a bacterial periodontal pathogen that can localize to atherosclerotic plaques, traffics to autophagosomes as a way of evading the conventional endocytic trafficking to lysosomes [40], After intracellular uptake, *P. gingivalis* transits from early autophagosomes to late autophagosomes and prevents the formation of autolysosomes, a mechanism not yet well elucidated [41]. On the other hand, a delay in the delivery of lysosomal enzymes to phagosomes was initially described for dimorphic bacteria and named "the pregnant pause". The dimorphic life cycles of these pathogens have dramatic consequences for phagosome traffic. In the transmissible state, *C. burnetii, L. pneumophila* and others, such as *Leishmania sp.*, block phagosome maturation; after a pregnant pause that includes the bacterial differentiation process, replicative forms emerge and thrive in lysosomes [42]. Autophagy is one of the mechanisms activated by these intracellular pathogens for delaying lysosomal fusion. At late stages of cellular infection, both *Coxiella burnetii* and *Legionella pneumophila* develop vacuoles that have characteristics of phagolysosomes and are also decorated with LC3 [43,44]. In the case of *C. burnetii,* acquisition of LC3 is even an early event in the transit of phagosomes containing bacterium (Cph) and depends on bacterial protein synthesis because chloramphenicol avoid this LC3 recruitment [45]. Interactions with autophagic and also late endocytic compartments are maintained during the transit of Cph and in the development of the Coxiella replicative vacuoles (CRV) [46]. Indeed, autophagy induction or the overexpression of autophagic proteins LC3 and Rab 24 favor the generation and maturation of this CRV [47]. Taking together, these results demonstrated that *C. burnetii* transits through the normal endo/phagocytic pathway but actively interacts with autophagosomes at early times after infection. This intersection delays fusion with the lysosomal compartment, possibly favoring the intracellular differentiation and survival of the bacteria. In this period, *C. burnetii* differentiates from the transmissible forms (named small cell variant) to the replicative forms (large cell variant) (Figure 2B). In the case of *L. pneumophila,* it was recently demonstrated that this bacterium produced several Type IV effector proteins that control the timing of bacteria during intracellular transport. The early secretion of DrrA/SidM, LidA, and RalF factors, prolong association with the ER and permit the persistence of the bacteria in immature autophagosomal vacuoles for a period sufficient to differentiate into an acid-resistant, replicative form. Subsequent secretion of LepB releases the block of autophagosome maturation, and the adapted progeny continue to replicate within autophagolysosomes [48].

- *Evasion of pathogen recognition by the autophagic machinery*: This strategy is especially important in intracytoplasmic pathogens such as *Shigella flexneri, L. monocytogenes,* and *Burkholderia pseudomallei.* Shigella VirG, a protein required for intracellular actin-based motility, induced autophagy and favored the microorganism trap by autophagosomes, after binding between VirG and Atg5. However, *Shigella,* encoding Type III secretion effector, IcsB, competitively binds to Atg5, thereby camouflaging its own bacterial target molecule VirG from autophagic capture [49]. Furthermore, BopA, the counterpart of IcsB in Burkholderia pseudomallei, have similar autophagy-evading properties [50]. *Listeria monocytogenes* is a classic example of a "cytosol-adapted pathogen"; it can rapidly escape from the phagosome of macrophages and other non-phagocytic cells and replicate rapidly in the cytosol. Phagosome escape also enables cell-to-cell spread by the bacteria through a bacterial driven actin-based motility mechanism. Besides Act A, that as was shown plays a critical role in autophagic escape by polymerizing actin which favors bacteria movements [51], another virulence factor of *L. monocytogenes,* InlK, was recently shown to counteract the autophagic process. InlK interacts with the Major Vault Protein (MVP), the main component of cytoplasmic ribonucleoproteic particules named vaults. The recruitment of MVP to bacterial surface disguises intracytosolic bacteria from the autophagic recognition system leading to an increased survival rate [52].

3.3. Autophagy as a survival mechanism

A different type of host-microbial interaction belongs to the group of organisms that harness cell autophagy. Independently of the final localization within the cell, these particular organisms improve their intracellular cycle when interacting with autophagic compartments. Postulated benefits of host autophagy for microbes include the promotion of viral replication or morphogenesis via utilization of the autophagic machinery. Polyovirus localized in double-membrane autophagosome-like structures positive for LC3 serve as lipid membrane-scaffolds that enhance viral replication [53]. In a similar way, the rotavirus NSP4 protein colocalizes with LC3-positive vesicular compartments and is postulated to play a role in the formation of viroplasms and/or the packaging or transcription of the rotavirus genome [54].

Another mechanism is the utilization of autophagosomes as a protective intracellular niche to enhance the survival and growth of bacteria. As described above, dimorphic bacteria such as *C. burnetii* or *L. pneumophila* follow this method. In contrast, *Francisella tularensis,* enters LC3-positive compartments to allow cytoplasmic bacteria to regain access to the endocytic compartment to finally promote bacteria egress through exocytosis [55]. Autophagy could also favor intracellular pathogen survival by providing nutrients to pathogens, particularly those that reside in sequestered vacuoles that lack access to cytoplasmic nutrients. That is the case of *T. gondii,* which establishes its vacuole in the vicinity of autophagic compartments and that displays an impaired growth in *Atg5*-deficient MEF cells, leading to the conclusion that host cell autophagy plays a role in promoting parasite growth through nutrient recovery [56].

A special type of pathogen-autophagy interaction is produced by the protozoan parasite *Trypanosoma cruzi.* This pathogen exploits the autophagic pathway to efficiently colonize host cells, as will be described in detail in the next section.

3.4. Autophagy as an invasion strategy: The *Trypanosoma cruzi* case

The protozoan parasite *Trypanosoma cruzi* can invade a wide range of phagocytic and non-phagocytic cells in the infective, non-proliferative trypomastigote form. Inside the cell, trypomastigotes are temporarily contained in a membrane vesicle, the parasitophorous vacuole. Subsequently, the parasites escape to the cytosol, differentiate into the amastigote form, and replicate by binary division [57,58]. This replication process culminates after 9 cycles, followed by a new differentiation period where the parasites undergo a transition back into trypomastigotes. After that the parasites are released from the cell and infect the neighboring cells, maintaining the infection process.

The characteristics of the *T. cruzi* parasitophorous vacuole (TcPV) are directly related to the parasite invasion mechanism. Previously published data showed that two main invasion processes involving different signaling pathways are participating: the calcium dependent fusion of lysosomes with the host plasma membrane [59,60], and the activation of class I PI3K that produces a plasma membrane-derived vacuole initially devoid of lysosomal markers [61]. Although both pathways require the disruption of the host cell actin cytoskeleton, the lysosomal independent *T. cruzi* entry model appears to be more significant early after internalization (50% *versus* 20% lysosome-dependent entry process). However, lysosomal fusion is essential for the establishment of a productive infection [62] and for the progression and completion of the *T. cruzi* intracellular cycle [63,64], since the parasite Tc-Tox is activated in the acid environment provided by the lysosomes [57,65]. These strategies of cell invasion indicate that *T. cruzi* entry is a complex process that employs different components from the plasma membrane and the endocytic pathway to finally produce the intracellular infection. More recently, studies provided by our laboratory demonstrated that the *T. cruzi* parasitophorous vacuole (TcPV) is decorated with LC3 protein and that the autophagic inhibitors wortmannin, 3-methyladenine or vinblastine suppress this recruitment and also significantly reduce the intracellular infection. In contrast, induction of autophagy before infection by starvation or other means significantly increased the percentage of infected cells. Interestingly, infection was diminished in the absence of the specific autophagy genes Beclin1 or Atg5, which are required for initiation of autophagy, indicating that autophagic-derived compartments are required for efficient entry of *T. cruzi* into the host cell [66]. Live imaging using confocal microscopy showed that GFP-LC3 positive vesicles move towards plasma membrane and contact the sites where trypomastigotes, the *T. cruzi* invasive forms, bind the membrane [66,67] (Figure 2C). The pro-pathogen effects of autophagy on *T. cruzi* infection were observed in different classes of cells and *T. cruzi* strains, demonstrating that this interaction is a wide-spread phenomenon [68]. The autophagy modulation of host cells during the following stages of the *T. cruzi* intracellular cycle -trypomastigote to amastigote differentiation, amastigote replication and amastigote differentiation back to trypomastigote- seems to suffer no mdification compared to cells maintained in control conditions [66]. Since *T. cruzi* is an unicellular eukaryotic organism that also has its own autophagic pathway [69], other experimental procedures will be necessary to decipher the possible dual action of autophagy modulation on *T. cruzi* infected cells. Indeed, unpublished data from our laboratory show that classical inducers and inhibitors of mammalian autophagy have similar effects on *T. cruzi* and that protozoan autophagy is activated during *T. cruzi* metacyclogenesis, a process that renders metacyclic trypomastigotes from epimastigotes and that takes place in the digestive apparatus of the triatomine vectors.

3.5. Autophagy in action: *in vivo* infections studies

To date the outcome of the pathogen/autophagy relationship on *in vivo* infections models, with the exception of a few cases, remains little understood. Actions of autophagy as an innate immune component are easier to understand, particularly with the use of knockout mice. In this way, studies on *in vivo M. tuberculosis* murine infections showed that the most susceptible mice are those deficient in either IFN-γ or IFN-γ receptors [70]. These results clearly demonstrate that macrophage activation, and the macrophage autophagic pathway [30], are required as a critical components for controlling infection.

The main concerns with the *in vivo* models arise from the cases of pathogens that *in vitro* studies show to be favored by autophagy induction. At the moment, no current evidence demonstrates that autophagy gene deletion in the host attenuates microbial disease. Therefore, the physiological significance of microbial utilization of autophagy for "promicrobial" effects remains to be established [71]. The discrepant conclusions between *in vitro* and *in vivo* studies in *T. gondii* infection models exemplify this concept. Although *T. gondii* has impaired growth in *Atg5*-deficient cells, leading to the conclusion that host cell autophagy plays a role in promoting parasite growth through nutrient recovery [56], this parasite has increased virulence in mice with macrophage-specific deletion of Atg5 [72]. In agreement with these results, unpublished data from our laboratory show that autophagy-impaired mice are more susceptible to *T. cruzi* infection, while our previously results on cell cultures clearly demonstrate that decreased autophagic levels in Atg5 KO cells or in Beclin 1 KD cells significantly decreased *T. cruzi* infection [66].

One possible explanation for these discrepancies is the different effects of autophagy (or autophagic proteins) on phagocytic and non-phagocytic cells. When *T. gondii* infected macrophages were stimulated with CD40 receptor agonists, the parasite vacuole fuses with endosomes and lysosomes in an autophagy-dependent process leading to parasite destruction [35,36]. In contrast, in the non-phagocytic HeLa cells, Wang and colleagues reported the beneficial effects of autophagic induction for parasite survival and growth [56]. Considering this possibility, the comparative analysis of autophagic modulation on the course of a specific pathogen infection in phagocytic and non-phagocytic cells *in vitro* prior to mice infection would be productive in the future. However, this simplistic point of view will never replace the conclusions obtained from mice experiments, especially when immune responses are implicated.

Acknowledgements

I want to especially thank Dr. María Isabel Colombo, who introduced me in the world of pathogens and also shared with me her great knowledge about cell biology and autophagy. Our laboratory was supported by grants from Universidad Juan A. Maza, Comisión Nacional Salud Investiga (Ministerio de Salud de la Nación), CONICET-Fundación Fulbright, and CONICET-Fundación Bunge y Born. Our current grants are provided by Secretaría de Ciencia, Técnica y Posgrado (Sectyp, Universidad Nacional de Cuyo), CONICET (PIP 2011-2013) and Agencia Nacional de Promoción Científica y Tecnológica (PICT# 2008 2235) to Patricia S. Romano.

Author details

Patricia Silvia Romano*

Address all correspondence to: promano@fcm.uncu.edu.ar

Laboratory of Trypanosoma cruzi and the Host Cell, Institute of Histology and Embryology, National Council of Scientific and Technical Research (IHEM-CONICET), School of Medicine-National University of Cuyo. Mendoza, Argentina

References

[1] Winterbourn, C. C, & Kettle, A. J. (2012). Redox Reactions and Microbial Killing in the Neutrophil Phagosome. Antioxid Redox Signal

[2] Vieira, O. V, Botelho, R. J, & Grinstein, S. (2002). Phagosome maturation: aging gracefully. Biochem J , 366, 689-704.

[3] Hackam, D. J, Rotstein, O. D, Zhang, W. J, Demaurex, N, Woodside, M, Tsai, O, & Grinstein, S. (1997). Regulation of phagosomal acidification. Differential targeting of Na+/H+ exchangers, Na+/K+-ATPases, and vacuolar-type H+-atpases. J Biol Chem , 272, 29810-29820.

[4] Duclos, S, Diez, R, Garin, J, Papadopoulou, B, Descoteaux, A, Stenmark, H, & Desjardins, M. (2000). Rab5 regulates the kiss and run fusion between phagosomes and endosomes and the acquisition of phagosome leishmanicidal properties in RAW 264.7 macrophages. J Cell Sci 113 Pt , 19, 3531-3541.

[5] Hackstadt, T. (2000). Redirection of host vesicle trafficking pathways by intracellular parasites. Traffic , 1, 93-99.

[6] Meresse, S, Steele-mortimer, O, Moreno, E, Desjardins, M, Finlay, B, & Gorvel, J. P. (1999). Controlling the maturation of pathogen-containing vacuoles: a matter of life and death. Nat Cell Biol 1: EE188., 183.

[7] Meijer, A. J, & Codogno, P. (2004). Regulation and role of autophagy in mammalian cells. Int J Biochem Cell Biol , 36, 2445-2462.

[8] Noda, T, Suzuki, K, & Ohsumi, Y. (2002). Yeast autophagosomes: de novo formation of a membrane structure. Trends Cell Biol , 12, 231-235.

[9] Axe, E. L, Walker, S. A, Manifava, M, Chandra, P, Roderick, H. L, Habermann, A, Griffiths, G, & Ktistakis, N. T. (2008). Autophagosome formation from membrane compartments enriched in phosphatidylinositol 3-phosphate and dynamically connected to the endoplasmic reticulum. J Cell Biol , 182, 685-701.

[10] Matsunaga, K, Morita, E, Saitoh, T, Akira, S, Ktistakis, N. T, Izumi, T, Noda, T, & Yoshimori, T. (2010). Autophagy requires endoplasmic reticulum targeting of the PI3-kinase complex via Atg14L. J Cell Biol , 190, 511-521.

[11] Hailey, D. W, Rambold, A. S, Satpute-krishnan, P, Mitra, K, Sougrat, R, Kim, P. K, & Lippincott-schwartz, J. (2010). Mitochondria supply membranes for autophagosome biogenesis during starvation. Cell , 141, 656-667.

[12] Moreau, K, Ravikumar, B, Puri, C, & Rubinsztein, D. C. (2012). Arf6 promotes autophagosome formation via effects on phosphatidylinositol 4,5-bisphosphate and phospholipase D. J Cell Biol , 196, 483-496.

[13] Yamamoto, H, Kakuta, S, Watanabe, T. M, Kitamura, A, Sekito, T, Kondo-kakuta, C, Ichikawa, R, Kinjo, M, & Ohsumi, Y. (2012). Atg9 vesicles are an important mem-

brane source during early steps of autophagosome formation. J Cell Biol , 198, 219-233.

[14] Baba, M, Takeshige, K, Baba, N, & Ohsumi, Y. (1994). Ultrastructural analysis of the autophagic process in yeast: detection of autophagosomes and their characterization. J Cell Biol , 124, 903-913.

[15] Suzuki, K, Kubota, Y, Sekito, T, & Ohsumi, Y. (2007). Hierarchy of Atg proteins in pre-autophagosomal structure organization. Genes Cells , 12, 209-218.

[16] Kabeya, Y, Mizushima, N, Ueno, T, Yamamoto, A, Kirisako, T, Noda, T, Kominami, E, Ohsumi, Y, & Yoshimori, T. (2000). LC3, a mammalian homologue of yeast Apg8p, is localized in autophagosome membranes after processing. EMBO J , 19, 5720-5728.

[17] Cheong, H, Yorimitsu, T, Reggiori, F, Legakis, J. E, Wang, C. W, & Klionsky, D. J. (2005). Atg17 regulates the magnitude of the autophagic response. Mol Biol Cell , 16, 3438-3453.

[18] Fujita, N, Itoh, T, Omori, H, Fukuda, M, Noda, T, & Yoshimori, T. (2008). The Atg16L complex specifies the site of LC3 lipidation for membrane biogenesis in autophagy. Mol Biol Cell , 19, 2092-2100.

[19] Gutierrez, M. G, Munafo, D. B, Beron, W, & Colombo, M. I. (2004). Rab7 is required for the normal progression of the autophagic pathway in mammalian cells. J Cell Sci , 117, 2687-2697.

[20] Fader, C. M, Sanchez, D. G, Mestre, M. B, & Colombo, M. I. and VAMP3/cellubrevin: two v-SNARE proteins involved in specific steps of the autophagy/multivesicular body pathways. Biochim Biophys Acta , 1793, 1901-1916.

[21] Moreau, K, Ravikumar, B, Renna, M, Puri, C, & Rubinsztein, D. C. (2011). Autophagosome precursor maturation requires homotypic fusion. Cell , 146, 303-317.

[22] Furuta, N, Fujita, N, Noda, T, Yoshimori, T, & Amano, A. (2010). Combinational soluble N-ethylmaleimide-sensitive factor attachment protein receptor proteins VAMP8 and Vti1b mediate fusion of antimicrobial and canonical autophagosomes with lysosomes. Mol Biol Cell , 21, 1001-1010.

[23] Aguilera, M. O, Beron, W, & Colombo, M. I. (2012). The actin cytoskeleton participates in the early events of autophagosome formation upon starvation induced autophagy. Autophagy 8:

[24] Klionsky DJ, Abdalla FC, Abeliovich H, Abraham RT, cevedo-Arozena A, Adeli K, Agholme L, Agnello M, Agostinis P, guirre-Ghiso JA, Ahn HJ, it-Mohamed O, it-Si-Ali S, Akematsu T, Akira S, Al-Younes HM, Al-Zeer MA, Albert ML, Albin RL, egre-Abarrategui J, Aleo MF, Alirezaei M, Almasan A, monte-Becerril M, Amano A, Amaravadi R, Amarnath S, Amer AO, ndrieu-Abadie N, Anantharam V, Ann DK, noopkumar-Dukie S, Aoki H, Apostolova N, Arancia G, Aris JP, Asanuma K, Asare NY, Ashida H, Askanas V, Askew DS, Auberger P, Baba M, Backues SK, Baehrecke

http://dx.doi.org/10.5772/55179

EH, Bahr BA, Bai XY, Bailly Y, Baiocchi R, Baldini G, Balduini W, Ballabio A, Bamber BA, Bampton ET, Banhegyi G, Bartholomew CR, Bassham DC, Bast RC, Jr., Batoko H, Bay BH, Beau I, Bechet DM, Begley TJ, Behl C, Behrends C, Bekri S, Bellaire B, Bendall LJ, Benetti L, Berliocchi L, Bernardi H, Bernassola F, Besteiro S, Bhatia-Kissova I, Bi X, Biard-Piechaczyk M, Blum JS, Boise LH, Bonaldo P, Boone DL, Bornhauser BC, Bortoluci KR, Bossis I, Bost F, Bourquin JP, Boya P, Boyer-Guittaut M, Bozhkov PV, Brady NR, Brancolini C, Brech A, Brenman JE, Brennand A, Bresnick EH, Brest P, Bridges D, Bristol ML, Brookes PS, Brown EJ, Brumell JH, Brunetti-Pierri N, Brunk UT, Bulman DE, Bultman SJ, Bultynck G, Burbulla LF, Bursch W, Butchar JP, Buzgariu W, Bydlowski SP, Cadwell K, Cahova M, Cai D, Cai J, Cai Q, Calabretta B, Calvo-Garrido J, Camougrand N, Campanella M, Campos-Salinas J, Candi E, Cao L, Caplan AB, Carding SR, Cardoso SM, Carew JS, Carlin CR, Carmignac V, Carneiro LA, Carra S, Caruso RA, Casari G, Casas C, Castino R, Cebollero E, Cecconi F, Celli J, Chaachouay H, Chae HJ, Chai CY, Chan DC, Chan EY, Chang RC, Che CM, Chen CC, Chen GC, Chen GQ, Chen M, Chen Q, Chen SS, Chen W, Chen X, Chen X, Chen X, Chen YG, Chen Y, Chen Y, Chen YJ, Chen Z, Cheng A, Cheng CH, Cheng Y, Cheong H, Cheong JH, Cherry S, Chess-Williams R, Cheung ZH, Chevet E, Chiang HL, Chiarelli R, Chiba T, Chin LS, Chiou SH, Chisari FV, Cho CH, Cho DH, Choi AM, Choi D, Choi KS, Choi ME, Chouaib S, Choubey D, Choubey V, Chu CT, Chuang TH, Chueh SH, Chun T, Chwae YJ, Chye ML, Ciarcia R, Ciriolo MR, Clague MJ, Clark RS, Clarke PG, Clarke R, Codogno P, Coller HA, Colombo MI, Comincini S, Condello M, Condorelli F, Cookson MR, Coombs GH, Coppens I, Corbalan R, Cossart P, Costelli P, Costes S, Coto-Montes A, Couve E, Coxon FP, Cregg JM, Crespo JL, Cronje MJ, Cuervo AM, Cullen JJ, Czaja MJ, D'Amelio M, rfeuille-Michaud A, Davids LM, Davies FE, De FM, de Groot JF, de Haan CA, De ML, De MA, De T, V, Debnath J, Degterev A, Dehay B, Delbridge LM, Demarchi F, Deng YZ, Dengjel J, Dent P, Denton D, Deretic V, Desai SD, Devenish RJ, Di GM, Di PG, Di PC, az-Araya G, az-Laviada I, az-Meco MT, az-Nido J, Dikic I, nesh-Kumar SP, Ding WX, Distelhorst CW, Diwan A, Djavaheri-Mergny M, Dokudovskaya S, Dong Z, Dorsey FC (2012) Guidelines for the use and interpretation of assays for monitoring autophagy. Autophagy 8: 445-544.

[25] Morselli, E, Marino, G, Bennetzen, M. V, Eisenberg, T, Megalou, E, Schroeder, S, Cabrera, S, Benit, P, Rustin, P, Criollo, A, Kepp, O, Galluzzi, L, Shen, S, Malik, S. A, Maiuri, M. C, Horio, Y, Lopez-otin, C, Andersen, J. S, Tavernarakis, N, Madeo, F, & Kroemer, G. (2011). Spermidine and resveratrol induce autophagy by distinct pathways converging on the acetylproteome. J Cell Biol , 192, 615-629.

[26] Eisenberg, T, Knauer, H, Schauer, A, Buttner, S, Ruckenstuhl, C, Carmona-gutierrez, D, Ring, J, Schroeder, S, Magnes, C, Antonacci, L, Fussi, H, Deszcz, L, Hartl, R, Schraml, E, Criollo, A, Megalou, E, Weiskopf, D, Laun, P, Heeren, G, Breitenbach, M, Grubeck-loebenstein, B, Herker, E, Fahrenkrog, B, Frohlich, K. U, Sinner, F, Tavernarakis, N, Minois, N, Kroemer, G, & Madeo, F. (2009). Induction of autophagy by spermidine promotes longevity 1. Nat Cell Biol , 11, 1305-1314.

[27] Yuk, J. M, Yoshimori, T, & Jo, E. K. (2012). Autophagy and bacterial infectious diseases 2. Exp Mol Med , 44, 99-108.

[28] Chua, J, Vergne, I, Master, S, & Deretic, V. (2004). A tale of two lipids: Mycobacterium tuberculosis phagosome maturation arrest. Curr Opin Microbiol , 7, 71-77.

[29] Sturgill-koszycki, S, Schlesinger, P. H, Chakraborty, P, Haddix, P. L, Collins, H. L, Fok, A. K, Allen, R. D, Gluck, S. L, Heuser, J, & Russell, D. G. (1994). Lack of acidification in Mycobacterium phagosomes produced by exclusion of the vesicular proton-ATPase2. Science , 263, 678-681.

[30] Gutierrez, M. G, Master, S. S, Singh, S. B, Taylor, G. A, Colombo, M. I, & Deretic, V. (2004). Autophagy is a defense mechanism inhibiting BCG and Mycobacterium tuberculosis survival in infected macrophages. Cell , 119, 753-766.

[31] Sakurai, A, Maruyama, F, Funao, J, Nozawa, T, Aikawa, C, Okahashi, N, Shintani, S, Hamada, S, Ooshima, T, & Nakagawa, I. (2010). Specific behavior of intracellular Streptococcus pyogenes that has undergone autophagic degradation is associated with bacterial streptolysin O and host small G proteins Rab5 and Rab7. J Biol Chem , 285, 22666-22675.

[32] Nakagawa, I, Amano, A, Mizushima, N, Yamamoto, A, Yamaguchi, H, Kamimoto, T, Nara, A, Funao, J, Nakata, M, Tsuda, K, Hamada, S, & Yoshimori, T. (2004). Autophagy defends cells against invading group A Streptococcus. Science , 306, 1037-1040.

[33] Yoshimori, T, & Amano, A. (2009). Group a Streptococcus: a loser in the battle with autophagy5. Curr Top Microbiol Immunol , 335, 217-226.

[34] Mauthe, M, Yu, W, Krut, O, Kronke, M, Gotz, F, Robenek, H, & Proikas-cezanne, T. (2012). WIPI-1 Positive Autophagosome-Like Vesicles Entrap Pathogenic Staphylococcus aureus for Lysosomal Degradation. Int J Cell Biol 2012: 179207.

[35] Andrade, R. M, Wessendarp, M, Gubbels, M. J, Striepen, B, & Subauste, C. S. (2006). CD40 induces macrophage anti-Toxoplasma gondii activity by triggering autophagy-dependent fusion of pathogen-containing vacuoles and lysosomes. J Clin Invest , 116, 2366-2377.

[36] Subauste, C. S. (2009). Autophagy in immunity against Toxoplasma gondii. Curr Top Microbiol Immunol , 335, 251-265.

[37] Orvedahl, A, Alexander, D, Talloczy, Z, Sun, Q, Wei, Y, Zhang, W, Burns, D, Leib, D. A, & Levine, B. (2007). HSV-1 ICP34.5 confers neurovirulence by targeting the Beclin 1 autophagy protein. Cell Host Microbe , 1, 23-35.

[38] Chaumorcel, M, Lussignol, M, Mouna, L, Cavignac, Y, Fahie, K, Cotte-laffitte, J, Geballe, A, Brune, W, Beau, I, Codogno, P, & Esclatine, A. (2012). The human cytomegalovirus protein TRS1 inhibits autophagy via its interaction with Beclin 1. J Virol , 86, 2571-2584.

[39] Sinha, S, Colbert, C. L, Becker, N, Wei, Y, & Levine, B. (2008). Molecular basis of the regulation of Beclin 1-dependent autophagy by the gamma-herpesvirus 68 Bcl-2 homolog M11. Autophagy , 4, 989-997.

[40] Dorn, B. R, & Dunn, W. A. Jr., Progulske-Fox A ((2001). Porphyromonas gingivalis traffics to autophagosomes in human coronary artery endothelial cells. Infect Immun , 69, 5698-5708.

[41] Belanger, M, Rodrigues, P. H, & Dunn, W. A. Jr., Progulske-Fox A ((2006). Autophagy: a highway for Porphyromonas gingivalis in endothelial cells. Autophagy , 2, 165-170.

[42] Swanson, M. S, & Fernandez-moreira, E. (2002). A microbial strategy to multiply in macrophages: the pregnant pause. Traffic , 3, 170-177.

[43] Amer, A. O, & Swanson, M. S. (2005). Autophagy is an immediate macrophage response to Legionella pneumophila. Cell Microbiol , 7, 765-778.

[44] Ghigo, E, Colombo, M. I, & Heinzen, R. A. (2012). The Coxiella burnetii Parasitophorous Vacuole.. Adv Exp Med Biol , 984, 141-169.

[45] Romano, P. S, Gutierrez, M. G, Beron, W, Rabinovitch, M, & Colombo, M. I. (2007). The autophagic pathway is actively modulated by phase II Coxiella burnetii to efficiently replicate in the host cell. Cell Microbiol , 9, 891-909.

[46] Beron, W, Gutierrez, M. G, Rabinovitch, M, & Colombo, M. I. (2002). Coxiella burnetii localizes in a Rab7-labeled compartment with autophagic characteristics. Infect Immun , 70, 5816-5821.

[47] Gutierrez, M. G, Vazquez, C. L, Munafo, D. B, Zoppino, F. C, Beron, W, Rabinovitch, M, & Colombo, M. I. (2005). Autophagy induction favours the generation and maturation of the Coxiella-replicative vacuoles. Cell Microbiol , 7, 981-993.

[48] Joshi, A. D, & Swanson, M. S. (2011). Secrets of a successful pathogen: legionella resistance to progression along the autophagic pathway. Front Microbiol 2: 138.

[49] Ogawa, M, Yoshimori, T, Suzuki, T, Sagara, H, Mizushima, N, & Sasakawa, C. (2005). Escape of intracellular Shigella from autophagy. Science , 307, 727-731.

[50] Kayath, C. A, & Hussey, S. El hN, Nagra K, Philpott D, Allaoui A ((2010). Escape of intracellular Shigella from autophagy requires binding to cholesterol through the type III effector, IcsB. Microbes Infect , 12, 956-966.

[51] Dortet, L, Mostowy, S, & Cossart, P. (2012). Listeria and autophagy escape: involvement of InlK, an internalin-like protein. Autophagy , 8, 132-134.

[52] Dortet, L, Mostowy, S, Samba-louaka, A, Gouin, E, Nahori, M. A, Wiemer, E. A, Dussurget, O, & Cossart, P. (2011). Recruitment of the major vault protein by InlK: a Listeria monocytogenes strategy to avoid autophagy. PLoS Pathog 7: e1002168.

[53] Jackson, W. T, & Giddings, T. H. Jr., Taylor MP, Mulinyawe S, Rabinovitch M, Kopito RR, Kirkegaard K ((2005). Subversion of cellular autophagosomal machinery by RNA viruses. PLoS Biol 3: e156.

[54] Berkova, Z, Crawford, S. E, Trugnan, G, Yoshimori, T, Morris, A. P, & Estes, M. K. (2006). Rotavirus NSP4 induces a novel vesicular compartment regulated by calcium and associated with viroplasms. J Virol , 80, 6061-6071.

[55] Checroun, C, Wehrly, T. D, Fischer, E. R, Hayes, S. F, & Celli, J. (2006). Autophagy-mediated reentry of Francisella tularensis into the endocytic compartment after cytoplasmic replication. Proc Natl Acad Sci U S A , 103, 14578-14583.

[56] Wang, Y, Weiss, L. M, & Orlofsky, A. (2009). Host cell autophagy is induced by Toxoplasma gondii and contributes to parasite growth. J Biol Chem , 284, 1694-1701.

[57] Andrews, N. W. (1994). From lysosomes into the cytosol: the intracellular pathway of Trypanosoma cruzi. Braz J Med Biol Res , 27, 471-475.

[58] Dvorak, J. A, & Howe, C. L. (1976). The attraction of Trypanosoma cruzi to vertebrate cells in vitro 4. J Protozool , 23, 534-537.

[59] Rodriguez, A, Samoff, E, Rioult, M. G, Chung, A, & Andrews, N. W. (1996). Host cell invasion by trypanosomes requires lysosomes and microtubule/kinesin-mediated transport. J Cell Biol , 134, 349-362.

[60] Tardieux, I, Webster, P, Ravesloot, J, Boron, W, Lunn, J. A, Heuser, J. E, & Andrews, N. W. (1992). Lysosome recruitment and fusion are early events required for trypanosome invasion of mammalian cells. Cell , 71, 1117-1130.

[61] Woolsey, A. M, Sunwoo, L, Petersen, C. A, Brachmann, S. M, Cantley, L. C, & Burleigh, B. A. kinase-dependent mechanisms of trypanosome invasion and vacuole maturation. J Cell Sci , 116, 3611-3622.

[62] Andrade, L. O, & Andrews, N. W. (2004). Lysosomal fusion is essential for the retention of Trypanosoma cruzi inside host cells. J Exp Med , 200, 1135-1143.

[63] Andrews, N. W, & Whitlow, M. B. (1989). Secretion by Trypanosoma cruzi of a hemolysin active at low pH. Mol Biochem Parasitol , 33, 249-256.

[64] Ley, V, Robbins, E. S, Nussenzweig, V, & Andrews, N. W. (1990). The exit of Trypanosoma cruzi from the phagosome is inhibited by raising the pH of acidic compartments. J Exp Med , 171, 401-413.

[65] Andrews, N. W. (1994). From lysosomes into the cytosol: the intracellular pathway of Trypanosoma cruzi. Braz J Med Biol Res , 27, 471-475.

[66] Romano, P. S, Arboit, M. A, Vazquez, C. L, & Colombo, M. I. (2009). The autophagic pathway is a key component in the lysosomal dependent entry of Trypanosoma cruzi into the host cell. Autophagy , 5, 6-18.

[67] Romano, P. S, Arboit, M. A, Vazquez, C. L, & Colombo, M. I. (2009). The autophagic pathway is a key component in the lysosomal dependent entry of Trypanosoma cruzi into the host cell. Autophagy , 5, 6-18.

[68] Romano, P. S, Cueto, J. A, Casassa, A. F, Vanrell, M. C, Gottlieb, R. A, & Colombo, M. I. (2012). Molecular and cellular mechanisms involved in the Trypanosoma cruzi/host cell interplay. IUBMB Life , 64, 387-396.

[69] Alvarez, V. E, & Kosec, G. Sant'Anna C, Turk V, Cazzulo JJ, Turk B ((2008). Autophagy is involved in nutritional stress response and differentiation in Trypanosoma cruzi. J Biol Chem , 283, 3454-3464.

[70] Cooper, A. M, Dalton, D. K, Stewart, T. A, Griffin, J. P, Russell, D. G, & Orme, I. M. (1993). Disseminated tuberculosis in interferon gamma gene-disrupted mice. J Exp Med , 178, 2243-2247.

[71] Deretic, V, & Levine, B. (2009). Autophagy, immunity, and microbial adaptations. Cell Host Microbe , 5, 527-549.

[72] Zhao, Z, Fux, B, Goodwin, M, Dunay, I. R, Strong, D, Miller, B. C, Cadwell, K, Delgado, M. A, Ponpuak, M, Green, K. G, Schmidt, R. E, Mizushima, N, Deretic, V, Sibley, L. D, & Virgin, H. W. (2008). Autophagosome-independent essential function for the autophagy protein Atg5 in cellular immunity to intracellular pathogens. Cell Host Microbe , 4, 458-469.

Chapter 14

Autophagic Balance Between Mammals and Protozoa: A Molecular, Biochemical and Morphological Review of Apicomplexa and Trypanosomatidae Infections

Thabata Lopes Alberto Duque,
Xênia Macedo Souto,
Valter Viana de Andrade-Neto,
Vítor Ennes-Vidal and
Rubem Figueiredo Sadok Menna-Barreto

Additional information is available at the end of the chapter

http://dx.doi.org/10.5772/54970

1. Introduction

Protozoa are unicellular eukaryotes that are able to live as parasites or as free-living organisms and interact with a great variety of environments and organisms, from bacteria to man; in addition, they represent one of most important sources of parasitic diseases. Every year, more than one million people die from complications from protozoal infections worldwide [1-5]. Of the medically relevant protozoa, Trypanosomatidae and Apicomplexa constitute a substantial group including the causative agents of several human diseases such as Chagas disease, sleeping sickness, leishmaniasis, malaria and toxoplasmosis [1,5,6]. The life cycles of these parasites are highly complex, involving different hosts and different specific interactions with a variety of cells and tissues [7- 11]. Some of these parasites live in the extracellular matrix or blood of host mammals, but the majority of them infect host cells to complete their cycle. Despite the high infection and mortality rates of these protozoa, especially in low-income populations of developing regions such as Africa, Asia and the Americas, current therapies for these parasitic diseases are very limited and unsatisfactory. The development of efficient drugs is urgently necessary, as are serious public health initiatives to improve patients' quality of life [12-16].

The Trypanosomatidae family belongs to the order Kinetoplastida and is comprised of flagellated protists characterised by the presence of the kinetoplast, a DNA-enriched portion of the mitochondrion localised close to the flagellar pocket. The most studied pathogenic trypanosomatids are the following: (a) *Trypanosoma brucei*, which is responsible for sleeping sickness in Africa; (b) *T. cruzi*, which is the causative agent of Chagas disease in Latin America; and (c) a variety of *Leishmania* species that cause leishmaniasis in tropical and subtropical areas worldwide. These illnesses have been classified by the World Health Organization as neglected diseases, which affect people living in poverty in developing countries and for which no efficient therapy is available [17-19].

The Apicomplexa family encompasses a large group of protists, including approximately 5,000 known parasitic species, which are characterised by the presence of an apical complex containing a set of organelles involved in the infection process. Apicomplexan parasites infect invertebrate and vertebrate hosts, including humans and other mammals. The most serious parasitic disorder is caused by apicomplexan *Plasmodium* species, the etiological agent of malaria, which causes more than one million deaths annually [1]. Toxoplasmosis is another important disease caused by the apicomplexan parasite *Toxoplasma gondii;* it has been estimated that almost half of the human population worldwide is infected with this protozoa [20]. The life cycle of the apicomplexan parasites generally consists of complex asexual and sexual reproduction, but some differences are observable among distinct genera. Malaria transmission occurs during the blood feeding of the *Anopheles* mosquito, whereas toxoplasmosis is mainly transmitted by the ingestion of raw meat or contaminated cat feces.

Autophagy is a physiological self-degradative pathway essential for the maintenance of the metabolic balance in eukaryotes, leading to the turnover of cellular structures during both the normal cell cycle and during conditions of stress, such as starvation [21,22]. This process depends on double-membrane vesicles known as autophagosomes, which are responsible for the engulfment of macromolecules and organelles and the recycling of their components without an inflammatory response [23]. In eukaryotic cells, proteins known as Atgs contribute to the formation of autophagosomes and their targeting to lysosomes [24]. The autophagic machinery interfaces with many cellular pathways, such as that of the immune response and the inflammatory process, and acts as an inductor or suppressor of these processes [25]. Some molecules and organelles can undergo autophagy by specific proteins, such as in the selective pathway known as xenophagy, which is also observed in the degradation of intracellular pathogens [26,27]. The involvement of autophagy in this process has been demonstrated in the interactions of different pathogens with the host cells [28-30]. In protozoan infections, the role of autophagy has been debated in light of conflicting evidence presented in the literature, which tends to vary with the experimental model. Some studies suggest that parasites evade host cell defences using autophagy, while others suggest that the host uses autophagy to eliminate the pathogen [31-35]. However, there is no doubt that the autophagic machinery decisively influences the pathogenesis and virulence of protozoan infections; this machinery may therefore represent a promising target for drug discovery [36]. The autophagic process also occurs in the protozoa [37,38] and could occur in parallel to the host cell pathway, thus increasing the complexity of the phenomena. In the following sub-sections, the biology of

Trypanosomatidae and Apicomplexa protozoa will be reviewed in relation to the role of autophagy during the infection of the host cells.

2. Trypanosomatids and autophagy

As previously mentioned, the transmission of neglected diseases caused by trypanosomatids (sleeping sickness, Chagas disease and leishmaniasis) depends on an insect vector, and the environmental change from one host to another is a drastic event for the protozoa. To complete its life cycle, many metabolic and morphological changes must occur for the parasite to survive in a new host [39-42]. In addition to the kinetoplast, other characteristic ultrastructural structures are present in these parasites, including a single mitochondrion, unique flagella, sub-pellicular microtubules, glycosomes, acidocalcisomes and reservosomes (the last one is present exclusively in *T. cruzi*) [8]. In the context of the remodelling of sub-cellular structures, autophagy is greatly involved in eukaryotic homeostasis (including in that of trypanosomatids). However, the deregulation of this pathway, which is induced by conditions of stress, also leads to the parasite's death (Table 1). The sequencing of the complete genome of trypanosomatids has enabled the identification of parasitic genes [43-45]. Blast analysis comparing the trypanosome genome with yeast and mammalian genomes, with a particular emphasis on genes encoding autophagic machinery, has indicated the presence of some *ATG* genes in trypanosomatids [46,47]. However, the partial lack of a ubiquitin-like system, which is crucial for autophagosome formation, and the absence of cytoplasm-to-vacuole-targeting pathway orthologs suggest that these parasites have alternative autophagic features.

3. *T. brucei*

T. brucei is the etiological agent of sleeping sickness (or African trypanosomiasis) and is transmitted by the infected tsetse fly (*Glossina* sp.). After a blood feeding, procyclic trypomastigotes migrate from the insect midgut to the salivary gland where they undergo differentiation to infective metacyclic forms. Subsequently, these metacyclic trypomastigotes are inoculated into the mammalian host during the blood meal of the fly and differentiate into a proliferative bloodstream slender form. Interestingly, after a new differentiation, adapted short-stumpy forms evade the host immune system and disseminate the infection to the whole body; these forms are also able to cross the blood-brain barrier, which causes severe behavioural abnormalities, such as somnolence during daytime [48] (Figure 1). Unlike all other pathogenic trypanosomatids, which have an intracellular life-stage, *T. brucei* remains in the bloodstream of the mammalian host throughout the process of infection and, as such, is exposed to different environmental conditions that can trigger autophagy.

3.1. Role of autophagy in *T. brucei*

The first report on this parasite and autophagy was published in the 1970s by Vickerman and colleagues. These authors described the presence of myelin-like structures in different forms of the parasite observed by transmission electron microscopy [49, 50]. Many years later, it was suggested that the autophagic pathway is involved in the turnover of glycosomes during protozoan differentiation [51]. Glycosomes are peroxysome-like organelles that perform early glycolytic steps and are also involved in lipid metabolism. It was demonstrated that glycosome contents are altered depending on the form of the parasite, with many of these organelles being close to glysosomes during the differentiation process. A similar phenomenon was observed after nutrient deprivation of the parasite, reinforcing the fact that differentiation may cause the degradation of glycosomes by pexophagy.

Further genomic and bioinformatic analyses were performed that identified in *T. brucei* many *ATG* orthologs to those of yeasts and mammals [47,52]. These genes are involved in different steps of the autophagic pathway, such as induction (*ATG24, PEX14, TOR1* and *TOR2, VAC8*), vesicle nucleation (*ATG6, VPS15* and *VPS34*) and vesicle expansion and completion (*ATG3, ATG7, ATG9*, two isoforms of *ATG4* and *ATG8*). Two isoforms of Atg4 and two of Atg8 were recently characterised structurally [53], and it was postulated that Atg8.2 is essential for autophagosome formation and that Atg8 depletion is associated with delayed cell death [54].

It is thought that many drugs may trigger autophagy in African trypanosomes. Dihydroxyacetone (DHA), spermine (snake venom) and vasoactive intestinal peptide (VIP – a neuropeptide secreted by the immune system) induce the appearance of morphological features of autophagy in *T. brucei* [55-58]. DHA is an interesting compound to be used in therapy for sleeping sickness because its phosphorylation is DHA kinase-dependent, and DHA kinase is present in mammals and other eukaryotes but not in trypanosomes. After DHA uptake, this compound is not eliminated, leading to typical morphological characteristics of autophagy similar to those found in rapamycin treatment. In another report [59], the authors showed that hydrogen peroxide can produce the appearance of autophagic profiles, suggesting that the release of reactive oxygen species acts as a signal in the autophagic pathway in *T. brucei,* as it does in other eukaryotic cells [60-62].

4. *T. cruzi*

T. cruzi is the causative agent of Chagas disease. It is mainly transmitted by triatomine bugs, which are commonly known as "kissing bugs". In the insect midgut, proliferative forms of the parasite called epimastigotes differentiate to metacyclic trypomastigotes after migration to the posterior intestine. During the blood meal, triatomines eliminate urine and feces with infective trypomastigotes that then gain access to the vertebrate bloodstream. After internalisation in the host cell, trypomastigotes remain in parasitophorous vacuoles (PV) that fuse with lysosomes, allowing an acidification of this compartment, which is an essential step towards differentiation into proliferative amastigotes. In the cytosol, successive parasite cycles occur until a new intracellular differentiation to trypomastigotes occurs; it is these forms that are responsible for the infection and dissemination to other cells and tissues [8] (Figure 2).

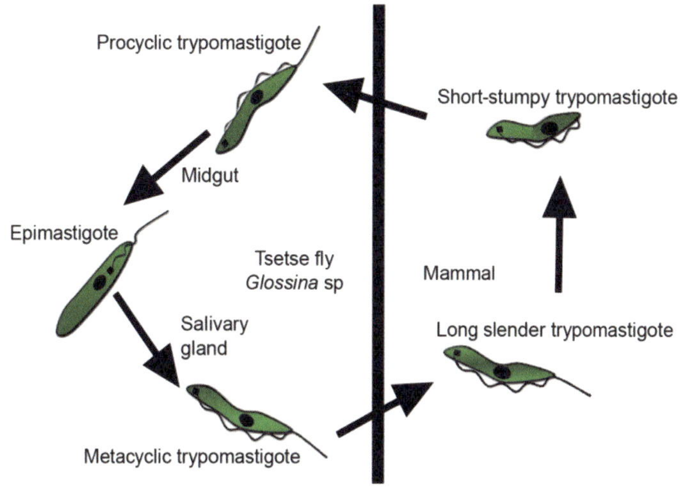

Figure 1. *T. brucei* life cycle.

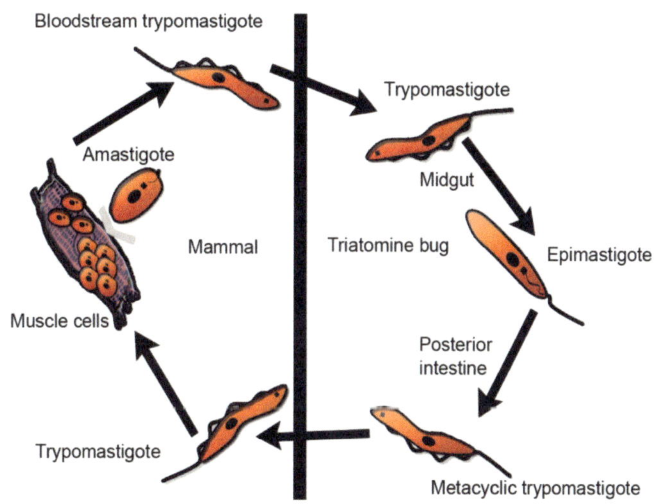

Figure 2. *T. cruzi* life cycle.

4.1. Role of autophagy in *T. cruzi*

Ultrastructural evidence of autophagy in *T. cruzi* was observed after the treatment of epimastigotes and bloodstream trypomastigotes with drugs; the appearance of myelin-like figures was the most recurrent feature detected [63-67]. Recently, the synergistic combination of amiodarone and posoconazole was able to trigger autophagy in replicative amastigotes [68].

In this way, different classes of therapeutic agents are able to induce the formation of autophagosomes, an event associated with parasite-related autophagic cell death, being the interplay between other programmed cell death as apoptosis or necrosis not discarded [69]. Due to the limitations of cell models, previous studies of different parasite forms have employed alternative techniques, such as monodansylcadaverine (MDC) staining and ATG gene expression, to demonstrate autophagy in the parasite [66,67]. Unfortunately, *T. cruzi* molecular machinery does not allow the use of double-stranded RNA to knock down target RNAs [70]; in addition, the lack of recognition of protozoan proteins by anti-Atg commercial antibodies hampers the evaluation of autophagy in this parasite. In spite of the advances in molecular and cellular biology, transmission electron microscopy remains a gold standard for autophagy analysis [71,72].

Aside from the description of autophagosomes in all *T. cruzi* life stages, description of the Atg cascade involved in autophagosome formation is not complete. Almost all *T. brucei ATG* genes have ortholog genes in *T. cruzi* [37,47]. In this parasite, two isoforms of Atg8 were described, with only Atg8.1 localised in autophagosomes as expected. These data suggest that there is only partially shared autophagic machinery, as is observed in human Atg8 orthologs [37]. In another study [37], the authors described the participation of *T. cruzi* Atg4 and Atg8 isoforms under conditions of nutritional stress and in the differentiation process from epimastigotes to metacyclic trypomastigotes, a process known as metacyclogenesis. The authors observed a remarkable expression of Atg8.1 by immunofluorescence microscopy, which was suggestive of intense autophagy in differentiating epimastigotes. Moreover, Atg8 co-localised with reservosomes, which are pre-lysosomal compartments related to energy supply that are present only in epimastigotes [73,74]. The reservosomal content consumed during metacyclogenesis and the presence of Atg8 in this organelle strongly suggest that there is crosstalk between autophagy and reservosomes [75,76]. Transmission electron microscopy studies have produced images from endoplasmic reticulum profiles surrounding reservosomes that indicate the possible origin of preautophagosomal structures [66]. It is well known that PI3K inhibitors, such as 3-methyladenine and wortmannin, prevent autophagy in different experimental models [54,66]; however, these data are controversial due to a previous report demonstrating that treatment with kinase inhibitors staurosporine, genistein, 3-methyladenine and wortmannin led to the formation of autophagosomes [77]. The data indicate the necessity of careful use of PI3K inhibitors to block autophagy and the urgent need for the development of new specific autophagic inhibitors [78].

4.2. Host cell autophagy and *T. cruzi* infection

Though thought to be essential for parasite success, lysosomal fusion could be involved in autophagy during host cell interaction and might contribute to the process of degradation and elimination of *T. cruzi*. In 2009, the role of autophagy in parasite entry and co-localisation with the PV was described, resulting in increased infection of Chinese hamster ovary cells; this observation was subsequently confirmed in macrophage and heart cell lineages [34,79]. Starvation conditions and the addition of rapamycin led to an increase in the scale of the infection; this increase was partially reversed by 3-methyladenine, wortmannin and vinblas-

http://dx.doi.org/10.5772/54970

tine, suggesting that autophagy favours the parasite during *T. cruzi*-host cell interactions. However, other groups demonstrated that classical autophagic stimuli (nutritional stress and rapamycin) did not produce an increase in parasite proliferation or even in the number of infected cells [33]. Recently, studies have emphasised role of autophagy in the control of *T. cruzi* infection using different cells and parasite strains (Figure 3) [80,81]. Once more, the conflicting data presented in the literature need to be further debated in light of the complexity of the protozoal strains and host cell models employed.

5. *Leishmania* species

The other medically important trypanosomatids are *Leishmania* species. Leishmaniasis is transmitted to mammals by sandflies, mainly of the *Phlebotomus* and *Lutzomia* genuses. Amastigotes differentiate into replicative procyclic promastigotes in the digestive tract of these sandflies, proliferate in the Phlebotominae gut, and then migrate to the proboscis where a new differentiation occurs to metacyclic promastigotes, the infective forms of the parasite. During the sandflies' blood meals, metacyclic promastigotes are inoculated into mammalian tissue and are phagocytised by macrophages. Inside the host cells, promastigotes differentiate into amastigotes that replicate and are responsible for cell lysis and dissemination in the organism (Figure 4). Currently, more than 20 species of *Leishmania* are known, each causing different clinical manifestations of the disease, including cutaneous leishmaniasis and visceral leishmaniasis (or Kala-azar). The pathogenicity depends on the *Leishmania* species and the host's immune response [8].

5.1. Role of autophagy in *Leishmania* sp.

Many groups have investigated autophagy cell death induced by drugs or antimicrobial peptides in various *Leishmania* species using electron microscopy and MDC staining [82-89]. Bioinformatics analysis has been a crucial checkpoint in the characterisation of *ATG* and *TOR* pathways in trypanosomatids [38,47,90]. In 2006, the role of autophagy in the differentiation process of *L. major* and *L. mexicana* was first evaluated [38,90]. The authors developed *L. major* VPS4, a mutant that could not complete the differentiation to the infective forms due to interference in autophagosome formation during conditions of starvation. The increase in Atg8 expression in differentiating forms supports the hypothesis that autophagy plays a pivotal role in metacyclogenesis [38,91]. In *L. mexicana*, the lack of cysteine peptidases CPA and CPB impairs autophagosomes formation and parasite differentiation; this finding is corroborated by the results of wortmaninn treatment and *ATG* deletion [90].

Recently, a subunit of protein kinase A in *L. donovani* that interferes with autophagy and protozoa differentiation was identified [92]. As observed in other trypanosomatids, the presence of Atg8-like proteins and their association with Atg4 in *Leishmania* species indicates that these proteins play a role in vesicle expansion [93]. Interestingly, the Atg5-Atg12 complex involved in autophagosome elongation was not previously detected [47], but recent studies have demonstrated its existence. It has also been shown that Atg5 deletion severally affects

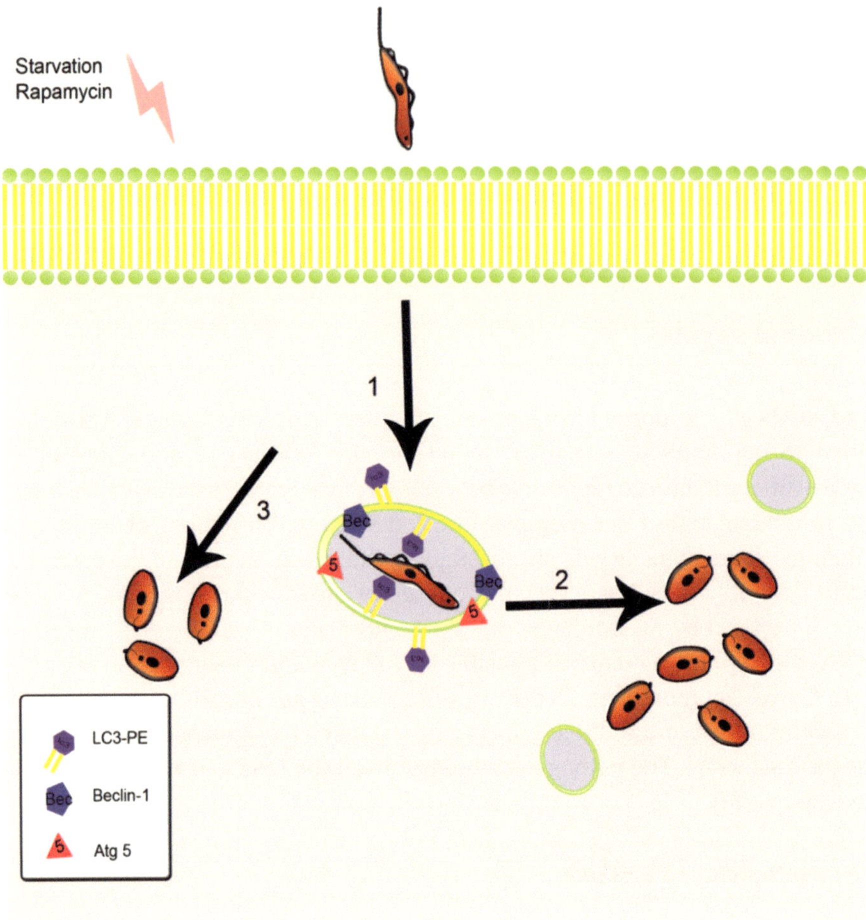

Figure 3. Autophagy in *T.cruzi*–host cell interaction. Romano et al [34] showed the co-localization of parasite vacuole with Atg proteins in the beginning of infection (1). Moreover, the replication of amastigotes is the same with or without autophagy induction (2) [33,34]. Rapamycin and starvation control infection reducing the number of amastigotes per cell (3) [80,81].

parasite homeostasis, producing a phenotype characterised by mitochondrial disruption, phospolipid accumulation and abnormal promastigote morphology [93,94]. Table 1 summarises the autophagic events in the three pathogenic trypanosomatids described in this chapter.

5.2. Host cell autophagy and *L. amazonensis* infection

The connection between the endosomal/lysosomal pathway and the PV results in macromolecules being taken up by the parasite, as demonstrated in *T. cruzi* infection [96]. In this context, a notable increase in the proliferation of *L. amazonensis* amastigotes was observed after autophagic induction by nutritional deprivation, rapamycin treatment or interferon-gamma. This mechanism was partially reversed by the autophagic inhibitors wortmaninn or 3-methyladenine, which significantly reduced amastigote replication (Figure 5) [33]. However,

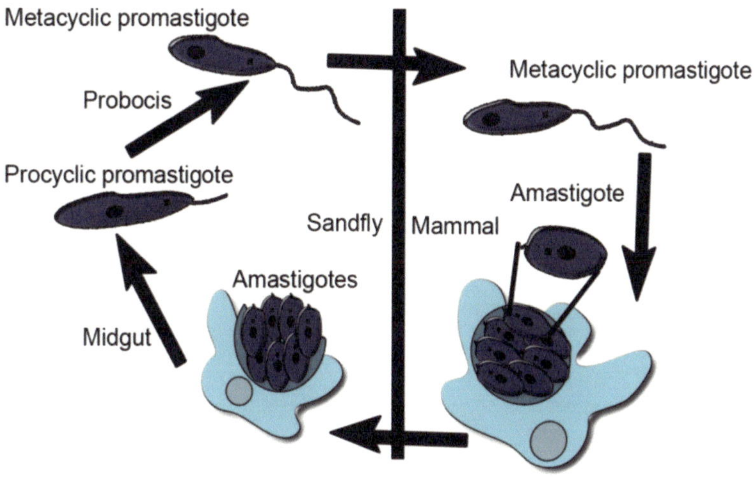

Figure 4. *Leishmania* sp. life cycle.

a recent report presented no correlation between the increase in LC3 expression and heightened *L. amazonensis* infection after treatment with autophagy inducers and inhibitors. In addition, macrophage autophagy was observed in inflammatory infiltrates of *L. amazonensis*-infected mice [97] and in natural human *L. donovani* infection [98].

6. Apicomplexa and autophagy

The phylum Apicomplexa comprises one of the most medically relevant groups of protists, which cause serious health and economic problems. Among these parasites, *Toxoplasma gondii* and *Plasmodium* species are well-known apicomplexans; it is estimated that malaria caused by *P. falciparum* kills over a million people annually. Another widespread disease is toxoplasmosis, which is caused by the apicomplexan parasite *T. gondii;* the severity of disease caused by this organism is directly related to patients′ immunosuppression and is characterised by congenital transmission. In this context, knowledge of the detailed mechanisms involved in parasite infection and survival, including the role of autophagy, could contribute important information to the development of novel strategies for controlling Apicomplexa infections. Autophagy is an evolutionarily conserved pathway found in all eukaryotes, from unicellular organisms to metazoans; orthologs for approximately 30% of autophagy-related genes have been detected in apicomplexan sequenced genomes [99].

Among the key molecules involved in early autophagy steps, Atg1/ULK complex, Atg8 and Atg9 play crucial roles in cargo selectivity and in autophagosome formation [100,101]. Unlike other cell models, in Apicomplexa protozoa, the Atg8 C-terminal appears to not undergo processing before its association with phosphatidylethanolamine (PE) in the membrane of

Parasite	Life-stage	Phenotype	Stimuli	References
T. brucei	bloodstream trypomastigotes	autophagic cell death	DHA, neuropeptides, rapamycin, starvation	[55,58, 59]
	procyclic trypomastigotes	autophagic cell death	spermine (snake venom)	[57]
		Autophagy-induced differentiation	rapamycin, starvation	[54,56]
		unfolded protein response in endoplasmic reticulum associate with autophagy	DTT	[95]
T. cruzi	epimastigotes, trypomastigotes	autophagic cell death	SBIs; LPAs and cetoconazole; naphthoquinones; naphthoimidazoles; MBHA; posoconazole and amiodarone	[63-65,67, 68,71,72,]
	metacyclic trypomastigotes	Autophagy-induced differentiation	starvation; differentiation medium	[37]
L. amazonensis	promastigotes, amastigotes	autophagic cell death	amiodarone; elatol; lipophilic diamine	[83,86,89]
L. chagasi	promastigotes	autophagic cell death	yangambin	[87]
L. donovani	promastigotes	autophagic cell death	antimicrobial peptides; cryptolepine	[82,88]
L. major	promastigotes, amastigotes	autophagic cell death	cathepsin inhibitors	[85]
	metacyclic promastigotes	autophagy induces differentiation	differentiation medium; starvation	[38,91]
L. donovani	metacyclic promastigotes	autophagy induces differentiation	differentiation medium; starvation	[90]

Table 1. Summary of autophagic events in trypanosomatids. DHA: Dihydroxyacetone; DTT: dithiothretiol; SBIs: sterol biosynthesis inhibitors; LPAs: lysophospholipid analogues; MBHA: Morita–Baylis–Hillman adduct.

autophagosomes, suggesting a different regulation of this Atg protein in these organisms than in mammals and fungi [102]. Using a technique to detect lipidated Atg8 in *Plasmodium* species, only a single band corresponding to ATG8 was observed, suggesting that this parasite's Atg8 exists predominantly in the PE-conjugated form [22].

Two important kinases have opposing roles in the autophagic process: TOR (target of rapamycin) and class III phosphatidylinositol3-kinase (PI3K) [78,103]. In well-established autophagic models, TOR and class III PI3K represent negative and positive regulators, respectively, that act through complexes with regulatory subunits orchestrated by signalling cascades.

http://dx.doi.org/10.5772/54970

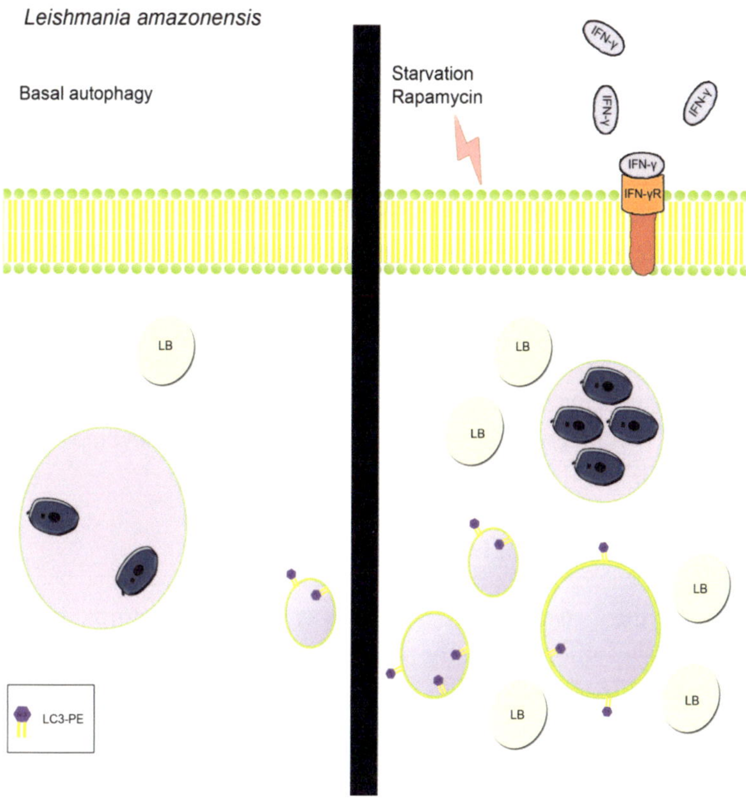

Figure 5. Autophagy in *L. amazonensis*-host cell interaction. When autophagy is induced, more amastigotes replicate and PV is smaller than in basal autophagic cells. Also, more lipid bodies are present, increasing infection and signaling to replication [33,97].

Analysis of the *T. gondii* genome revealed the presence of TOR and PI3K but not of other proteins crucial to the formation of these complexes [99]. Curiously, no genes for TOR complex machinery were found in the *Plasmodium* genome. Thus, it is possible that these unicellular eukaryotes have specific unknown proteins for several steps of the autophagic pathway instead of an absence of key proteins [22,104].

7. *T. gondii*

T. gondii is an obligate intracellular parasite with a complex life cycle involving one definitive feline host where the sexual phase occurs and intermediate hosts, such as birds, other mammals and man [105]. The main transmission routes to humans are the following: (i) the ingestion of raw meat containing tissue cysts (essentially bradyzoites forms); (ii) the ingestion of water and food contaminated with feline feces residue containing oocysts; and (iii) transplacentary pathway of tachyzoites [106]. After oral ingestion, tissue cysts or oocysts rupture, liberating the slow-replicating forms known as bradyzoites and sporozoites, respectively, which then invade intestinal epithelial cells. In the intracellular environment, the parasites differentiate into the fast-replicating tachyzoites that proliferate inside the host cell PV. The sustained infection depends on the modification of the PV membrane by the insertion of *T. gondii* secreted proteins, which prevent the fusion to lysosomes and, consequently, the elimination of the parasite (Figure 6) [20,107].

In healthy adults, *T. gondii* cysts are established in the host cells mainly in the eyes, brain and muscles during the chronic phase of toxoplasmosis [108]; however, in immunocompromised patients, such as HIV-positive patients, or in congenital toxoplasmosis, the disease becomes much more severe, and its complications could lead to death [20,109,110]. Despite the high percentage of people infected, the available therapy for toxoplasmosis is effective only in the tachyzoite stage and presents limited efficacy against the tissue cyst, which is the latent form of the parasite [111]. In this context, many efforts are necessary to develop new drugs to treat *T. gondii* infection [17].

7.1. Role of autophagy in *T. gondii* infection

Only a few studies on the *T. gondii* autophagic pathway have been performed, and these studies suggest opposing roles of autophagy in the parasite infection [102,112]. The presence of TgAtg8 in autophagic vesicles was observed in tachyzoites during their intracellular replication; similarly, severe parasite growth arrest due to TgAtg3 knockdown and recent identifications of the presence of TgAtg1 and TgAtg4 in the parasite suggest a role for autophagy in *T. gondii* homeostasis, although long-term exposure to autophagic stimuli was found to be harmful to the parasite (Figure 7) [112; 113].

Tachyzoites divide by a process called endodyogeny, whereby two daughter cells are developed inside a mother cell and leave residual material at the end of division. During this process, autophagy might be involved in recycling the mother cell organelles, such as micronemes and rhoptries, which are synthesised de novo in the daughter cells; however the accumulation of organelles after endodyogeny has not been observed in TgATG3 knockout organisms, making other experiments necessary to confirm this hypothesis [113]. One important phenotype detected in autophagic mutants is the loss of mitochondrial integrity [102,112]. Mitophagy, which is the autophagy of mitochondria, regulates the mitochondrial number to match metabolic demand; this process represents a quality control that is necessary for the removal of damaged organelles [114]. Autophagic stimuli are able to direct the mitochondrial network of tachyzoites towards their autophagic pathway, but the molecular machinery involved in selective targeting of the organelle remains unclear [102,112]. Nutrient deprivation has been shown to be a classic stimulus for the autophagic pathway activation in a large variety of organisms [37,115]. In *T. gondii* tachyzoites, starvation induces autophagy in extracellular and intracellular parasites [102,112]. Furthermore, autophagosomes were observed in parasites after a long extracellular nutritional restriction, suggesting that autophagy can act as a mechanism of resistance to starvation for nutrient recycling until the infection of a new host cell [102].

http://dx.doi.org/10.5772/54970

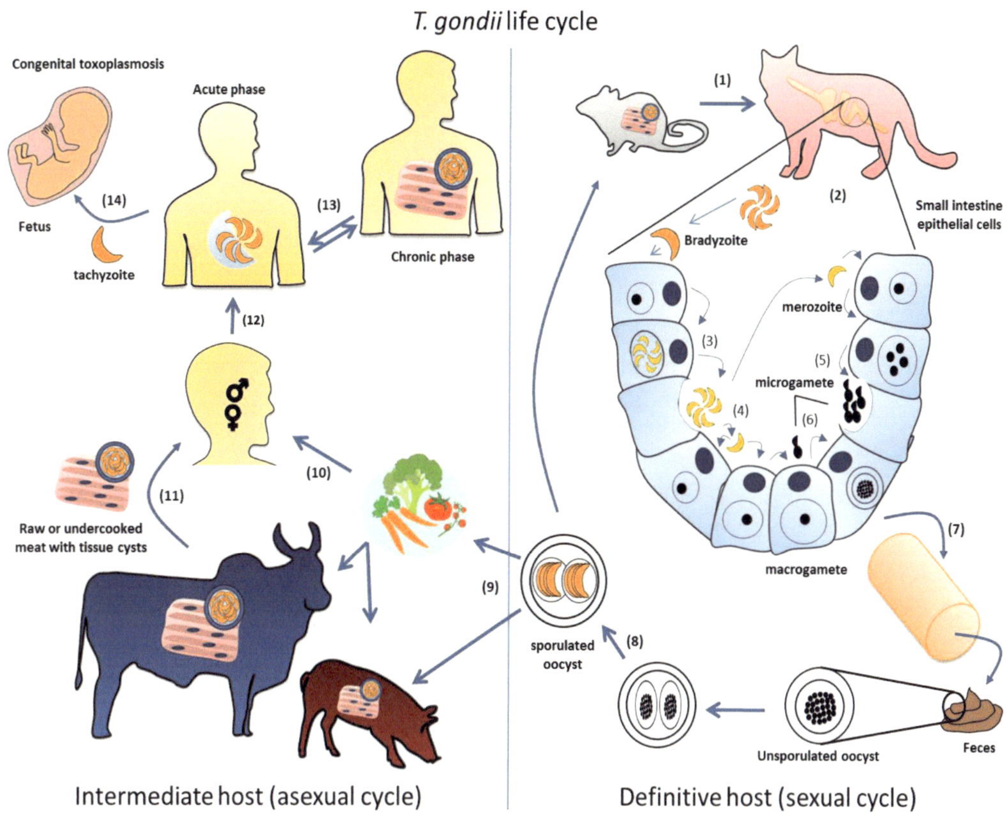

Figure 6. *T. gondii* life cycle. (1) Definitive host infection; (2) Cyst disruption and intestinal epithelial cell infection; (3) Formation of merozoites; (4,5) Start of sexual phase with the formation of macrogametes and flagellate microgametes from merozoites; (6) Fusion of microgamete and macrogamete; (7) Oocyst release to the environment in the faeces; (8) The unsporulated oocysts become infective and contaminate the environment [116-118]; (9) The sporulated oocysts can cause infection of animals via consumption of contaminated food and water. (10,11) Human infection occurs by the ingestion of raw or undercooked meat of infected animals containing *T. gondii* cysts; (12) *T. gondii* tachyzoite multiplication in the intermediate host; (13) Tachyzoite-bradyzoite differentiation and formation of tissue cysts; (14) Transplacentary transmission of tachyzoites.

The data presented here demonstrate possible functions of *T. gondii* autophagy in parasite homeostasis. However, it has been proposed that, when strongly induced, the autophagic pathway represents a self-destructive mechanism leading to protozoal death. The molecular pathway of autophagic cell death is still unknown, and it is debated whether the pathway is a type of programmed cell death or a survival response to death stimuli [119]. Intracellular starved tachyzoites showed systematic mitochondrial fragmentation and a defect in host cell internalisation. As *T. gondii* is an obligate intracellular protozoa, the loss of invasion capacity leads to parasite death. The impairment in infective ability was related to the loss of mitochondrial integrity because organelles from apical complexes, such as rhoptries and micronemes, which are usually associated with the invasion process, are intact in these parasites [112]. Interestingly, these authors also demonstrated that autophagic inhibitor 3-methylade-

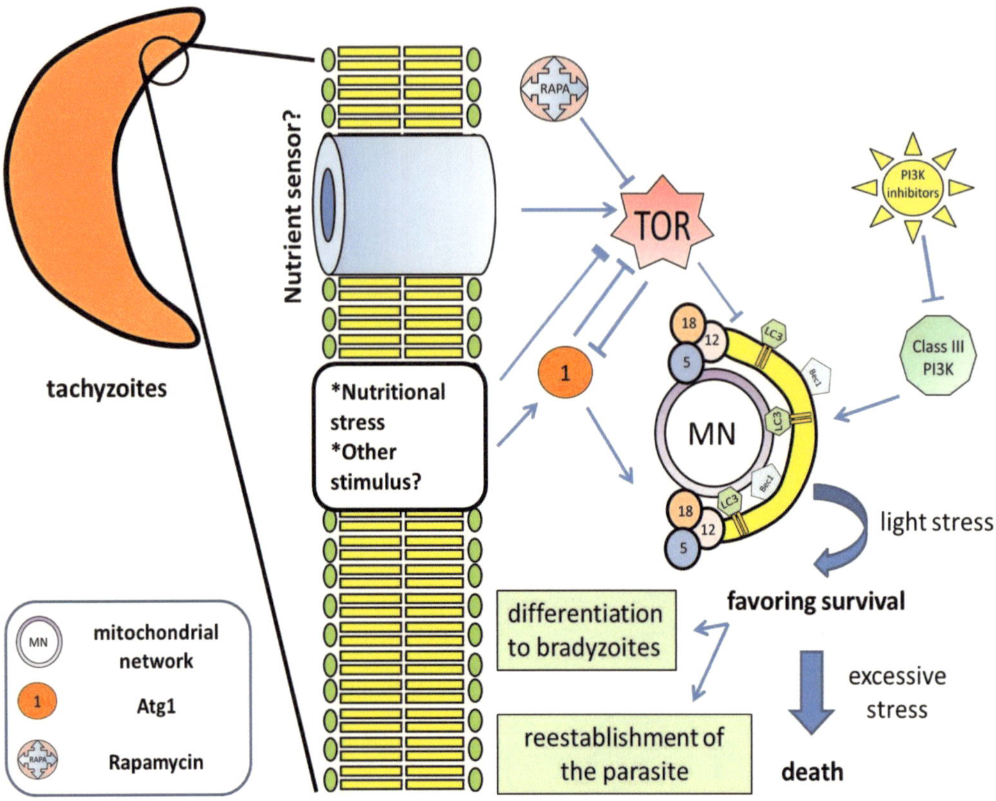

Figure 7. *T. gondii* tachyzoites response to autophagic stress. Autophagy acts in survival or death mechanisms in apicomplexan parasites depending on the environmental stress conditions. Arrows: activation; Headless arrows: inhibition.

nine prevented mitochondrial fragmentation, suggesting autophagic involvement in *T. gondii* death.

While nutritional stress has been extensively used as a model for autophagy, this condition is not easily encountered in the host cells and tissues *in vivo*. However, parasites could be exposed to nutritional restriction in the extracellular environment. The viability of tachyzoites kept in an axenic medium for periods of up to 12 hours drastically decreases, but a significant number of parasites nevertheless differentiate into bradyzoites [120]. Indeed, these observations raise the hypothesis that autophagy could be an adaptive mechanism of *T. gondii* to survive for short periods in starvation conditions, allowing the parasite to recover when favourable conditions occur or even to differentiate into a cystic form. Another interesting point for discussion is the correlation between mitochondrial fragmentation in intracellular tachyzoites and the depletion of amino acids in the culture medium [112]. Activated macrophages infected with the parasite showed low availability of the essential amino acid tryptophan, a condition that directly contributes to the protozoa's death in these cells [121,122]. In this context, TOR kinase is a vital component of the amino acid sensing mechanism in eukaryotic cells, as suggested by the detection of TgTOR by bioinformatic approaches and the evaluation of the activity of the classical TOR inhibitor rapamycin. This inhibitor triggered mitochondrial fragmentation of intracellular tachyzoites in starved parasites, and this phenotype was reversed by adding 3-methyladenine [112].

7.2. Host cell autophagy and *T. gondii* infection

As previously mentioned, *T. gondii* can infect any nucleated cell, but the parasite tropism principally involves nervous and muscular cells where the establishment of cystic forms is observed in chronic toxoplasmosis [111,123]. As was observed for *T. cruzi*-host cell interactions, controversial data on the importance of autophagy during *T. gondii* infection have been described in the literature; indeed, it has been suggested that autophagy can either control or facilitate parasite internalisation and proliferation [32,35,124-128]. Despite the relevance of muscular and nervous cells for the establishment of infection and for the course of the disease, very little has been reported on the role of autophagy in the progression of infection. As we will discuss in the next paragraphs, previous studies on the connection between the autophagic pathway and *T. gondii* infection were performed in macrophages, which are cells that play an important role in the immune response against this parasite [129].

Previous reports have shown that cellular immunity mediated by CD40 stimulation redirects the *T. gondii* to a lysosomal compartment via the autophagic route, resulting in the antimicrobial activity of the macrophage *in vitro* and *in vivo* [124,125]. *In vivo*, parasite elimination was dependent on GTPase p-47, IFN-γ, IGTP, and PI3K and culminated in the rupture of the parasite's membrane [125] (Figure 8). Additionally, the relationship between autophagy and the fusion of lysosomes with the *T. gondii* PV seems to be dependent on the synergy between TRAF6 signalling downstream of CD40 and TNF-α [126]. However, the IFN-γ/p47 GTPase-dependent elimination of the parasite by macrophages is independent of CD40/TNF signalling *in vitro*, demonstrating the primary role of IFN-γ in immunity against *T. gondii* in mice [127]. As observed in astrocytes, autophagy is activated to eliminate intracellular parasite debris and thus prevent the host cell death. Investigations in macrophages also indicated that the CD40-p21-Beclin 1 pathway is a CD40-dependent immunity route to mediating *in vivo* protection [128]. Similarly, Atg5 is required for damage to the PV membrane and removal of the parasite in primary macrophages stimulated by IFN-γ, despite the fact that no autophagosomes involving *T. gondii* have been detected. Atg5 also appeared crucial for *in vivo* p47 GTPase IIGP1 recruitment to the vacuole membrane induced by IFN-γ, suggesting an additional autophagy-independent role for Atg5 in the GTPase trafficking process [32]. In *T. gondii* infected astrocytes, the participation of autophagy has been shown to be indirect. The IFN-γ-stimulation of astrocytes infected with tachyzoites triggers the recruitment of p47 GTPases to the PV and usually leads to rupture of the vacuole and parasite membrane. In this case, autophagy acts by removing protozoal debris that accumulates in the cytoplasm and causes cell injury. Additionally, autophagy assists in antigen presentation through MHC class II in astrocytes, allowing an intracerebral immune response to parasite [130].

So far, little has been described regarding the involvement of autophagy in the interaction of *T. gondii* with nonprofessional phagocytes. In primary fibroblasts or Hela cells, infection with tachyzoites induced LC3 conjugation to PE, accumulation of LC3-containing vesicles close to the PV and an overexpression of beclin-1 and phosphatidylinositol-3-phosphate in the host cells in the mTOR-independent pathway. The infection of Atg5-deficient fibroblasts was reduced in physiological concentrations of amino acids, reinforcing the host cell autophagic role in the recovery of nutrients by the parasite. Because the classical function of autophagy involves recycling of various cellular components and because *T. gondii* depends on the uptake of many nutrients from the host cell, it has been proposed that the parasite may take advantage of the mammalian autophagic machinery to achieve successful infection [35]. Table 2 shows the host autophagic roles during *T. gondii* infection.

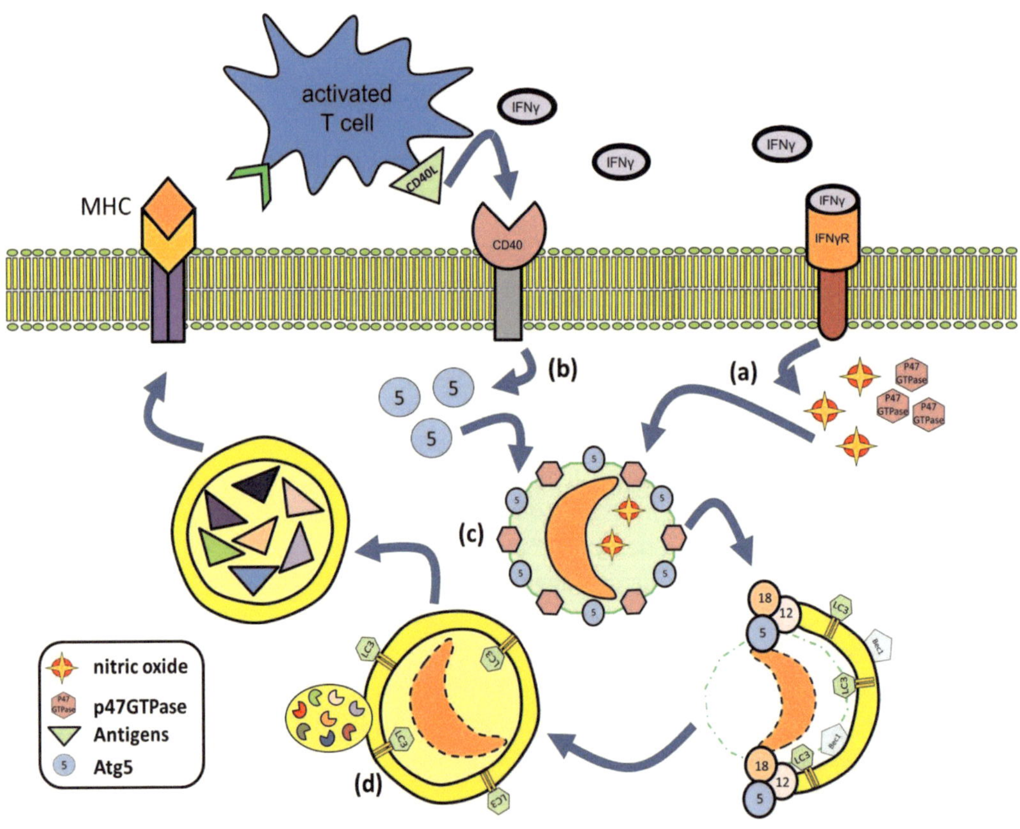

Figure 8. Autophagic role in *T. gondii* interaction with professional phagocytic cells. (a) INF-Y recruits P47GTPases to the PV membrane and induce nitric oxide production which limits the parasite replication. (b) CD40L activates Atg5 and recruits the autophagic machinery to the PV membrane. (c) PV and parasite membrane degradation by P47GTPase and Atg5. (d) Elimination of *T. gondii* debris by autophagolysosomal fusion and possible contribution of this process in antigen presentation through class II MHC.

http://dx.doi.org/10.5772/54970

Host cell	Induction	Phenotype	Reference
Peritoneal Macrophages and RAW264.7 lineage	CD40 stimulation and rapamycin	accumulation of LC3 around PV and low parasite load	[124]
Peritoneal macrophages	INF-γ stimulation	autophagy- dependent elimination of intracellular parasite debris	[125]
Peritoneal macrophages	INF-γ stimulation	Atg5-dependent PV membrane disruption	[32]
bone marrow Macrophages	CD40-p21-Beclin 1 pathway	stimulation of autophagy for protection against *T. gondii*	[128]
astrocytes	INF-γ stimulation	*T. gondii* debris removal by autophagy after vacuole and parasite membrane rupture by p47 GTPases	[130]
primary fibroblasts and Hela cells	*T. gondii* infection	Induction of LC3 conjugation to PE, accumulation of vesicles containing LC3 close to PV, beclin-1 and PI3K inside the cell	[35]

Table 2. Autophagy in *T. gondii*-host cell interactions

8. *Plasmodium* sp.

Plasmodium species are causative agents of malaria, the illness with the highest morbidity rate among human parasitic diseases. Currently, 5 species of *Plasmodium* sp. (*P. falciparum, P. vivax, P. malariae, P. ovale* and *P. knowlesi*) can infect humans, and lethality is associated with *P. falciparum* [131-133]. Sporozoites are transmitted by *Anopheles* sp. mosquitoes (definitive hosts) to the mammals (intermediate hosts), where they migrate primarily to the liver. After internalisation in hepatocytes, the parasites convert from elongated sporozoites (invasion competent and motile) to round proliferative trophozoites (metabolically active), which start the asexual reproduction process known as schizogony. At the end of the reproductive process, the daughter cells (merozoites) initiate maturation for erythrocyte invasion. When the merozoites become mature, they are enclosed in a membrane (the merosome) and released from hepatocytes to invade red blood cells, causing clinical symptoms of malaria (Figure 9). [135-137].

8.1. Role of autophagy in *Plasmodium* sp. infection

Recent publications have suggested that autophagy is involved in the differentiation of sporozoites to merosomes in hepatocytes [137,138]. The sporozoite-to-trophozoite differentiation is accompanied by the elimination of organelles unnecessary for schizogony and the production of merozoites in liver cells [137]. For example, micronemes and rhoptries are compartmentalised in the cytoplasm of sporozoites and sequestered in double-membrane structures resembling autophagosomes. In axenic conditions, the treatment of parasites with 3-methyladenine resulted in significant delay of the sporozoite differentiation process [139].

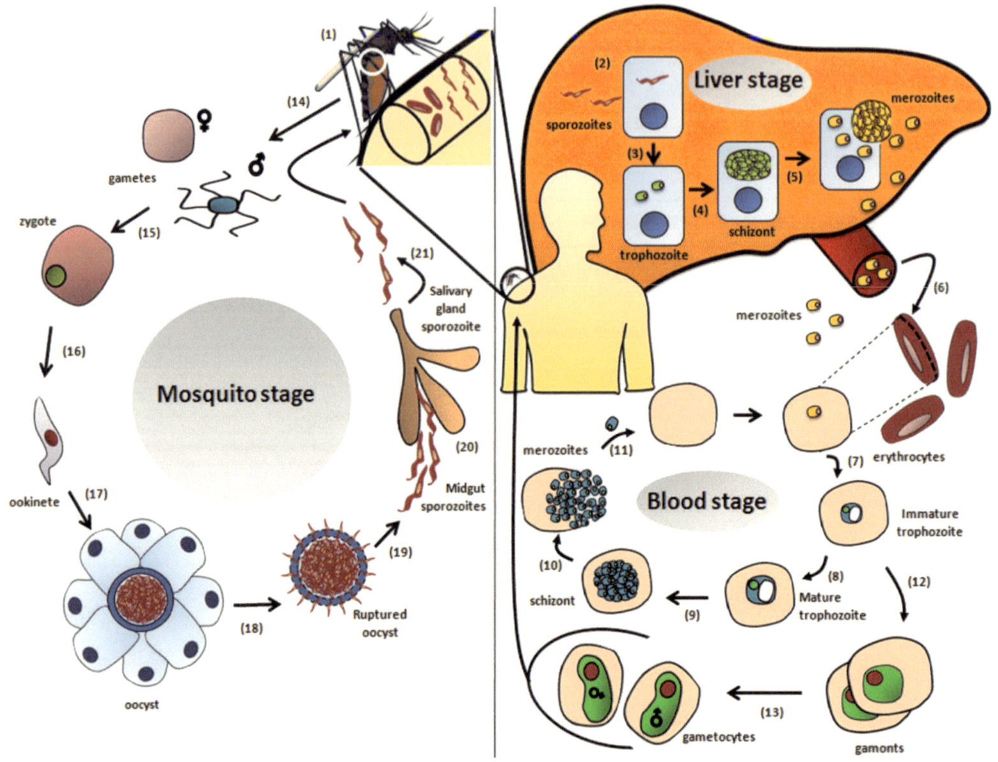

Figure 9. *Plasmodium* sp. life cycle. (1) Inoculation of sporozoites by malaria-infected female *Anopheles* mosquito into the human host. (2) Sporozoites infect hepatocytes. (3) Sporozoite-trophozoite differentiation. (4) Schizont formation. (5) Schizont rupture and release of merozoites. (6) Merozoites infect red blood cells. (7,8) Trophozoite maturation. (9) Schizont formation in red blood cells. (10) Schizont rupture and release of merozoites. (11) Infection of new red blood cells by the merozoites. (12,13) Differentiation of some parasites in gametocytes (sexual erythrocytic stages). (14) Ingestion of gametocytes by the mosquito during a blood meal. (15) Zygote formation in the mosquito´s stomach when the microgametes penetrate the macrogametes. (16) Zygote-ookinete differentiation. (17) Ookinetes invade the midgut wall of the mosquito where they develop into oocysts. (18,19) Oocysts rupture and release sporozoites. (20) Sporozoites migrate to the mosquito's salivary glands. (21) Mosquito inoculates sporozoites into a new human, perpetuating the parasite cycle.

After sporozoite differentiation, Atg8 is present in autophagosomes during the replication phase, suggesting an additional independent role for this protein in autophagy [137, 138,140].

The involvement of autophagy in *Plasmodium*-infected red blood cells has been poorly studied. One study demonstrated that erythrocytes infected with *P. falciparum* trophozoites and maintained in supplemented culture medium expressed Atg8 in the parasite cytosol. However, when these infected cells are submitted to restriction of glucose and amino acids, an increase in the number of autophagosomes labelled by Atg8 was observed, and these vesicles were found close to red blood cell membranes. Once erythrocytes no longer have organelles in the cytoplasm, the potential targets of autophagosomes in this cell model are debated. One hypothesis suggested that these autophagosomes target haemoglobin and blood nutrients to

http://dx.doi.org/10.5772/54970

favour nutrient uptake by the parasite (Gaviria and colleagues, unpublished results). Surprisingly, no TOR ortholog was found in the *P. falciparum* genome (Sinai & Roepe, unpublished results), suggesting that under normal growth conditions, *P. falciparum* autophagy is configured as a survival process that is constitutively regulated by the acquisition of nutrients, although this route is unusual. Table 3 summarises the published reports on autophagic features in apicomplexans.

Little is known about the involvement of autophagy in the *Plasmodium sp.*-host cell interactions. So far, *Plasmodium ATG8* knock-out resulted in a lethal phenotype, indicating that this gene is essential for the mammalian life-stage of the parasite [22]. However, there have been no studies on the importance of the host cell autophagic machinery during the infection.

Parasite	Localisation	Induction	Phenotype	Reference
T. gondii	extracellular	Amino acid starvation	Basal: maintenance of life	[102]
	intracellular	Amino acid starvation and rapamycin	mitochondrial fragmentation	[111]
		Glucose and/or pyruvate starvation	Arrested mitochondrial fragmentation	
Plasmodium sp.	intracellular	sporozoite to trophozoite conversion in the liver	recycling of secretory organelles	[136]

Table 3. Autophagy in Apicomplexan parasites

9. Conclusion

The present chapter addresses the positive and negative regulations of the autophagic process of infected mammalian cells and the possible effects of these regulations on the *in vitro* and *in vivo* modulation of this process. This review also describes the autophagy pathway in pathogenic trypanosomatids and apicomplexans responsible for some of the most relevant neglected illnesses worldwide. The pivotal role of autophagy in pathogenicity and virulence was demonstrated in T. cruzi, T. brucei, Leishmania sp., T. gondii and Plasmodium sp., which suggests that autophagic machinery is a possible target for anti-parasitic intervention.

Acknowledgements

This work was supported with grants from CNPq (Universal), FAPERJ (APQ1) and IOC/FIOCRUZ.

Author details

Thabata Lopes Alberto Duque[1,2], Xênia Macedo Souto[1,3], Valter Viana de Andrade-Neto[4], Vítor Ennes-Vidal[5] and Rubem Figueiredo Sadok Menna-Barreto[1*]

*Address all correspondence to: rubemb@ioc.fiocruz.br

1 Laboratory of Cell Biology, Oswaldo Cruz Institute, Oswaldo Cruz Foundation, Rio de Janeiro, RJ, Brazil

2 Laboratory of Cell Biology, Department of Biology, Federal University of Juiz de Fora, MG, Brazil

3 Laboratory of Structural Biology, Oswaldo Cruz Institute, Oswaldo Cruz Foundation, Rio de Janeiro, RJ, Brazil

4 Laboratory of Biochemistry of Trypanosomatids, Oswaldo Cruz Institute, Oswaldo Cruz Foundation, Rio de Janeiro, RJ, Brazil

5 Laboratory of Molecular Biology and Endemic Diseases, Oswaldo Cruz Institute, Oswaldo Cruz Foundation, Rio de Janeiro, RJ, Brazil

Thabata Lopes Alberto Duque and Xênia Macedo Souto equally contributed to this work

References

[1] Nayyar GML, Breman JG, Newton PN, Herrington J. Poor-quality antimalarial drugs in southeast Asia and sub-Saharan Africa. Lancet Infectious Diseases 2012;12(6): 488-96.

[2] Soeiro MN, De Castro SL. *Trypanosoma cruzi* targets for new chemotherapeutic approaches. Expert Opinion on Therapeutic Targets 2009;13(1):105-21.

[3] Kobets T, Grekov I, Lipoldova M. Leishmaniasis: prevention, parasite detection and treatment. Current Medicinal Chemistry 2012;19(10): 1443-74.

[4] Welburn SC, Maudlin I. Priorities for the elimination of sleeping sickness. Advances in Parasitology 2012;79:299-337.

[5] Centers for Disease Control and Prevention. Toxoplasmosis. http://www.cdc.gov/parasites/toxoplasmosis/ (accessed 17 october 2012).

[6] World Health Organization. Working to overcome the global impact of neglected tropical diseases - First WHO report on neglected tropical diseases. Switzerland. 2010.

[7] Bañuls AL, Hide M, Prugnolle F. Leishmania and the leishmaniases: a parasite genetic update and advances in taxonomy, epidemiology and pathogenicity in humans. Advances in Parasitology 2007;64:1-109.

[8] Stuart K, Brun R, Croft S, Fairlamb A, Gürtler RE, McKerrow J, et al. Kinetoplastids: related protozoan pathogens, different diseases. The Journal of Clinical Investigation 2008;118(4): 1301-10.

[9] Greenwood BM, Fidock DA, Kyle DE, Kappe SH, Alonso PL, Collins FH, et al. Malaria: progress, perils, and prospects for eradication. Journal of Clinical Investigation 2008;118(4): 1266-76.

[10] Boyle JP, Radke JR. A history of studies that examine the interactions of *Toxoplasma* with its host cell: Emphasis on in vitro models. International Journal of Parasitology 2009;39(8): 903-14.

[11] Teixeira AR, Gomes C, Lozzi SP, Hecht MM, Rosa AeC, Monteiro PS, et al. Environment, interactions between *Trypanosoma cruzi* and its host, and health. Cadernos de Saúde Pública 2009;25 (1): S32-44.

[12] Nwaka S, Hudson A. Innovative lead discovery strategies for tropical diseases. Nature Reviews Drug Discovery 2006;5(11): 941-55.

[13] Hotez PJ, Bottazzi ME, Franco-Paredes C, Ault SK, Periago MR. The neglected tropical diseases of Latin America and the Caribbean: a review of disease burden and distribution and a roadmap for control and elimination. PLoS Neglected Tropical Diseases 2008;2(9):e300.

[14] Le Pape P. Development of new antileishmanial drugs--current knowledge and future prospects. Journal of Enzyme Inhibition and Medicinal Chemistry 2008;23(5): 708-18.

[15] Nissapatorn V, Sawangjaroen N. Parasitic infections in HIV infected individuals: diagnostic & therapeutic challenges. The Indian Journal of Medical Research 2011;134(6): 878-97.

[16] Hotez PJ, Savioli L, Fenwick A. Neglected tropical diseases of the Middle East and North Africa: review of their prevalence, distribution, and opportunities for control. PLoS Neglected Tropical Diseases 2012;6(2):e1475.

[17] Lindoso JA, Lindoso AA. Neglected tropical diseases in Brazil. Revista do Instituto de Medicina Tropical de São Paulo 2009;51(5): 247-53.

[18] Feasey N, Wansbrough-Jones M, Mabey DC, Solomon AW. Neglected tropical diseases. British Medical Bulletin 2010;93: 179-200.

[19] World Health Organization. http://www.who.int/neglected_diseases/en/ access in November 8, 2012.

[20] Montoya JG, Liesenfeld, O. Toxoplasmosis. Lancet 2004;*363*(9425): 1965-76.

[21] Kiel JAKW. Autophagy in unicellular eukaryotes. Philosophical Transactions of the Royal Society 2010;365: 819-830.

[22] Brennand A, Gualdrón-López M, Coppens I, Rigden DJ, Ginger ML, Michels PAM. Autophagy in parasitic protists: Unique features and drug targets. Molecular and Biochemical Parasitology 2011;177 (2): 83-99.

[23] Levine B, Yuan J. Autophagy in cell death: an innocent convict? Clinical Investigation 2005;115(10):2679-88.

[24] Mitzushima N, Yoshimori T, Ohsumi Y. The role of Atg proteins in autophagosome formation. Annual Review of Cell Developmental Biology 2011; 27: 107-32.

[25] Levine B, Mizushima N, Virgin HW. Autophagy in immunity and inflammation. Nature 2011;469(7330): 323-35.

[26] Sumpter R, Levine B. Autophagy and innate immunity: triggering, targeting and tuning. Seminars in Cell & Developmental Biology 2010;21(7): 699-711.

[27] Kuballa P, Nolte WM, Castoreno AB, Xavier RJ. Autophagy and the immune system. Annual Review of Immunology 2012;30: 611-46.

[28] Gutierrez MG, Master SS, Singh SB, Taylor GA, Colombo MI, Deretic V. Autophagy is a defense mechanism inhibiting BCG and *Mycobacterium tuberculosis* survival in infected macrophages. Cell. 2004;119(6): 753-66.

[29] Deretic V, Levine B. Autophagy, immunity, and microbial adaptations. Cell Host & Microbe 2009;5(6): 527-49.

[30] Skendros P, Mitroulis I. Host cell autophagy in immune response to zoonotic infections. Clinical & Development Immunology 2012;2012: 91052. doi: 10.1155/2012/910525. (accessed 17 October 2012).

[31] Picazarri K, Nakada-Tsukui K, Nozaki T. Autophagy during proliferation and encystation in the protozoan parasite *Entamoeba invadens*. Infection & Immunity. 2008;76(1): 278-88.

[32] Zhao Z, Fux B, Goodwin M, Dunay IR, Strong D, Miller BC, et al. Autophagosome-independent essential function for the autophagy protein Atg5 in cellular immunity to intracellular pathogens. Cell host and Microbe 2008;4(5): 458-69.

[33] Pinheiro RO, Nunes MP, Pinheiro CS, D'Avila H, Bozza PT, Takiya CM, et al. Induction of autophagy correlates with increased parasite load of *Leishmania amazonensis* in BALB/c but not C57BL/6 macrophages. Microbes and Infection 2009;11(2): 181-90.

[34] Romano PS, Arboit MA, Vázquez CL, Colombo MI. The autophagic pathway is a key component in the lysosomal dependent entry of *Trypanosoma cruzi* into the host cell. Autophagy 2009;5(1): 6-18.

[35] Wang Y, Weiss LM, Orlofsky A. Host cell autophagy is induced by *Toxoplasma gondii* and contributes to parasite growth. The Journal of biological chemistry 2009;284(3): 1694-1701.

[36] Duszenko M, Ginger ML, Brennand A, Gualdrón-López M, Colombo MI, et al. Autophagy in protists. Autophagy. 2011;7(2): 127-58.

[37] Alvarez VE, Kosec G, Sant'Anna C, Turk V, Cazzulo JJ, Turk B. Autophagy is involved in nutritional stress response and differentiation in *Trypanosoma cruzi*. The Journal of Biological Chemistry. 2008; 283(6): 3454-64.

[38] Besteiro S, Williams RA, Morrison LS, Coombs GH, Mottram JC. Endosome sorting and autophagy are essential for differentiation and virulence of *Leishmania major*. The Journal of Biological Chemistry 2006;281(16): 11384-96.

[39] Saraiva EM, Pimenta PF, Brodin TN, Rowton E, Modi GB, Sacks DL. Changes in lipophosphoglycan and gene expression associated with the development of *Leishmania major* in *Phlebotomus papatasi*. Parasitology. 1995;111 (Pt 3): 275-87.

[40] Nolan DP, Rolin S, Rodriguez JR, Van Den Abbeele J, Pays E. Slender and stumpy bloodstream forms of *Trypanosoma brucei* display a differential response to extracellular acidic and proteolytic stress. European Journal of Biochemistry 2000;267(1): 18-27

[41] Gonçalves RL, Barreto RF, Polycarpo CR, Gadelha FR, Castro SL, Oliveira MF. A comparative assessment of mitochondrial function in epimastigotes and bloodstream trypomastigotes of *Trypanosoma cruzi*. Journal of Bioenergetics and Biomembranes 2011;43(6): 651-61.

[42] Castro DP, Moraes CS, Gonzalez MS, Ratcliffe NA, Azambuja P, Garcia ES. *Trypanosoma cruzi* immune response modulation decreases microbiota in *Rhodnius prolixus* gut and is crucial for parasite survival and development. PLoS One. 2012;7(5):e36591.

[43] Berriman M, Ghedin E, Hertz-Fowler C, Blandin G, Renauld H, Bartholomeu DC, et al. The genome of the African trypanosome *Trypanosoma brucei*. Science 2005;309(5733): 416-22.

[44] El-Sayed NM, Myler PJ, Bartholomeu DC, Nilsson D, Aggarwal G, Tran AN, et al. The genome sequence of *Trypanosoma cruzi*, etiologic agent of Chagas disease. Science 2005;309(5733): 409-15.

[45] Ivens AC, Peacock CS, Worthey EA, Murphy L, Aggarwal G, Berriman M, et al. The genome of the kinetoplastid parasite, *Leishmania major*. Science 2005;309(5733): 436-42.

[46] Rigden DJ, Herman M, Gillies S, Michels PA. Implications of a genomic search for autophagy-related genes in trypanosomatids. Biochemical Society Transactions 2005;33(Pt 5): 972-4.

[47] Herman M, Gillies S, Michels PA, Rigden DJ. Autophagy and related processes in trypanosomatids: insights from genomic and bioinformatic analyses. Autophagy 2006;2(2): 107-18.

[48] Hidron A, Vogenthaler N, Santos-Preciado JI, Rodriguez-Morales AJ, Franco-Paredes C, Rassi A. Cardiac involvement with parasitic infections. Clinical Microbiology Reviews 2010;23(2): 324-49.

[49] Brown RC, Evans DA, Vickerman K. Developmental changes in ultrastructure and physiology of *Trypanosoma brucei*. Transactions of Royal Society of Tropical Medicine Hygiene 1972;66(2): 336-7.

[50] Vickerman K, Tetley L. Recent ultrastructural studies on trypanosomes. Annales de la Société Belge de Médecine Tropicale 1977;57(4-5): 441-57.

[51] Herman M, Pérez-Morga D, Schtickzelle N, Michels PA. Turnover of glycosomes during life-cycle differentiation of *Trypanosoma brucei*. Autophagy 2008;4(3): 294-308.

[52] Barquilla A, Crespo JL, Navarro M. Rapamycin inhibits trypanosome cell growth by preventing TOR complex 2 formation. Proceedings of National Academy of Sciences U S A 2008;105(38): 14579-84.

[53] Koopmann R, Muhammad K, Perbandt M, Betzel C, Duszenko M. *Trypanosoma brucei* ATG8: structural insights into autophagic-like mechanisms in protozoa. Autophagy 2009;5(8): 1085-91.

[54] Li FJ, Shen Q, Wang C, Sun Y, Yuan AY, He CY. A role of autophagy in *Trypanosoma brucei* cell death. Cellular Microbiology 2012;14(8): 1242-56.

[55] Uzcátegui NL, Carmona-Gutiérrez D, Denninger V, Schoenfeld C, Lang F, Figarella K, et al. Antiproliferative effect of dihydroxyacetone on *Trypanosoma brucei* bloodstream forms: cell cycle progression, subcellular alterations, and cell death. Antimicrobial Agents and Chemotherapy 2007;51(11): 3960-8.

[56] Uzcátegui NL, Denninger V, Merkel P, Schoenfeld C, Figarella K, Duszenko M. Dihydroxyacetone induced autophagy in African trypanosomes. Autophagy 2007;3(6): 626-9.

[57] Merkel P, Beck A, Muhammad K, Ali SA, Schönfeld C, Voelter W, et al. Spermine isolated and identified as the major trypanocidal compound from the snake venom of *Eristocophis macmahoni* causes autophagy in *Trypanosoma brucei*. Toxicon 2007;50(4): 457-69.

[58] Delgado M, Anderson P, Garcia-Salcedo JA, Caro M, Gonzalez-Rey E. Neuropeptides kill African trypanosomes by targeting intracellular compartments and inducing autophagic-like cell death. Cell Death and Differetiation 2009;16(3): 406-16.

[59] Denninger V, Koopmann R, Muhammad K, Barth T, Bassarak B, Schönfeld C, et al. Kinetoplastida: model organisms for simple autophagic pathways? Methods in Enzymology 2008;451: 373-408.

[60] Zhang Y, Qi H, Taylor R, Xu W, Liu LF, Jin S. The role of autophagy in mitochondria maintenance: characterization of mitochondrial functions in autophagy-deficient *S. cerevisiae* strains. Autophagy 2007;3(4): 337-46.

[61] Chen Y, Gibson SB. Is mitochondrial generation of reactive oxygen species a trigger for autophagy? Autophagy 2008;4(2): 246-8.

[62] Scherz-Shouval R, Elazar Z. Regulation of autophagy by ROS: physiology and pathology. Trends in BiochemicalSciences 2011;36(1): 30-8.

[63] Braga MV, Magaraci F, Lorente SO, Gilbert I, de Souza W. Effects of inhibitors of Delta24(25)-sterol methyl transferase on the ultrastructure of epimastigotes of Trypanosoma *cruzi*. Microscopy and Microanalysis 2005;11(6): 506-15.

[64] Santa-Rita RM, Lira R, Barbosa HS, Urbina JA, de Castro SL. Anti-proliferative synergy of lysophospholipid analogues and ketoconazole against *Trypanosoma cruzi* (Kinetoplastida: Trypanosomatidae): cellular and ultrastructural analysis. The Journal of Antimicrobial Chemotherapy 2005;55(5): 780-4.

[65] Menna-Barreto RF, Corrêa JR, Pinto AV, Soares MJ, de Castro SL. Mitochondrial disruption and DNA fragmentation in *Trypanosoma cruzi* induced by naphthoimidazoles synthesized from beta-lapachone. Parasitology Research 2007;101(4): 895-905.

[66] Menna-Barreto RF, Corrêa JR, Cascabulho CM, Fernandes MC, Pinto AV, Soares MJ, et al. Naphthoimidazoles promote different death phenotypes in *Trypanosoma cruzi*. Parasitology 2009;136(5): 499-510.

[67] Fernandes MC, Da Silva EN, Pinto AV, De Castro SL, Menna-Barreto RF. A novel triazolic naphthofuranquinone induces autophagy in reservosomes and impairment of mitosis in *Trypanosoma cruzi*. Parasitology 2012;139(1): 26-36.

[68] Veiga-Santos P, Barrias ES, Santos JF, de Barros Moreira TL, de Carvalho TM, Urbina JA, et al. Effects of amiodarone and posaconazole on the growth and ultrastructure of Trypanosoma cruzi. International Journal of Antimicrobial Agents 2012;40(1): 61-71.

[69] Menna-Barreto RF, Salomão K, Dantas AP, Santa-Rita RM, Soares MJ, Barbosa HS, et al. Different cell death pathways induced by drugs in *Trypanosoma cruzi*: an ultrastructural study. Micron 2009;40(2): 157-68.

[70] DaRocha WD, Otsu K, Teixeira SM, Donelson JE. Tests of cytoplasmic RNA interference (RNAi) and construction of a tetracycline-inducible T7 promoter system in *Trypanosoma cruzi*. Molecular and Biochemical Parasitolology 2004;133(2): 175-86.

[71] Sandes JM, Borges AR, Junior CG, Silva FP, Carvalho GA, Rocha GB, et al. 3-Hydroxy-2-methylene-3-(4-nitrophenylpropanenitrile): A new highly active compound

against epimastigote and trypomastigote form of *Trypanosoma cruzi*. Bioorganic Chemistry 2010;38(5): 190-5.

[72] Benitez D, Pezaroglo H, Martínez V, Casanova G, Cabrera G, Galanti N, et al. Study of *Trypanosoma cruzi* epimastigote cell death by NMR-visible mobile lipid analysis. Parasitology 2012 139(4): 506-15.

[73] Soares MJ, Souto-Padrón T, De Souza W. Identification of a large pre-lysosomal compartment in the pathogenic protozoon Trypanosoma cruzi. Journal of Cell Science 1992;102 (Pt 1): 157-67.

[74] Soares MJ. The reservosome of *Trypanosoma cruzi* epimastigotes: an organelle of the endocytic pathway with a role on metacyclogenesis. Memórias do Instituto Oswaldo Cruz 1999;94 (1): 139-41.

[75] Figueiredo RC, Rosa DS, Soares MJ. Differentiation of *Trypanosoma cruzi* epimastigotes: metacyclogenesis and adhesion to substrate are triggered by nutritional stress. The Journal of Parasitology 2000;86(6): 1213-8.

[76] Alvarez VE, Kosec G, Sant Anna C, Turk V, Cazzulo JJ, Turk B. Blocking autophagy to prevent parasite differentiation: a possible new strategy for fighting parasitic infections? Autophagy 2008;4(3): 361-3

[77] Braga MV, de Souza W. Effects of protein kinase and phosphatidylinositol-3 kinase inhibitors on growth and ultrastructure of *Trypanosoma cruzi*. FEMS Microbiology Letters 2006;256(2): 209-16.

[78] Petiot A, Ogier-Denis E, Blommaart EF, Meijer AJ, Codogno P. Distinct classes of phosphatidylinositol 3'-kinases are involved in signaling pathways that control macroautophagy in HT-29 cells. The Journal of Biological Chemistry. 2000; 275(2): 992-8.

[79] Romano PS, Cueto JA, Casassa AF, Vanrell MC, Gottlieb RA, Colombo MI. Molecular and cellular mechanisms involved in the *Trypanosoma cruzi*/host cell interplay. IUBMB Life 2012;64(5): 387-96.

[80] Martins RM, Alves RM, Macedo S, Yoshida N. Starvation and rapamycin differentially regulate host cell lysosome exocytosis and invasion by *Trypanosoma cruzi* metacyclic forms. Cellular Microbiology 2011;13(7): 943-54.

[81] Maeda FY, Alves RM, Cortez C, Lima FM, Yoshida N. Characterization of the infective properties of a new genetic group of *Trypanosoma cruzi* associated with bats. Acta Tropica 2011;120(3): 231-7.

[82] Bera A, Singh S, Nagaraj R, Vaidya T. Induction of autophagic cell death in *Leishmania donovani* by antimicrobial peptides. Molecular and Biochemichal Parasitology 2003;127(1): 23-35.

[83] Dos Santos AO, Veiga-Santos P, Ueda-Nakamura T, Filho BP, Sudatti DB, Bianco EM, et al. Effect of elatol, isolated from red seaweed *Laurencia dendroidea*, on *Leishmania amazonensis*. Marine Drugs 2010;8(11): 2733-43.

[84] Santos AO, Santin AC, Yamaguchi MU, Cortez LE, Ueda-Nakamura T, Dias-Filho BP, et al. Antileishmanial activity of an essential oil from the leaves and flowers of *Achillea millefolium*. Annals of Tropical Medicine and Parasitology 2010;104(6): 475-83.

[85] Schurigt U, Schad C, Glowa C, Baum U, Thomale K, Schnitzer JK, et al. Aziridine-2,3-dicarboxylate-based cysteine cathepsin inhibitors induce cell death in Leishmania major associated with accumulation of debris in autophagy-related lysosome-like vacuoles. Antimicrobial Agents and Chemotherapy 2010;54(12): 5028-41.

[86] de Macedo-Silva ST, de Oliveira Silva TL, Urbina JA, de Souza W, Rodrigues JC. Antiproliferative, Ultrastructural, and Physiological Effects of Amiodarone on Promastigote and Amastigote Forms of *Leishmania amazonensis*. Molecular Biology International 2011; doi: 10.4061/2011/876021 (accessed 17 October 2012).

[87] Monte Neto RL, Sousa LM, Dias CS, Barbosa Filho JM, Oliveira MR, Figueiredo RC. Morphological and physiological changes in *Leishmania* promastigotes induced by yangambin, a lignan obtained from *Ocotea duckei*. Experimental Parasitology 2011;127(1):215-21.

[88] Sengupta S, Chowdhury S, Bosedasgupta S, Wright CW, Majumder HK. Cryptolepine-Induced Cell Death of *Leishmania donovani* Promastigotes Is Augmented by Inhibition of Autophagy. Molecular Biological International 2011;2011: 187850.

[89] Silva AL, Adade CM, Shoyama FM, Neto CP, Padrón TS, de Almeida MV, et al. In vitro leishmanicidal activity of N-dodecyl-1,2-ethanediamine. Biomedicine and Pharmacotherapy 2012;66(3): 180-6.

[90] Williams RA, Tetley L, Mottram JC, Coombs GH. Cysteine peptidases CPA and CPB are vital for autophagy and differentiation in *Leishmania mexicana*. Molecular Microbiology 2006;61(3): 655-74.

[91]] Besteiro S, Williams RA, Coombs GH, Mottram JC. Protein turnover and differentiation in *Leishmania*. International Journal for Parasitology 2007;37(10): 1063-75.

[92] Bhattacharya A, Biswas A, Das PK. Identification of a protein kinase A regulatory subunit from *Leishmania* having importance in metacyclogenesis through induction of autophagy. Molecular Microbiology 2012;83(3): 548-64.

[93] Williams RAM, Woods KL, Juliano L, Mottram JC, Coombs GH. Characterisation of unusual families of ATG8-like proteins and ATG12 in the protozoan parasite *Leishmania major*. Autophagy 2009;5(2): 159-172.

[94] Williams RAM, Smith TK, Cull B, Mottram JC, Coombs GH. ATG5 is Essential for ATG8-Dependent Autophagy and Mitochondrial Homeostasis in *Leishmania major*. Plos Pathogens 2012; 8 (5): 1-14.

[95] Goldshmidt H, Matas D, Kabi A, Carmi S, Hope R, Michaeli S. Persistent ER stress induces the spliced leader RNA silencing pathway (SLS), leading to programmed cell death in *Trypanosoma brucei*. PLoS Pathogens 2010;6(1):e1000731.

[96] Schaible UE, Schlesinger PH, Steinberg TH, Mangel WF, Kobayashi T, Russell DG. Parasitophorous vacuoles of *Leishmania mexicana* acquire macromolecules from the host cell cytosol via two independent routes. Journal of Cell Science 1999;112(Pt 5): 681-93.

[97] Cyrino LT, Araújo AP, Joazeiro PP, Vicente CP, Giorgio S. *In vivo* and *in vitro Leishmania amazonensis* infection induces autophagy in macrophages. Tissue & Cell 2012; doi: http://dx.doi.org/10.1016/j.tice.2012.08.003 (accessed 17 October 2012).

[98] Mitroulis I, Kourtzelis I, Papadopoulos VP, Mimidis K, Speletas M, Ritis K. *In vivo* induction of the autophagic machinery in human bone marrow cells during *Leishmania donovani* complex infection. Parasitology International 2009;58(4):475-7.

[99] Besteiro S. Which roles for autophagy in *Toxoplasma gondii* and related apicomplexan parasites? Molecular and Biochemical Parasitology 2012;184: 1-8.

[100] Xie Z, Nair U, Klionsky DJ. Atg8 controls phagophore expansion during autophagosome formation. Molecular biology of the cell 2008;19(8): 3290-8.

[101] Shvets E, Fass E, Scherz-Shouval R, Elazar Z. The N-terminus and Phe52 residue of LC3 recruit p62/SQSTM1 into autophagosomes. Journal of cell science 2008;121(Pt 16): 2685-95.

[102] Besteiro S, Brooks CF, Striepen B, Dubremetz JF. Autophagy protein Atg3 is essential for maintaining mitochondrial integrity and for normal intracellular development of *Toxoplasma gondii* tachyzoites. PLoS pathogens 2011;7(12): e1002416.

[103] Diaz-Troya S, Perez-Perez ME, Florencio FJ, Crespo JL. The role of TOR in autophagy regulation from yeast to plants and mammals. Autophagy 2008;4(7): 851-65.

[104] Rigden DJ, Michels PA, Ginger ML. Autophagy in protists: Examples of secondary loss, lineage-specific innovations, and the conundrum of remodeling a single mitochondrion. Autophagy 2009;5(6): 784-94.

[105] Dubey JP, Jones JL. *Toxoplasma gondii* infection in humans and animals in the United States. International journal for parasitology 2008;38(11): 1257-78.

[106] Furtado JM, Smith JR, Belfort R Jr., Gattey D, Winthrop KL. Toxoplasmosis: a global threat. Journal of global infectious diseases 2011;3(3): 281-4.

[107] Mordue DG, Hakansson S, Niesman I, Sibley, LD. *Toxoplasma gondii* resides in a vacuole that avoids fusion with host cell endocytic and exocytic vesicular trafficking pathways. Experimental parasitology 1999;92(2): 87-99.

[108] Sullivan WJ Jr., Jeffers V. Mechanisms of *Toxoplasma gondii* persistence and latency. FEMS microbiology reviews 2012;36(3): 717-33.

[109] Ambroise-Thomas P, Pelloux H. Toxoplasmosis - congenital and in immunocompromised patients: a parallel. Parasitology today 1993;9(2): 61-3.

[110] Mele A, Paterson PJ, Prentice HG., Leoni P, Kibbler CC. Toxoplasmosis in bone marrow transplantation: a report of two cases and systematic review of the literature. Bone marrow transplantation 2002;29(8): 691-8.

[111] Innes EA. A brief history and overview of *Toxoplasma gondii*. Zoonoses and public health 2010;57(1): 1-7.

[112] Ghosh D, Walton JL, Roepe PD, Sinai AP. Autophagy is a cell death mechanism in *Toxoplasma gondii*. Cellular microbiology 2012;14(4): 589-607.

[113] Besteiro S. Role of ATG3 in the parasite *Toxoplasma gondii*: Autophagy in an early branching eukaryote. Autophagy 2012;8(3): 435-7.

[114] Youle RJ, Narendra DP. Mechanisms of mitophagy. Nature reviews Molecular cell biology 2011;12(1): 9-14.

[115] Meijer AJ, Codogno P. Regulation and role of autophagy in mammalian cells. The international journal of biochemistry and cell biology 2004;36(12): 2445-62.

[116] Dubey JP, Frenkel JK. Cyst-induced toxoplasmosis in cats. The Journal of protozoology 1972;19(1): 155-7.

[117] Dubey JP, Frenkel JK. Experimental toxoplasma infection in mice with strains producing oocysts. The Journal of parasitology 1973; 59(3): 505-12.

[118] Dubey JP. Feline toxoplasmosis and coccidiosis: a survey of domiciled and stray cats. Journal of the American Veterinary Medical Association 1973; 162(10): 873-7.

[119] Denton D, Nicolson S, Kumar S. Cell death by autophagy: facts and apparent artefacts. Cell death and differentiation 2012;19(1): 87-95.

[120] Yahiaoui B, Dzierszinski F, Bernigaud A, Slomianny C, Camus D, Tomavo S. Isolation and characterization of a subtractive library enriched for developmentally regulated transcripts expressed during encystation of *Toxoplasma gondii*. Molecular and biochemical parasitology 1999;99(2): 223-35.

[121] Pfefferkorn ER. Interferon gamma blocks the growth of *Toxoplasma gondii* in human fibroblasts by inducing the host cells to degrade tryptophan. Proceedings of the National Academy of Sciences of the United States of America 1984;81(3): 908-12.

[122] Pfefferkorn ER, Eckel M, Rebhun S. Interferon-gamma suppresses the growth of *Toxoplasma gondii* in human fibroblasts through starvation for tryptophan. Molecular and biochemical parasitology 1986;20(3): 215-24.

[123] Black MW, Boothroyd JC. Lytic cycle of *Toxoplasma gondii*. Microbiology and molecular biology reviews 2000;64(3): 607-23.

[124] Andrade RM, Wessendarp M, Gubbels MJ, Striepen B, Subauste, CS. CD40 induces macrophage anti-*Toxoplasma gondii* activity by triggering autophagy-dependent fu-

sion of pathogen-containing vacuoles and lysosomes. The Journal of clinical investigation 2006;116(9): 2366-77.

[125] Ling Y M, Shaw MH, Ayala C, Coppens I, Taylor GA, Ferguson DJ, et al.. Vacuolar and plasma membrane stripping and autophagic elimination of *Toxoplasma gondii* in primed effector macrophages. The Journal of experimental medicine 2006;203(9): 2063-271.

[126] Subauste CS, Andrade RM, Wessendarp M. CD40-TRAF6 and autophagy-dependent anti-microbial activity in macrophages. Autophagy 2007;3(3): 245-8.

[127] Zhao Y, Wilson D, Matthews S, Yap GS. Rapid elimination of *Toxoplasma gondii* by gamma interferon-primed mouse macrophages is independent of CD40 signaling. Infection and immunity 2007;75(10): 4799-803.

[128] Portillo JA, Okenka G, Reed E, Subauste A, Van Grol J, Gentil K, et al. The CD40-autophagy pathway is needed for host protection despite IFN-Gamma-dependent immunity and CD40 induces autophagy via control of P21 levels. PLoS One 2010;5(12): e14472.

[129] Bogdan C, Rollinghoff M. How do protozoan parasites survive inside macrophages? Parasitology Today 1999;15(1): 22-8.

[130] Halonen SK. Role of autophagy in the host defense against *Toxoplasma gondii* in astrocytes. Autophagy 2009;5(2): 268-9.

[131] Cox-Singh J, Davis TM, Lee KS, Shamsul SS, Matusop A, Ratnam S, et al. *Plasmodium knowlesi* malaria in humans is widely distributed and potentially life threatening. Clinical infectious diseases 2008;46(2): 165-71.

[132] Marchand RP, Culleton R, Maeno Y, Quang NT, Nakazawa S. Co-infections of *Plasmodium knowlesi, P. falciparum,* and *P. vivax* among Humans and Anopheles dirus Mosquitoes, Southern Vietnam. Emerging infectious diseases 2011;17(7): 1232-9.

[133] William T, Menon J, Rajahram G, Chan L, Ma G, Donaldson S, et al. Severe *Plasmodium knowlesi* malaria in a tertiary care hospital, Sabah, Malaysia. Emerging infectious diseases 2011;17(7): 1248-55.

[134] Frevert U, Engelmann S, Zougbede S, Stange J, Ng B, Matuschewski K, et al. Intravital observation of *Plasmodium berghei* sporozoite infection of the liver. PLoS biology 2005;3(6): e192.

[135] Amino R, Thiberge S, Martin B, Celli S, Shorte S, Frischknecht F, et al. Quantitative imaging of *Plasmodium* transmission from mosquito to mammal. Nature Medicine 2006;12(2): 220-4.

[136] Sturm A, Amino R, van de Sand C, Regen T, Retzlaff S, Rennenberg A, et al. Manipulation of host hepatocytes by the malaria parasite for delivery into liver sinusoids. Science 2006;313(5791): 1287-90.

[137] Jayabalasingham B, Bano N, Coppens I. Metamorphosis of the malaria parasite in the liver is associated with organelle clearance. Cell Research 2010; 20: 1043-1059.

[138] Coppens I. Metamorphoses of malaria: the role of autophagy in parasite differentiation. Essays in biochemistry 2011;51: 127-36.

[139] Vaid A, Ranjan R, Smythe WA, Hoppe HC, Sharma P. PfPI3K, a phosphatidylinositol-3 kinase from *Plasmodium falciparum,* is exported to the host erythrocyte and is involved in hemoglobin trafficking. Blood 2010;115(12): 2500-7.

[140] Kitamura K, Kishi-Itakura C, Tsuboi T, Sato S, Kita K, Ohta N, Mizushima N. Autophagy-Related Atg8 Localizes to the Apicoplast of the Human Malaria Parasite *Plasmodium falciparum.* Plos one 2012;7(8): 1-10.

Chapter 15

Induction of Autophagy by Anthrax Lethal Toxin

Aiguo Wu, Yian Kim Tan and Hao A. Vu

Additional information is available at the end of the chapter

http://dx.doi.org/10.5772/55873

1. Introduction

Autophagy is an intracellular process whereby cells break down long-lived proteins and organelles and is morphologically characterized by the formation of many large autophagic vacuoles in cytoplasm [1]. This evolutionary process is conserved across all eukaryotic cells and is fundamentally important in normal and pathological cell physiology and development [2, 3]. Autophagy occurs constitutively at a basal level in quiescent cells but the process may be up-regulated during periods of starvation [4] and in response to other stress stimuli [5]. Many recent studies also suggest the increasing association of autophagy in numerous physiological and pathological conditions such as neurodegeneration, death of cancer cells, tissue formation and host cells response to pathogens [5].

The process of autophagy begins with the formation of isolation membrane or phagophore followed by sequestration of organelles or part of the cytoplasm to form autophagosome. The double-membrane autophagosome subsequently fuses with lysosome to form autolysosome where its content are degraded and released into the cytoplasm [6]. Several important autophagy related genes (*ATG*) that are critical for autophagosome formation have been identified recently. Microtubule-associated protein 1 light chain 3 (LC3) is the mammalian orthologue of yeast Atg8 that is required for autophagosome formation [7]. During autophagy, cytosolic form of LC3-I is processed into a lipidated LC3-II which is tightly associated with autophagosome membranes [8]. In addition, Atg8 has also been identified for its involvement in the expansion of isolation membrane [9]. The other protein complex that is essential for elongation of the isolation membrane is Atg5-Atg12 complex [10].

Vegetative *Bacillus anthracis* generates two essential virulence factors: the anthrax toxin and the poly-γ-D glutamic acid capsule [11]. The primary virulence factor is a secreted zinc-dependent metalloprotease toxin known as lethal factor (LF), which is introduced into the cytosol by protective antigen (PA) through its receptors on the cells [12]. LF exerts its toxic effect through the disruption of mitogen-activated protein kinase kinase (MAPKK) signalling pathway, which is essential in mounting an efficient and prompt immune response against the invading pathogen [13]. LF is also a potent inhibitor on many functions of immune cells such as macrophages, dendritic cells, neutrophils, T cells and B cells [14].

2. Anthrax lethal toxin induces autophagy

Our study has provided evidence that autophagy was involved in anthrax pathogenesis. These results are briefly described below.

2.1. Cells and induction of autophagy

RAW 264.7 murine macrophage cells were transfected with pEGFP-LC3. The transfected cells were treated with anthrax LT. The existence of autophagy was identified by immunoblotting, fluorescent punctuate counting, formation of acidic vacuoles, and viability of the LT treated cells.

2.2. Acridine orange staining showed increased Acidic Vacuoles (AVO)

Increased in AVO formation is a typical feature observed in cells undergoing autophagy [16]. Hence, we examined the effect of LT on AVO formation in RAW 264.7 cells by using lysosomotropic agent acridine orange (AO). RAW 264.7 cells treated with LT displayed a dose-dependent increase in AVO formation [17].

2.3. Increased GFP-LC3 punctuate when cells were treated with LT

Atg8 is an ubiquitin-like protein that undergoes conjugation process during autophagy and is determined to be an essential component for autophagy [7]. LC3 is the human orthologue of Atg8 and is also the most widely used protein marker for detecting autophagic organelles. During autophagy, cytosolic LC3-I is linked to phosphatidylethanolamine (PE) to form LC3-II and remain tightly bound to the autophagosomal membranes [8]. This process can be indirectly monitored through the use of reporter protein GFP conjugated to LC3 [18]. In order to determine if LT induces autophagy, we overexpressed GFP-LC3 in cells and observed for fluorescent punctuate distribution of GFP-LC3, which represent autophagosome formation.

Transfection of cells with GFP-LC3 for fluorescence microscopy analysis is widely used to detect autophagosome. Stable GFP-LC3 expressing RAW 264.7 cells were treated with anthrax LT for 2 hours and exhibited increased GFP-LC3 punctuates distribution whereas untreated cells displayed a diffuse GFP-LC3 appearance (Figure 1) [17]. These punctuate fluorescent dots indicate autophagosomes formation. Most of these fluorescent dots were probably autophagosomes as autolysosomes had weaker or no fluorescence signals due to the presence of lower LC3-II proteins [8]. The reduced LC3-II level in autolysosome may be possibly due to degradation or recycling back to cytosolic LC3-I [8]. The punctuate distribution of GFP-LC3 in LT treated cells were similar to those treated with rapamycin which serve as positive control for autophagy induction. Rapamycin binds to and inhibits mammalian target of rapamycin (mTOR), a negative regulator of autophagy [5]. Nutrients starvation is also able to trigger autophagy. Accordingly, cells incubated in nutrient free salt solutions, EBSS, for 2 hours showed punctuate distribution of GFP-LC3. Autophagy induced by nutrients starvation produced more intense fluorescence punctuates compared to rapamycin or LT treated cells [17].

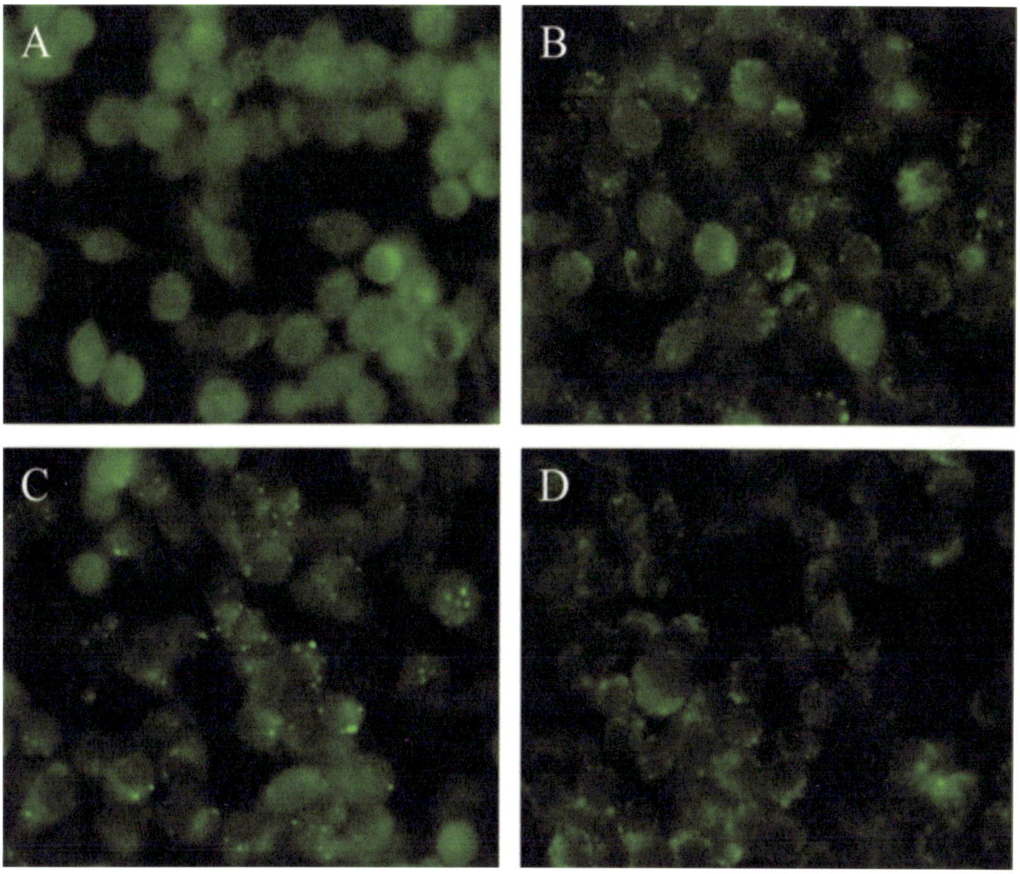

Figure 1. Lethal toxin induced punctuates EGFP-LC3 distribution in cells. Stably transfected RAW 264.7 cells expressing EGFP-LC3 were treated for 2 hours with (A) PBS, (B) 500ng/ml PA + 50ng/ml LF, (C) EBSS, (D) 4μM rapamycin. PBS, PA, LF and rapamycin were added directly into medium and EBSS treated cells were washed 3 times with PBS before incubation in EBSS. Images (40x) were taken from specimens under fluorescence microscope and are representative of 3 experiments.

2.4. Conversion of LC3-I to LC3-II

During autophagy, processing of cytosolic LC3-I to LC3-II permits autophagosomal membrane recruitment through an autophagic specific conjugation. As the amount of LC-II correlates with the extent of autophagosome formation [8], immunoblotting of LC3-II can be used to determine autophagy induction. To further corroborate that the GFP-LC3 punctuate observed was indeed autophagy induction by LT, we therefore examined the endogenous LC3-II levels in LT-treated cells.

RAW 264.7 cells were pre-treated with E64d and pepstatin A for 1 hour to inhibit lysosomal proteases followed by incubation with LT for 1 and 3 hours. At both time points, increase of LC3-II level from LT treated cells were detected, although the ratio differs from 1 to 3 hours after incubation (Figure 2A) [17]. It is not unusual to observe fluctuation of LC3-II level across various time point during autophagy induction [20].

We further determined if LT components could also induce autophagy individually. Treatment of cells with PA alone showed moderate increase of LC3-II while LF alone produced similar ratio of LC3-II / actin as control cells (Figure 2B) [17]. RAW 264.7 cells treated with LT for 1 hour showed elevated amount of endogenous LC3-II (Figure 2B). These observations are generally consistent with the fluorescent punctuate count in GFP-LC3 transfected RAW 264.7 cells. It may appear obvious that LF did not induce autophagy simply because it is not able to cross cell membrane in the absence of PA although it was reported that a small fragment of LF can enter into the cell cytoplasm without the assistance from PA [21]. Apparently, this mechanism of PA-independent insertion of LF into cytosol did not have an observable effect on autophagy induction at the concentration tested.

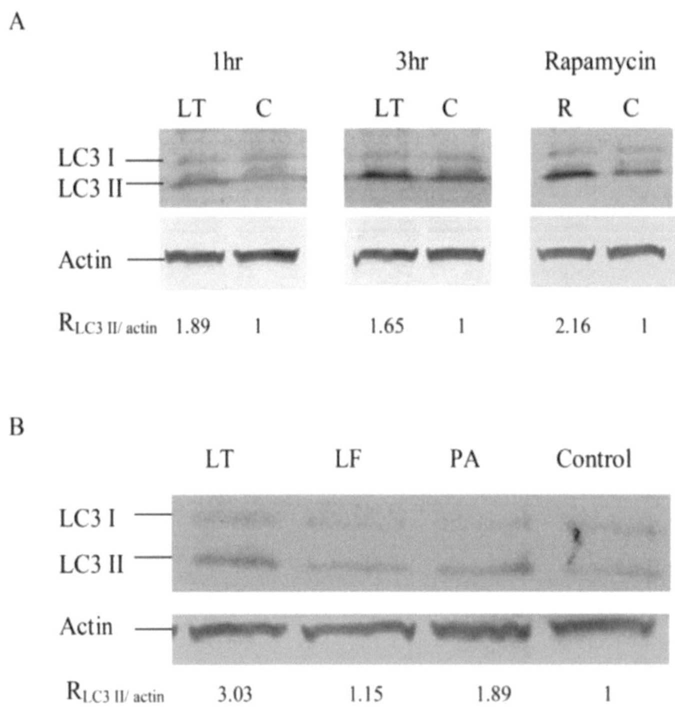

Figure 2. Immunoblot analysis of endogenous LC3-II conversion in RAW 264.7 cells. (A) Cells were pre-treated with 10μg/ml E64d and 10μg/ml Pepstatin A for 1 hour followed by incubation with LT (PA 500μg/ml + LF 100μg/ml) for 1 and 3 hours. Cells treated with 4μM rapamycin were used as positive control for autophagy induction. (B) Cells were pre-treated with E64d 10μg/ml and Pepstatin A 10μg/ml for 1 hour followed by incubation with 500ng/ml PA, 50ng/ml LF and LT (PA 500ng/ml + LF 50μg/ml) for another 1 hour. Total proteins were analysed by using anti-LC3 and anti-actin antibodies. Ratio of LC3 II/actin is shown under the blot.

2.5. Autophagy inhibitor may increase cell death

Inhibition of autophagy process can be used to investigate the role of autophagy in cellular response to toxins, bacteria or viruses. Depending on the interaction between autophagy mechanism and stimulus, the induction of autophagy may sometimes be beneficial or detrimental to the cells. Autophagy protects cells against *Vibrio cholera* cytolysin intoxication [15] but has an opposite effect when autophagy is activated in response to diphtheria toxin treatment [22]. Hence, we attempt to study the effect of autophagy on LT intoxication.

RAW 264.7 cells were treated with 10mM 3MA for 1 hour to inhibit autophagy followed by 2 and 3 hours of incubation with LT. Cells viability as determined by MTS assay showed no differences in 3MA treated and untreated cells (data not shown). This could be attributed to rapid lysis of RAW 264.7 cells when subject to LT treatment. Thus, we decided to use another cell line that is also susceptible to LT induced cell death but at a slower lysis rate than RAW 264.7. LT does not appear to cause instant lysis on human promyelocytic leukemia cell line HL-60 but is cytotoxic when HL-60 cells are differentiated into macrophage-like cells with PMA [23]. Differentiated HL-60 cells were pre-treated with 3MA for 1 hour followed by introduction of LT. Cells pre-treated with 3MA showed accelerated cell death compared to control cells at all the time points tested [17]. This suggests that autophagy may function as a defense mechanism against LT intoxication. While 3MA is often used as a specific inhibitor of autophagy [15], it also has effects on various aspects of metabolism that is unrelated to autophagy [24]. More studies need to be conducted to further understand the role of autophagy in LT intoxication.

3. Discussion

LT is recognized as a critical virulence factor in *B. anthracis* pathogenesis. Having been extensively researched for numerous years, LT pleiotropic actions on many cellular mechanisms have been described. Autophagy is activated during periods of physiological stress such as starvation as a means to sustain cell viability in a nutrient limiting environment [4]. In addition, autophagy is also implicated as a protective cellular response for the elimination of infectious agents [25]. However, certain pathogens are able to manipulate autophagy by altering certain processes for its survival and proliferation [25]. Recently, autophagy has become a rapidly growing biomedical marker as more studies unravel the role of autophagy in many physiological and pathological processes [6].

During autophagy, isolation membranes or phagophores elongate to sequester cytoplasmic components and become enclosed to form a double membrane autophagosome. Herein, we reported LT induced autophagosome formation in cells as demonstrated by the punctuate GFP-LC3 distribution in the cytoplasm and the corresponding increase in the punctuate counts. Another frequently used method as an indicator of autophagy is the monitoring of LC3-II conversion. LC3-II protein associates tightly to autophagosome and was determined to be correlated with autophagosome in cells [8]. Indeed, LT-treated cells displayed enhanced LC3-II conversion, which is a typical representative of autophagosome formation. As expected, PA was determined to be a critical component for autophagy induction. By itself, PA caused a moderate increase in LC3-II levels compared with non-treated controls. This could be attributed to a self-protection response of the host cells upon PA exposure. However, cells treated with LT (PA + LF) caused a dramatic increase in LC3-II levels [17]. This could be mainly the result of cellular stress and defence mechanism against the rapid toxic effects of LT. LT activity is believed to persist longer in the cells than PA alone, as indicated by its continuous enzymatic cleavage of substrate in the cells for 4-5 days [12, 26]. The prolonged presence of active LF in the cytoplasm may possibly play a contributing role in the dramatic increase of autophagy.

Autophagy may function as a defensive mechanism against toxins or invading pathogens but may also be exploited by microbes for survival/replication or even leading to death of host cells. In our study, autophagy was determined to be beneficial to differentiated human promyelocytic leukemia HL-60 cells exposed to LT as cells blocked from autophagy expressed accelerated cell death [17]. Probably similar to the cellular response to *V. cholerae* cytolysin intoxication [15], autophagy was presumably activated to enhance LT clearance from cytoplasm by diverting them to autophagosome and eventually eliminated by lysosomal degradation. As this study involved the use of cell lines, it is integral that the defensive role of autophagy be further determined on human primary macrophages or other immune cells. Other more specific autophagy gene knockdown/knockout studies can be carried out to confirm the results obtained from the commonly used autophagy inhibitor 3MA.

Meanwhile, circumstantial evidence from other non-autophagy related LT studies also suggests a possible link between lethal toxin and autophagy [27, 28]. As described earlier, autophagy proceeds from nascent vacuoles to become degradative autophagosomes by acquiring lysosomal proteins, including lysosome associated membrane protein (LAMP)-1 [29]. The maturation culminates with the subsequent fusion of the autophagome with lysosome to form autolysosome where it then degrades and releases its contents into the cytoplasm. LAMP-1 protein is also a major component of lysosome [30]. Kuhn et al analysed the proteomic profile of macrophages treated with LT and reported that LAMP1 protein was one of the highly upregulated protein [27], conceivably to increase lysosome capacity for fusing with autophagosomes and binding to late autophagosomes. In another separate study, several compounds were tested for its ability to modulate LT-induced cell death in macrophages [28]. Interestingly, the presence of rapamycin, an autophagy inducer, protected macrophages from LT-induced cell death. In contrast, macrophages co-treated with autophagy inhibitors, wortmannin or LY294002, exhibited accelerated cell death upon treatment with LT. Although autophagy was not part of their experimental design [28], it is worthy to note that the only compound tested in that study that protected macrophage from LT death in that experiment is a well known autophagy inducer, rapamycin. The results from these studies are in agreement with our current findings that LT activate autophagy and it may function as a cellular defense mechanism against LT intoxication.

Taken together, this study provides new insights into a hitherto undescribed effect of LT on cells; the induction of autophagic response in cells by PA and LT and the plausible role of autophagy in *B. anthracis* infection. Looking beyond, modulation of autophagy may potentially counter the detrimental effects of LT exposure in cells and remains a subject for further investigation.

Author details

Aiguo Wu[1], Yian Kim Tan[2] and Hao A. Vu[3]

1 Department of Molecular Microbiology, George Mason University, Manassas VA, USA

2 DSO National Laboratories, Singapore, Singapore

3 Biosecurity Research Institute, Kansas State University, Manhattan, USA

References

[1] R.A. Lockshin, Z. Zakeri, Apoptosis, autophagy, and more, Int. J. Biochem. Cell Biol. 2004;36(12) 2405-2419.

[2] B. Djehiche, J. Segalen, Y. Chambon, Inhibition of autophagy of fetal rabbit gonoducts by puromycin, tunicamycin and chloroquin in organ culture, Tissue Cell 1996;28(1) 115-121.

[3] S. Tsukamoto, A. Kuma, M. Murakami, C. Kishi, A. Yamamoto, N. Mizushima, Autophagy is essential for preimplantation development of mouse embryos, Science 2008;321(5885) 117-120.

[4] W. Martinet, G.R. De Meyer, L. Andries, A.G. Herman, M.M. Kockx, In situ detection of starvation-induced autophagy, J. Histochem. Cytochem. 2006;54(1) 85-96.

[5] V. Deretic, Autophagy: an emerging immunological paradigm, J Immunol. 2012 189(1):15-20

[6] D.J. Klionsky, Autophagy: from phenomenology to molecular understanding in less than a decade, Nat Rev Mol Cell Biol 2007;8(11) 931-937.

[7] D.J. Klionsky, A.M. Cuervo, P.O. Seglen, Methods for monitoring autophagy from yeast to human, Autophagy 2007;3(3) 181-206.

[8] Y. Kabeya, N. Mizushima, T. Ueno, A. Yamamoto, T. Kirisako, T. Noda, E. Kominami, Y. Ohsumi, T. Yoshimori, LC3, a mammalian homologue of yeast Apg8p, is localized in autophagosome membranes after processing, EMBO J. 2000;19(21) 5720-5728.

[9] Z. Xie, U. Nair, D.J. Klionsky, Atg8 controls phagophore expansion during autophagosome formation, Mol. Biol. Cell 2008;19(8) 3290-3298.

[10] N. Mizushima, H. Sugita, T. Yoshimori, Y. Ohsumi, A new protein conjugation system in human. The counterpart of the yeast Apg12p conjugation system essential for autophagy, J. Biol. Chem. 1998;273(51) 33889-33892.

[11] M. Mock, A. Fouet, Anthrax, Annu. Rev. Microbiol. 2001;55 647-671.

[12] L. Abrami, S. Liu, P. Cosson, S.H. Leppla, F.G. van der Goot, Anthrax toxin triggers endocytosis of its receptor via a lipid raft-mediated clathrin-dependent process, J. Cell Biol. 2003;160(3) 321-328.

[13] J.F. Bodart, A. Chopra, X. Liang, N. Duesbery, Anthrax, MEK and cancer, Cell Cycle 2002;1(1) 10-15.

[14] D.J. Banks, S.C. Ward, K.A. Bradley, New insights into the functions of anthrax toxin, Expert Rev Mol Med 2006;8(7) 1-18.

[15] M.G. Gutierrez, H.A. Saka, I. Chinen, F.C. Zoppino, T. Yoshimori, J.L. Bocco, M.I. Colombo, Protective role of autophagy against Vibrio cholerae cytolysin, a pore-forming toxin from V. cholerae, Proc. Natl. Acad. Sci. U. S. A. 2007;104(6) 1829-1834.

[16] Y. Xu, S.O. Kim, Y. Li, J. Han, Autophagy contributes to caspase-independent macrophage cell death, J. Biol. Chem. 2006;281(28) 19179-19187.

[17] Y.K. Tan, C.M. Kusuma, L.J. St John , H.A. Vu HA, K. Alibek , A. Wu, Induction of autophagy by anthrax lethal toxin. Biochem Biophys Res Commun. 2009;379(2):293-7.

[18] N. Mizushima, Methods for monitoring autophagy, Int. J. Biochem. Cell Biol. 2004;36(12) 2491-2502.

[19] K.E. Beauregard, R.J. Collier, J.A. Swanson, Proteolytic activation of receptor-bound anthrax protective antigen on macrophages promotes its internalization, Cell Microbiol 2000; 2(3) 251-258.

[20] M.A. Delgado, R.A. Elmaoued, A.S. Davis, G. Kyei, V. Deretic, Toll-like receptors control autophagy, EMBO J. 2008;27(7) 1110-1121.

[21] N. Kushner, D. Zhang, N. Touzjian, M. Essex, J. Lieberman, Y. Lu, A fragment of anthrax lethal factor delivers proteins to the cytosol without requiring protective antigen, Proc. Natl. Acad. Sci. U. S. A. 2003;100(11) 6652-6657.

[22] K. Sandvig, B. van Deurs, Toxin-induced cell lysis: protection by 3-methyladenine and cycloheximide, Exp. Cell Res. 1992;200(2) 253-262.

[23] A. Kassam, S.D. Der, J. Mogridge, Differentiation of human monocytic cell lines confers susceptibility to Bacillus anthracis lethal toxin, Cell Microbiol 2005;7(2) 281-292.

[24] L.H. Caro, P.J. Plomp, E.J. Wolvetang, C. Kerkhof, A.J. Meijer, 3-Methyladenine, an inhibitor of autophagy, has multiple effects on metabolism, Eur. J. Biochem. 1988;175(2) 325-329.

[25] D. Schmid, C. Munz, Innate and adaptive immunity through autophagy, Immunity 2007;27(1) 11-21.

[26] S.D. Ha, D. Ng, J. Lamothe, M.A. Valvano, J. Han, S.O. Kim, Mitochondrial proteins Bnip3 and Bnip3L are involved in anthrax lethal toxin-induced macrophage cell death, J. Biol. Chem. 2007;282(36) 26275-26283.

[27] J.F. Kuhn, P. Hoerth, S.T. Hoehn, T. Preckel, K.B. Tomer, Proteomics study of anthrax lethal toxin-treated murine macrophages, Electrophoresis 2006;27(8) 1584-1597.

[28] M. Tsuneoka, T. Umata, H. Kimura, Y. Koda, M. Nakajima, K. Kosai, T. Takahashi, Y. Takahashi, A. Yamamoto, c-myc induces autophagy in rat 3Y1 fibroblast cells, Cell Struct. Funct. 2003;28(3) 195-204.

[29] W.A. Dunn, Jr., Studies on the mechanisms of autophagy: maturation of the autophagic vacuole, J. Cell Biol. 1990;110(6) 1935-1945.

[30] E.L. Eskelinen, Roles of LAMP-1 and LAMP-2 in lysosome biogenesis and autophagy, Mol. Aspects Med. 2006;27(5-6) 495-502.

Chapter 16

Up-Regulation of Autophagy Defense Mechanisms in Mouse Mesenchymal Stromal Cells in Response to Ionizing Irradiation Followed by Bacterial Challenge

Nikolai V. Gorbunov, Thomas B. Elliott,
Dennis P. McDaniel, K. Lund, Pei-Jyun Liao,
Min Zhai and Juliann G. Kiang

Additional information is available at the end of the chapter

http://dx.doi.org/10.5772/55397

1. Introduction

Mesenchymal stroma along with epithelium, endothelium, reticuloendothelium, and lymphoid components is an essential constituent of tissue barriers that sustain immunochemical homeostatic interactions of tissue with internal and external environments. Thus, mesenchymal stroma protects the body from infections [1-8]. A breach of immune and structural integrity of tissue barriers under patho-physiological conditions such as complicated injury can lead to translocation of bacteria from different host-associated microbiomes and colonization of vital organs that can ultimately result in multiple organ failure and sepsis [9].

It is well documented that suppression of radiosensitive lymphoid and epithelial cells by ionizing irradiation results in impairment of tissue barriers and provokes bacterial translocation and sepsis leading to lethal outcome [10-12]. Under these circumstances one would expect increasing stress impact to the ubiquitously present and relatively radioresistant mesenchymal stromal components and their implication in host defense response [13]. This idea is supported by experimental and clinical observations indicating that injury can induce recruitment of mesenchymal stromal cells (MSCs) from bone marrow and promote their proliferative activity in order to re-constitute integrity of fractured tissue and mediate natural debridement [2, 5, 14, 15]. Moreover, it has been shown recently that transplanted MSCs can suppress experimental sepsis and can promote healing of radiation-induced cutaneous injury and survival from acute radiation syndrome [16-20]. All of this evidence suggests that MSCs play a crucial role in mitigation of systemic and local effects of tissue injury under different pathophysiological conditions. However, in the case of total body irradiation, the dynamics of MSC response to inflammatory stimuli can be skewed due to the cytotoxic effects of irradiation; but the mechanisms of remodeling of irradiated MSCs and their antimicrobial barrier capacity are not known and need to be delineated. *In vivo* assessment of stromal cell responses against bacteria is nearly impossible, because of (i) complexity of the architecture of the mesenchymal network in tissues and (ii) the fact of lethal complications in the hematopoietic radiation syndrome occurring at so low a level of microorganisms that the responses are difficult to detect [3,10].

From this perspective our attention was attracted by the macroautophagy-lysosomal (autolysosomal) mechanism described recently *in vitro* in cultured mesenchymal fibroblastic stromal cells [13]. The autophagy/autolysosomal mechanism mediates cell secretory functions and biodegradation mechanisms implicated in phagocytosis and cell remodeling activated in response to damage to cell constituents, endoplasmic reticular (ER) stress, and protein misfolding [21-23]. Thus, the autolysosomal pathway is responsible for decomposition of damaged proteins and organelles as well as phagocytized bacteria and viruses and is considered to be a part of the innate defense mechanism [23- 25].

The dynamics of macroautophagy (hereafter referred to as autophagy) in mammalian cells are well described in recent reviews [22, 26-28]. It has been proposed that autophagy is initiated by the formation of the phagophore, followed by a series of steps, including the elongation and expansion of the phagophore, closure and completion of a double-membrane autophagosome (which surrounds a portion of the cytoplasm), autophagosome maturation through docking and fusion with an endosome (the product of fusion is defined as an amphisome) and/ or lysosome (the product of fusion is defined as an autolysosome), breakdown and degradation of the autophagosome inner membrane and cargo through acid hydrolases inside the autolysosome, and, finally, release of the resulting macromolecules through permeases [22]. These processes, along with the drastic membrane traffic, are mediated by factors known as autophagy-related proteins (i.e., ATG-proteins) and the lysosome-associated membrane proteins (LAMPs) that are conserved in evolution [29]. The autophagic pathway is complex. To date there are over 30 ATG genes identified in mammalian cells as regulators of various steps of autophagy such as cargo recognition, autophagosome formation, etc. [22, 30]. The core molecular machinery is comprised of (i) components of signaling cascades, such as the ULK1 and ULK2 complexes and class III PtdIns3K complexes, (ii) autophagy membrane processing components such as mammalian Atg9 (mAtg9) that contributes to the delivery of membrane to the autophagosome as it forms, and (iii) two conjugation systems: the microtubule-associated protein 1 (MAP1) light chain 3 (i.e., LC3) and the Atg12–Atg5–Atg16L complex. The two conjugation systems are proposed to function during elongation and expansion of the phagophore membrane [22, 27, 30]. A conservative estimate of the autophagy network counts over 400 proteins, which, besides the ATG-proteins, also including stress-response factors, cargo adaptors, and chaperones such as p62/SQSTM1 and heat shock protein 70 (HSP70) [23, 27, 30, 32, 33-35].

Autophagy is considered as a cytoprotective process leading to tissue remodeling, recovery, and rejuvenation. However, under circumstances leading to mis-regulation of the autolysosomal pathway, autophagy can eventually cause cell death, either as a precursor of apoptosis in apoptosis-sensitive cells or as a result of destructive self-digestion [36].

We hypothesized that: (i) MSCs enable activation of the autophagy pathway in response to ionizing irradiation; (ii) this mechanism is a part of adaptive remodeling essential for recovery of MSCs from the radiation-induced injury; and (iii) activation of autophagy in the irradiated MSCs can be potentiated by a challenge with Gram-negative or Gram-positive bacteria, e.g., *Escherichia coli* or *Staphylococcus epidermidis,* in order to sustain the MSC phagocytic antibacterial functions. The objective of the current chapter is to provide evidence to substantiate the proposed hypothesis.

2. Hypothesis test: Experimental procedures and technical approach

2.1. Mouse bone marrow Mesenchymal Stromal Cells (MSCs)

The cultures of MSCs were established and expanded as described previously [13]. Phenotype, proliferative activity, and colony-forming ability of the cells were monitored by flow cytometry and immunofluorescence imaging using positive markers for MSCs, i.e., CD44 and Sca1 [13].

2.2. Irradiation of MSCs and challenge with bacteria

2.2.1. MSC irradiation

MSC irradiation with gamma-photons was conducted using the ^{60}Co source in the Armed Forces Radiobiology Research Institute. The range of the applied doses was from 1 Gy through 12 Gy at a dose rate of 0.4 Gy/min. Dosimetry was performed using the alanine/electron paramagnetic resonance system. Calibration of the dose rate with alanine was traceable to the National Institute of Standards and Technology and the National Physics Laboratory of the United Kingdom. The irradiated cells were given a 24 h rest and then were subjected to either analyses or a challenge with *S. epidermidis* or *E. coli.*

2.2.2. Challenge of MSCs with bacteria

Irradiated and non-irradiated MSC cultures (~90% confluency) were challenged with either *S. epidermidis* or *E. coli* ($5x10^7$ bacteria/ml) for 1-3 h in antibiotic-free medium. For assessment of the cellular alteration during a period ≥ 3 h, the incubation medium was replaced with fresh medium containing penicillin and streptavidin antibiotics.

2.3. Cell analyses

Cell analysis for (i) the radiation-induced DNA double-strand breaks, viability, pro-apoptotic alterations, MSC proliferative activity, integrity of cell monolayers, and colony-forming activity; (ii) bacterial growth suppression, (iii) bacterial phagocytosis and autophagy (ATG), and (iv) response of stress-proteins, were conducted using flow cytometry techniques, fluorescence confocal imaging, protein immunoblotting, bright-field microscopy, and transmission electron microscopy (TEM).

The flow cytometry–based assessments of (i) the radiation-induced DNA double-strand breaks; (ii) proliferative activity; and (iii) cell viability were conducted using, respectively, the H2A.X phosphorylation assay kit (Cell Signaling Solutions, Temecula, CA); Click-iT® EdU Cell Proliferation Assay Kit, which utilizes a modified nucleoside, EdU (5-ethynyl-2′-deoxyuridine) that, in turn, is incorporated during *de novo* DNA synthesis in a quick-click chemistry reaction] (Life Technologies Corp., Grand Island, NY); and the CYTOX® Blue stain (Life Technologies Corp., Grand Island, NY).

The radiation-induced apoptotic response in MSCs was determined by immunoblot analysis of caspase-3, a marker of apoptosis.

The data presented in Fig. 1 indicate that the MSC cultures displayed a high integrity and survival from damage produced by irradiation with doses 1 Gy – 12 Gy. The irradiated cells challenged with bacteria were also able to sustain integrity of confluent monolayers (Fig. 1). The treated cells did not manifest signs of pro-apoptotic alterations. Moreover, MSCs challenged with *E. coli* and *S. epidermidis* at 24 h following irradiation (8 Gy) were able to suppress the bacterial growth by 1.4-fold and 1.8-fold, respectively (not shown).

2.4. Analysis of the cell proteins

Proteins from MSCs were extracted in accordance with the protocol described previously [12]. Aliquots of proteins were resolved on SDS-polyacrylamide slab gels (NuPAGE 4-12% Bis-Tris; Invitrogen, Carlsbad, CA). After electrophoresis, proteins were blotted onto a PDVF membrane and the blots were incubated with antibodies (1 μg/ml) raised against MAP LC3, Lamp1, p65(NFκB), HSP70, Sirt3a, SUMO1, and actin (Abcam, Santa Cruz Biotechnology Inc., LifeSpan Biosciences, Inc., eBiosciences) followed by incubation with species-specific IgG peroxidase conjugate.

2.5. Immunofluorescent staining and image analysis

MSCs (5 specimens per group) were fixed in 2% paraformaldehyde, processed for immunofluorescence analysis and analyzed with fluorescence confocal microscopy (30). Normal donkey serum and antibody were diluted in phosphate-buffered saline (PBS) containing 0.5% BSA and 0.15% glycine. Any nonspecific binding was blocked by incubating the samples with purified normal donkey serum (Santa Cruz Biotechnology, Inc., Santa Cruz, CA) diluted 1:20. Primary antibodies were raised against MAP LC3, Lamp1, p62/SQSTM1, p65(NFκB), FoxO3a, Tom 20. That was followed by incubation with secondary fluorochrome-conjugated antibody and/or streptavidin-AlexaFluor 610 conjugate (Molecular Probes, Inc., Eugene OR), and with Heochst 33342 (Molecular Probes, Inc., Eugene OR) diluted 1:3000. Secondary antibodies used were AlexaFluor 488 and AlexaFluor 594 conjugated donkey IgG (Molecular Probes Inc., Eugene OR). Negative controls for nonspecific binding included normal goat serum without primary antibody or with secondary antibody alone. Five confocal fluorescence and DIC images of crypts (per specimen) were captured with a Zeiss LSM 710 microscope. The immunofluorescence image analysis was conducted as described previously [12].

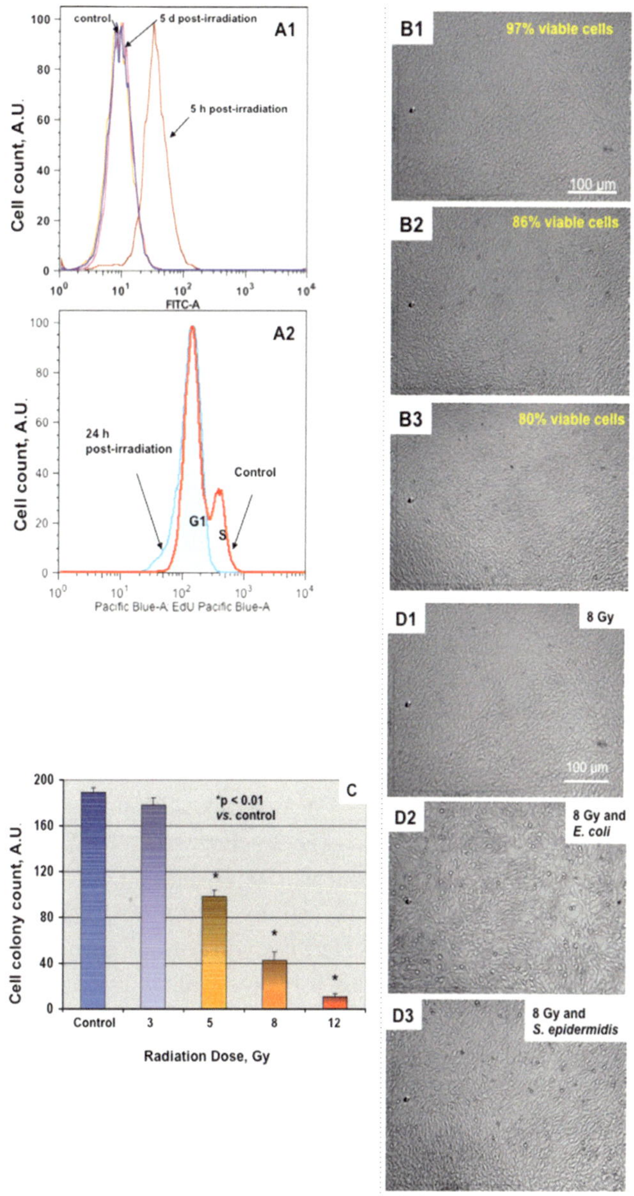

Figure 1. Functional ability of MSCs subjected to ionizing irradiation and bacterial challenge. A1. Analysis of radiation-induced DNA double-strand breaks with flow cytometry assay of the phosphorylated H2A.X (γ-H2A.X). Conditions: Control (in blue line) and irradiated MSCs were analyzed at 5 h (brown line) and 5 d (pink line) after 8-Gy irradiation. A2. Analysis of the radiation-induced suppression of MSC proliferative activity with flow cytometry assay of incorporated EdU, a modified nucleoside. The S-phase cell population was absent after irradiation Conditions: Control (red line) and irradiated (8 Gy, blue line) MSCs were analyzed 24 h after 8-Gy irradiation. A.U. is % of maximal cell count per channel. B. Bright-field microscopy analysis of effect of ionizing irradiation on ability of MSCs to form confluent

monolayers. Panel B1, control; Panel B2, 8-Gy irradiation; Panel B3, 12-Gy irradiation. Conditions: Images were captured 24 h after irradiation. Panel C. Analysis of the radiation-induced suppression of MSC colony-forming ability. Conditions: MSCs were harvested 24 h after irradiation and 200 MSCs from each radiation dose sample were aliquoted to Petri dishes and cultivated for 10 days. Panels D. Bright-field microscopy analysis of effects of bacterial challenge on ability of the irradiated MSCs shown in panel "B2" to form confluent monolayers. Panel D1 is after irradiation only (8 Gy), Panel D2 is same as "Panel D1" but after challenge with *E. coli*; Panel D3 is same as "Panel D1" but after challenge with *S. epidermidis*. Conditions: Images were captured 24 h after the bacterial challenge.

2.6. Transmission Electron Microscopy (TEM)

MSCs in culture were fixed in 4% formaldehyde and 4% glutaraldehyde in PBS overnight, post-fixed in 2% osmium tetroxide in PBS, dehydrated in a graduated series of ethanol solutions, and embedded in Spurr's epoxy resin. Blocks were processed as described previously [12,13]. The sections of embedded specimens were analyzed with a Philips CM100 electron microscope.

2.7. Statistical analysis

Statistical significance was determined using Student's *t*-test for independent samples. Significance was reported at a level of $p<0.05$.

3. Role of autophagy in adaptive response of MSCs to radiation injury and phagocytosis of *S. epidermidis* and *E. coli*

3.1. Alterations in the MSC stress-response-proteins following irradiation and bacterial challenge

The 8-Gy irradiation resulted in substantial DNA double-strand breaks in MSCs detectable with the γ-H2AX assay at 5 h post-exposure (Fig. 1A, brown line). This effect disappeared at 5 d post-irradiation recovery (Fig. 1A1, pink line). The observed DNA damage was accompanied by suppression of the cell proliferative activity determined with Click-iT® EdU Cell Proliferation Assay. As shown in Fig. 1A2 (red line), in control groups the cell populations were represented by the cells in both G1 and S phases. Following irradiation, the entire cell population was in G1 phase. The data presented in Fig. 1B indicate that the MSC cultures displayed a high integrity and survival from damage produced by irradiation at doses ranging from 1 Gy to 12 Gy, but that their ability to form colonies was reduced in a radiation dose-dependent manner (Fig. 1C).

The irradiated cells challenged with bacteria were also able to sustain integrity of confluent monolayers (Fig. 1D). These cells did not manifest signs of pro-apoptotic alterations. Moreover, MSCs challenged with *E. coli* and *S. epidermidis* 24 h after irradiation (8 Gy) were able to suppress the bacterial growth by 1.4-fold and 1.8-fold, respectively (not shown).

The data presented in Fig. 1 indicate that, despite radiation-produced damage and suppression of proliferative activity, MSCs demonstrated substantial radioresistance and absence of significant apoptotic and necrotic transformations in a wide range of radiation doses, i.e., 1-12

http://dx.doi.org/10.5772/55397

Gy. Interactive investigation of the stress-response factors implicated in cell survival may be important for the development of effective therapies for radiation injury (RI).

According to a current paradigm, the general stress responses involve conserved signaling modules that, in turn, are interconnected to the cellular adaptive mechanisms [37]. It is suggested that the stress due to molecular and organelle damage, impact of pro-oxidants, and infections triggers a cascade of responses attributed to specific sensitive transcriptional and post-transcriptional mechanisms mediating inflammation, antioxidant response, adaptation, and remodeling [36-40]. Ionizing radiation (IR) *per se* stimulates signaling cascades mediated by transcription factors and pathways that are believed to play a central role in protective response(s) to the molecular and subcellular damage and the oxidative stress. They include (but are not limited to) a battery of thiol-containing redox-response elements, redox-sensitive transcription factors such as nuclear factor-kappa B (NFκB) and forkhead box O3a (FoxO3a), stress-response adaptors such as the chaperone heat-shock protein 70 (HSP70) and NAD^+-dependent deacetylase sirtuin-3 (Sirt3), and activators of the autolysosomal degradation. Overall, these effector systems are crucial in maintaining homeostasis, which is altered due to damage to the cell constituents [33, 40-46]. It should be noted that, while the role of the IR-induced NFκB response in cell survival is well communicated, HSP70, the mitochondrial Sirt3, and FoxO3a are relatively newly-determined players implicated in adaptive mechanisms [43-48]. Thus, it has recently been observed that HSP70 and Sirt3 can sustain cell radioresistance and antioxidant capacity of mitochondria respectively; and that FoxO3a can promote cell survival by inducing the expression of antioxidant enzymes, autophagy, and factors involved in cell cycle withdrawal, such as the cyclin-dependent kinase inhibitor (CKI) p27 [33, 44-48].

Although the transcription factors NFκB and FoxO3a are normally sequestered in the cytoplasm, ionizing irradiation, bacterial products, pro-inflammatory effectors, and oxidative stress can stimulate their nuclear translocation and DNA-binding activity [13, 42, 43]. NFκB and FoxO3a are known to regulate numerous genes, including autophagy genes, and therefore, could link responses to IR and bacterial challenge with up-regulation of autolysosomal activity [13, 40, 42, 43, 45]. We do not, however, exclude implication of stress-induced adaptors and chaperones such as the heat-shock proteins (HSPs). HSP70, in particular, was shown to promote cell radioresistance and can regulate autophagy [33, 46]. Therefore, we assumed that a battery of stress-sensitive mechanisms mediated by survival factors such as NFκB, FoxO3a, Sirt3, and HSP70 are involved in an adaptive response of MSCs to IR and bacterial challenge.

Immunoblotting analysis of stress-response proteins presented in Fig. 2A indicates that control MSCs had relatively high amounts of constitutively expressed HSP70 and (p65)NFκB and a detectable amount of Sirt3. These basal levels did not significantly change at 24 h following 8-Gy irradiation. A slight increase in HSP70, (p65)NFκB, and Sirt3 occurred only after 12-Gy irradiation. Up-regulation of Casp-3 was not detected in the irradiated MSCs (Fig. 2A), which suggested the absence of pro-apoptotic alterations. Additional bacterial challenge of the 8-Gy irradiated MSCs did not compromise their viability (Fig. 1D) and did not affect the profile of the stress-proteins, except that IR induced a significant increase in Sirt3, a mitochondrial stress-response protein, and MMP3, the type 3 matrix metalloproteinase, essential for remodeling of extracellular matrix (Fig. 2B).

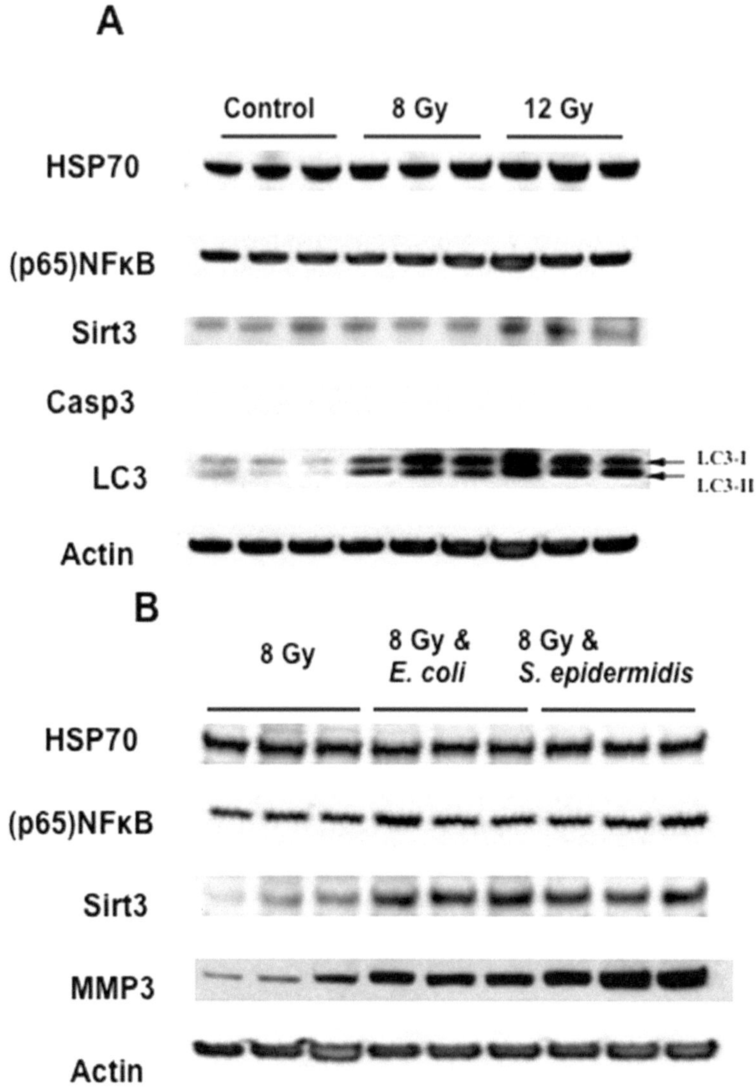

Figure 2. Immunoblot analysis of stress-response proteins in MSCs subjected to irradiation and bacterial challenge. A. MSCs, control and irradiated with 8 Gy and 12 Gy. Conditions: MSCs were harvested 24 h after irradiation then lysed and subjected to immunoblot analysis for stress-response proteins (HSP70, NFκB, and Sirt3), autolysosomal proteins (LC3-1 and LC3-II), and pro-apoptotic protein Caspase-3. B. MSCs irradiated with 8 Gy were challenged with either *E. coli* or *S. epidermidis*. Conditions: Irradiated MSCs were challenged with approximately $5x10^7$ bacteria /ml for 3 h in MesenCult Medium (without antibiotics). The cells were harvested and lysed 24 h after challenge. The protein lysates were subjected to immunoblot analysis for stress-response proteins (HSP70, NFκB, Sirt3, and MMP3).

Although in these experiments we did not observe a significant alteration of the amount of (p65)NFκB (Fig. 2B), the response of NFκB to IR was characterized by re-compartmentalization of (p65)NFκB resulting in an increase in its nuclear fraction (Fig. 3A). It should be noted that pre-incubation of the cells with pyrrolidine dithiocarbamate, an inhibitor of NFκB translocation, or wortmannin, an inhibitor of autophagy, resulted in development of pro-apoptotic alterations and loss of confluency after irradiation (not shown). Immunofluorescence imaging of spacial localization of FoxO3a in MSCs (Fig. 3B) indicated that FoxO3a response to IR was associated with an increase in its nuclear fraction in a manner similar to (p65)NFκB.

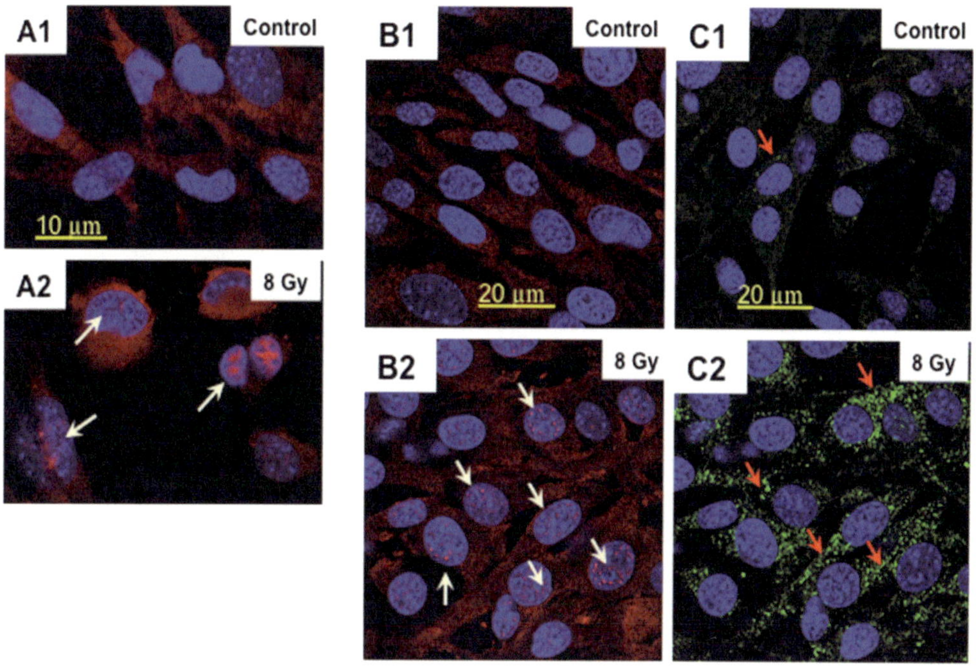

Figure 3. Confocal immunofluorescence imaging of nuclear translocation of NFκB and FoxO3a and activation of autophagy in MSCs subjected to 8-Gy irradiation. A. Projections of NFκB (red) in MSCs: A1, control; A2, 24 h after irradiation. Increase of nuclear fraction of p65 subunit of NFκB was observed in the irradiated cells due to transactivation of NFκB (indicated with white arrows). B. Projections of FoxO3a (red) in MSCs: B1, control; B2, 24 h after irradiation. Increase of nuclear fraction of FoxO3a was observed in the irradiated cells due to transactivation of FoxO3a (indicated with white arrows). C. Projections of LC3-positive autophagy vacuoles (green) in MSCs: C1, control; C2, 24 h after irradiation. A massive accumulation of autophagosomes occurred in irradiated MSCs (indicated with red arrows). Counterstaining of nuclei was with Hoechst 33342 (blue channel). The confocal images were taken with pinhole setup to obtain 0.5 µm Z-sections.

3.2. Autophagy—Autolysosomal response and secretory-activity in the irradiated MSCs subjected to bacterial challenge

The autophagy-autolysosomal pathway is considered to be an evolutionarily developed pro-survival mechanism, the purpose of which is to remove damaged and misfolded proteins, compromised organelles, and pathogens including bacteria [12, 21, 29, 34, 40]. A key step in autophagosome biogenesis is the conversion of light-chain protein 3 type I (LC3-I, also known as ubiquitin-like protein, Atg8) to type II (LC3-II). The conversion occurs via the cleavage of the LC3-I carboxyl terminus by a redox-sensitive Atg4 cysteine protease. The subsequent binding of the modified LC3-I to phosphatidylethanolamine, i.e., process of lipidation of LC3-1, on the isolation membrane, as it forms, is mediated by E-1- and E-2-like enzymes Atg7 and Atg3 [22, 26, 27, 40, 49]. Therefore, conversion of LC3-I to LC3-II and formation of LC3-positive vesicles are considered to be a marker of activation of autophagy [22, 26, 27, 33, 40]. Notably, a growing body of reports suggests implication of FoxO3a and HSP-70 in regulation of LC3 expression and translocation [32, 43, 45].

A line of evidence suggests that autophagy is a more selective process than the "bulk process" as it was originally defined [40, 49]. The discovery and characterization of autophagic adapters like p62/sequestrosome 1 (SQSTM1) and NBR1 (neighbor of BRCA1 gene 1), and target-ubiquitination with small ubiquitin-like modifier 1 (SUMO1) has provided mechanistic insight into this process. p62/SQSTM1 and NBR1 are both selectively degraded by autophagy and are able to act as cargo receptors for degradation of ubiquitinated/sumoylated substrates. A direct interaction between these autophagic adapters and the autophagosomal marker protein LC3-II, mediated by a so-called LIR (LC3-interacting region) motif, and their inherent ability to polymerize or aggregate, as well as their ability to specifically recognize substrates, are required for efficient selective autophagy [40, 49].

We hypothesized that autophagy and xenophagy (i.e., selective degradation of foreign pathogens by autophagy) can be implicated in the pro-survival response of MSCs to IR-related damage and bacterial challenge. To address this hypothesis we conducted immunoblotting and immunofluorescence confocal imaging of autophagy MAP (LC3) protein, lysosomal LAMP1 and SUMO1 in MSCs after irradiation and challenge with *E. coli* and *S. epidermidis*.

The immunoblotting analysis of MSC proteins revealed a drastic increase in LC3-I and LC3-II (compared to control) at 24 h following 8-Gy and 12-Gy irradiation (Fig. 2A). These results suggested that, indeed, the autophagy MAP (LC3) pathway is implicated in that MSC response to IR-induced injury. In contrast to MAP (LC3), we did not observe a substantial increase in HSP-70. This was most likely due to relatively high background expression of this stress-response protein in the cells (Fig. 2A).

The above immunoblotting results were corroborated by the immunofluorescence confocal image analysis of the LC3 protein in MSCs. Thus, the data presented in Fig. 3C suggest that up-regulation of LC3-I/LC3-II proteins in the 8-Gy-irradiated cells was associated with massive formation of the LC3-positive vesicles which are well-documented to be features of autophagy [12, 13, 33]. The further TEM-assessment of the 8-Gy irradiated MSCs (in comparison with controls) revealed the presence of multiple vacuoles, which were formed by double-layer membrane and sequestered constituents of different densities (Figs. 4 A-C). Some of these vacuoles can be identified as secretory autolysosomes by the presence of multilamellar structures (most likely fibers of collagen) released extracellularly (Figs. 4 D and E), while others contained fractured organelles including mitochondria (Figs. 4 C and F).

http://dx.doi.org/10.5772/55397

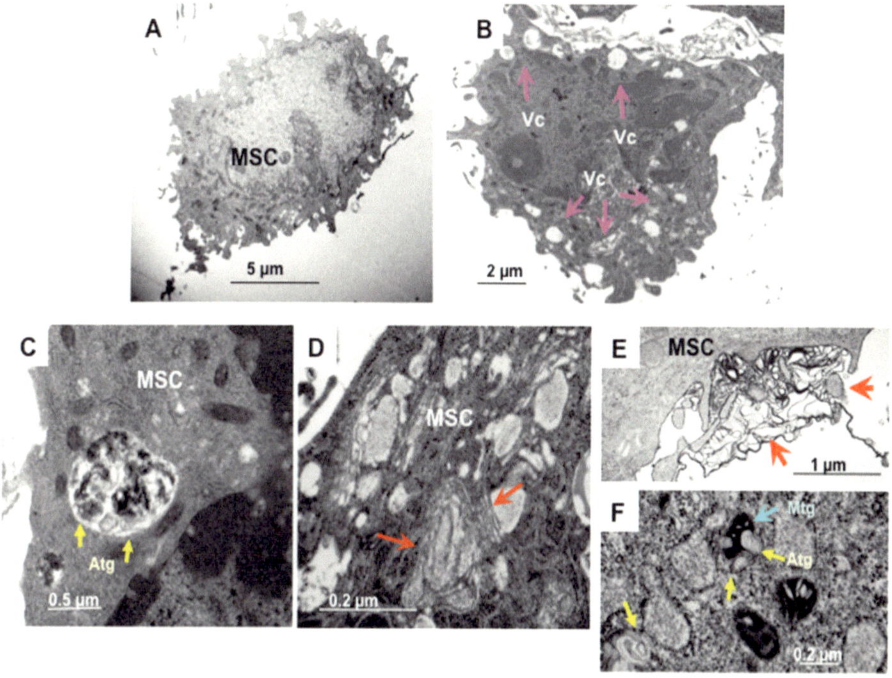

Figure 4. Transmission electron (TEM) analysis of autolysosomal vacuoles in irradiated MSCs. Panel A: A control MSC. Panel B: Irradiated MSC. A massive formation of different density autolysosomal vacuoles occurred after irradiation (indicated with pink arrows). Specimens were fixed 24 h after irradiation with 8 Gy gamma-photons. Panel C: Autolysosome sequestering cellular constituents (indicated with yellow arrows) in an irradiated MSC. Panel D: Formation of secretory autolysosomes containing multilamellar structures (indicated with red arrows) in an irradiated MSC. Panel E: Extracellular secretion of multilamellar structures from an irradiated cell. Panel F: Autolysosome sequestering mitochondria, *e.g.*, mitophagy, (indicated with blue arrow) in an irradiated MSC. Abbreviations: "Vc", vacuoles; "Atg", autophagosomes/autolysosomes; "Mtg", mitophagy.

In our recent research we demonstrated that intact MSCs are able to up-regulate autophagy in response to challenge with *E. coli* and employ this mechanism for inactivation of the microorganisms [13]. The data presented in this report showed that the irradiated MSCs retained their ability to phagocytise bacteria in a manner similar to that of non-irradiated MSCs (Fig. 5). Indeed, sequestration and degradation of *E. coli* and *S. epidermidis* in the MSC vacuoles, constituted by characteristic autophagosomal membranes, was observed at 5 h after bacterial challenge of both non-irradiated and irradiated MSCs (Figs. 5 A, B, D, and E).

The immunofluorescence confocal image analysis of the irradiated MSCs challenged with bacteria showed that the vacuoles containing bacteria were LC3-positive and that this LC3 immunoreactivity was co-localized with immunoreactivity to LAMP1, a marker of lysosomes, indicating presence of fusion of autophagosomes with lysosomes, i.e., formation of autolysosomes (Fig. 6). This increase in autolysosomal activity was accompanied by accumulation of the proteins LC3-II, a marker of up-regulation of autophagy, LAMP1, and p62/SQSTM1, a target adaptor (Figs. 7A and C). Meanwhile, the level of SUMO1, a target modifier protein, in the cells decreased after bacterial challenge (Fig. 7A). Interestingly, irradiation and bacterial challenge resulted in up-regulation of the factors responsible for modification of extracellular matrix, such as collagen III, MMP3, and MMP13, i.e., the collagenase-3, (Fig. 7C), indicating that the stress-response aimed at multiple targets including extracellular ones.

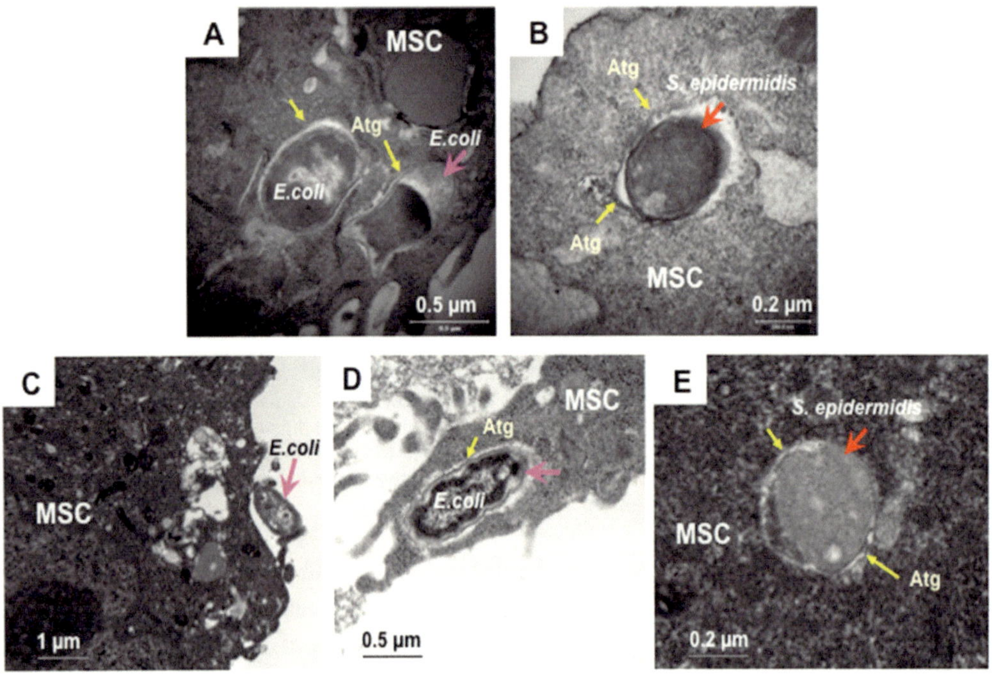

Figure 5. Assessment of phagocytosis and autophagy/autolysosomal processing of *E. coli* or *S. epidermidis* in irradiated MSCs with transmission electron microscopy. Panels A and B: Autolysosomal degradation of phagocytized *E. coli* and *S. epidermidis* in control MSCs. Autophagosome (ATG) membranes are indicated with yellow arrows. Conditions: Control MSCs were challenged with ~$5x10^7$ bacteria /ml for 3 h as indicated in *Methods*. The cells were harvested and fixed for TEM 5 h after challenge. Panels C, D, and E: TEM micrographs obtained from the 8-Gy irradiated cells. C - Engulfing and up-take of *E. coli* (pink arrows) by the cell plasma membrane extrusions (black arrows). D - Autolysosomal degradation of phagocytized *E. coli.* E - Autolysosomal degradation of phagocytized *S. epidermidis.* Atg, autophagosomes/autolysosomes. Conditions: 8-Gy irradiated MSCs were challenged with ~$5x10^7$ bacteria/ml for 3 h as indicated in *Methods*. The cells were harvested and fixed for TEM 5 h after challenge.

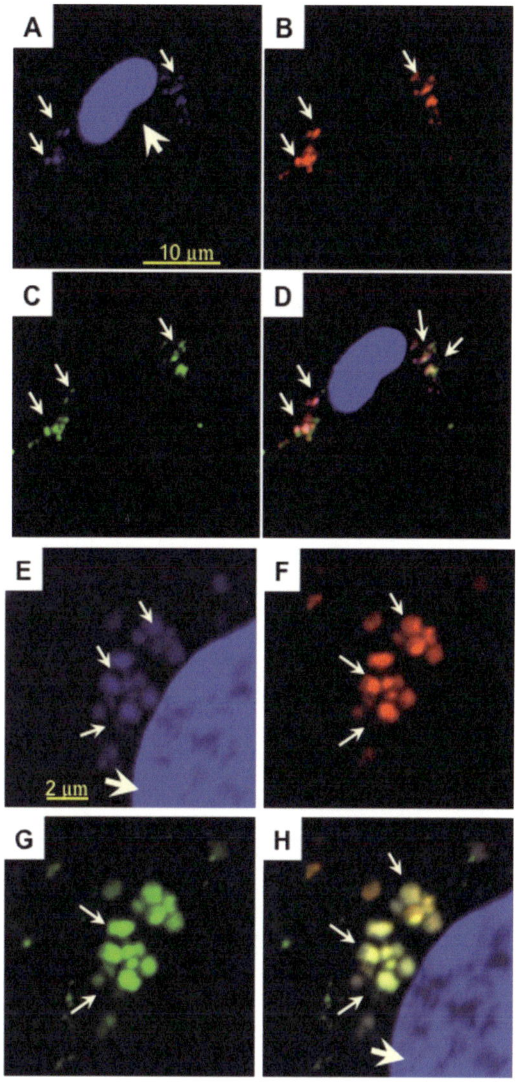

Figure 6. Confocal immunofluorescence imaging of autolysosomal sequestration of *E. coli* and *S. epidermidis* microorganisms phagocytized by MSCs irradiated at 8-Gy. Panel A – (Blue channel). *E. coli* (small arrows) and MSC nuclear DNA (large arrow) are indicated. Panel B – (Red channel). Spatial localization of LAMP1 is indicated with arrows. Panel C – (Green channel). Spatial localization of LC3 is indicated with arrows. Panel D – Overlay of images appeared in the blue, red, and green channels and presented in panels A, B, and C, respectively. Spatial co-localization of LAMP1, LC3, and *E. coli* DNA is indicated with arrows. Conditions: 8-Gy irradiated MSCs were challenged with ~5x10^7 *E. coli*/ml for 3 h. The cells were fixed 24 h after challenge. The fixed cells were subjected to immunofluorescence analysis for autolysosomal proteins. Panel E – (Blue channel). *S. epidermidis* (small arrows) and MSC nuclear DNA (large arrow) are indicated. Panel F – (Red channel). Spatial localization of LAMP1 is indicated with arrows. Panel G – (Green channel). Spatial localization of LC3 is indicated with arrows. Panel H – Overlay of images appeared in the blue, red, and green channels and presented in panels E, F, and G, respectively. Spatial co-localization of LAMP1, LC3, and *S. epidermidis* DNA is indicated with arrows. Conditions: 8-Gy irradiated MSCs were challenged with ~5x10^7 *S. epidermidis*/ml for 3 h. The cells were fixed 24 h after challenge. The fixed cells were subjected to immunofluorescence analysis for autolysosomal proteins. Counterstaining of nuclei was with Hoechst 33342 (blue channel). The confocal images were taken with pinhole setup to obtain 0.5 µm Z-sections.

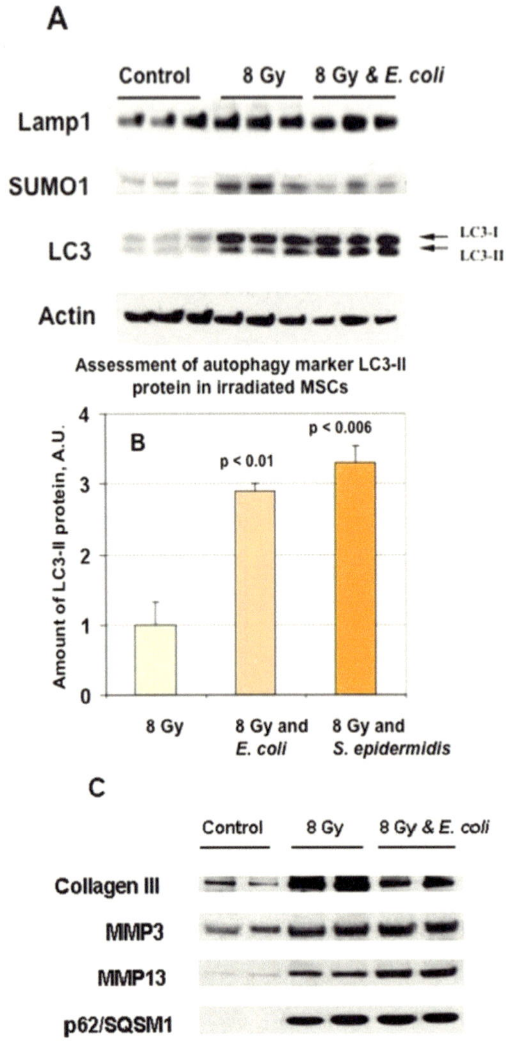

Figure 7. Immunoblot assessment of autolysosomal response in MSCs subjected to 8-Gy irradiation followed by challenge with either *E. coli* or *S. epidermidis.* Panel A. Representative immunoblotting bands of SUMO1, LC3, and LAMP1. Note that irradiated MSCs were challenged with *E. coli.* Panel B. Densitometry histograms of LC3-II bands of the immunoblots of proteins from the irradiated MSCs subjected to challenge with *E. coli* or *S. epidermidis*. The presented bars indicate the relative density of LC3-II protein (normalized to density of actin bands). The statistical significance was determined by Student's *t*-test (n=3). Panel C. Representative immunoblots of collagen III, MMP3, MMP13, and p62/SQSM1. Note that irradiated MSCs were challenged with *E. coli.* Conditions: Irradiated MSCs were challenged with approximately $5x10^7$ bacteria/ml for 3 h as indicated in *Methods*. The cells were harvested and lysed 24 h after challenge.

Various cells eliminate bacterial microorganisms by autophagy, and this elimination is in many cases crucial for host resistance to bacterial translocation. Targeting of microorganisms can occur outside of the host cells in extracellular matrix by different defense mechanisms, such as the cell-produced oxidative burst, nitric oxide, antibacterial peptides, and extracellular traps [50]. The data presented in the present report (Figs. 2, 4, 5, 7) suggest that MSCs can employ

http://dx.doi.org/10.5772/55397

the autophagy mechanism to modify extracellular matrix by releasing collagen and matrix metalloproteases in order to increase efficacy of extracellular entrapment, uptake, and further phagocytosis of the microorganisms.

Recent observations suggest that autophagosomes do not form randomly in the cytoplasm, but rather sequester the bacteria selectively [23, 49, 51]. Therefore, autophagosomes that engulf microbes are sometimes much larger than those formed during degradation of cellular organelles, suggesting that the elongation step of the autophagosome membrane is involved in bacteria-surrounding autophagy [13, 32]. This effect could be observed by comparison of profiles of the autophagosomes, which appeared in the irradiated MSCs before and after challenge with bacteria (Figs. 3 and 6). The mechanism underlying selective induction of autophagy at the site of microbe phagocytosis remains unknown. However, it is likely mediated by pattern recognition receptors, stress-response elements, adaptor proteins, e.g., p62/SQSTM1, and ubiquitin-like modifiers, which can target bacteria and ultimately recruit factors essential for the formation of autophagosomes [21, 22, 52].

4. Conclusion

Survival of multicellular organisms in a non-sterile environment requires a network of host defense mechanisms. The initial contact of pathogenic and opportunistic microorganisms with a host usually takes place at internal or external body surfaces. Microbial growth and translocation are controlled by multi-layer integrative tissue barriers that mediate innate defense mechanisms. Tissue injury compromises barrier function, and increases risk of infection and sepsis. Recent observations from our laboratory indicate that ubiquitous MSCs can modulate systemic responses to bacterial infection and support tissue repair and healing when recruited at sites of injury [2, 5, 9, 14-18]. However, these "compensatory" responses of MSCs can be skewed and suppressed after irradiation. To elucidate the role of autophagy in response of stromal cells to radiation injury and bacterial infection we irradiated cultured MSCs and challenged them with *S. epidermidis* or *E. coli.* Using this cell model we showed that (i) irradiation induced translocation of cytosol NF-kB and FoxO3a to the nucleus; (ii) irradiation and bacterial challenge induced increases in Sirt3 stress-response factors, LC3, MMP3, MMP13, collagen III, SUMO1, and p62/SQSM1 proteins; and (iii) the antibacterial defense response of the irradiated MSCs was characterized by extensive phagocytosis and inactivation of both *S. epidermidis* or *E. coli* in autolysosomes.

Our communication is the first report demonstrating a potential role of MSCs in sustaining antibacterial barrier functions of irradiated tissues. We postulate that effector mechanisms expressed by MSCs can contribute to the innate defense response to IR injury alone or, especially, when IR is combined with trauma.

Grants

This work was supported by AFRRI Intramural RAB2CF (to JGK) and NIAID YI-AI-5045-04 (to JGK). There are no ethical and financial conflicts in the presented work.

Disclaimer

The opinions or assertions contained herein are the authors' private views and are not to be construed as official or reflecting the views of the Uniformed Services University of the Health Sciences, Armed Forces Radiobiology Research Institute, the U.S. Department of Defense or the National Institutes of Health.

Acknowledgements

The authors thank Prof. J.S . Greenberger (UPMC/University of Pittsburgh Schools of the Health Sciences) for his criticism and encouragement in conducting of our investigation, and Ms. Dilber Nurmemet for her technical support.

Author details

Nikolai V. Gorbunov[1*], Thomas B. Elliott[2], Dennis P. McDaniel[3], K. Lund[3], Pei-Jyun Liao[1], Min Zhai[1] and Juliann G. Kiang[2,3*]

*Address all correspondence to: nikolaiv.gorbunov@gmail.com

1 The Henry M. Jackson Foundation, USA

2 Armed Forces Radiobiology Research Institute, USA

3 School of Medicine, Uniformed Services University of the Health Sciences, Bethesda, Maryland, USA

References

[1] Phan, S. H. Biology of fibroblasts and myofibroblasts. Proc Am Thorac Soc (2008). , 5, 334-7.

[2] Krebsbach, P. H, Kuznetsov, S. A, Bianco, P, & Robey, P. G. Bone marrow stromal cells: characterization and clinical application. Crit Rev Oral Biol Med (1999). , 10, 65-81.

[3] Winkler, I. G, Barbier, V, Wadley, R, Zannettino, A. C, Williams, S, & Lévesque, J. P. Positioning of bone marrow hematopoietic and stromal cells relative to blood flow *in vivo*: serially reconstituting hematopoietic stem cells reside in distinct nonperfused niches. Blood (2010). , 116, 375-85.

[4] Zusman, I, Gurevich, P, & Ben-hur, H. Two secretory immune systems (mucosal and barrier) in human intrauterine development, normal and pathological. Int J Mol Med (2005). , 16, 127-33.

[5] Powell, D. W, Pinchuk, I. V, & Saada, J. I. Chen Xin, and Mifflin RC. Mesenchymal Cells of the Intestinal Lamina Propria. Annu Rev Physiol (2011). , 73, 13-237.

[6] Turner, H. L, & Turner, J. R. Good fences make good neighbors: Gastrointestinal mucosal structure. Gut Microbes (2010). , 1, 22-29.

[7] Hooper, L. V, & Macpherson, A. J. Immune adaptations that maintain homeostasis with the intestinal microbiota. Nat Rev Immunol (2010). , 10, 159-69.

[8] Louis, N. A, & Lin, P. W. The Intestinal Immune Barrier. Neoreviews (2009). ee190., 180.

[9] Kirkup BC Jr, Craft DW, Palys T, Black C, Heitkamp R, Li C, Lu Y, Matlock N, McQueary C, Michels A, Peck G, Si Y, Summers AM, Thompson M, Zurawski DV.Traumatic wound microbiome workshop. Microb Ecol (2012). , 64, 37-50.

[10] Ledney, G. D, & Elliott, T. B. Combined injury: factors with potential to impact radiation dose assessments. Health Physics (2010). , 98, 145-152.

[11] Kiang, J. G, Jiao, W, Cary, L. H, Mog, S. R, Elliott, T. B, Pellmar, T. C, & Ledney, G. D. Wound trauma increases radiation-induced mortality by activation of iNOS pathway and elevation of cytokine concentrations and bacterial infection. Radiat Res (2010). , 173, 319-32.

[12] Gorbunov, N. V, Garrison, B. R, & Kiang, J. G. Response of crypt Paneth cells in the small intestine following total-body gamma-irradiation. Int J Immunopathol Pharmacol (2010). , 23, 1111-23.

[13] Gorbunov, N. V, Garrison, B. R, Zhai, M, Mcdaniel, D. P, Ledney, G. D, Elliott, T. B, & Kiang, J. G. Autophagy-Mediated Defense Response of Mouse Mesenchymal Stromal Cells (MSCs) to Challenge with *Escherichia coli*. In: Cai J and Wang RE (eds) Protein Interactions. Rijeka: InTech; (2012). , 23-44.

[14] Jackson, W. M, Alexander, P. G, Bulken-hoover, J. D, Vogler, J. A, Ji, Y, Mckay, P, Nesti, L. J, & Tuan, R. S. Mesenchymal progenitor cells derived from traumatized muscle enhance neurite growth. J Tissue Eng Regen Med (2012). doi:10.1002/term. 539.Epub ahead of print].

[15] Jackson, W. M, Aragon, A. B, Djouad, F, Song, Y, Koehler, S. M, Nesti, L. J, & Tuan, R. S. Mesenchymal progenitor cells derived from traumatized human muscle. J Tissue Eng Regen Med (2009). , 3, 129-38.

[16] Le Blanc K, Ringdén O.Immunomodulation by mesenchymal stem cells and clinical experience. J Intern Med (2007). , 262, 509-25.

[17] Lee, J. W, Fang, X, Gupta, N, Serikov, V, & Matthway, M. A. Allogeneic human mesenchymal stem cells for treatment of *E. coli* endotoxin-induced acute lung injury in the *ex vivo* perfused human lung. Proc Natl Acad Sci USA (2009). , 106, 16357-62.

[18] Nemeth, K, Mayer, B, & Mezey, E. Modulation of bone marrow stromal cell functions in infectious diseases by toll-like receptor ligands. J Mol Med (2010). , 88, 5-10.

[19] Akita, S, Akino, K, Hirano, A, Ohtsuru, A, & Yamashita, S. Mesenchymal stem cell therapy for cutaneous radiation syndrome. Health Phys (2010). , 98, 858-62.

[20] Lange, C, Brunswig-spickenheier, B, Cappallo-obermann, H, Eggert, K, Gehling, U. M, Rudolph, C, Schlegelberger, B, Cornils, K, Zustin, J, Spiess, A. N, & Zander, A. R. Radiation rescue: mesenchymal stromal cells protect from lethal irradiation. PLoS One (2011). e14486.

[21] Levine, B, Mizushima, N, & Virgin, H. W. Autophagy in immunity and inflammation. Nature (2011). , 469, 323-35.

[22] Yang, Z, & Klionsky, D. J. Eaten alive: a history of macroautophagy. Nat Cell Biol (2010). , 12, 814-22.

[23] Yano, T, & Kurata, S. Intracellular recognition of pathogens and autophagy as an innate immune host defence. J Biochem (2011). , 150, 143-9.

[24] Mizushima, N, Levine, B, Cuervo, A. M, & Klionsky, D. J. Autophagy fights disease through cellular self-digestion. Nature (2008). , 451, 1069-75.

[25] Klionsky, D. J. The Autophagy Connection. Dev Cell. (2010). , 19, 11-2.

[26] Tooze, S. A, & Yoshimori, T. The origin of the autophagosomal membrane. Nat Cell Biol (2010). , 12, 831-5.

[27] Weidberg, H, Shvets, E, & Elazar, Z. Biogenesis and cargo selectivity of autophagosomes. Annu Rev Biochem (2011). , 80, 125-56.

[28] Eskelinen, E. L. New insights into the mechanisms of macroautophagy in mammalian cells. Int Rev Cell Mol Biol (2008). , 266, 207-47.

[29] Eskelinen, E. L, & Saftig, P. Autophagy: a lysosomal degradation pathway with a central role in health and disease. Biochim Biophys Acta (2009). , 1793, 664-73.

[30] Mizushima, N, & Levine, B. Autophagy in mammalian development and differentiation. Nat Cell Biol (2010). , 12, 823-30.

[31] Kabeya, Y, Mizushima, N, Yamamoto, A, Oshitani-okamoto, S, Ohsumi, Y, & Yoshimori, T. LC3, GABARAP and GATE16 localize to autophagosomal membrane depending on form-II formation. J Cell Sci (2004). , 117, 2805-12.

[32] Behrends, C, Sowa, M. E, Gygi, S. P, & Harper, J. W. Network organization of the human autophagy system. Nature (2010). , 466, 68-76.

[33] Viiri, J, Hyttinen, J. M, Ryhänen, T, Rilla, K, Paimela, T, Kuusisto, E, Siitonen, A, Urtti, A, Salminen, A, & Kaarniranta, K. p/62 sequestosome 1 as a regulator of proteasome inhibitor-induced autophagy in human retinal pigment epithelial cells. Mol Vis (2010)., 16, 1399-414.

[34] Ryhänen, T, Hyttinen, J. M, Kopitz, J, Rilla, K, Kuusisto, E, Mannermaa, E, Viiri, J, Holmberg, C. I, Immonen, I, Meri, S, Parkkinen, J, Eskelinen, E. L, Uusitalo, H, Salminen, A, & Kaarniranta, K. Crosstalk between Hsp70 molecular chaperone, lysosomes and proteasomes in autophagy-mediated proteolysis in human retinal pigment epithelial cells. J Cell Mol Med (2009)., 13, 3616-31.

[35] Behl, C. BAG3 and friends: co-chaperones in selective autophagy during aging and disease. Autophagy (2011)., 7, 795-8.

[36] Sridhar, S, Botbol, Y, Macian, F, & Cuervo, A. M. Autophagy and Disease: always two sides to a problem. J Pathol (2012)., 226, 255-73.

[37] Kültz, D. Molecular and evolutionary basis of the cellular stress response. Annu Rev Physiol (2005)., 67, 225-57.

[38] Burhans, W. C, & Heintz, N. H. The cell cycle is a redox cycle: linking phase-specific targets to cell fate. Free Radic Biol Med (2009)., 47, 1282-93.

[39] Baltimore, D. NF-κB is 25. Nat Immunol (2011)., 12, 683-5.

[40] Murrow, L, & Debnath, J. Autophagy as a Stress-Response and Quality-Control Mechanism: Implications for Cell Injury and Human Disease. Annu Rev Pathol. (2012). Oct 15. [Epub ahead of print]

[41] Wei, S. J, Botero, A, Hirota, K, Bradbury, C. M, Markovina, S, Laszlo, A, Spitz, D. R, Goswami, P. C, Yodoi, J, & Gius, D. Thioredoxin nuclear translocation and interaction with redox factor-1 activates the activator protein-1 transcription factor in response to ionizing radiation. Cancer Res (2000)., 60, 6688-95.

[42] Li, N, & Karin, M. Ionizing radiation and short wavelength UV activate NF-κB through two distinct mechanisms. Proc Natl Acad Sci USA (1998)., 95, 13012-7.

[43] Tsai, W. B, Chung, Y. M, Takahashi, Y, Xu, Z, & Hu, M. C. Functional interaction between FOXO3a and ATM regulates DNA damage response. Nat Cell Biol (2008)., 10, 460-7.

[44] Aquila, D, Rose, P, Panno, G, Passarino, M. L, Bellizzi, G, & , D. Sirt3 gene expression: a link between inherited mitochondrial DNA variants and oxidative stress. Gene (2012)., 497, 323-9.

[45] Rodriguez-rocha, H, Garcia-garcia, A, Panayiotidis, M. I, & Franco, R. DNA damage and autophagy. Mutat Res (2011)., 711, 158-66.

[46] Graner, M. W, Raynes, D. A, Bigner, D. D, & Guerriero, V. Heat shock protein 70-binding protein 1 is highly expressed in high-grade gliomas, interacts with multiple

heat shock protein 70 family members, and specifically binds brain tumor cell surfaces. Cancer Sci (2009). , 100, 1870-9.

[47] Miyamoto, K, Araki, K. Y, Naka, K, Arai, F, Takubo, K, Yamazaki, S, Matsuoka, S, Miyamoto, T, Ito, K, Ohmura, M, Chen, C, Hosokawa, K, Nakauchi, H, Nakayama, K, Nakayama, K. I, Harada, M, Motoyama, N, Suda, T, & Hirao, A. Foxo3a is essential for maintenance of the hematopoietic stem cell pool. Cell Stem Cell (2007). , 1, 101-12.

[48] Burhans, W. C, & Heintz, N. H. The cell cycle is a redox cycle: linking phase-specific targets to cell fate. Free Radic Biol Med (2009). , 47, 1282-93.

[49] Reggiori, F, Komatsu, M, Finley, K, & Simonsen, A. Selective types of autophagy. Int J Cell Biol (2012). Epub 2012 May 15.

[50] von Köckritz-Blickwede, M, & Nizet, V. Innate immunity turned inside-out: antimicrobial defense by phagocyte extracellular traps. J Mol Med (Berl) (2009). , 87, 775-83.

[51] Nakagawa, I, Amano, A, Mizushima, N, Yamamoto, A, Yamaguxhi, H, Kamimoto, T, Nara, A, Funao, J, Nakata, M, Tsuda, K, Hamada, S, & Yoshimori, T. Autophagy defends cells against invading Group A *Streptococcus*. Science (2004). , 306, 1037-40.

[52] Ichimura, Y, & Komatsu, M. Selective degradation of by autophagy. Semin Immunopathol (2010). , 62.

Section 7

Autophagy in Neurodegenerative Diseases

Chapter 17

Role of Autophagy in Parkinson's Disease

Grace G.Y. Lim, Chengwu Zhang and
Kah-Leong Lim

Additional information is available at the end of the chapter

http://dx.doi.org/10.5772/54927

1. Introduction

Parkinson's disease (PD) is a prevalent neurodegenerative movement disorder whose occurrence crosses geographic, racial and social boundaries affecting 1-2% of the population above the age of 65 (Dorsey et al., 2007). Clinically, the disease is attended by a constellation of motoric deficits that progressively worsen with age, which ultimately leads to near total immobility. Although pathological changes are distributed in the PD brain (Braak et al., 2003), the principal lesion that underlies the characteristic motor phenotype of PD patients is unequivocally the loss of dopaminergic neurons in the *substantia nigra pars compacta* (SNpc) of the midbrain. This neuronal loss results in a severe depletion of striatal dopamine (DA) and thereby an impaired nigrostriatal system that otherwise allows an individual to execute proper, coordinated movements. Accordingly, pharmacological replacement of brain DA via L-DOPA administration represents an effective symptomatic recourse for the patient (especially during the initial stages of the disease) and remains a clinical gold standard treatment for PD. However, neither L-DOPA nor any currently available therapies could slow or stop the insidious degenerative process in the PD brain. Thus, PD remains an incurable disease. Invariably, the debilitating nature and morbidity of the disease present significant healthcare, social, emotional and economic problems. As the world population rapidly ages, these problems undoubtedly would also increase. According to a recent report, more than 4 million individuals in Europe's five most and the world's ten most populous countries are currently afflicted with PD (Dorsey et al., 2007). In less than 20 years' time, the number of PD sufferers is projected to increase to close to 10 million (i.e. in 2030). This is definitely a worrying trend, and one that aptly emphasizes the urgency to develop more effective treatment modalities for the PD patient. Towards this endeavour, a better understanding of the molecular mechanism(s) that underlies the pathogenesis of PD would definitely be helpful, as the illumination of which would allow the identification and therapeutic exploitation of key molecules/events involved in the pathogenic process.

Although a subject of intense research, the etiology of PD unfortunately remains incompletely understood. However, a broad range of studies conducted over the past few decades, including epidemiological, genetic and post-mortem analysis, as well as *in vitro* and *in vivo* modelling, have contributed significantly to our understanding of the pathogenesis of the disease. In particular, the recent identification and functional characterization of several genes, including *α-synuclein, parkin, DJ-1, PINK1* and *LRRK2,* whose mutations are causative of rare familial forms of PD have provided tremendous insights into the molecular pathways underlying dopaminergic neurodegeneration (Lim and Ng, 2009; Martin et al., 2011). Collectively, these studies implicate aberrant protein and mitochondrial homeostasis as key contributors to the development of PD, with oxidative stress likely acting an important nexus between the two pathogenic events.

2. Aberrant protein homeostasis & PD

Perhaps the most glaring evidence suggesting that protein homeostasis has gone awry in the PD brain is the presence of intra-neuronal inclusions, known as Lewy Bodies (LBs), in affected regions of the diseased brain in numbers that far exceed their occasional presence in the normal brain (Lewy, 1912). These signature inclusions of PD comprise of a plethora of protein constituents that include several PD-linked gene products such as α-synuclein, parkin, DJ-1, PINK1 and LRRK2. In a recent report, Wakabayashi and colleagues have documented more than 90 components of the LB and have grouped them into 13 functional groups (Table 1) (Wakabayashi et al., 2012). Among these, α-synuclein is recognized as the major component of LB and thought to be the key initiator of LB biogenesis.

However, whether LB biogenesis represents a cytoprotective or pathogenic mechanism in PD remains debatable. Notwithstanding this, how proteins aggregate to form LB is intriguing in the first place, as the cell is endowed with several complex surveillance machineries to detect and repair faulty proteins, and also destroy those are beyond repair rapidly (Fig. 1). In this surveillance system, the chaperones (comprising of members of the heat-shock proteins) represent the first line of defense in ensuring the correct folding and refolding of proteins (Liberek et al., 2008). When a native folding state could not be attained, the chaperones will direct the misfolded protein for proteolyic removal typically by the proteasome. Proteins that are destined for proteasome-mediated degradation are usually added a chain of ubiquitin via a reaction cascade that involves the ubiquitin-activating (E1), -conjugating (E2) and -ligating (E3) enzymes, whereby successive iso-peptide linkages are formed between the terminal residue (G76) of one ubiquitin molecule and a lysine (K) residue (most commonly K48) within another. The (G76-K48) polyubiquitinated substrate is then recognized by the 26S proteasome as a target for degradation (Pickart and Cohen, 2004). It is noteworthy to mention that although the G76-K48 chain linkage is the most common form of polyubiquitin, ubiquitin self-assembly can occur at any lysine residues within the molecule (at positions 6, 11, 27, 29, 33, 48 and 63)

(Pickart, 2000; Peng et al., 2003). In addition, proteins can also be monoubiquitinated. Notably, both K63-linked polyubiquitination and monoubiquitination of proteins are not typically associated with proteasome-mediated degradation (Pickart, 2000; Peng et al., 2003).

Group	Components	Remarks
1	α-synuclein; Neurofilaments	Structural Elements
2	Agrin; 14-3-3; Synphilin-1; Tau	α-synuclein-binding proteins
3	Dorfin; GSK-3β; NUB1; Parkin; Pin1; SIAH-1	Synphilin-1-binding proteins
4	Ubiquitin; E1; UbcH7; TRAF6; TRIM9; Proteasome subunits; PA700; PA28; β-TrCP; Cullin-1; HDAC4; NEDD8; p38; p62 (Sequestosome 1); ROC1; UCHL1	UPS-related proteins
5	LC3; GABARAP; GATE-16; Glucocerebrosidase; NBR-1	Autophagosome-lysosome system
6	γ-tubulin; HDAC6; Peri-centrin	Aggresome-related proteins
7	DJ-1; CHIP; Clusterin/apolipoprotein J; DnaJB6; Heat Shock Proteins; Torsin A, SOD1 & 2; FOXO3a	Stress response-related proteins
8	CaMKII; Casein Kinase II; CDK5, G-Protein Coupled Receptor Kinase 5; LRRK2; PINK1; IκBα; NFκB; p35; phospho-lipase C-δ; Tissue Transglutaminase	Signal transduction-related proteins
9	MAP1B; MAP2; Sept4/H5	Cytoskeletal proteins
10	Cox IV; Cytochrome C; Omi/HtrA2	Mitochondria-related proteins
11	Cyclin B; Retinoblastoma Protein	Cell cycle proteins
12	Amyloid Precursor Protein; Calbindin; Choline Acetyltransferase; Chromogranin A; Synaptophysin; Synaptotagmin; Tyrosine Hydroxylase; VMAT2	Cytosolic Proteins
13	Complement Proteins; Immunoglobulin	Immune-related proteins

Table 1. Components of Lewy Body (Wakabayashi et al., 2012)

Whilst the coupling of chaperone and ubiquitin protein system (UPS) provides an efficient way for the cell to deal with protein misfolding, there are times when the capacity of these systems may be exceeded by the production of misfolded proteins (e.g. under conditions of cellular stress). In such cases, aggregation-prone proteins that failed to be degraded may be transported along microtubules in a retrograde fashion to the microtubule organizing center to form an "aggresome", a term originally coined by Johnston and Kopito more than a decade ago (Johnston et al., 1998). According to the model, aggresome formation represents a cellular response towards proteasome impairments and their localization to the juxta-nuclear region is to facilitate their capture by lysosomes and thereby their clearance by macroautophagy (hereafter referred to as autophagy). Consistent with this, aggregation-prone proteins often generate aggresome-like structures when ectopically expressed in cultured cells in the presence of proteasome inhibition (Wong et al., 2008). Moreover, several groups including ours have demonstrated that autophagy induction promotes the clearance of aggresomes whereas the reverse is true when the bulk degradation system is inhibited (Fortun et al., 2003; Iwata et al., 2005b; Opazo et al., 2008; Wong et al., 2008).

Together, the chaperone, ubiquitin-proteasome and autophagy systems thus function in synergism to effectively counterbalance the threat of protein misfolding and aggregation. Accordingly, aberrations in one or more of these systems would be expected to promote protein aggregation and inclusion body formation, as in the case of affected neurons in the PD brain where LBs occur.

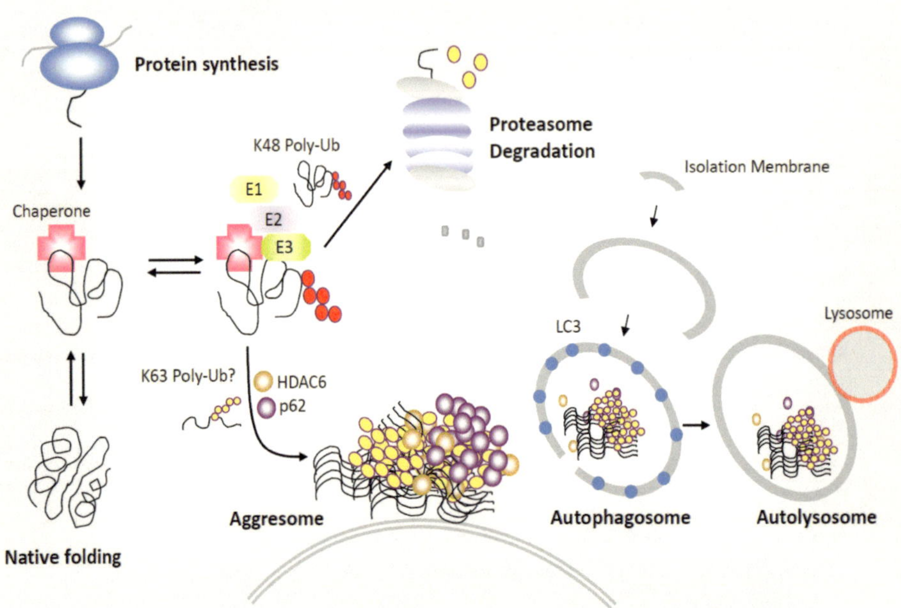

Figure 1. Schematic depiction of the collaboration among the chaperone, ubiquitin-proteasome and autophagy systems in the maintenance of intracellular protein homeostasis.

3. Biogenesis of Lewy bodies – An aggresome-related process reflecting failed autophagy?

As mentioned earlier, α-synuclein is a major component of LBs, suggesting that aberrant α-synuclein homeostasis contributes to the biogenesis of these inclusion bodies in the PD brain. The presynaptic terminal-enriched α-synuclein protein is an interesting molecule in that it is typically unfolded (or intrinsically disordered) in its native state, although the protein is extremely sensitive to its environment and can be moulded into an assortment of structurally unrelated conformations including a fibrillization-prone partially folded structure as well as various α-helical and β-sheet species occurring in both monomeric and oligomeric states (Uversky, 2007). Along with this conformation flexibility, α-synuclein also tends to misfold and becomes aggregated in the process. PD-associated mutations, including missense substitutions (A53T, A30P and E46K), duplication or triplication are known to enhance α-synuclein accumulation and aggregation (Giasson et al., 1999; Narhi et al., 1999; Conway et al., 2000; Uversky, 2007). Further, several groups have demonstrated in different experimental models that various exogenous neurotoxicants linked to PD, including pesticides, herbicides and metal ions, significantly accelerate the aggregation of α-synuclein (Manning-Bog et al., 2002; Uversky et al., 2002; Sherer et al., 2003). Not surprisingly, α-synuclein accumulation and aggregation can lead to impairments of the chaperone and UPS systems [For a recent review, see (Tan et al., 2009)]. Under such conditions, the isolation of α-synuclein aggregates into an aggresome would represent an alternative way by which the protein could be cleared, i.e. via autophagy. Indeed, emerging evidence suggest that LB biogenesis may be an aggresome-related process (Olanow et al., 2004). Because the protofibrillar, oligomeric forms of α-synuclein are thought to be more toxic than fibrillar, aggregated α-synuclein species, aggresome formation may also be regarded as a "protective" response that serves as a trap to immobilize soluble toxic forms of α-synuclein. However, this process has to be coupled to the active removal of the aggresomes by autophagy, as the unregulated growth of an inclusion body could conceivably affect cellular functions, physically or otherwise.

The relevance of aggresome formation to LB biogenesis in PD is exemplified by their striking similarities to each other in terms of structural organization, protein composition and intracellular localization (Olanow et al., 2004). For example, aggresome-related proteins such as γ-tubulin and HDAC6 can be found in LB (Table 1). HDAC6 plays an important role during aggresome formation by facilitating the retrograde transport of ubiquitinated misfolded proteins along the microtubule network to the γ-tubulin-positive MTOC by the dynein motor complex (Kawaguchi et al., 2003). Moreover, LBs are also immunopositive for p62 and NBR1, which are autophagy adapter proteins capable of binding to ubiquitinated substrates and the autophagosome protein LC3 (Bjorkoy et al., 2005; Pankiv et al., 2007; Kirkin et al., 2009). By virtue of this binding property, p62 and NBR1 may provide a link between aggresome-related proteins and their clearance by the autophagy machinery. Interestingly, all the three ubiquitin-binding autophagy receptors, i.e. p62, HDAC6 and NBR1, show preference for K63-linked polyubiquitin chains (Olzmann et al., 2007; Tan et al., 2008a; Kirkin et al., 2009), suggesting that this form of ubiquitin modification may underlie the formation as well as autophagic degradation of protein aggregates. Consistent with this, we found that K63-linked ubiquitination promotes the formation of inclusion bodies associated with PD and other neurodegenerative diseases and importantly, acts as a cargo selection signal for their subsequent removal by autophagy (Tan et al., 2008a; Tan et al., 2008b). As per our original proposal (Lim et al., 2006), it is tempting to think that the cell may switch to an alternative, proteasome-independent form of ubiquitination under conditions of proteasome-related stress that could help divert cargo proteins away from an otherwise overloaded proteasome. All these would culminate to the ultimate clearance of these proteins by autophagy (Fig. 1).

What remains curious about LB biogenesis is that it apparently takes place in the presence of constitutive autophagy, which is a characteristic of post-mitotic neurons (Wong and Cuervo, 2010). Moreover, α-synuclein is itself a substrate for autophagy (Webb et al., 2003).

Although α-synuclein can also be degraded by the proteasome, the aggregates of which appear to be preferentially cleared by the autophagy system (Petroi et al., 2012). Consistent with this, autophagy is recruited as the primary removal system in transgenic mice over-expressing oligomeric species of α-synuclein (Ebrahimi-Fakhari et al., 2011). Further, the protein can also be removed via chaperone-mediated autophagy (CMA), a specialized form of lysosomal degradation by which proteins containing a particular pentapeptide motif related to KFERQ are transported across the lysosomal membrane via the action of the integral membrane protein LAMP-2A and both cytosolic and lumenal hsc70 (Klionsky et al., 2011). Notably, the intralysosomal level of α-synuclein is significantly increased along with LAMP-2A and hsc70 in mice treated with the herbicide paraquat (which induces parkinsonism) or expressing α-synuclein as a transgene (Mak et al., 2010). Thus in theory, the level of α-synuclein, whether present as soluble or aggregated species, should be effectively managed in neurons under normal conditions or even when they are undergoing stress. Indeed, even in the PD brain, LB takes a significant length of time to develop. Given this, and that the autophagy system arguably represents the final line of cellular defense against the buildup of protein aggregates, the simplest explanation that could account for the presence of LB in PD is that the autophagy system has either become suboptimal in its function or is otherwise impaired altogether during the disease pathogenesis process.

4. Autophagy and PD

Morphological evidence of autophagic vacuole (AV) accumulation is certainly evident in PD as well as in several other neurodegenerative disorders (Anglade et al., 1997). However, whether the phenomenon represents attempts by the neuron to clean up its cobwebs of aggregated proteins, or a prelude to cell death, or simply a failure in AV consumption remains poorly understood. Notwithstanding this, two elegant studies conducted in 2006 aptly illustrated the importance of competent autophagy function to neuronal homeostasis (Hara et al., 2006; Komatsu et al., 2006). By means of targeted genetic disruption of essential components of the autophagy process (Atg5 or Atg7), these studies demonstrated that ablation of autophagy function in neural cells of mice results in extensive neurodegeneraion that is accompanied by widespread inclusion pathology, suggesting that autophagy failure can precipitate protein aggregation and subsequent cell death in affected neurons.

Supporting a role for failed autophagy in PD in the face of α-synuclein accumulation, α-synuclein was recently demonstrated to inhibit autophagy when over-expressed, both *in vitro* and *in vivo* (Winslow et al., 2010). The inhibition apparently occurs at a very early stage of autophagosome formation, which is likely a result of disrupted localization and mobilization of Atg9, a multi-spanning membrane protein whose associated vesicles are important sources of membranes for the synthesis of early autophagosomes (Yamamoto et al., 2012). Interesting, the reverse, i.e. autophagy enhancement, was observed when α-synuclein is depleted via RNAi-mediated knockdown (Winslow et al., 2010), suggesting that the protein might play a regulatory role in the synthesis of autophagosome. Moreover, targeted disruption of autophagy (via *Atg7* deletion) in midbrain dopaminergic neurons results in abnormal presynaptic accumulation of α-synuclein that is accompanied by dendritic and axonal dystrophy, reduced striatal DA content, and the formation of somatic and dendritic ubiquitinated inclusions (Friedman et al., 2012). Significant age-dependent loss of nigral dopaminergic neurons were also recorded in these *Atg7* conditionally knockout mice (*Atg7*-cKO^{TH}), with 9 month old *Atg7*-cKO^{TH} mice exhibiting about 40% reduction in the number of SN neurons that is accompanied by markedly decreased spontaneous motor activity and coordination relative to controls (Friedman et al., 2012). Together, these results suggest that failure in autophagy function precipitates inclusions formation in dopaminergic neurons that leads to their demise.

Besides macroautophagy, α-synuclein can also affect the function of CMA. For example, disease-associated α-synuclein mutants bind to the CMA lysosomal receptor with high affinity but are poorly translocated, resulting in the blockage of uptake and degradation of CMA substrates (Cuervo et al., 2004). The increase in cytosolic α-synuclein levels that ensued could favour its aggregation and concomitantly, amplify the burden of misfolded protein load for the cell. Interestingly, DA modification of α-synuclein also impairs CMA-mediated degradation by a similar mechanism (Martinez-Vicente et al., 2008). In this case, membrane-bound DA-α-synuclein monomers appear to seed the formation of oligomeric complexes, which consequently placed the translocation complex under siege. Consistent with this, CMA inhibition following L-DOPA treatment is more pronounced in ventral midbrain cultures containing dopaminergic neurons than in non-DA producing cortical neurons. Importantly, α-synuclein appears to be the principal mediator of DA-induced blockage of CMA, as ventral midbrain cultures derived from α-synuclein null mice are relatively spared from the inhibitory effects of DA on CMA (Martinez-Vicente et al., 2008). More recently, Malkus and Ischiropoulos demonstrated that CMA activity in the adult brain of A53T α-synuclein-expressing transgenic mice varies across different regions, with brain regions vulnerable to α-synuclein aggregation displaying marked deficiencies in CMA (Malkus and Ischiropoulos, 2012). Their results support an integral role for the lysosome in maintaining α-synuclein homeostasis and at the same time, provides an explanation to why certain brain regions are vulnerable to inclusion formation and cellular dysfunction while others are spared.

Perhaps the most direct evidence linking lysosomal dysfunction to PD is the demonstration that loss-of-function mutations in a gene encoding for the lysosomal P-type ATPase named ATP13A2 cause a juvenile and early-onset form of parkinsonism that is also characterized by pyramidal degeneration and dementia (Ramirez et al., 2006). In patient-derived fibroblasts as well as in ATP13A2-silenced primary mouse neurons, deficient ATPase function results in impaired lysosomal degradation capacity that concomitantly enhanced the accumulation and toxicity of α-synuclein (Usenovic et al., 2012). Importantly, silencing of endogenous α-synuclein ameliorated the toxicity in neurons depleted of ATP13A2, suggesting that ATP13A2-induced parkinsonism may be contributed by α-synuclein accumulation amid functional impairments of the lysosome. Supporting this, overexpression of wild type ATP13A2 suppresses α-synuclein-mediated toxicity in *C. elegans* while knockdown of ATP13A2 expression promotes the accumulation of misfolded α-synuclein in the animal (Rappley et al., 2009). Together, these studies demonstrate a functional link between ATP13A2-related lysosomal dysfunction and α-synuclein in promoting neurodegeneration.

Besides *α-synuclein* and *ATP13A2*, several other PD-linked genes have also been associated directly or indirectly with the autophagic process. For example, emerging evidence suggest that mutations in LRRK2 promote dysregulation in autophagy, although the role of LRRK2 in controlling autophagy-lysosome pathway is likely to be complex (discussed further in section 6). In the case of parkin, which has the ability to promote K63-linked ubiquitination, we and others have shown that the ubiquitin ligase is involved in aggresome formation and thereby their removal via autophagy (at least indirectly) (Lim et al., 2005; Olzmann et al., 2007). Consistent with its role as an "aggresome-promoter", parkin-related cases are frequently (although not exclusively) devoid of classic LBs, as revealed by a number of autopsy studies (Takahashi et al., 1994; Mori et al., 1998; Hayashi et al., 2000). In recent years, the attention to parkin-autophagy axis has however shifted towards its ability to remove damaged mitochondria via a specialized form of autophagy known as "mitophagy", a term originally coined by Lemasters (Lemasters, 2005). Accordingly, impairment in mitochondrial quality control due to failed mitophagy in parkin-deficient neurons is now thought to be a key mechanism that predisposes them to degeneration.

5. Mitophagy and PD

A role for mitochondria dysfunction in the pathogenesis of PD has long been appreciated. Through post-mortem analysis performed as early as 1989, several groups have recorded a significant reduction in the activity of mitochondrial complex I as well as ubiquinone (co-enzyme Q10) in the SN of PD brains (Schapira et al., 1989; Shults et al., 1997; Keeney et al., 2006). Moreover, mitochondrial poisoning recapitulates PD features in humans and represents a popular strategy to model the disease in animals (Dauer and Przedborski, 2003). Similarly, impairment of mitochondrial homeostasis via genetic ablation of TFAM, a mitochondrial transcription factor, in dopaminergic neurons of mice (named MitoPark mouse) results in energy crisis and neurodegeneration (Sterky et al., 2011).

Rather than being solitary and static structures as depicted in many textbooks, mitochondria are now recognized to be dynamic and mobile organelles that constantly undergo membrane remodeling through repeated cycles of fusion and fission as well as regulated turnover via mitophagy. These processes help to maintain a steady pool of healthy mitochondrial essential for energy production and beyond (e.g. calcium homeostasis). Following the seminal discovery by Youle group that identified parkin as a key mammalian regulator of mitophagy (Narendra et al., 2008), intensive research is now focused on elucidating the precise mechanism underlying parkin-mediated mitophagy and whether impaired clearance of damaged mitochondria may trigger the demise of dopaminergic neurons in the PD brain.

Mechanistically, the picture regarding parkin-mediated mitophagy that has emerged thus far is depicted in Figure 2.

http://dx.doi.org/10.5772/54927

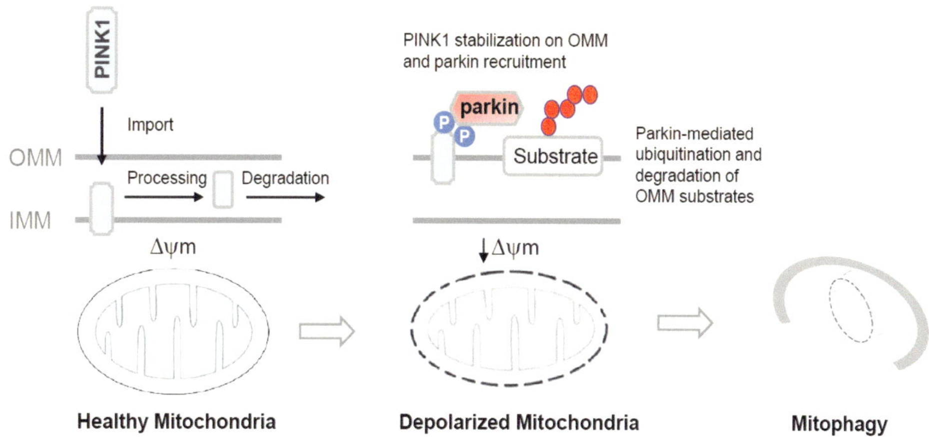

Figure 2. Model of Parkin/PINK1-mediated mitophagy

In this model, another PD-linked gene known as PINK1, which is a mitochondrial serine/threonine kinase, collaborate closely with parkin to bring about the mitophagy process. Briefly, a key initial event that occurs upon mitochondrial depolarization is the selective accumulation of PINK1 in the outer membrane of the damaged organelle. Normally, PINK1 accumulation in healthy mitochondria is prevented by the sequential proteolytic actions of mitochondrial processing peptidase (MPP) and presenilin-associated rhomboid-like protease (PARL) that rapidly cleaves the protein to generate an unstable 53 kDa PINK1 species that is usually degraded by the proteasome or by an unknown "proteasome-like" protease (Becker et al., 2012; Greene et al., 2012). In depolarized mitochondria, PINK1 stabilization on the outer membrane enables the protein to recruit parkin to the organelle, a process that is apparently dependent on PINK1 autophosphorylation at Ser228 and Ser402 (Okatsu et al., 2012). Once recruited onto the mitochondria, parkin becomes activated and promotes the ubiquitination and subsequent degradation of many outer membrane proteins (Chan et al., 2011; Yoshii et al., 2011) including the pro-fusion mitofusin proteins (Poole et al., 2010; Ziviani et al., 2010), the elimination of which is thought to prevent unintended fusion events involving damaged mitochondria and thereby their re-entry into undamaged mitochondrial network from occurring. Mitophagy induction then occurs, which likely involves parkin-mediated K63 ubiquitination that will help recruit the autophagy adaptors HDAC6 and p62 that subsequently lead to mitochondrial clustering around the peri-nucleus region. By virtue of their association with the autophagy process, the concerted actions of p62 and HDAC6 will presumably facilitate the final removal of damaged mitochondria by the lysosome (Ding et al., 2010; Geisler et al., 2010; Lee et al., 2010). However, a recent study from Mizushima's lab revealed that the initial cargo recognition step of mitophagy does not involves the interaction between LC3 and the adaptor molecules. Rather, parkin recruitment on the mitochondria induces the formation of ULK1 (Atg1) puncta and Atg9 structures (Itakura et al., 2012). Because ULK1 complex functions as an essential upstream nucleation step of the hierachical autophagy cascade, their results suggest that mitophagosome is generated in a de novo fashion on damaged mitochondria. Autophagosomal LC3 is however important for the efficient incorporation of damaged mitochondria into the autophagosome at a later stage. Notwithstanding this, how parkin participates in the de novo synthesis of isolation membrane awaits further clarifications. Interestingly, the whole mitophagy process bears striking resemblance to the formation and autophagic clearance of aggresomes. Indeed, we have termed the mitochondrial clustering phenomenon as (formation of) "mito-aggresomes" (Lee et al., 2010). Importantly, several groups, including ours, have demonstrated that PD-associated parkin mutants are defective in supporting mitophagy due to distinct problem at recognition, transportation or ubiquitination of impaired mitochondria (Lee et al., 2010; Matsuda et al., 2010), thereby implicating dysfunctional mitophagy in the development of parkin-related parkinsonism.

Given the pivotal role of parkin/PINK1 pathway in mitochondrial quality control, it is perhaps not surprising to note that deficiency in parkin or PINK1 function results in the accumulation of abnormal mitochondria in several parkin/PINK1-related PD models. This defect is perhaps most prominently observed in *Drosophila* parkin or PINK1 mutants, especially in their flight musculature, which is plagued by pronounced mitochondrial lesions and muscle degeneration (Greene et al., 2003; Clark et al., 2006; Park et al., 2006; Wang et al., 2007). Importantly, parkin over expression in pink1-/- flies significantly ameliorates all the mutant phenotypes, although the reverse, does not happen, i.e. pink1 over expression in parkin null flies does not compensate for the loss of parkin function. These results suggest that parkin acts in the same pathway but downstream of pink1 (Clark et al., 2006; Park et al., 2006). The hierachy is consistent with the proposed model of parkin/PINK1 pathway in the regulation of mitochondrial quality control, although parkin in this case can apparently do the job in the complete absence of pink1. Notably, several other studies also suggested that mitophagy can take place in a PINK1-deficient background (Dagda et al., 2009; Cui et al., 2010; Dagda et al., 2011). Conversely, Seibler and colleagues found PINK1 to be essential for parkin-mediated mitophagy. They demonstrated that parkin recruitment to depolarized mitochondria is impaired in human dopaminergic neurons derived via the induced pluripotent stem cells route from PINK1-related PD patients, a defect that can be rescued by the re-introduction of wild-type PINK1 into PINK1-deficient neurons (Seibler et al., 2011).

As with the case with virtually all the biological models initially proposed, the parkin/PINK1 mitophagy model is currently less than perfect and clearly needs be continually updated with each new piece of significant data. The relevance of mitophagy to sporadic PD is also debatable, although we and others have previously shown that parkin dysfunction (presumably triggering mitophagy deficiency) may also underlie the pathogenesis of sporadic PD (Pawlyk et al., 2003; LaVoie et al., 2005; Wang et al., 2005a). Perhaps one of most challenging tasks at hand is to demonstrate unequivocally that mitophagy impairment, instead of a generalized impairment in the autophagy process, contributes directly to neurodegeneration *in vivo*. This would require the genetic differentiation of targeted components that are exclusively involved in mitophagy. Currently, key components of mitophagy and autophagy tend to overlap. Even parkin appear to subserve both types of autophagy processes (and more). Thus, although mitochondrial quality control is invariably important for neuronal survival, whether failure in the removal of damaged mitochondria is in itself a driver of disease pathogenesis or is a consequence of a progressive and general decline in autophagy function in the PD brain remains to be clarified.

6. Autophagy induction as therapeutic strategy for PD?

If failure in autophagy function were to underlie PD pathogenesis, it follows intuitively that stimulation of autophagy in the PD brain might be beneficial for the patient. Indeed, work from Rubinsztein lab and others have demonstrated that autophagy enhancement promotes beneficial outcomes in several experimental models of PD, supporting that such an approach could represent a viable therapeutic strategy (Rubinsztein et al., 2012).

Notably, most neurodegenerative disease-associated proteins, including α-synuclein, that are prone to aggregation are substrates of autophagy. Accordingly, pharmacological or genetic enhancement of autophagy can in theory help remove these aggregation-prone proteins and concomitantly reduce their associated toxicity. Rapamycin, an inducer of mTOR (mammalian Target of Rapamycin), is widely established to be a potent autophagy inducer. Expectedly, rapamycin treatment of cellular or animal models of α-synucleinopathies reduces the levels of both soluble and aggregated species of α-synuclein in an autophagy-dependent manner (Crews et al., 2010). Similarly, trehalose also accelerates the clearance of α-synuclein by means of its ability to induce autophagy, albeit in an mTOR-independent manner (Sarkar et al., 2007). Further, trehalose-treated cells are protected against subsequent pro-apoptotic insults. Together, trehalose and rapamycin exert an additive effect in the clearance of aggregate-prone proteins (Sarkar et al., 2007). Perhaps unsurprisingly, rapamycin can also rescue failed mitophagy in parkin deficient cells and result in improved mitochondrial function (Siddiqui et al., 2012), suggesting that generalized autophagy activation can help clean up all the cellular "cobwebs" be it protein aggregates or damaged organelles. More recently, Steele and colleagues showed that latrepirdine, a neuroactive compound associated with enhanced cognition and neuroprotection, also stimulates the degradation of α-synuclein and concomitantly protects against α-synuclein-induced toxicity in 3 model systems: yeast, differentiated SH-SY5Y cells and wild type mouse (Steele et al., 2012). The beneficial effects of latrepirdine again appear to be related to autophagy induction, as evident by the elevation of several autophagy markers in mouse brain following chronic administration of the compound. Using a genetic approach, Spencer and colleagues demonstrated via lentivirus-mediated gene transfer of beclin 1, a key promoter of autophagy, that genetic enhancement of autophagy in α-synuclein overexpressing mice ameliorates the synaptic and dendritic pathology in these transgenic animals and reduces the accumulation of the protein *in vivo* (Spencer et al., 2009). Taken together, these studies support the therapeutic applications of autophagy induction in PD, particularly in preventing the accumulation of α-synuclein.

Notwithstanding the above promising findings regarding the protective effects of autophagy induction, it is important to recognize that autophagy induction is a "double-edge sword" that can cut both ways, i.e. being protective or pro-death under different conditions. One therefore have to consider this caveat in considering autophagy induction as a therapeutic strategy for PD. Notably, the parkinsonian neurotoxin MPP+ that induces selective loss of dopaminergic neurons has been demonstrated by several groups to activate autophagy (Zhu et al., 2007; Xilouri et al., 2009; Wong et al., 2011), a process that appears to act through the dephosphorylation of LC3 (which enhances its recruitment into autophagosomes) (Cherra et al., 2010) and/or CDK5-mediated phosphorylation of endophilin B1 (which promotes its dimerization and recruitment of the UVRAG/Beclin 1 complex to induce autophagy) (Wong et al., 2011). In this case, autophagy induction is apparently harmful to dopaminergic neurons. Moreover, stimulation of autophagy also contributes to neuronal death induced by overexpression of α-synuclein (Xilouri et al., 2009). Conversely, inhibition of autophagy pharmacologically with 3-methylalanine (3-MA) or genetically via Atg5 or Atg12 gene silencing significantly attenuates neuronal loss associated with MPP + treatment or mutant α-synuclein expression, as is the case with knockdown of CDK5 or endophilin B1 (Wong et al., 2011). Along these lines, we found that mutant α-synuclein-associated toxicity is aggravated by the accumulation of iron, which act together to trigger autophagic cell death. The toxicity that α-synuclein-iron elicits can be ameliorated by pharmacological inhibition of autophagy (Chew et al., 2011). Interestingly, autophagy activation elicited by mutant α-synuclein overexpression can also result in excessive mitophagy and thereby unintended loss of mitochondria, which in turn promotes bioenergetics deficit and neuronal degeneration (Choubey et al., 2011). Further supporting a "pathological" role for autophagy, loss of DJ-1 function associated with recessive parkinsonism has been found to increase (instead of decrease) autophagic flux, although it is currently unclear how this relates to neuronal death in the context of DJ-1 deficiency (Irrcher et al., 2010).

Finally, mutations in LRRK2, which currently represent the most prevalent genetic contributor to PD, are also implicated in aberrant autophagy induction. For example, transgenic mice expressing disease-associated LRRK2 mutants (R1441C and G2019S) frequently exhibit increased incidence of autophagic vacuoles in their brain (Ramonet et al., 2011). Similarly, cells expressing G2019S LRRK2 mutant show increase autophagosome content and autophagy-dependent shortening of neurites (Plowey et al., 2008). Conversely, ablation of LRRK2 in mice promotes impairment of the autophagy pathway as evident by the accumulation of p62, lipofuscin granules, ubiquitinated proteins and α-synuclein-positive inclusions in their kidneys (Tong et al., 2010). The relationship between LRRK2 and autophagy is however complicated. For example, Gomez-Suaga and colleagues have recently demonstrated that LRRK2-induced accumulation of autophagosome is related to the ability of the kinase to activate a calcium-dependent protein kinase kinase-beta (CaMKK-beta)/adenosine monophosphate (AMP)-activated protein kinase (AMPK) pathway via modulation of NAADP-dependent Ca2+ channel on lysosomal membrane (Gomez-Suaga et al., 2012). However, they also detected at the same time a reduction in the acidification of lysosomes that can compromise autophagosome turnover and thereby autophagy (Gomez-Suaga et al., 2012), suggesting that autophagy is actually impaired rather than activated in LRRK2-expressing cells. Consistent with this, another study revealed that the expression of LRRK2 R1441C mutant leads to impaired autophagic balance that is characterized by AV accumulation containing incompletely degraded materials and increased levels of p62 (Alegre-Abarrategui et al., 2009).

http://dx.doi.org/10.5772/54927

Accordingly, siRNA-mediated knockdown of LRRK2 expression results in increased autophagic activity and prevented cell death caused by inhibition of autophagy in starvation conditions. Thus, the precise role of autophagy in LRRK2-related parkinsonism is anybody's guess at this moment, begging again caution in the proposed use of autophagy inducers as a therapeutic recourse.

In a related development, we recently found that disease-associated LRRK2 G2019S mutant can trigger marked mitochondrial abnormalities when overexpressed in *Drosophila*, a phenotype that can be rescued by parkin co-expression (Ng et al., 2012). Given the role of parkin in promoting mitophagy, it is tempting to speculate that the LRRK2 mutant may retard the clearance of damaged mitochondria via mitophagy in the absence of parkin overexpression. Indeed, the mitochondrial phenotype LRRK2 G2019S mutant induces in the flight muscle is reminiscent of that brought about by the loss of parkin function. Alternatively, this could also be a result of LRRK2-induced impairment in autophagy in general. Importantly, we further found that pharmacological or genetic activation of AMPK can effectively compensate for parkin deficiency to bring about a significant suppression of dopaminergic and mitochondrial dysfunction in mutant LRRK2 flies (Ng et al., 2012). Our results suggest a neuroprotective role for AMPK that might be related to mitophagy/autophagy modulation. AMPK is an evolutionarily conserved cellular energy sensor that is activated by ATP depletion or glucose starvation (Hardie, 2011). When activated, AMPK switches the cell from an anabolic to a catabolic mode and in so doing, helps to regulate diverse cellular processes that impact on cellular energy demands. Interestingly, like parkin, AMPK can also regulate mitophagy and also autophagy through its ability to phosphorylate the autophagy initiator ATG1 (Egan et al., 2011; Kim et al., 2011). Lending relevance to our findings, a recent report demonstrated that AMPK is activated in mice treated with MPTP and that inhibition of AMPK function by compound C enhances MPP(+)-induced cell death (Choi et al., 2010). More recently, a PD cohort-based study revealed that Metformin-inclusive sulfonylurea therapy reduces the risk for the disease occurring with Type 2 diabetes in a Taiwanese population (Wahlqvist et al., 2012). Metformin is a direct activator of AMPK. Together, these findings suggest that AMPK activation may protect against the development of PD, presumably via its ability to maintain energy balance via the modulation of autophagy as well as a range of other cellular processes. Given that caveats associated with direct autophagy induction, and that excessive autophagy can result in energy crisis especially in the aged brain, perhaps AMPK activation, through its ability to maintain both protein and energy homeostasis, would represent a better approach than direct autophagy induction as a therapeutic strategy for PD.

7. Conclusion

In essence, the case for autophagy dysfunction as a contributor to the pathogenesis of PD is rather compelling. As evident from the above discussion, virtually all the major PD-associated gene products have some direct or indirect relationship with the autophagy-lysosome axis. What is less clear is whether autophagy induction is neuroprotective or is a key driver of neurodegeneration. One can envisage that the activation of autophagy may be beneficial in the short term (particularly when the induction is transient and timely), but deleterious when it is becomes chronic or excessive. Finding the tipping autophagy threshold point between neuroprotection and neurodegeneration would therefore be an important endeavour, the clarification of which has important implications for the future development of autophagy-related therapeutics for the PD patient.

Acknowledgements

This work was supported by grants from Singapore Millennium Foundation, National Medical Research Council and A*STAR Biomedical Research Council (LKL). G.L. is supported by a graduate scholarship from the Singapore Millennium Foundation.

Author details

Grace G.Y. Lim[1], Chengwu Zhang[2] and Kah-Leong Lim[1,2,3]

1 Department of Physiology, National University of Singapore, Singapore

2 National Neuroscience Institute, Singapore

3 Duke-NUS Graduate Medical School, Singapore

References

[1] Alegre-abarrategui, J, Christian, H, Lufino, M. M, Mutihac, R, Venda, L. L, Ansorge, O, & Wade-martins, R. (2009). LRRK2 regulates autophagic activity and localizes to specific membrane microdomains in a novel human genomic reporter cellular model. Hum Mol Genet , 18, 4022-4034.

[2] Anglade, P, Vyas, S, Javoy-agid, F, Herrero, M. T, Michel, P. P, Marquez, J, Mouatt-prigent, A, Ruberg, M, Hirsch, E. C, & Agid, Y. (1997). Apoptosis and autophagy in nigral neurons of patients with Parkinson's disease. Histol Histopathol , 12, 25-31.

[3] Becker, D, Richter, J, Tocilescu, M. A, Przedborski, S, & Voos, W. (2012). Pink1 kinase and its membrane potential (Deltapsi)-dependent cleavage product both localize to outer mitochondrial membrane by unique targeting mode. J Biol Chem , 287, 22969-22987.

[4] Bjorkoy, G, Lamark, T, Brech, A, Outzen, H, Perander, M, Overvatn, A, Stenmark, H, & Johansen, T. forms protein aggregates degraded by autophagy and has a protective effect on huntingtin-induced cell death. J Cell Biol , 171, 603-614.

[5] Braak, H. Del Tredici K, Rub U, de Vos RA, Jansen Steur EN, Braak E ((2003). Staging of brain pathology related to sporadic Parkinson's disease. Neurobiol Aging , 24, 197-211.

[6] Chan, N. C, Salazar, A. M, Pham, A. H, Sweredoski, M. J, Kolawa, N. J, Graham, R. L, Hess, S, & Chan, D. C. (2011). Broad activation of the ubiquitin-proteasome system by Parkin is critical for mitophagy. Hum Mol Genet , 20, 1726-1737.

[7] Cherra, S. J. rd, Kulich SM, Uechi G, Balasubramani M, Mountzouris J, Day BW, Chu CT ((2010). Regulation of the autophagy protein LC3 by phosphorylation. J Cell Biol , 190, 533-539.

[8] Chew, K. C, Ang, E. T, Tai, Y. K, Tsang, F, Lo, S. Q, Ong, E, Ong, W. Y, Shen, H. M, Lim, K. L, Dawson, V. L, Dawson, T. M, & Soong, T. W. (2011). Enhanced autophagy from chronic toxicity of iron and mutant A53T alpha-synuclein: implications for neuronal cell death in Parkinson disease. J Biol Chem , 286, 33380-33389.

[9] Choi, J. S, Park, C, & Jeong, J. W. (2010). AMP-activated protein kinase is activated in Parkinson's disease models mediated by 1-methyl-4-phenyl-1,2,3,6-tetrahydropyridine. Biochem Biophys Res Commun , 391, 147-151.

[10] Choubey, V, Safiulina, D, Vaarmann, A, Cagalinec, M, Wareski, P, Kuum, M, Zharkovsky, A, & Kaasik, A. induces neuronal death by increasing mitochondrial autophagy. J Biol Chem , 286, 10814-10824.

[11] Clark, I. E, Dodson, M. W, Jiang, C, Cao, J. H, Huh, J. R, Seol, J. H, Yoo, S. J, Hay, B. A, & Guo, M. (2006). Drosophila pink1 is required for mitochondrial function and interacts genetically with parkin. Nature , 441, 1162-1166.

[12] Conway, K. A, Lee, S. J, Rochet, J. C, Ding, T. T, Williamson, R. E, & Lansbury, P. T. Jr. ((2000). Acceleration of oligomerization, not fibrillization, is a shared property of both alpha-synuclein mutations linked to early-onset Parkinson's disease: implications for pathogenesis and therapy. Proc Natl Acad Sci U S A , 97, 571-576.

[13] Crews, L, Spencer, B, Desplats, P, Patrick, C, Paulino, A, Rockenstein, E, Hansen, L, Adame, A, Galasko, D, & Masliah, E. (2010). Selective molecular alterations in the autophagy pathway in patients with Lewy body disease and in models of alpha-synucleinopathy. PLoS One 5:e9313.

[14] Cuervo, A. M, Stefanis, L, Fredenburg, R, Lansbury, P. T, & Sulzer, D. (2004). Impaired degradation of mutant alpha-synuclein by chaperone-mediated autophagy. Science , 305, 1292-1295.

[15] Cui, M, Tang, X, Christian, W. V, Yoon, Y, & Tieu, K. (2010). Perturbations in mitochondrial dynamics induced by human mutant PINK1 can be rescued by the mitochondrial division inhibitor mdivi-1. J Biol Chem , 285, 11740-11752.

[16] Dagda, R. K, & Cherra, S. J. rd, Kulich SM, Tandon A, Park D, Chu CT ((2009). Loss of PINK1 function promotes mitophagy through effects on oxidative stress and mitochondrial fission. J Biol Chem , 284, 13843-13855.

[17] Dagda, R. K, Gusdon, A. M, Pien, I, Strack, S, Green, S, Li, C, Van Houten, B, & Cherra, S. J. rd, Chu CT ((2011). Mitochondrially localized PKA reverses mitochondrial pathology and dysfunction in a cellular model of Parkinson's disease. Cell Death Differ , 18, 1914-1923.

[18] Dauer, W, & Przedborski, S. (2003). Parkinson's disease: mechanisms and models. Neuron , 39, 889-909.

[19] Ding, W. X, Ni, H. M, Li, M, Liao, Y, Chen, X, Stolz, D. B, & Dorn, G. W. nd, Yin XM ((2010). Nix is critical to two distinct phases of mitophagy, reactive oxygen species-mediated autophagy induction and Parkin-ubiquitin-mitochondrial priming. J Biol Chem 285:27879-27890., 62.

[20] Dorsey, E. R, Constantinescu, R, Thompson, J. P, Biglan, K. M, Holloway, R. G, Kieburtz, K, Marshall, F. J, Ravina, B. M, Schifitto, G, Siderowf, A, & Tanner, C. M. (2007). Projected number of people with Parkinson disease in the most populous nations, 2005 through 2030. Neurology , 68, 384-386.

[21] Egan, D. F, Shackelford, D. B, Mihaylova, M. M, Gelino, S, Kohnz, R. A, Mair, W, Vasquez, D. S, Joshi, A, Gwinn, D. M, Taylor, R, Asara, J. M, Fitzpatrick, J, Dillin, A, Viollet, B, Kundu, M, Hansen, M, & Shaw, R. J. (2011). Phosphorylation of ULK1 (hATG1) by AMP-activated protein kinase connects energy sensing to mitophagy. Science , 331, 456-461.

[22] Fortun, J, & Dunn, W. A. Jr., Joy S, Li J, Notterpek L ((2003). Emerging role for autophagy in the removal of aggresomes in Schwann cells. J Neurosci , 23, 10672-10680.

[23] Friedman, L. G, Lachenmayer, M. L, Wang, J, He, L, Poulose, S. M, Komatsu, M, Holstein, G. R, & Yue, Z. (2012). Disrupted autophagy leads to dopaminergic axon and dendrite degeneration and promotes presynaptic accumulation of alpha-synuclein and LRRK2 in the brain. J Neurosci , 32, 7585-7593.

[24] Geisler, S, Holmstrom, K. M, Skujat, D, Fiesel, F. C, Rothfuss, O. C, Kahle, P. J, & Springer, W. (2010). PINK1/Parkin-mediated mitophagy is dependent on VDAC1 and SQSTM1. Nat Cell Biol 12:119-131., 62.

[25] Giasson, B. I, Uryu, K, Trojanowski, J. Q, & Lee, V. M. (1999). Mutant and wild type human alpha-synucleins assemble into elongated filaments with distinct morphologies in vitro. J Biol Chem , 274, 7619-7622.

[26] Gomez-suaga, P, Luzon-toro, B, Churamani, D, Zhang, L, Bloor-young, D, Patel, S, Woodman, P. G, Churchill, G. C, & Hilfiker, S. (2012). Leucine-rich repeat kinase 2 regulates autophagy through a calcium-dependent pathway involving NAADP. Hum Mol Genet , 21, 511-525.

[27] Greene, A. W, Grenier, K, Aguileta, M. A, Muise, S, Farazifard, R, Haque, M. E, Mcbride, H. M, & Park, D. S. Fon EA ((2012). Mitochondrial processing peptidase regulates PINK1 processing, import and Parkin recruitment. EMBO Rep , 13, 378-385.

[28] Greene, J. C, Whitworth, A. J, Kuo, I, Andrews, L. A, Feany, M. B, & Pallanck, L. J. (2003). Mitochondrial pathology and apoptotic muscle degeneration in Drosophila parkin mutants. Proc Natl Acad Sci U S A , 100, 4078-4083.

[29] Hara, T, Nakamura, K, Matsui, M, Yamamoto, A, Nakahara, Y, Suzuki-migishima, R, Yokoyama, M, Mishima, K, Saito, I, Okano, H, & Mizushima, N. (2006). Suppression of basal autophagy in neural cells causes neurodegenerative disease in mice. Nature , 441, 885-889.

[30] Hardie, D. G. (2011). AMP-activated protein kinase: an energy sensor that regulates all aspects of cell function. Genes Dev , 25, 1895-1908.

[31] Hayashi, S, Wakabayashi, K, Ishikawa, A, Nagai, H, Saito, M, Maruyama, M, Takahashi, T, Ozawa, T, Tsuji, S, & Takahashi, H. (2000). An autopsy case of autosomal-recessive juvenile parkinsonism with a homozygous exon 4 deletion in the parkin gene. Mov Disord , 15, 884-888.

[32] Irrcher, I, et al. (2010). Loss of the Parkinson's disease-linked gene DJ-1 perturbs mitochondrial dynamics. Hum Mol Genet , 19, 3734-3746.

[33] Itakura, E, Kishi-itakura, C, Koyama-honda, I, & Mizushima, N. (2012). Structures containing Atg9A and the ULK1 complex independently target depolarized mitochondria at initial stages of Parkin-mediated mitophagy. J Cell Sci , 125, 1488-1499.

[34] Iwata, A, Riley, B. E, Johnston, J. A, & Kopito, R. R. and microtubules are required for autophagic degradation of aggregated huntingtin. J Biol Chem , 280, 40282-40292.

[35] Johnston, J. A, Ward, C. L, & Kopito, R. R. (1998). Aggresomes: a cellular response to misfolded proteins. J Cell Biol , 143, 1883-1898.

[36] Kawaguchi, Y, Kovacs, J. J, Mclaurin, A, Vance, J. M, Ito, A, & Yao, T. P. (2003). The deacetylase HDAC6 regulates aggresome formation and cell viability in response to misfolded protein stress. Cell , 115, 727-738.

[37] Keeney, P. M, Xie, J, Capaldi, R. A, & Bennett, J. P. Jr. ((2006). Parkinson's disease brain mitochondrial complex I has oxidatively damaged subunits and is functionally impaired and misassembled. J Neurosci , 26, 5256-5264.

[38] Kim, J, Kundu, M, Viollet, B, & Guan, K. L. (2011). AMPK and mTOR regulate autophagy through direct phosphorylation of Ulk1. Nat Cell Biol , 13, 132-141.

[39] Kirkin, V, Lamark, T, Sou, Y. S, Bjorkoy, G, Nunn, J. L, Bruun, J. A, Shvets, E, Mcewan, D. G, Clausen, T. H, Wild, P, Bilusic, I, Theurillat, J. P, Overvatn, A, Ishii, T, Elazar, Z, Komatsu, M, Dikic, I, & Johansen, T. (2009). A role for NBR1 in autophagosomal degradation of ubiquitinated substrates. Mol Cell , 33, 505-516.

[40] Klionsky, D. J, et al. (2011). A comprehensive glossary of autophagy-related molecules and processes (2nd edition). Autophagy , 7, 1273-1294.

[41] Komatsu, M, Waguri, S, Chiba, T, Murata, S, Iwata, J, Tanida, I, Ueno, T, Koike, M, Uchiyama, Y, Kominami, E, & Tanaka, K. (2006). Loss of autophagy in the central nervous system causes neurodegeneration in mice. Nature , 441, 880-884.

[42] LaVoie MJOstaszewski BL, Weihofen A, Schlossmacher MG, Selkoe DJ ((2005). Dopamine covalently modifies and functionally inactivates parkin. Nat Med , 11, 1214-1221.

[43] Lee, J. Y, Nagano, Y, Taylor, J. P, Lim, K. L, & Yao, T. P. (2010). Disease-causing mutations in Parkin impair mitochondrial ubiquitination, aggregation, and HDAC6-dependent mitophagy. J Cell Biol , 189, 671-679.

[44] Lemasters, J. J. (2005). Selective mitochondrial autophagy, or mitophagy, as a targeted defense against oxidative stress, mitochondrial dysfunction, and aging. Rejuvenation Res , 8, 3-5.

[45] Lewy, F. H. (1912). Paralysis agitans. I. Pathologische Anatomie Ed. M Lewandowski:, 920-933.

[46] Liberek, K, Lewandowska, A, & Zietkiewicz, S. (2008). Chaperones in control of protein disaggregation. Embo J , 27, 328-335.

[47] Lim, K. L, & Ng, C. H. (2009). Genetic models of Parkinson disease. Biochim Biophys Acta , 1792, 604-615.

[48] Lim, K. L, Dawson, V. L, & Dawson, T. M. (2006). Parkin-mediated lysine 63-linked polyubiquitination: a link to protein inclusions formation in Parkinson's and other conformational diseases? Neurobiol Aging , 27, 524-529.

[49] Lim, K. L, Chew, K. C, Tan, J. M, Wang, C, Chung, K. K, Zhang, Y, Tanaka, Y, Smith, W, Engelender, S, Ross, C. A, Dawson, V. L, & Dawson, T. M. (2005). Parkin mediates nonclassical, proteasomal-independent ubiquitination of synphilin-1: implications for Lewy body formation. J Neurosci , 25, 2002-2009.

[50] Mak, S. K, Mccormack, A. L, Manning-bog, A. B, & Cuervo, A. M. Di Monte DA ((2010). Lysosomal degradation of alpha-synuclein in vivo. J Biol Chem , 285, 13621-13629.

[51] Malkus, K. A, & Ischiropoulos, H. (2012). Regional deficiencies in chaperone-mediated autophagy underlie alpha-synuclein aggregation and neurodegeneration. Neurobiol Dis , 46, 732-744.

[52] Manning-bog, A. B, Mccormack, A. L, Li, J, Uversky, V. N, & Fink, A. L. Di Monte DA ((2002). The herbicide paraquat causes up-regulation and aggregation of alpha-synuclein in mice: paraquat and alpha-synuclein. J Biol Chem , 277, 1641-1644.

[53] Martin, I, Dawson, V. L, & Dawson, T. M. (2011). Recent advances in the genetics of Parkinson's disease. Annu Rev Genomics Hum Genet , 12, 301-325.

[54] Martinez-vicente, M, Talloczy, Z, Kaushik, S, Massey, A. C, Mazzulli, J, Mosharov, E. V, Hodara, R, Fredenburg, R, Wu, D. C, Follenzi, A, Dauer, W, Przedborski, S, Ischiropoulos, H, Lansbury, P. T, Sulzer, D, & Cuervo, A. M. (2008). Dopamine-modified alpha-synuclein blocks chaperone-mediated autophagy. J Clin Invest , 118, 777-788.

[55] Matsuda, N, Sato, S, Shiba, K, Okatsu, K, Saisho, K, Gautier, C. A, Sou, Y. S, Saiki, S, Kawajiri, S, Sato, F, Kimura, M, Komatsu, M, Hattori, N, & Tanaka, K. (2010). PINK1 stabilized by mitochondrial depolarization recruits Parkin to damaged mitochondria and activates latent Parkin for mitophagy. J Cell Biol , 189, 211-221.

[56] Mori, H, Kondo, T, Yokochi, M, Matsumine, H, Nakagawa-hattori, Y, Miyake, T, Suda, K, & Mizuno, Y. (1998). Pathologic and biochemical studies of juvenile parkinsonism linked to chromosome 6q. Neurology , 51, 890-892.

[57] Narendra, D, Tanaka, A, Suen, D. F, & Youle, R. J. (2008). Parkin is recruited selectively to impaired mitochondria and promotes their autophagy. J Cell Biol , 183, 795-803.

[58] Narhi, L, Wood, S. J, Steavenson, S, Jiang, Y, Wu, G. M, Anafi, D, Kaufman, S. A, Martin, F, Sitney, K, Denis, P, Louis, J. C, Wypych, J, Biere, A. L, & Citron, M. (1999). Both familial Parkinson's disease mutations accelerate alpha-synuclein aggregation. J Biol Chem , 274, 9843-9846.

[59] Ng, C. H, Guan, M. S, Koh, C, Ouyang, X, Yu, F, Tan, E. K, Neill, O, Zhang, S. P, Chung, X, & Lim, J. KL ((2012). AMP Kinase Activation Mitigates Dopaminergic Dysfunction and Mitochondrial Abnormalities in Drosophila Models of Parkinson's Disease. J Neurosci , 32, 14311-14317.

[60] Okatsu, K, Oka, T, Iguchi, M, Imamura, K, Kosako, H, Tani, N, Kimura, M, Go, E, Koyano, F, Funayama, M, Shiba-fukushima, K, Sato, S, Shimizu, H, Fukunaga, Y, Taniguchi, H, Komatsu, M, Hattori, N, Mihara, K, Tanaka, K, & Matsuda, N. (2012). PINK1 autophosphorylation upon membrane potential dissipation is essential for Parkin recruitment to damaged mitochondria. Nat Commun 3:1016.

[61] Olanow, C. W, Perl, D. P, Demartino, G. N, & Mcnaught, K. S. (2004). Lewy-body formation is an aggresome-related process: a hypothesis. Lancet Neurol , 3, 496-503.

[62] Olzmann, J. A, Li, L, Chudaev, M. V, Chen, J, Perez, F. A, Palmiter, R. D, & Chin, L. S. linked polyubiquitination targets misfolded DJ-1 to aggresomes via binding to HDAC6. J Cell Biol , 178, 1025-1038.

[63] Opazo, F, Krenz, A, Heermann, S, Schulz, J. B, & Falkenburger, B. H. (2008). Accumulation and clearance of alpha-synuclein aggregates demonstrated by time-lapse imaging. J Neurochem , 106, 529-540.

[64] Pankiv, S, Clausen, T. H, Lamark, T, Brech, A, Bruun, J. A, Outzen, H, Overvatn, A, Bjorkoy, G, & Johansen, T. Binds Directly to Atg8/LC3 to Facilitate Degradation of Ubiquitinated Protein Aggregates by Autophagy. J Biol Chem , 282, 24131-24145.

[65] Park, J, Lee, S. B, Lee, S, Kim, Y, Song, S, Kim, S, Bae, E, Kim, J, Shong, M, Kim, J. M, & Chung, J. (2006). Mitochondrial dysfunction in Drosophila PINK1 mutants is complemented by parkin. Nature , 441, 1157-1161.

[66] Pawlyk, A. C, Giasson, B. I, Sampathu, D. M, Perez, F. A, Lim, K. L, Dawson, V. L, Dawson, T. M, Palmiter, R. D, Trojanowski, J. Q, & Lee, V. M. (2003). Novel monoclonal antibodies demonstrate biochemical variation of brain parkin with age. J Biol Chem , 278, 48120-48128.

[67] Peng, J, Schwartz, D, Elias, J. E, Thoreen, C. C, Cheng, D, Marsischky, G, Roelofs, J, Finley, D, & Gygi, S. P. (2003). A proteomics approach to understanding protein ubiquitination. Nat Biotechnol , 21, 921-926.

[68] Pickart, C. M. (2000). Ubiquitin in chains. Trends Biochem Sci , 25, 544-548.

[69] Pickart, C. M, & Cohen, R. E. (2004). Proteasomes and their kin: proteases in the machine age. Nat Rev Mol Cell Biol , 5, 177-187.

[70] Plowey, E. D, & Cherra, S. J. rd, Liu YJ, Chu CT ((2008). Role of autophagy in G2019S-LRRK2-associated neurite shortening in differentiated SH-SY5Y cells. J Neurochem , 105, 1048-1056.

[71] Poole, A. C, Thomas, R. E, Yu, S, Vincow, E. S, & Pallanck, L. (2010). The mitochondrial fusion-promoting factor mitofusin is a substrate of the PINK1/parkin pathway. PLoS One 5:e10054.

[72] Ramirez, A, Heimbach, A, Grundemann, J, Stiller, B, Hampshire, D, Cid, L. P, Goebel, I, Mubaidin, A. F, Wriekat, A. L, Roeper, J, Al-din, A, Hillmer, A. M, Karsak, M, Liss, B, Woods, C. G, Behrens, M. I, & Kubisch, C. (2006). Hereditary parkinsonism with dementia is caused by mutations in ATP13A2, encoding a lysosomal type 5 P-type ATPase. Nat Genet , 38, 1184-1191.

[73] Ramonet, D, et al. (2011). Dopaminergic neuronal loss, reduced neurite complexity and autophagic abnormalities in transgenic mice expressing G2019S mutant LRRK2. PLoS One 6:e18568.

[74] Rappley, I, Gitler, A. D, & Selvy, P. E. LaVoie MJ, Levy BD, Brown HA, Lindquist S, Selkoe DJ ((2009). Evidence that alpha-synuclein does not inhibit phospholipase D. Biochemistry , 48, 1077-1083.

[75] Rubinsztein, D. C, Codogno, P, & Levine, B. (2012). Autophagy modulation as a potential therapeutic target for diverse diseases. Nat Rev Drug Discov , 11, 709-730.

[76] Sarkar, S, Davies, J. E, Huang, Z, Tunnacliffe, A, & Rubinsztein, D. C. (2007). Trehalose, a novel mTOR-independent autophagy enhancer, accelerates the clearance of mutant huntingtin and alpha-synuclein. J Biol Chem , 282, 5641-5652.

[77] Schapira, A. H, Cooper, J. M, Dexter, D, Jenner, P, Clark, J. B, & Marsden, C. D. (1989). Mitochondrial complex I deficiency in Parkinson's disease. Lancet 1:1269.

[78] Seibler, P, Graziotto, J, Jeong, H, Simunovic, F, Klein, C, & Krainc, D. (2011). Mitochondrial Parkin Recruitment Is Impaired in Neurons Derived from Mutant PINK1 Induced Pluripotent Stem Cells. J Neurosci , 31, 5970-5976.

[79] Sherer, T. B, Kim, J. H, Betarbet, R, & Greenamyre, J. T. (2003). Subcutaneous rotenone exposure causes highly selective dopaminergic degeneration and alpha-synuclein aggregation. Exp Neurol , 179, 9-16.

[80] Shults, C. W, Haas, R. H, Passov, D, & Beal, M. F. levels correlate with the activities of complexes I and II/III in mitochondria from parkinsonian and nonparkinsonian subjects. Ann Neurol , 42, 261-264.

[81] Siddiqui, A, Hanson, I, & Andersen, J. K. (2012). Mao-B elevation decreases parkin's ability to efficiently clear damaged mitochondria: protective effects of rapamycin. Free Radic Res , 46, 1011-1018.

[82] Spencer, B, Potkar, R, Trejo, M, Rockenstein, E, Patrick, C, Gindi, R, Adame, A, Wyss-coray, T, & Masliah, E. (2009). Beclin 1 gene transfer activates autophagy and ameliorates the neurodegenerative pathology in alpha-synuclein models of Parkinson's and Lewy body diseases. J Neurosci , 29, 13578-13588.

[83] Steele, J. W, et al. (2012). Latrepirdine stimulates autophagy and reduces accumulation of alpha-synuclein in cells and in mouse brain. Mol Psychiatry.

[84] Sterky, F. H, Lee, S, Wibom, R, Olson, L, & Larsson, N. G. (2011). Impaired mitochondrial transport and Parkin-independent degeneration of respiratory chain-deficient dopamine neurons in vivo. Proc Natl Acad Sci U S A , 108, 12937-12942.

[85] Takahashi, H, Ohama, E, Suzuki, S, Horikawa, Y, Ishikawa, A, Morita, T, Tsuji, S, & Ikuta, F. (1994). Familial juvenile parkinsonism: clinical and pathologic study in a family. Neurology , 44, 437-441.

[86] Tan, J. M, Wong, E. S, & Lim, K. L. (2009). Protein misfolding and aggregation in Parkinson's disease. Antioxid Redox Signal , 11, 2119-2134.

[87] Tan, J. M, Wong, E. S, Dawson, V. L, Dawson, T. M, & Lim, K. L. linked polyubiquitin potentially partners with to promote the clearance of protein inclusions by autophagy. Autophagy 4:251-253., 62.

[88] Tan, J. M, Wong, E. S, Kirkpatrick, D. S, Pletnikova, O, Ko, H. S, Tay, S. P, Ho, M. W, Troncoso, J, Gygi, S. P, Lee, M. K, Dawson, V. L, Dawson, T. M, & Lim, K. L. linked

ubiquitination promotes the formation and autophagic clearance of protein inclusions associated with neurodegenerative diseases. Hum Mol Genet , 17, 431-439.

[89] Tong, Y, Yamaguchi, H, Giaime, E, Boyle, S, Kopan, R, & Kelleher, R. J. rd, Shen J ((2010). Loss of leucine-rich repeat kinase 2 causes impairment of protein degradation pathways, accumulation of alpha-synuclein, and apoptotic cell death in aged mice. Proc Natl Acad Sci U S A , 107, 9879-9884.

[90] Usenovic, M, Tresse, E, Mazzulli, J. R, Taylor, J. P, & Krainc, D. (2012). Deficiency of ATP13A2 leads to lysosomal dysfunction, alpha-synuclein accumulation, and neurotoxicity. J Neurosci , 32, 4240-4246.

[91] Uversky, V. N. (2007). Neuropathology, biochemistry, and biophysics of alpha-synuclein aggregation. J Neurochem , 103, 17-37.

[92] Uversky, V. N, Li, J, Bower, K, & Fink, A. L. (2002). Synergistic effects of pesticides and metals on the fibrillation of alpha-synuclein: implications for Parkinson's disease. Neurotoxicology , 23, 527-536.

[93] Wahlqvist, M. L, Lee, M. S, Hsu, C. C, Chuang, S. Y, Lee, J. T, & Tsai, H. N. (2012). Metformin-inclusive sulfonylurea therapy reduces the risk of Parkinson's disease occurring with Type 2 diabetes in a Taiwanese population cohort. Parkinsonism Relat Disord.

[94] Wakabayashi, K, Tanji, K, Odagiri, S, Miki, Y, Mori, F, & Takahashi, H. (2012). The Lewy Body in Parkinson's Disease and Related Neurodegenerative Disorders. Mol Neurobiol.

[95] Wang, C, Lu, R, Ouyang, X, Ho, M. W, Chia, W, Yu, F, & Lim, K. L. (2007). Drosophila overexpressing parkin R275W mutant exhibits dopaminergic neuron degeneration and mitochondrial abnormalities. J Neurosci , 27, 8563-8570.

[96] Wang, C, Ko, H. S, Thomas, B, Tsang, F, Chew, K. C, Tay, S. P, Ho, M. W, Lim, T. M, Soong, T. W, Pletnikova, O, Troncoso, J, Dawson, V. L, Dawson, T. M, & Lim, K. L. alterations in parkin solubility promote parkin aggregation and compromise parkin's protective function. Hum Mol Genet , 14, 3885-3897.

[97] Webb, J. L, Ravikumar, B, Atkins, J, Skepper, J. N, & Rubinsztein, D. C. (2003). Alpha-Synuclein is degraded by both autophagy and the proteasome. J Biol Chem , 278, 25009-25013.

[98] Winslow, A. R, Chen, C. W, Corrochano, S, Acevedo-arozena, A, Gordon, D. E, Peden, A. A, Lichtenberg, M, Menzies, F. M, Ravikumar, B, Imarisio, S, Brown, S, Kane, O, Rubinsztein, C. J, & Alpha-synuclein, D. C. impairs macroautophagy: implications for Parkinson's disease. J Cell Biol , 190, 1023-1037.

[99] Wong, A. S, Lee, R. H, Cheung, A. Y, Yeung, P. K, Chung, S. K, Cheung, Z. H, & Ip, N. Y. (2011). Cdk5-mediated phosphorylation of endophilin B1 is required for induced autophagy in models of Parkinson's disease. Nat Cell Biol , 13, 568-579.

[100] Wong, E, & Cuervo, A. M. (2010). Autophagy gone awry in neurodegenerative diseases. Nat Neurosci , 13, 805-811.

[101] Wong, E. S, Tan, J. M, Soong, W. E, Hussein, K, Nukina, N, Dawson, V. L, Dawson, T. M, Cuervo, A. M, & Lim, K. L. (2008). Autophagy-mediated clearance of aggresomes is not a universal phenomenon. Hum Mol Genet , 17, 2570-2582.

[102] Xilouri, M, Vogiatzi, T, Vekrellis, K, Park, D, & Stefanis, L. (2009). Abberant alpha-synuclein confers toxicity to neurons in part through inhibition of chaperone-mediated autophagy. PLoS One 4:e5515.

[103] Yamamoto, H, Kakuta, S, Watanabe, T. M, Kitamura, A, Sekito, T, Kondo-kakuta, C, Ichikawa, R, Kinjo, M, & Ohsumi, Y. (2012). Atg9 vesicles are an important membrane source during early steps of autophagosome formation. J Cell Biol , 198, 219-233.

[104] Yoshii, S. R, Kishi, C, Ishihara, N, & Mizushima, N. (2011). Parkin mediates proteasome-dependent protein degradation and rupture of the outer mitochondrial membrane. J Biol Chem.

[105] Zhu, J. H, Horbinski, C, Guo, F, Watkins, S, Uchiyama, Y, & Chu, C. T. (2007). Regulation of autophagy by extracellular signal-regulated protein kinases during 1-methyl-4-phenylpyridinium-induced cell death. Am J Pathol , 170, 75-86.

[106] Ziviani, E, Tao, R. N, & Whitworth, A. J. (2010). Drosophila parkin requires PINK1 for mitochondrial translocation and ubiquitinates mitofusin. Proc Natl Acad Sci U S A , 107, 5018-5023.

Chapter 18

Neuronal Autophagy and Prion Proteins

Audrey Ragagnin, Aurélie Guillemain,
Nancy J. Grant and Yannick J. R. Bailly

Additional information is available at the end of the chapter

http://dx.doi.org/10.5772/55646

1. Introduction

Protein and organelle turnover is essential to maintain cellular homeostasis and survival. Removing and recycling cell constituents is achieved by autophagy in all cells, including neurons. Autophagy contributes to various physiological processes, such as intracellular cleansing, cellular homeostasis, development, differentiation, longevity, tumor suppression, elimination of invading pathogens, antigen transport to the innate and adaptative immune systems, and counteracting endoplasmic reticulum (ER) stress and diseases characterized by the accumulation of protein aggregates [1]. Autophagy plays a role in a number of infectious and inflammatory diseases, in addition to protein unfolding and misfolding diseases that lead to neuron, muscle and liver degeneration or heart failure [2-4]. Whereas autophagy has long been defined as a form of non-apoptotic, programmed cell death [5], recent findings suggest that autophagy functions primarily to sustain cells, and only defects in autophagy lead to cell death [6].

2. Autophagy in neuronal physiology

Autophagy was initially identified and characterized in a few cell types including neurons. The distinct vacuoles which feature this self-eating process were originally described at the ultrastructural level [7, 8]. The formation of autophagosomes was associated with chromatolysis of a restricted neuroplasm area, free of organelles, but filled with various types of vesicles [9]. The function of autophagy in mature neurons , however, is still debated. In comparison with other organs, rodent brains show high expression levels of autophagy-related (Atg) proteins and low levels of autophagy markers such as autophagosome number and LC3-II. Indeed, even under prolonged fasting conditions, the number of autophagosomes does not increase in neurons, probably because their nutrient supply from peripheral organs is maintained [10]. However, mice with CNS defects in their autophagic machinery exhibit neurological deficits, such as abnormal limb-clasping reflexes, locomotor ataxia, and lack of motor coordination, in addition to a significant loss of large pyramidal neurons in the cerebral cortex and Purkinje cells (PCs) in the cerebellar cortex [11-13].

Macroautophagy (hereafter referred to as autophagy) is initiated when a portion of the cytoplasm is sequestrated within a double-membrane organelle, the so-called autophagosome [14]. The autophagic machinery has been extensively detailed at the molecular level in a number of reviews including several chapters of this book [14-17]. Atg and several non-Atg proteins have been identified as regulators of key steps leading to the degradation of cytosolic components in lysosomes: initiation and nucleation of phagophores, expansion of autophagosomes, maturation of autophagosomes into amphisomes/autolysosomes, and execution of autophagic degradation [18]. The endosomal sorting complex required for transport (ESCRT) pathway functions in the sorting of transmembrane proteins into the inner vesicles of multivesicular bodies (MVB) during endocytosis. Also it is conceivably an essential part of the basal autophagy process in neurons because ubiquitin- or p62/SQSTM1 (p62)-labelled inclusions and autophagosomes accumulate in neurons deficient in ESCRT components [19]. Increasing evidence suggests that autophagy is regulated in a cell type-specific manner and as such autophagy may serve a distinct function in neurons and may show difference in the molecular machinery underlying basal autophagy (Fig. 1).

2.1. Axonal autophagy

In neurons, autophagy occurs in axons, suggesting that it may be uniquely regulated in this compartment and specifically adapted to local axonal physiology [20]. In primary dorsal root ganglion neurons, autophagosomes initiate distally in nerve terminals and mature during their transport toward the cell soma [21, 22]. In non-neuronal cells, the autophagosomal membrane has multiple possible origins, including endocytosed plasma membrane (amphisome), ER, mitochondria, and trans-Golgi membranes [18, 21, 23- 32]. In contrast, the origin of autophagosomes in the axons is likely to be restricted to the sources of membrane available in the terminals such as smooth ER and plasma membrane [33, 34], excluding rough ER and Golgi dictyosomes.

As observed in ultrastructural studies of axotomized neurons [9, 35], analysis of Purkinje cell (PC) degeneration in lurcher mutant $GluR\delta2^{Lc}$ [36] demonstrates autophagosomes in their axonal compartment. In this study, an excitotoxic insult mediated by $GluR\delta2^{Lc}$ triggered a rapid and robust accumulation of autophagosomes in dystrophic axonal swellings providing evidence that autophagy is induced in dystrophic terminals and that autophagosome biogenesis occurs in axons [37]. The molecular scenario underlying the initiation of this axonal autophagy is unclear. Liang *et al.* [38] suggested that autophagy in lurcher PCs could be directly activated by an interaction between the postsynaptic $GluR\delta2^{Lc}$, nPIST and beclin 1 an important regulator of autophagy. Nevertheless, how activation of this postsynaptic signaling pathway in dendrites initiates autophagosome formation in axon compartments is uncertain. PC death that is correlated with early signs of autophagy appears to be independent of depolarization

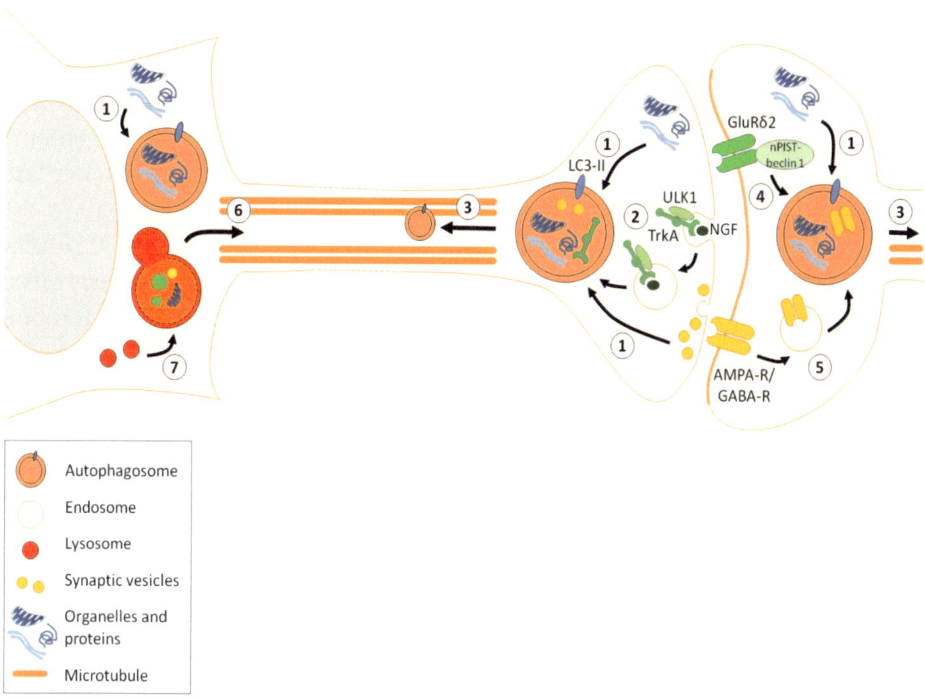

Figure 1. Physiological neuronal autophagy. Autophagy recycles synaptic components to sustain neuronal homeostasis and regulate synaptic plasticity and growth. **1.** Autophagic degradation of organelles, synaptic vesicles and proteins. **2.** ULK1-mediated autophagy of endocytosed NGF-bound TrkA receptors. **3.** Dynein-mediated retrograde transport of autophagosomes. **4.** GluRδ2 activation of beclin 1-dependent autophagy via nPIST. **5.** Targeting of postsynaptic receptors to autophagosomes via endocytosis. **6.** Kinesin-dependent anterograde transport of autophagosomes. **7.** Formation of autophagolysosomes by fusion of autophagosomes with lysosomes.

in the heteroallelic mutant Lurcher/hotfoot bearing only one copy of the lurcher allele and no wild-type GRID2 [39]. However, in the lurcher mutant bearing only one copy of the lurcher allele and one copy of the wild-type GRID2 allele, the leaky channel of GluRδ2^{Lc} depolarizes the neuron and this could transduce an electrical signal to the distal ends causing rapid physiological changes within axons. This effect combined with the local changes in postsynaptic signaling in dendrites may promote autophagosome biogenesis [37, 40].

2.2. Microtubule-dependent dynamics of neuronal autophagy

Previous data indicate that autophagy is a microtubule-dependent process. In cultured sympathetic neurons, autophagosomes formed in the distal ends of axon undergo retrograde transport along microtubules to the cell body where lysosomes that are necessary for the degradation step of autophagy are usually located [21]. Consistent with these observations, prominent retrograde transport of GFP-LC3-labelled autophagosomes has been observed in the axons of primary cerebellar granule cells [36]. In serum-deprived PC12 cells, autolysosomes formed by fusion of autophagosomes with lysosomes move in both anterograde and retrograde directions in neurites, and this trafficking requires microtubules [41]. Furthermore, both pharmacological and siRNA-based inhibition of directional microtubule motor proteins kinesin and dynein partially block respectively, anterograde and retrograde neuritic transport of autophagosomes, indicating that they participate in this transport. Recent observations in primary dorsal root ganglionic neurons support a maturation model in which autophagosomes initiate distally, engulfing mitochondria and ubiquitinated cargo, and move bidirectionally along microtubules driven by bound anterograde kinesin and retrograde dynein motors [22]. Fusion with late endosomes or lysosomes may then allow autophagosomes to escape from the early distal pool by robust retrograde dynein-driven motility. The involvement of the dynein-dynactin complex in the movement of autophagosomes along microtubules to lysosomes has also been demonstrated in non-neuronal cells [42]. Consistent with the formation of an autolysosomal compartment, autophagosomes increasingly acidify as they approach the cell soma, thereby fueling the catalysis of the degradation of their engulfed contents. Fully acidified autolysosomes undergo bidirectional motility suggesting reactivation of kinesin motors [22, 41].

The interaction of the autophagic membrane marker Atg8/LC3 with the microtubule-associated protein 1B (MAP1B) [43] implicates microtubule-dependent, axon-specific regulation of autophagosomes. Overexpression of MAP1B in non-neuronal cells reduces the number of LC3-associated autophagosomes without impairing autophagic degradation. The scarcity of LC3-labelled autophagosomes in CNS neurons under normal conditions may be explained by their high expression levels of MAP1B [10, 36]. By modifying microtubule function, the LC3-MAP1B interaction has been proposed to accelerate the delivery of LC3-autophagosomes to lysosomes, thereby promoting efficient autophagic turnover [37]. The exact mechanism underlying the involvement of microtubule in autophagosome formation, as well as targeting and fusion with lysosomes is open to debate [44, 45]. Based on (*i*) the absence of obvious changes in LC3 autophagosomes when they are associated with phosphorylated MAP1B-P, (*ii*) the elevated level of MAP1B-P bound to LC3 in dystrophic terminals containing a large number of autophagosomes [36] and, (*iii*) the conserved role of MAP1B-P in axonal growth and repair during development or injury (which implicates autophagy in remodelling axonal terminals during regeneration) [46], the interactions of LC3 with MAP1B and MAP1B-P have been proposed to represent a regulatory determinant of autophagy in axons under normal and pathological conditions respectively [37].

2.3. Functions of neuronal autophagy

Neurons, as non-dividing cells, are more sensitive to toxic components than dividing cells. Therefore, their survival and the maintenance of their specialized functions under physiological and pathological conditions is crucial requiring a tight quality control of cytoplasmic components and their degradation. Autophagy is believed to be of particular importance in the synaptic compartments of neurons where high energy requirements and protein turnover are necessary to sustain synaptic growth and activity. The CNS displays relatively low levels of autophagosomes under normal conditions, even after starvation, but requires an indispensable turnover of cytosolic contents by autophagy even in the absence of any disease-associated mutant proteins [10, 47, 48]. The scarcity of immature autophagosomes in neurons is likely to reflect a highly efficient autophagic degradation in the healthy brain. Accordingly, inhibition of autophagy causes neurodegeneration in mature neurons suggesting that autophagy may regulate neuronal homeostasis [11, 12]. For example, abnormal protein accumulation and eventual neurodegeneration are observed in the CNS of mice lacking the *atg5* or the *atg7* genes. This implies that basal autophagy is normally highly active and required for neuronal survival [11, 12]. The cardinal importance of autophagy in central neurons is further supported by recent studies showing a rapid accumulation of autophagosomes in cortical neurons when lysosomal degradation is inhibited. Thus, constitutive autophagy apparently plays an active role in neurons even under nutrient-rich conditions [49, 50].

2.3.1. Axonal homeostasis

Constitutive autophagy is probably essential for axonal homeostasis. Suppression of basal autophagy by either deleting an *atg* gene or inhibiting autophagic clearance in neurons disrupts axonal transport of vesicles destined for lysosomal degradation, and causes axonal swelling and dystrophy [11, 12, 37, 50]. For examples, Atg1/Unc-51 mutants in *C. elegans* show defaults in axonal membranes [51], and Unc-51.1, the murine homologue of Unc-51 is necessary for axonal extension, suggesting a possible role for these proteins in axonal membrane homeostasis [20, 52, 53]. In the cerebellum, neuron-specific deletion of FIP200, a protein implicated in autophagosome biogenesis, causes axon degeneration and neuronal death [13]. Altogether, these data suggest that autophagy is essential to maintain axonal structure and function through retrograde axonal transport [16]. The degree of vulnerability and the formation of intracellular inclusions vary significantly among the different types of CNS neurons in mutant brains deficient in Atg5 or Atg7 suggesting disparate intrinsic requirements for autophagy and relative levels of basal autophagy [20]. For example, while ubiquitinated inclusions are rare in the Atg5- or Atg7-deficient PCs, these cells are among the most susceptible neurons to *Atg 5/7* gene deletion [54, 55]. ULK1, the human homologue of Atg1 is incorporated into the active NGF-TrkA complex after its K-63 polyubiquitination and association with p62 [52, 56]. The subsequent interaction of ULK1 with endocytosis regulators allows trafficking of NGF-bound TrkA receptors into endocytic vesicles [57] providing a possible mechanism of crosstalk between autophagy and endocytosis. By fusing with autophagosomes, some membrane compartments, including endosomes, can be removed from axons and degraded in lysosomes. This process maintains the homeostasis of the axonal membrane networks and as such is essential for axonal physiology [20, 53].

Indeed, dysfunctional autophagy has been implicated in axonal dystrophy. Axonal swellings occur in autophagy-deficient mouse brains [11, 12] and genetic ablation of *Atg7* provokes cell-autonomous axonal dystrophy and degeneration, inferring that autophagy is crucial for membrane trafficking and turnover in axons [53]. In Atg5- or Atg7-deficient PCs, axonal endings exhibit an accumulation of abnormal organelles and membranous profiles much earlier than the somato-dendritic compartment [53, 54]. Axonal degeneration is increasingly believed to precede somatic death by a non-apoptotic auto-destructive mechanism [58, 59]. The "dying-back" progressive retrograde degeneration of the distal axon is a likely model of the chronic injury observed in neurodegenerative diseases [59]. NGF-deprivation induces autophagosome accumulation in the distal tips of neurites of PC12 cells, and knocking down *Atg7* or *beclin 1* expression delays neurite degeneration of NGF-deprived sympathetic neurons [60]. This suggests that overactive or deficient autophagy contributes to axonal degeneration in a dying-back manner due to the fragility of the axonal tips [20].

2.3.2. Dendritic autophagy

Early autophagosomes have also been observed in dendrites and the cell body of neurons suggesting that axon terminals are not be the only sites where neuronal autophagosomes form, and that autophagy may play a regulatory function in dendrites under physiological and pathological conditions [19]. Along this line, mTOR a key regulator of the autophagic pathway, modulates postsynaptic long-term potentiation and depression, suggesting that autophagy may critically control synaptic plasticity at the postsynaptic, dendritic compartment [61]. Further investigations are required to determine the specific roles of autophagy in dendrites and axons.

Since autophagosomes can fuse with endosomes and form amphisomes, there is a link between autophagy and endocytosis [62]. ESCRT proteins have recently been implicated in normal autophagy [19, 63, 64]. The endocytic pathways, in particular multi-vesicular bodies (MVBs) may serve as critical routes for autophagosomes to reach lysosomes, because defects in ESCRT function prevents fusion or maturation of autophagosomes. The ESCRT-MVB pathway could represent the primary, if not the only, route for delivering autophagosomes to lysosomes in some cell types [20]. In neurons, a large part of the endocytosed cargo merges with the autophagic pathway prior to being degraded by lysosomes [65]. Alterations in ESCRT function have also been linked to autophagy-deficiency in fronto-temporal dementia (FTD) and amyotrophic lateral sclerosis (ALS). In these cases, the particular vulnerability of the neurons appears to be associated with a dysfunction in the autophagosome-MVB pathway in the dendritic compartment [19].

2.3.3. Protein homeostasis

Neurons deficient in Atg5 or Atg7 exhibit an accumulation of polyubiquitinated proteins in inclusion bodies even though the proteasome function is normal, suggesting that basal autophagy prevents spontaneous protein aggregation and plays an essential role in protein clearance and homeostasis in neurons. Such a function is even more critical in neurons expressing disease-related proteins like the aggregate-prone mutant α-synuclein and poly-glutamine-containing proteins [66-68], although how autophagy selectively degrades these disease-related proteins is unclear. The ubiquitin-associated protein p62 is a likely candidate, providing a link between autophagy and selective protein degradation. Indeed, p62 binds numerous proteins through multiple protein-protein interacting motifs, including one for LC3 [55, 56] and the ubiquitin-associated C-terminal domain which binds ubiquitinated proteins. The relationship between p62 and autophagy is further supported by the observation that a marked accumulation of p62 and LC3 occurs only when lysosomes, but not proteasomes, are blocked. Furthermore, p62 protein levels are elevated in autophagy-deficient neurons [36, 55].

This argues that p62 is a specific substrate of autophagic degradation rather than a molecule involved in autophagosome formation since p62-knockout mice display intact autophagosomes and slower protein degradation. Autophagy-deficient cells and neurons accumulate ubiquitin- and p62-positive inclusions, and this accumulation is greatly reduced by ablating p62 [55, 69]. p62 with mutations in the LC3 recognition sequence escape autophagic degradation, leading to the formation of inclusions, whereas those with mutations in the self-oligomerizing domain PB1 are poorly degraded, but no protein inclusions form. Thus increased levels and oligomerization of p62 are required for the formation of inclusion bodies, and their degradation is facilitated by oligomerization. Ubiquitinated aggregates induced by proteasome inhibition are also reduced in p62-deficient cells suggesting that p62 is a general mediator of inclusion formation and normally functions as an adaptator targeting proteins for autophagic degradation [20].

2.3.4. Neuronal autophagy in synapse development, function and remodeling

Neuronal autophagy has been recently shown to play an important role in synapse development. The ubiquitin-proteasome system negatively regulates growth of the neuromuscular junction (NMJ) in *Drosophila* [70] whereas NMJ is positively regulated by neuronal autophagy; a decrease or an increase in autophagy correspondingly affects synapse size [71]. Indeed, an overexpression or a mutation of *Atg1*, a gene involved in autophagy induction, respectively enhanced or decreased NMJ growth. Furthermore, this positive effect of autophagy on NMJ development is mediated by downregulating Hiw, an E3 ubiquitin ligase which negatively regulates synaptic growth by downregulating Wallenda (Wnd), a MAP kinase kinase [70- 72]. Although autophagy is considered as a nonselective bulk degradation process, it can regulate specific developmental events in a substrate-selective mode [73, 74]. In *C. elegans*, when presynaptic afferences are removed from postsynaptic cells, GABAA receptors are selectively targeted to autophagosomes [73]. Accordingly, Hiw could traffic to autophagosomes via a still unknown mechanism, although Hiw could be unselectively degraded by autophagy along with other presynaptic proteins. Interestingly, the synaptic density in mice carrying an atg1 mutation is decreased due to excessive activity of the MAP kinase ERK, suggesting that activated ERK negatively regulates synapse formation and that Atg1 regulates synaptic structure by downregulating ERK activity [75]. As pointed out by Shen and Ganetzky [71], autophagy is a perfect candidate to modulate synaptic growth and plasticity in function of environmental conditions, resulting in plausible consequences in learning and memory.

Autophagy has recently been shown to regulate neurotransmission at the presynaptic level [76]. Besides enhancing protein synthesis via the mTORC1 complex, mTOR activity inhibits autophagy by an Atg13 phosphorylation-induced blockade of Atg1 [77]. In the nervous system, mTORC1 promotes learning and synaptic plasticity dependent on protein synthesis [78- 80]. Conversely, the mTOR inhibitor rapamycin impedes protein synthesis and blocks cell injury-induced axonal hyperexcitability and synaptic plasticity, as well as learning and memory [81, 82]. In prejunctional dopaminergic axons, inhibition of mTOR induces autophagy as shown by an increase in autophagosome formation, and decreases axonal volume, synaptic vesicle number and evoked dopamine release. Similarly, non-dopaminergic striatal terminals also display more autophagosomes and fewer synaptic vesicles. Conversely, chronic autophagy deficiency in dopamine neurons increases dopaminergic axon size and evoked dopamine release, and promotes rapid presynaptic recovery. Thus mTOR-dependent axonal autophagy locally regulates presynaptic structure and function. In cultured brain slices, the occurrence of autophagosomes in presynaptic terminals isolated from their cell bodies confirms that autophagosomes are locally synthesized [83], and supports the view that this autophagy may serve to modulate presynaptic terminal function by sequestrating presynaptic components [76]. The global stimulating effect of chronic autophagy deficiency on dopaminergic neurons is consistent with the implication of autophagy in neurite retraction of sympathetic neurons *in vitro* [84] and neuritic growth in developing neurons [21].

There are only a few other reports indicating that autophagy may participate in synapse remodeling. In the cerebellar cortex of the ($Bax^{-/-}$;$GluR\delta2^{Lc}$) double mutant mouse (Fig. 2A), prominent autophagic profiles are evident in parallel fiber terminals subjected to intense remodeling in the absence of the PCs, their homologous target neurons [85]. As mentioned above, endocytosed GABAA receptors are selectively targeted to autophagosomes *in C. elegans* neurons [73], whereas autophagy promotes synapse outgrowth in *Drosophila* [71]. Autophagy may also modulate synaptic plasticity as recently demonstrated in mammalian hippocampal neurons [61]. Here, neuronal stimulation by chemical LTD induces NMDAR-dependent autophagy by inhibiting the PI3K-Akt-mTOR pathway. Enhanced autophagosome formation in the dendrites and spines of these neurons targets internalized AMPA receptors to lysosomes suggesting that autophagy contributes to the NMDAR-dependent synaptic plasticity required to maintain LTD and assure certain brain functions [61]. A possible mechanism for this formation of autophagosomes and autophagic degradation of AMPARs in dendritic shafts and spines may involve a change in endosome cycling. The formation of more amphisomes due to the fusion of endosomes with autophagosome [86, 87] would reduce the recycling endosome population, and direct more AMPAR-containing endosomes to autophagosomes for lysosomal degradation. Another alternative, but not exclusive actor is p62. This autophagosomal protein is important for LTP and spatial memory [88], interacts with AMPAR and is required for the trafficking of AMPAR [89]. AMPAR via its interaction with p62 would be trapped in autophagosomes as their number increased [61]. mTOR regulates protein turnover in neurons by functioning at the intersection between protein synthesis and degradation. During learning and reactivation in the amygdale and hippocampus, rapamycin inhibition of mTOR has recently been shown to impair object recognition memory [90], implicating signaling mechanisms involved in protein synthesis, synaptic plasticity and cell metabolism in this cognitive function.

2.4. Few autophagosomes, a feature of basal neuronal autophagy

Neurons are highly resistant to large-scale induction of autophagy in response to starvation, probably due to the multiple energy sources available to assure their function [48]. Interestingly, the activity of mTOR, a negative regulator of autophagy is significantly reduced in hypothalamic neurons from mice after a 48h starvation [91], although there are reports that autophagy in neurons can be regulated independently of mTOR [92, 93]. For example, insulin impairs the induction of neuronal autophagy *in vitro*, but in its absence induction of autophagy is mTOR-dependent. Furthermore, a potent Akt inhibitor provokes robust autophagy [92]. Thus insulin signaling maintains a low level of autophagosome biogenesis in healthy neurons constituting a critical mechanism for controlling basal autophagy in neurons. In addition to insulin signaling, multiple parallel signaling pathways including the mTOR pathway can regulate autophagy in neurons. From these data, Yue and collaborators [20] have proposed that basal autophagy in CNS neurons is regulated by at least two mechanisms: (1) a non-cell-autonomous mechanism whereby regulators (nutrients, hormones and growth factors) are supplied by extrinsic sources (glia, peripheral organs), (2) a cell-autonomous mechanism controlled by intrinsic nutrient-mediated signaling or specific factors expressed in neurons.

Neurons may depend less on autophagy to provide free amino acids and energy under physiological conditions given their quasi exclusive use of blood-born glucose as a source of carbon and energy for protein synthesis. Accordingly, the primary function of neuronal autophagy may be different than a primary response to starvation, and autophagy regulatory mechanisms are likely to be specific in neurons. Furthermore, gender differences in autophagic capacity have been suggested by the faster autophagic response to starvation of cultured neurons from male rats compared to those from females [94]. While *in vivo* evidence of neuronal autophagy mediated by nutrient signaling is still missing, a number of stress-related signals, neuron injuries and neuropathogenic conditions trigger prominent formation and accumulation of autophagosomes in neurons. During this process, neurons may undergo a significant change in autophagy regulation, involving a deregulation that allows neurons to switch from basal level (neuron-specific process featured by a low number of autophagosomes) to an activated state (well-conserved induced autophagy with large-scale biosynthesis of autophagosomes) [20]. Hypoxic-ischemia [95, 96], excitotoxicity [97-99], the dopaminergic toxins, methamphetamine and MPP+ [65, 100, 101], proteasome inhibition [102-104], lysosomal enzyme/lipid storage deficiencies [105-108] are examples of these pathological inducers of neuronal autophagy (see below).

3. Autophagy in neuronal physiopathology

Autophagy normally protects effect against neurodegeneration, but defects in the autophagy machinery are sufficient to induce neurodegeneration. Indeed, neuron-specific disruption of autophagy results in neurodegeneration [11, 12]; for example PC-specific Atg7 deficiency impedes axonal autophagy via an important p62-independent axonopathic mechanism associated with neurodegeneration [55]. Furthermore, specific defects in selective autophagic components or in the cargo selection process can induce neurodegeneration. This hypothesis is supported by the studies of cargo recognition and degradation components, such as p62, NBR1, or ALFY [109, 110]. Defects at any one of the autophagic steps can cause an abnormal accumulation of cytosolic components and lead to disease states. Therefore, each step of the autophagic process needs to be tightly regulated for efficient autophagic degradation.

The housekeeping role of neuronal autophagy is more evident when neurons are loaded with pathogenic proteins [67]. In many neurodegenerative disorders, cytoplasmic, nuclear and extracellular inclusions composed of aggregated and ubiquitinated proteins are believed to contribute to organelle damage, synaptic dysfunction and neuronal degeneration. The autophagic process in diseased neurons participates in the clearance of abnormal aggregate-prone proteins such as the expanded glutamine (polyQ)-containing proteins (e.g. mutant huntingtin in Huntington's disease (HD)), mutant forms of α-synuclein in familial Parkinson disease (PD), different forms of tau in Alzheimer's disease (AD), tauopathies and FTD, mutant forms of SOD1 in motor diseases such as ALS, and mutant forms of PMP22 in peripheral neuropathies are cleared from diseased neurons by autophagy [19, 20, 55, 56, 66, 67, 111-115]. However, accumulation of these intracellular aggregates is believed to play a significant role in the etiology of neurodegenerative diseases including prion diseases (PrD) [3, 67]. One common feature is the dramatic cyto-pathological accumulation of autophagosomes in injured and degenerating neurons [116-121]. Such signs of defects in autophagy have been interpreted as a result of an "autophagic stress", or in other words an imbalance between protein synthesis and degradation [116]. This has traditionally been viewed as a highly destructive cellular mechanism, driving the cell to death [117]. In these diseases, it is now accepted that autophagy eliminates aggregate-prone proteins and damaged organelles more efficiently than the proteasome machinery. Since the proteasome is unable to degrade them [122], the clearance of misfolded, aggregated proteins originating from neuropathologic deficits is highly dependent on autophagy. However, a blockade of the autophagic flux is likely to impede the clearance of these proteins. The accumulation of aggregated proteins and organelles within the diseased neurons then contributes to cell dysfunction and in the end results in cell death [16], (Fig. 3). Indeed, pharmacological upregulation of autophagy reduces neuronal aggregates and slows down the progression of neurological symptoms in animal models of tauopathy and HD [123], AD [41, 124, 125] and PrD [126, 127].

The mechanisms that determine the activation of autophagy for the removal of aggregated proteins are not clearly understood, but failure of the other proteolytic systems to handle the altered proteins seems to at least partly underlie autophagy activation. Thus oligomers and fibers of particular proteins can block the proteolytic activity of the ubiquitin-proteasome system and chaperone-mediated autophagy (CMA) that results in autophagy upregulation [128, 129]. In addition, sequestration of negative regulators of autophagy in the protein aggregates could also provoke activation of this pathway. Thus it has been shown that blockage of autophagy in neurons leads to the accumulation of aggregated proteins and neurodegeneration even in the absence of aggregate-prone proteins [11, 12]. Although the specific reasons for the failure of the proteolytic systems are unknown, factors such as enhanced oxidative stress and aging seem to precipitate entry into a late failure stage when the activity of all degradation systems are blocked or decreased, leading to accumulation of autophagic vacuoles and aggregates and finally cell death [130].

Autophagy protects against cell death in the case of growth factor withdrawal, starvation and neurodegeneration, but it is required for some types of autophagic cell death [131-134]. However, the role of autophagy as a positive mediator of cell death is not well understood in mammalian systems, although many studies suggest that impaired autophagy sensitizes cells and organisms to toxic insults. Atg1-dependent autophagy restricts cell growth [135]. Cells

http://dx.doi.org/10.5772/55646

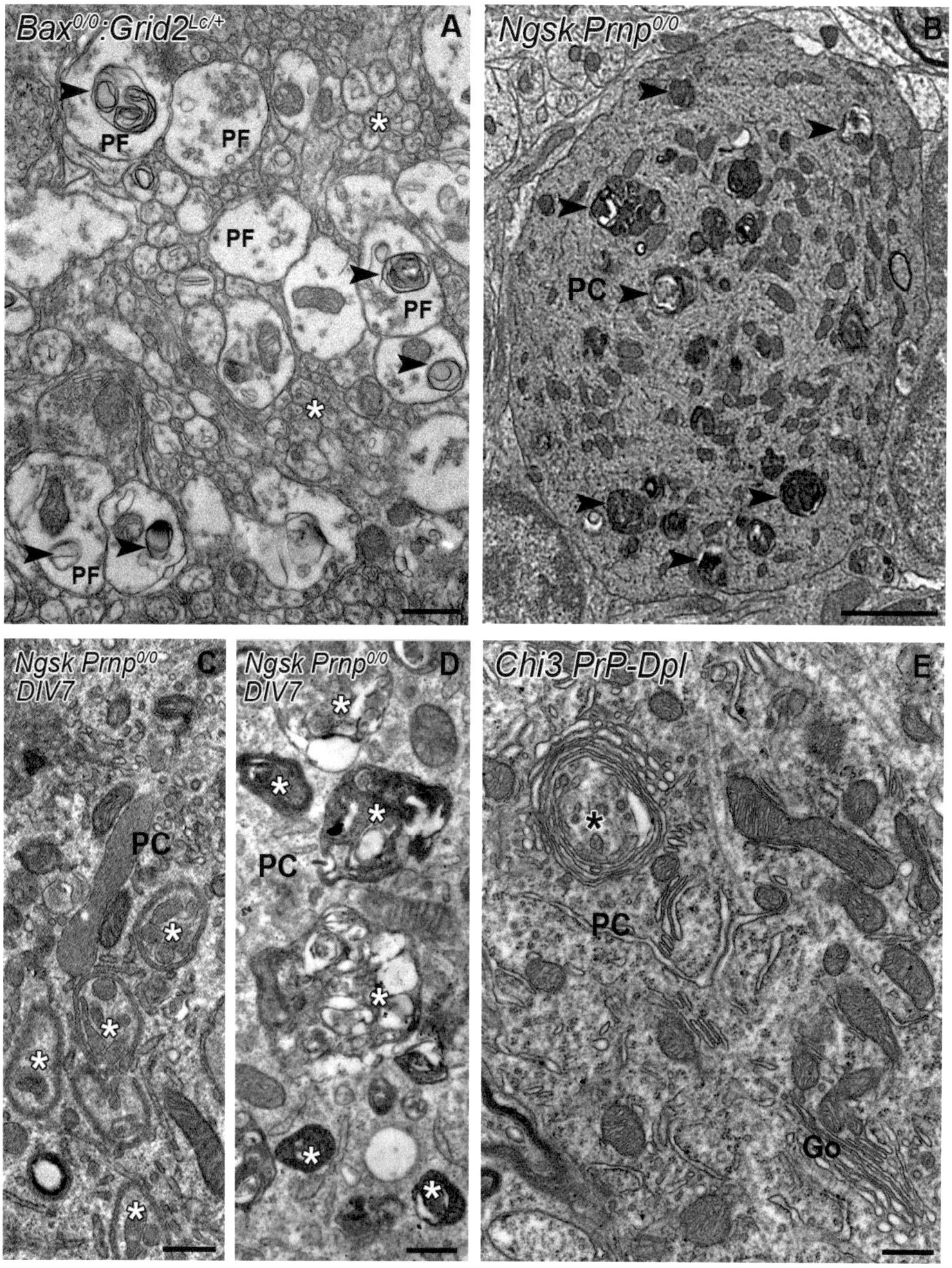

Figure 2. Autophagy in cerebellar neurons.A. Neuronal autophagy in the cerebellum of a Purkinje cell-deficient *Bax*$^{0/0}$;*Grid2*$^{Lc/+}$ double mutant mouse. Autophagic-like profiles (arrowheads) in presynaptic parallel fiber boutons (PF) in the cerebellar molecular layer. * intervaricose parallel fibers. **B.** Autophagolysosomes (arrowheads) characteristic of neuronal autophagy in the soma of a cerebellar Purkinje cell (PC) of a prion protein-deficient Ngsk *Prnp*$^{0/0}$ mouse. **C-D.** Phagophores, autophagosomes (* in C) and autophagolysosomes (* in D) in the soma of cerebellar Purkinje cells (PC) from prion protein-deficient Ngsk *Prnp*$^{0/0}$ mouse maintained 7 days *in vitro* (DIV7) in organotypic culture. **E.** Autophagosome (*) forming from a Golgi dictyosome in the Purkinje cell soma (PC) of a transgenic mouse expressing a neurotoxic Chi3 PrP-Dpl chimera. Go, normal Golgi dictyosome. Scale bars = 500 nm in A, C-E, 2 µm in B.

deficient in Pdk1, a positive regulator of mTOR pathway [136], display autophagy and reduced growth. The increased growth capacity that results from disrupting autophagy may contribute to the tumorigenicity of cells mutant for tumor suppressors [38, 137, 138]. Overexpression of Atg1 leads to apoptotic cell death [135]. Cells undergoing autophagic cell death display signs of apoptosis [139], as do Atg1-null cells [135]. Thus, elevated levels of autophagy promote cell death and the role of autophagy in cell death is likely to be context-dependent.

Neuronal autophagy is currently believed to constitute a protective mechanism that slows the advance of neurodegenerative disorders, and that its inhibition is associated with neurodegeneration [130]. Substantial attention is currently being focused on the molecular mechanisms underlying the autophagic fight against neurodegeneration, the role of autophagy in early stages of pathogenesis and therapeutical approaches to upregulate protective neuronal autophagy. It is unclear whether accumulation of autophagic vacuoles in degenerating neurons results from increased autophagic flux or impaired flux. A chronic imbalance between autophagosome formation and degradation causes "autophagic stress" [140]. Due to obvious therapeutic consequences, it is imperative to understand how autophagic stress occurs in each autophagy-associated neurodegenerative condition: either a cellular incapacity to support an excessive autophagic demand or a defective degradation (lysosomal) step [141].

4. Autophagy in prion diseases

4.1. Prion diseases

4.1.1. Infectious and familial prion diseases

Prion diseases (PrD) are transmissible spongiform encephalopathies (TSEs) which are fatal neurodegenerative diseases in humans (Creutzfeldt-Jakob disease (CJD), Gerstmann-Sträussler-Scheinker syndrome (GSS), variant CJD (vCJD), fatal familial insomnia (FFI) and kuru) and in animals (bovine spongiform encephalopathy (BSE), transmissible mink encephalopathy (TME), chronic wasting disease (CWD) and scrapie). In humans, PrD manifest after a long incubation period free of symptoms as a rapid progressive dementia that leads inevitably to death. Severe loss of neurons with extensive astrogliosis and moderate microglial activation, characteristic of all TSEs, results in a progressive spongiform degeneration of the brain tissue which is reflected by ataxia, behavioral changes and, in humans, a progressive cognitive decline [142-145]. According to the protein-only hypothesis [146], TSEs are caused by prions that are believed to be proteinaceous infectious particles mainly consisting of PrP^{SC}, an abnormal isoform of the normal, host-encoded prion protein (PrP^{C}), [142]. Prions are able to catalyze a switch from PrP^{C} conformation into an aggregated misfolded conformer PrP^{SC} which collects throughout the brain according to a prion strain-specific anatomo-pathologic signature. These PrDs share a protein misfolding feature with other neurodegenerative diseases (e.g. AD, PD and HD), [147].

The central role played by PrP^{C} in the development of PrD was first illustrated by the observation that disruption of the PrP gene (PRNP) in mice confers resistance to PrD and impairs

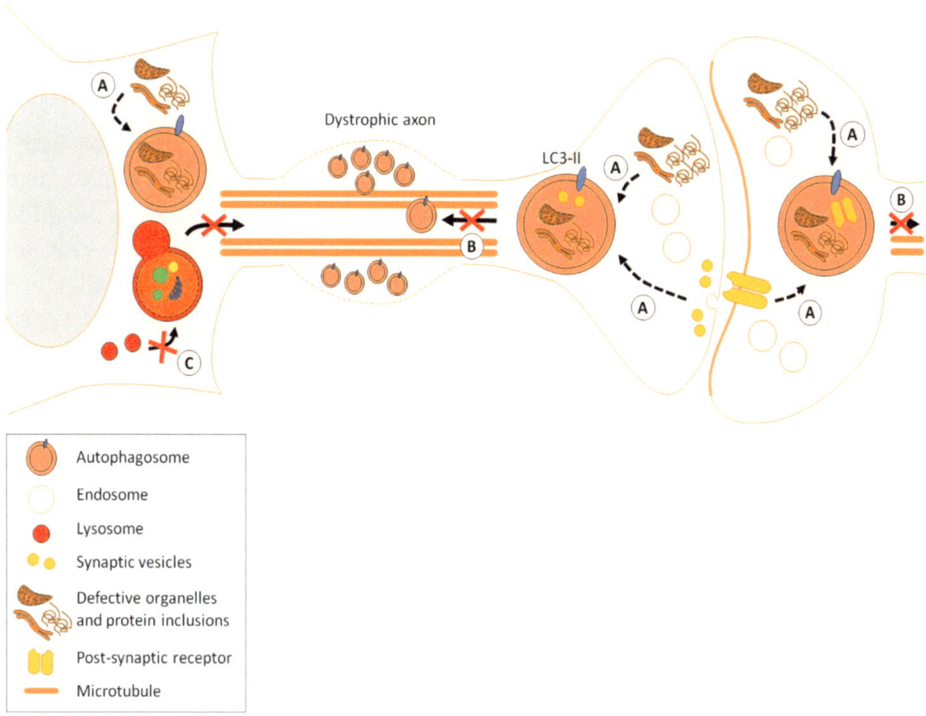

Figure 3. Impaired steps of neuronal autophagy in neurodegenerative disorders.A. Defective autophagosome biogenesis. **B.** Blockage of retrograde transport and accumulation of autophagosomes in dystrophic neurites. **C.** Failure of autophagosomes to fuse with lysosomes.

the propagation of infectious prions [148], while PrP-overexpressing (*tga20*) mice exhibit reduced incubation periods when compared with wild-type mice [149]. Overall, the current data argue for a primary role of the neuronal, GPI-anchored PrP^C in prion neuropathogenesis [150].The subversion of PrP^C function(s) as a result of its conversion into PrP^{SC} is assumed to account for prion-associated toxicity in neurons [151]. Whether PrP^{SC} triggers a loss of PrP^C physiological function (loss-of-function hypothesis) or promotes a gain of toxic activity (gain-of-function hypothesis), or both, is an ongoing debate in the TSE field [152]. Elucidating the roles of PrP^C in neurons should help to answer this question. Knockout experiments, however, have failed to reveal any obvious physiological role for PrP^C. Mice devoid of PrP^C are viable and display only minor phenotypic or behavioural alterations that vary according to the null strain, and hence, these results do not permit one to assign a specific function to PrP^C. *Ex vivo* studies support the involvement of PrP^C in copper homeostasis [153]. In addition, the localisation of PrP^C on the cell membrane and its affinity for the neuronal cell adhesion molecule (N-CAM), laminin, and the laminin receptor [154, 155] have implicated PrP^C in cell adhesion. Such properties may reflect the involvement of PrP^C in the outgrowth and maintenance of neurites, and even cell survival. Indeed, recent experimental evidence showing that PrP^C interacting with β1 integrin controls focal adhesion and turnover of actin microfilaments in neurons substantiates a role for PrP^C in neuritogenesis. Of note, integrins are well known inducers of autophagy (see review in chapter by Nollet and Miranti). Remarkably, during neuronal differentiation, the downregulation of Rho kinase (ROCK) activity by PrP^C is necessary for neurite sprouting [156]. A stress-protective activity has also been assigned to PrP^C based on results obtained with primary neuronal cultures. Neuronal cells derived from PrP-knockout mice are more sensitive to oxidative stress and serum deprivation than wild type cells [157-159]. Moreover, after ischemic brain injury, PrP^C-depleted mice revealed enlarged infarct volumes [160-162]. This neuroprotective role of PrP^C has been linked to cell signaling events. The interaction of PrP^C with the stress inducible protein (STI-1) generates neuroprotective signals that rescue cells from apoptosis [163]. Previous studies of both neuronal and non-neuronal cells substantiate the coupling of PrP^C to signaling effectors involved in cell survival, redox equilibrium and homeostasis (e.g. ERK1/2, NADPH oxidase [164], cyclic AMP-responsive element binding protein (CREB) transcription factor and metalloproteinases [165, 166]. According to these data, PrP^C has been proposed to function as a « dynamic cell surface platform for the assembly of signaling modules » [167]. Despite these overall advances, the sequence of cellular and molecular events that leads to neuronal cell demise in TSEs remains obscure [168, 169]. At present, one envisions that neuronal cell death results from several parallel, interacting or sequential pathways involving protein processing and proteasome dysfunction [170], oxidative stress [159, 171], apoptosis and autophagy [172].

4.1.2. Autophagy in prion-infected neurons

Prion propagation involves the endocytic pathway, specifically the endosomal and lysosomal compartments that are implicated in trafficking and recycling, as well as the final degradation of prions. Shifting the equilibrium between propagation and lysosomal clearance impairs the cellular prion load. This and the presence of autophagic vacuoles in prion diseased neurons [173, 174] suggest a role for autophagy in prion infection (reviewed in [172]). Indeed, the high numbers of autophagic vacuoles observed in neurons from experimentally prion-infected mice (Fig. 4A, B, Fig. 5) and hamsters is indicative of a robust activation of autophagy [175, 176]. Furthermore, autophagic vacuoles and multivesicular bodies have been detected in prion-infected neuronal cells in vitro [177].The formation of autophagic vacuoles has recently been observed in neuronal pericarya, neurites and synapses of neurons experimentally infected with scrapie, CJD and GSS [174], as well as in neuronal synaptic compartments in humans with certain PrD [173]. PrDs are further correlated with autophagy given that the Scrg1 protein (encoded by the scrapie responsive gene1, *Scrg1*) is upregulated in scrapie and BSE-infected brains, as well as in brains of patients with sporadic CJD [178-180] and is associated with neuronal autophagosomes [181, 182]. This Scrg1 protein is thus, a new marker for autophagic vacuoles in prion-infected neurons (Fig. 4A, B). In the brains of CJD and FFI patients and experimentally scrapie-infected hamsters, increased cytoplasmic levels of LC3-II-immunostained autophagosomes have been demonstrated in neurons, again indicating autophagy activation. In addition, the decreased p62 and polyubiquitinated proteins levels in hamster and human brains infected with prion suggest an upregulation of autophagy with enhanced autophagic flux and protein degradation. Downregulation of the mTOR pathway and upregulation of the beclin 1 pathway in these infected tissues provide further evidence of autophagy activation [183]. On the basis of these observations, Xu *et al.* [183] propose that neuronal autophagy is an intricate element of prion infections. They suggest that once PrP^{SC} enters host cells and is delivered to endosomes, it accumulates in amphisomes via fusion with autophagosomes and then with lysosomes. At this initial stage of infection, PrP^{SC} does not co-localize with autophagosomes, probably because PrP^{SC} levels are too low to be detected due to their rapid degradation in autophagolysosomes. In agreement with this explanation, blocking the fusion of autophagosomes with lysosomes using bafilomycin A1 permits the visualization of PrP-PG14 and PrP^{SC} in autophagosomes [183], as is the case for Aβ1-42 [184].

The role of lysosomes in PrDs is still controversial. Although autophagic lysosomal degradation of PrP^{SC} in infected neurons is supposed to clear prion aggregates and inhibit PrP^{SC} replication, there are indications that PrP^{SC} may subvert the autophagic-lysosomal system to promote the conversion of PrP^{C} into PrP^{SC}. Lysosomal inhibitors prevent the build-up of PrP^{SC} [126] and agonists of the autophagy-lysosome pathway enhance the clearance of PrP^{SC} [185, 186, 126]. However, as PrP^{SC} production increases, the accumulating PrP^{SC} may saturate the clearance capacity of the system causing lysosomal disruption and release of PrP^{SC} aggregates into the neuroplasm. In turn this would cause cell stress and over-activate autophagy, as has been reported in prion-diseased brain tissue [183].

The octapeptide repeats region of PrP^{C} has been shown to negatively influence autophagy. As measured by LC3-II expression, autophagy induced by serum deprivation occurs earlier and to a greater extent in hippocampal neurons from ZH-I *PrnP*$^{-/-}$ compared with those from wild type mice. Reintroduction of PrP^{C}, but not *PrP*^{C} lacking its N-t octapeptide region, into ZH-I PrnP$^{-/-}$neurons delays this upregulation of autophagy [187]. The transconformation of PrP^{C} into PrP^{SC} could interfere with the function of this domain and as a consequence, upregulate autophagy. It is conceivable that the activation of autophagy observed in PrD models reflects a defense mechanism designed to degrade prions and resist oxidative stress. A reduction in autophagy combined with endosomal/lysosomal dysfunction has indeed been proposed to contribute to the development of PrD [188]. Furthermore, the anti-cancer drug imatinib has been shown to activate lysosomal degradation of PrP^{SC} [186] and is a potent autophagy inducer [189]. When administered early during peripheral infection, imatinib delays both PrP^{SC} neuroinvasion and the onset of clinical disease in prion-infected mice [190]. Upregulation of autophagy has beneficial effects on the clearance of aggregate-prone proteins in PrD and other neurodegenerative diseases [66, 111-115, 191, 192]. Both lithium and trehalose enhance PrP^{SC} clearance from prion-infected cells by inducing autophagy, as demonstrated by increases in LC3-II protein and the number of GFP-LC3 puncta [193, 126]. Furthermore, PrP^{SC} can be cleared not only by mTOR-independent autophagy (lithium and trehalose), but also by the mTOR-dependent route because the mTOR inhibitor rapamycin also causes a decrease in cellular PrP^{SC}. Lithium-induced autophagy also reduces PrP^{C} levels. This treatment causes internalization of PrP^{C} [194], and the consequent reduction of available membrane-bound PrP^{C} is known to decrease its conversion into pathologic PrP^{SC} [195-198]. This would provide an additional, indirect way to reduce PrP^{SC} by reducing of PrP^{C} with lithium treatment.

Whether autophagy-inducing compounds are candidates for therapeutic approaches against prion infection has recently been investigated in prion-infected mice. Starting in the last third of the incubation periods, treatment with rapamycin and to a lesser extent with lithium significantly prolonged incubation times compared to mock-treated control mice [126, 172]. Along this line, activation of the class III histone deacetylase Sirtuin 1 (Sirt1) has been shown to mediate the neuroprotective effect of resveratrol against prion toxicity [199] and prevent prion protein-derived peptide 106-126 (PrP106-126) neurotoxicity via autophagy processing [200]. Moreover, Sirt1-induced autophagy protects against mitochondrial dysfunction induced by PrP106-126, whereas siRNA knockdown of Sirt1 sensitizes cells to PrP106-126-induced cell death and mitochondrial dysfunction. Finally, knockdown of Atg5 decreases LC3-II protein levels and blocks the effect of a Sirt1 activator against PrP106-126-induced mitochondrial dysfunction and neurotoxicity. Thus inducing Sirt1-mediated autophagy may be a principal neuroprotective mechanism against prion-induced mitochondrial apoptosis. Nevertheless, understanding the mechanisms underlying Sirt1-mediated autophagy against prion neurotoxicity and mitochondrial damage merits further investigation, in particular determining the Sirt1-mediated dowstream signaling network, including FOXOs, p53 and PGC-1α. More recently, the mTOR inhibitor and autophagy inducer rapamycin has been shown to delay disease onset and prevent PrP plaque deposition in a mouse model of the Gerstmann-Sträussler-Scheinker PrD [127]. Here, the reduction in symtom severity and prolonged survival correlate with increases in LC3-II levels in the brains of treated mice, suggesting that autophagy induction enhances elimination of misfolded PrP before plaques form. This is in agreement with the well known neuroprotective effects of rapamycin in various models of neurodegenerative diseases with misfolded aggregate-prone proteins (e.g. PD [111], ALS [201], HD [115], spinocerebellar ataxia [66, 202], FTD [203] and AD [41, 124, 125].

4.2. Doppel-expressing prion protein-deficient mice

Research efforts to determine the function of PrP^C using knockout mutant mice have revealed that large deletions in the PrP^C genome result in the ectopic neuronal expression of the prion-like protein Doppel (Dpl) causing late onset degeneration of PCs and ataxia in *PrnP*$^{-/-}$mouse lines, such as Ngsk [204], Rcm0 [205], ZH-II [206] and Rikn [207].

Similar PC degeneration is observed when the N-terminal truncated form of PrP is expressed (ΔPrP) in *Prnp*-ablated mouse lines [208] and when Dpl is overexpressed [209, 210]. Of note, full-length PrP^C antagonizes the neurotoxic effects of both Dpl and ΔPrP [208-212], but not PrP^C lacking the N-terminal residues 23-88 [213]. These results imply that Dpl and ΔPrP induce cell death by the same mechanism, likely by interfering with a cellular signaling pathway essential for cell survival and normally controlled by full-length PrP^C [209, 214]. The mechanism underlying Dpl-induced neurotoxicity is still under debate. PrP-deficient neurons undergo Dpl-induced apoptosis in a dose-dependent, cell autonomous manner [215]. Oxidative stress is a likely candidate to play a role in the death of these neurons because NOS activity is induced by Dpl both in vitro and in vivo [212, 216]. Endogenous, as well as exogenous PrP^C has been shown to inhibit Dpl-induced apoptosis, a neuroprotective function that has been attributed to its BCL-2-like properties [158]. Like BCL-2, PrP^C antagonizes mitochondrial apoptotic pathways, thereby protecting neurons from cell death [217- 219]. In BAX-induced apoptosis [220, 221], PrP^C probably acts by preventing the conformational changes in BAX that are necessary for its activation [222]. In primary cultures, Dpl-induced apoptosis of *Prnp*$^{+/+}$ as

http://dx.doi.org/10.5772/55646

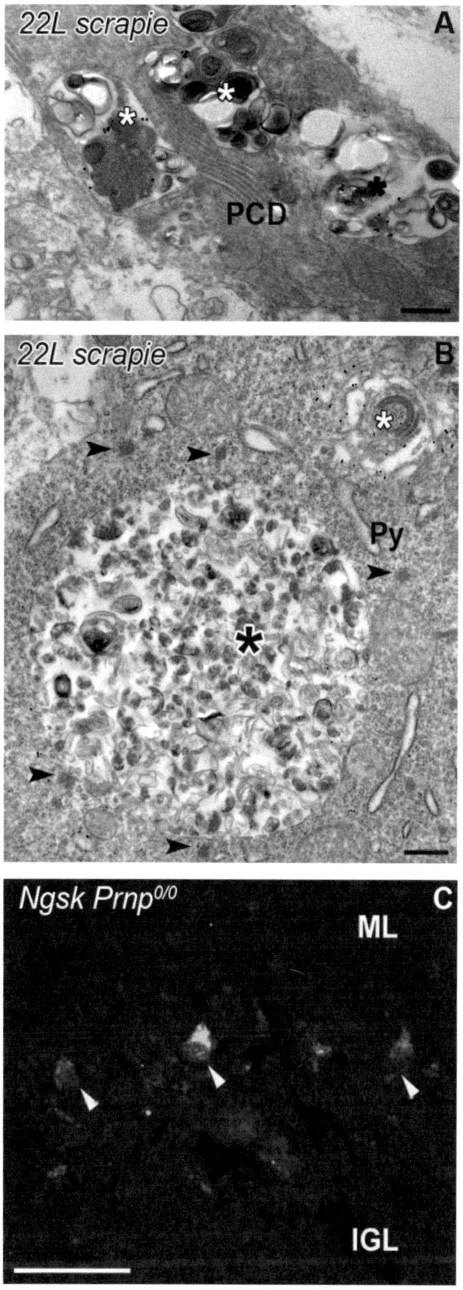

Figure 4. Scrapie responsive gene 1 (Scrg1)-immuno-cytochemistry in prion-infected and prion protein-deficient neurons.A-B. Scrg1 immunogold labeling in central neurons of a clinically ill 22L-scrapie-infected mouse. Scrg1-bound immunogold particles label autophagolysosomes (* in A) in a Purkinje cell dendrite (PCD) and an autophagosome forming from a Golgi dictyosome (white asterisk in B) in the somatic neuroplasm of a pyramidal neuron (Py) of the CA3 field of the hippocampus. In this neuron, lysosomes (arrowheads) and immunogold particles labelling Scrg1 surround a large autolysosome-like vacuole (black asterisk). **C.** Scrg1 immuno-fluorescent labeling of Purkinje cells (arrowheads) in the cerebellar cortex of a prion protein-deficient Ngsk *Prnp*$^{0/0}$ mouse. IGL, internal granular layer; ML, molecular layer. Scale bars = 500 nm in A-B and 50 μm in C.

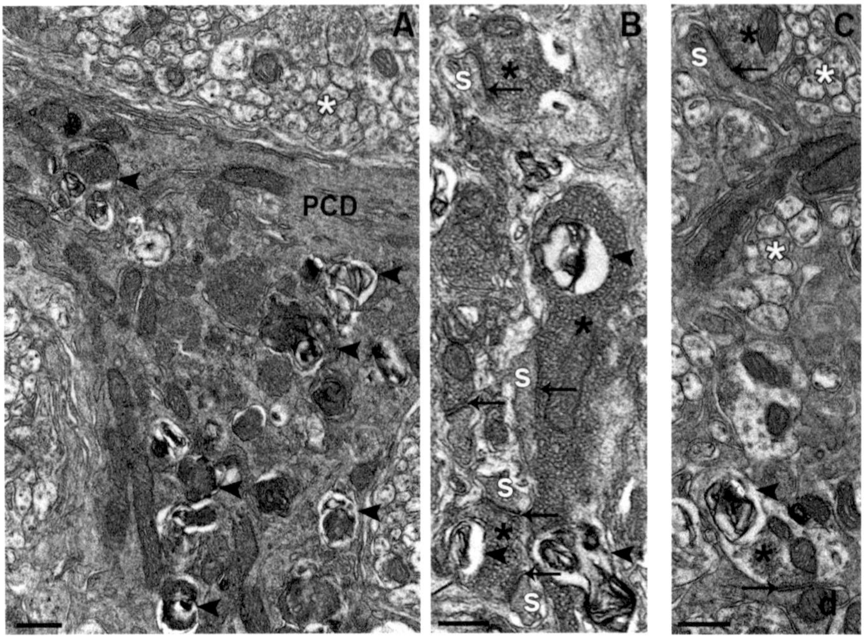

Figure 5. Neuronal autophagy in cerebellar neurons of a clinically ill 22L-scrapie-infected mouse. A. Accumulation of autophagosomes (arrowheads) in a main dendrite of a Purkinje cell (PCD) in the cerebellar molecular layer. *, parallel fibers. **B.** Autophagosomes (arrowheads) in presynaptic axon terminals (black asterisks) establishing synapses (arrows) on postsynaptic Purkinje cell dendritic spines (s). **C.** An intact parallel fiber bouton (black asterisk) makes a synapse (arrow) on a Purkinje cell spine (s) in the upper part of the picture and another parallel fiber bouton (black asterisk) containing an autophagosome (arrowhead) makes a synapse (arrow) on a putative interneuron dendrite (d) in the bottom of the picture. Scale bars = 500 nm.

well as *Prn*P$^{-/-}$ granule cells, has recently been shown to be inhibited by BAX deficiency or pharmacologically blocking caspase-3 suggesting that it is mediated by Bax and caspase-3 [223]. These results further confirm *in vivo* data concerning the effects of Bax expression on PC survival in the cerebellum of the Dpl-overexpressing Ngsk *PrnP*$^{-/-}$mouse that we reported several years ago [224]. In these mice, PC death is already significant as early as 6 months of age. During aging, quantification of PC populations shows that significantly more PCs survived in the Ngsk *PrnP*$^{-/-}$:*Bax*$^{-/-}$ double mutant mice than in the Ngsk *Prn*P$^{-/-}$mice. However, the number of surviving PCs is still lower than wild type levels and less than the number of surviving PCs in *Bax*$^{-/-}$ mutants. This suggests that neuronal expression of Dpl activates both BAX-dependent and BAX-independent pathways of cell death. Interestingly, a partial rescue of Ngsk *PrnP*$^{-/-}$PCs is observed in Ngsk *PrnP*$^{-/-}$-*Hu-bcl-2* double mutant mice, in a proportion similar to that found in Ngsk *PrnP*$^{-/-}$:*Bax*$^{-/-}$ mice, strongly supporting the involvement of BCL-2-dependent apoptosis in Dpl neurotoxicity [225]. The capacity of BCL-2 to apparently compensate for the deficit in PrP^{C} by partially rescuing PCs from Dpl-induced death suggests that the BCL-2-like property of PrP^{C} may counteract Dpl-like neurotoxic pathway in wild-type neurons. Although not exactly identical to BCL-2, PrP^{C} may functionally replace BCL-2 as it decreases in the aging brain [222]. The N-terminal domain of PrP^{C} which is partially homologous to the BH2 domain of BCL-2 family of proteins [226, 227] is probably responsible for the neuroprotective functions of PrP^C because BAX-induced apoptosis cannot be counteracted by N-terminally truncated PrP. BCL-2 antagonizes the pro-apoptotic effect of BAX by interacting directly with this BH2 domain [228-230], and this domain is missing in both Dpl and the neurotoxic mutated forms of PrP: ΔPrP [208, 214, 231] and Tg(PG14)PrP [232]. Interestingly, expression of Dpl fused to a BH2-containing octapeptide repeat and the N-terminal half of the hydrophobic region of PrP^C makes cells resistant to serum deprivation [233]. Furthermore, N-terminal deleted forms of PrP^C have been reported to activate both BAX-dependant and BAX-independant apoptotic pathways [231].

4.2.1. Autophagy in prion protein-deficient mice

The Dpl-activated, BAX-independent cell death mechanism may involve neuronal autophagy as we have detected the expression of Scrg1, a novel protein with a potential link to autophagy in the Ngsk *PrnP*$^{-/-}$PCs (Fig. 4C), [181]. Both neuronal Scrg1 mRNA and protein levels are increased in prion-diseased brains [179, 180], and Scrg1 is associated with dictyosomes of the Golgi apparatus and autophagic vacuoles in degenerating neurons of scrapie-infected Scrg1-overexpressing transgenic and WT mice (Fig. 4A, B), [181, 182]. Both before and during PC loss, protein levels of Scrg1 and the autophagic markers LC3-II and p62 are increased in Ngsk *PrnP*$^{-/-}$PCs, whereas their mRNA expression is stable, suggesting that the degradation of autophagic products is impaired in these neurons [234, 235]. Autophagic profiles collect in somato-dendritic and axonal compartments of Ngsk *PrnP*$^{-/-}$ (Figs. 2B, 6), but not wild-type PCs. The most robust autophagy occurs in dystrophic profiles of the PC axons in the cerebellar cortex (Fig. 6D) and at their preterminal and terminal levels in the deep cerebellar nuclei (Fig. 6A-C) suggesting that it initiates in these axons. Taken together, these data indicate that Dpl triggers autophagy and apoptosis in Ngsk *PrnP*$^{-/-}$ PCs. As reflected by the abundance of autophagosomes in the diseased Ngsk PCs, Dpl neurotoxicity induces a progressive dysfunction of autophagy, as well as apoptosis. Whether this autophagy dysfunction triggers apoptotic cascades or provokes autophagic cell death independent of apoptosis remains to be resolved. In the Ngsk *PrnP*$^{-/-}$ PCs, the increased expression of LC3-II and p62 at the protein level, without any change in mRNA levels, suggests that the ultimate steps of autophagic degradation are impaired. This is further confirmed by the prominence of autophagolysosomes in these neurons which indicate that the fusion of autophagosomes with lysosomes occurs normally, but downstream, the autophagic flux is blocked.

To further investigate the neurodegenerative mechanisms induced by Dpl in Ngsk cerebellar PCs, we are using an organotypic cerebellar culture system which allows an easier way to approach mechanistic questions than in vivo models [236]. For this purpose, we have assessed the growth and viability of PCs in cerebellar organotypic cultures from Ngsk and ZH-I *PrnP*$^{-/-}$ mice using morphometric methods to measure PC survival and development [237]. The timing and amplitude of PC growth impairment and neuronal death are similar in Ngsk and ZH-I *PrnP*$^{-/-}$ cultures (Fig. 7). In addition, increased amounts of autophagic (LC3-II, Fig. 8) and apoptotic (caspase-3, Fig. 9) markers are detected in protein extracts from both cultures indicating that both apoptosis and autophagy (Fig. 2C, D) contribute to PC death in Ngsk [235] and ZH-I cultures. This suggests that PrP^C -deficiency, rather than Dpl expression, is respon-

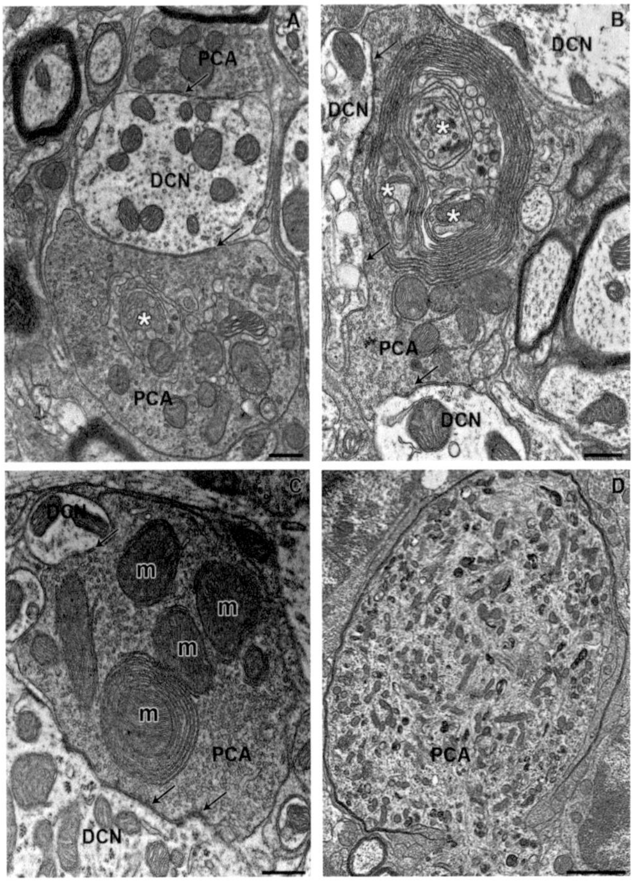

Figure 6. Neuronal autophagy in the cerebellar deep nuclei of a prion protein-deficient Ngsk *Prnp*$^{0/0}$ mouse.A. A presynaptic terminal of a Purkinje cell axon (PCA) establishes symmetric synapses on a postsynaptic dendrite from a neuron of the interpositus deep cerebellar nucleus (DCN) and contains an autophagosome (*). **B.** A double membrane wrap sequesters autophagosomes (*) in a Purkinje cell axon varicosity (PCA) symmetrically synapsing (arrows) on dendrites from neurons of the dentate deep cerebellar nucleus (DCN). **C.** Mitophagy by double membranes wrapping around mitochondria (m) in a Purkinje cell presynaptic axon terminal (PCA) making symmetric synapses (arrows) on postsynaptic dendrites from dentate deep nuclear neurons (DCN). **D.** Dystrophic Purkinje cell axon (PCA) filled with electron-dense autophagic profiles in the cerebellar internal granular layer. Scale bars = 500 nm in A-C, 2 µm in D.

sible for the neuronal growth deficit and loss in these cultures. For presently unknown reasons, the neurotoxic properties of Dpl do not seem to contribute to the degeneration of Ngsk PCs in these organotypic cultures. As the neurotoxicity induced by Dpl takes about 6 months to develop *in vivo*, it is possible that organotypic cultures are not mature enough to model 6-month-old cerebellar tissue. Nevertheless, *ex vivo* cerebellar organotypic cultures do provide a suitable system for analyzing the mechanisms underlying the neurotoxic effects of PrP^{C}-deficiency and prion infections [238] using pharmacological and siRNA-based approaches.

Our results have shown that PrP^{C} has a neuroprotective role in cerebellar PCs. As PCs survive *in vivo* in the cerebellum of the ZH-I mouse, the death of the ZH-I PCs in the organotypic

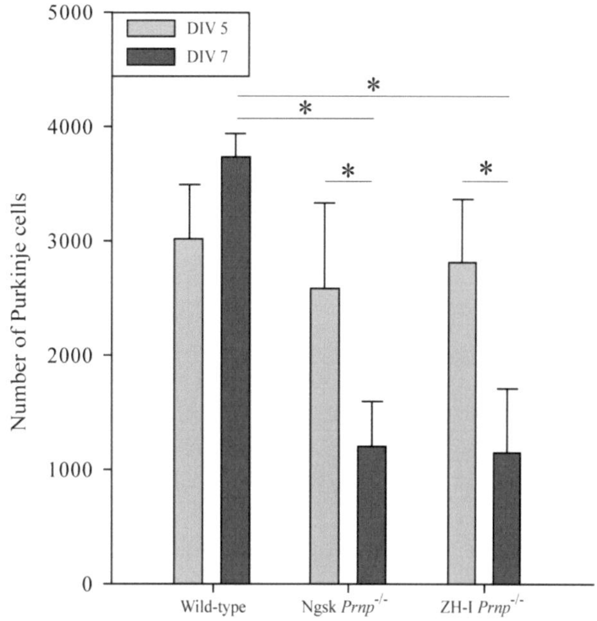

Figure 7. Purkinje cell loss in cerebellar organotypic cultures from wild-type and *Prnp*-deficient mice. PCs stained for calbindin by immuno-fluorescence were counted. This analysis reveals similar, significant reductions in the number of PCs between DIV5 and DIV7 for Ngsk *Prnp*$^{-/-}$ (53.5%) and ZH-I *Prnp*$^{-/-}$ (59%) cultures. During this period, the number of PCs in wild-type cultures is stable ($p > 0.05$). Although the number of PCs is not significantly different between genotypes at DIV5, by DIV7 there are similar decreases in mutant organotypic cultures (Ngsk: 67.8% and ZH-I: 69%) compared to wild-type cultures (two-way ANOVA followed by post-hoc Tukey test; * $p < 0.001$).

cultures is likely to stem from the inherent stress of the *ex vivo* conditions. As mentioned above, PrP^C negatively regulates autophagy as demonstrated by the upregulation of autophagy following serum deprivation in *PrnP*$^{-/-}$ hippocampal neurons when compared to PrP^C-expressing neurons [187]. Recent results suggest that PrP^C can directly modulate autophagic cell death. Using antisens oligonucleotides targeting the *Prnp* transcript, the downregulation of PrP^C expression in glial and non-glial tumor cells induces autophagy-dependent, apoptosis-independent cell death [239]. Previous data have shown that PrP^C acts as a SOD [240] and modulates the activity of Cu/Zn SOD by binding 5 Cu^{++} ions on its N-terminal octapeptide repeat domain [153, 157, 241]. A recent study of the effects of H_2O_2-induced oxidative stress on hippocampal neurons expressing PrP^C or deficient in PrP^C provides further support for the protective role of PrP^C against oxidative stress [242]. Although autophagy and apoptosis occur in both lines, the *Prnp*$^{-/-}$ neurons are less resistant to H_2O_2-induced oxidative stress than the *Prnp*$^{+/+}$ neurons confirming the anti-oxidant activity of PrP^C.

Furthermore, autophagy is more enhanced in *Prnp*$^{-/-}$ neurons than in *Prnp*$^{+/+}$ neurons. In the latter, this is due to H_2O_2-induced enhancement of autophagic flux, and in the former due to H_2O_2-induced impairment of autophagic flux. Similarly, experiments using Atg7 siRNA to inhibit autophagy have revealed that the increased autophagic flux in Prnp$^{+/+}$ neurons protects against H_2O_2 cytotoxicity. Thus a deficiency in Prnp may impair autophagic flux via H_2O_2-

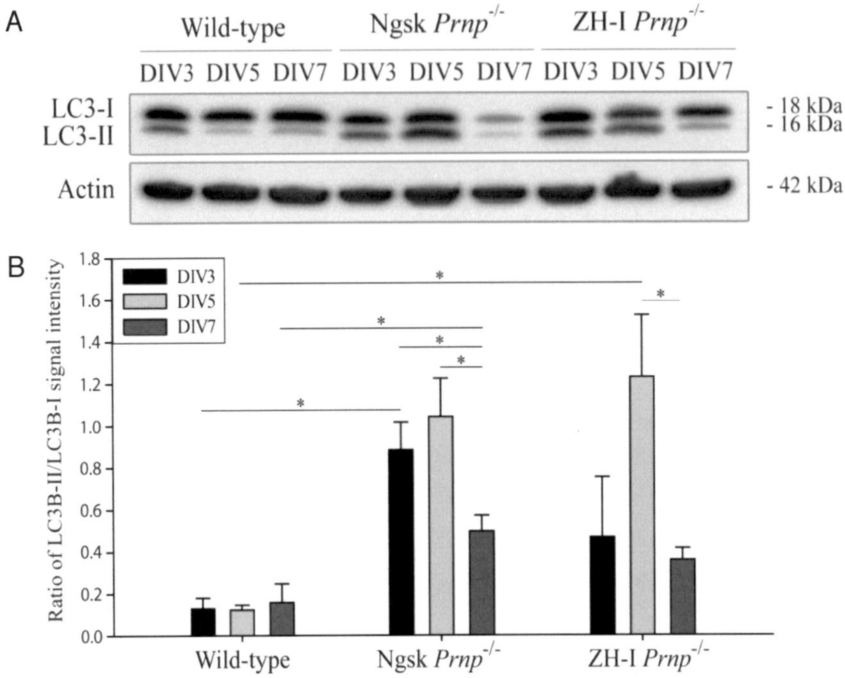

Figure 8. A. Western blot of the autophagic marker LC3B-II, in extracts prepared from organotypic cultures from wild-type, Ngsk and ZH-I *Prnp*$^{-/-}$ mouse cerebellum at DIV3, 5 and 7. Actin was used as a loading control. **B.** Autophagy was measured by quantifying the ratio of the band intensities of LC3B-II and LC3B-I ($n \geq 3$ mice) which reflects the amount of autophagosomes. Compared to wild-type cultures, this ratio increases in mutant cultures at DIV5 suggesting enhanced autophagy and then decreases at DIV7 probably as a result of either autophagic degradation or PC death (Kruskal-Wallis test followed by post-hoc Tukey test; * $p < 0.05$).

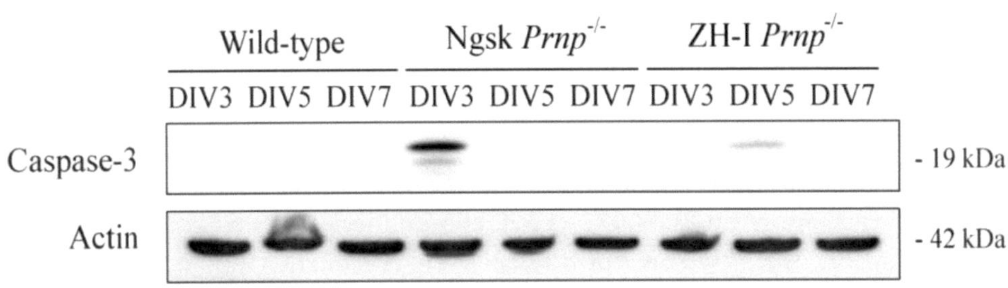

Figure 9. Western blot of the pro-apoptotic activated caspase-3. Activated caspase-3 is detected in extracts of organotypic cultures from Ngsk and ZH-I *Prnp*$^{-/-}$ mouse cerebellum, but not from wild-type mouse cerebellum. Actin serves as a loading control.

induced oxidative stress contributing to autophagic cell death [242]. Since autophagic flux is apparently normal in both *Prnp*$^{+/+}$ and *Prnp*$^{-/-}$ neurons in the absence of stress, the lack of PrP^C only seems to contribute to autophagy impairment under stress-induced conditions, such as H_2O_2 treatment [242], stress-inducing *in vitro* conditions, as well as Dpl-induced toxicity.

4.2.2. Prion protein PrP-doppel Dpl chimeras

When overexpressed ectopically in neurons, mutations within the central region of PrP^C are associated with severe neurotoxic activity, similar to that of Dpl [231, 243]. The absence of these segments, called central domains (CD) is believed to be responsible for neurodegeneration and ataxia. To understand the dual neurotoxicity *vs.* neuroprotective roles of PrP^C, transgenic mice expressing a fusion protein made of the CD of PrP^C inserted within the Dpl sequence have been generated [244]. These mice failed to develop typical Dpl-mediated neurological disorder indicating that this N-terminal portion of PrP^C reduces Dpl toxicity. To further investigate Dpl-like neurotoxicity, Lemaire-Vieille *et al.* recently generated lines of transgenic mice expressing three different chimeric PrP-Dpl proteins [245]. Chi1 (Dpl 1-57 replaced with PrP 1-125) and Chi2 (Dpl 1-66 replaced with PrP1-134) abrogates the pathogenecity of Dpl confirming the neuroprotective role of the PrP 23-134 N-terminal domain against Dpl toxicity. However, when Dpl 1-24 were replaced with PrP 1-124, these Chi3 transgenic mice that express a very low level of the chimeric protein develop ataxia, as early as 5 weeks of age. This phenotype is only rescued by overexpressing PrP^C, and not by a single copy of full-length PrP^C, indicating the strong toxicity of the chimeric protein Chi3. The Chi3 mice exhibit severe cerebellar atrophy with significant granule cell loss and prominent signs of autophagy in PCs (Fig. 2E). We conclude that the first 33 amino acids of Dpl, that are absent in Chi1 and Chi2 constructs, confer toxicity to the protein. This is confirmed *in vitro* by the highly neurotoxic effect of the 25-57 Dpl peptide on mouse embryo cortical neurons. Since this chimeric transgene is not expressed by PCs in the transgenic mice expressing Chi3, the signs of autophagy displayed by these neurons in vitro could result from the neurotoxic effect of the exogenous Chi3 chimeric protein, as well as the deleterious effect of losing their primary afferences (i.e. the granule cells).

5. Perspectives

The beneficial effects that autophagy has on prion infections is currently supported by a growing bulk of evidence from *in vivo* and *ex vivo* data and is strongly promising for future mid-term therapeutic approaches. To further understand the fascinating interplay between autophagy and PrDs, further investigations are necessary to decipher their molecular interactions. Important issues remain. How are the different phases of prion infection physiopathology i.e. propagation, trafficking, recycling and clearance connected with autophagy? Which autophagic pathways are activated by prions - the mTOR-dependent, mTOR-independent or both? The biological function of autophagy per se in prion infection is obscure as the cellular levels of autophagy can apparently modify cell susceptibility to prion infection, although changes in autophagy may be a pre-requisite or a consequence of a prion infection.

Overall, the results point to a need to counteract cell stress and to eliminate toxic aggregate-prone proteins that eventually saturate the usual degradation pathways, including autophagy. These are common features of prion disease and most of the other neurodegenerative diseases described in this review. Saturation of the autophagic machinery, loss or imbalance of autophagic flux is believed to lead to neurodegeneration. Understanding how autophagy relates to these diseases is a first step for developing autophagy modulation-based therapies for treating neurological disorders. This implies therapeutic consideration for each type of autophagic default at a precise step of the neurodegenerative disease concerned.

Acknowledgements

The authors are greatly indebted to Dr. Anne-Marie Haeberlé (CNRS UPR3212, Strasbourg) for excellent assistance in transmission electron microscopy and prion-infected tissue handling as well as to Dr. Catherine Vidal (Institut Pasteur, Paris) for intra-cerebellar inoculation of mice with 22L scrapie. A. R. is supported by a doctoral grant from the French Minister of Research and Technology and A. G. is supported by a grant from the French Centre National de la Recherche Scientifique and the AgroParisTech High School.

Author details

Audrey Ragagnin[1], Aurélie Guillemain[1], Nancy J. Grant[2] and Yannick J. R. Bailly[1]

1 Cytologie & Cytopathologie Neuronales, INCI CNRS UPR3212, Université de Strasbourg, Strasbourg, France

2 Trafic Membranaire Dans les Cellules Neurosécrétrices et Neuroimmunitaires, INCI CNRS UPR3212, Université de Strasbourg, Strasbourg, France

References

[1] Mizushima N, Komatsu M. Autophagy: renovation of cells and tissues. Cell (2011) 147:728-41.

[2] Levine B, Deretic V. Unveiling the roles of autophagy in innate and adaptive immunity. Nat Rev Immunol (2007) 7(10):767-77.

[3] Levine B, Kroemer G. Autophagy in the pathogenesis of disease. Cell (2008) 132:27-42.

[4] Mizushima N, Levine B, Cuervo AM, Klionsky DJ. Autophagy fights disease through cellular self-digestion. Nature (2008) 451:1069-75.

[5] Clarke PG. Developmental cell death: morphological diversity and multiple mechanisms. Anat Embryol (1990) 181:195-213.

[6] Kroemer G, Levine B. Autophagic cell death: the story of a misnomer. Nat Rev Mol Cell Biol (2008) 9:1004-10.

[7] Novikoff AB, Beaufay H, De Duve C. Electron microscopy of lysosomerich fractions from rat liver. J Biophys Biochem Cytol (1956) 2:S179-84.

[8] De Duve C. The significance of lysosomes in pathology and medicine. Proc Inst Med Chic (1966) 26:73-6.

[9] Dixon JS. "Phagocytic" lysosomes in chromatolytic neurones. Nature (1967) 215:657-658.

[10] Mizushima N, Yamamoto A, Matsui M, Yoshimori T, Ohsumi Y. In vivo analysis of autophagy in response to nutrient starvation using transgenic mice expressing a fluorescent autophagosome marker. Mol Biol Cell (2004) 15:1101-11.

[11] Hara T, Nakamura K, Matsui M, Yamamoto A, Nakahara Y, Suzuki-Migishima R, Yokoyama M, Mishima K, Saito I, Okano H, Mizushima N. Suppression of basal autophagy in neural cells causes neurodegenerative disease in mice. Nature (2006) 441:885–9.

[12] Komatsu M, Waguri S, Chiba T, Murata S, Iwata J, Tanida I, Ueno T, Koike M, Uchiyama Y, Kominami E, Tanaka K. Loss of autophagy in the central nervous system causes neurodegeneration in mice. Nature (2006) 441:880-4.

[13] Liang CC, Wang C, Peng X, Gan B, Guan JL. Neural-specific deletion of FIP200 leads to cerebellar degeneration caused by increased neuronal death and axon degeneration. J Biol Chem (2010) 285:3499-509.

[14] Xie Z, Klionsky DJ. Autophagosome formation: core machinery and adaptations. Nat Cell Biol (2007) 9:1102-9.

[15] He C, Klionsky DJ. Regulation mechanisms and signaling pathways of autophagy. Annu Rev Genet (2009) 43:67-93.

[16] Lee J-A. Neuronal autophagy: a housekeeper or a fighter in neuronal cell survival? Exp Neurobiol (2012) 21:1-8.

[17] Klionsky D, et 1268 al. Guidelines for the use and interpretation of assays for monitoring autophagy. Autophagy (2012) 8:445-544.

[18] Mizushima N, Yoshimori T, Ohsumi Y. The role of Atg proteins in autophagosome formation. Annu Rev Cell Dev Biol (2011) 27:107-32.

[19] Lee JA, Beigneux A, Ahmad ST, Young SG, Gao FB. ESCRT-III dysfunction causes autophagosome accumulation and neurodegeneration. Curr Biol (2007) 17:1561-1567.

[20] Yue Z, Friedman L, Komatsu M, Tanaka K. The cellular pathways of neuronal autophagy and their implication in neurodegenerative diseases. Biochim Biophys Acta (2009) 1793:1496-507.

[21] Hollenbeck PJ. Products of endocytosis and autophagy are retrieved from axons by regulated retrograde organelle transport. J Cell Biol (1993) 121:305-15.

[22] Maday S, Wallace KE, Holzbaur EL. Autophagosomes initiate distally and mature during transport toward the cell soma in primary neurons. J Cell Biol (2012) 196:407-17.

[23] Dunn WA Jr. Studies on the mechanisms of autophagy: formation of the autophagic vacuole. J Cell Biol (1990) 110:1923-33.

[24] Axe EL, Walker SA, Manifava M, Chandra P, Roderick HL, Habermann A, Griffiths G, Ktistakis NT. Autophagosome formation from membrane compartments enriched in phosphatidylinositol 3-phosphate and dynamically connected to the endoplasmic reticulum. J Cell Biol (2008) 182:685-701.

[25] Hayashi-Nishino M, Fujita N, Noda T, Yamaguchi A, Yoshimori T, Yamamoto A. A subdomain of the endoplasmic reticulum forms a cradle for autophagosome formation. Nat Cell Biol (2009) 11:1433-7.

[26] Ylä-Anttila P, Vihinen H, Jokitalo E, Eskelinen EL. 3D tomography reveals connections between the phagophore and endoplasmic reticulum. Autophagy (2009) 5:1180-5.

[27] Simonsen A & Tooze SA. Coordination of membrane events during autophagy by multiple class III PI3-kinase complexes. J Cell Biol (2009) 186:773-82.

[28] Tooze SA, Yoshimori T. The origin of the autophagosomal membrane. Nat Cell Biol (2010) 12:831-5.

[29] Hailey DW, Rambold AS, Satpute-Krishnan P, Mitra K, Sougrat R, Kim PK, Lippincott-Schwartz J. Mitochondria supply membranes for autophagosome biogenesis during starvation. Cell (2010) 141:656-67.

[30] Van der Vaart A, Reggiori F. The Golgi complex as a source for yeast autophagosomal membranes. Autophagy (2010) 6:800-1.

[31] Yen WL, Shintani T, Nair U, Cao Y, Richardson BC, Li Z, Hughson FM, Baba M, Klionsky DJ. The conserved oligomeric Golgi complex is involved in double-membrane vesicle formation during autophagy. J Cell Biol (2010) 188:101-14.

[32] Ravikumar B, Moreau K, Jahreiss L, Puri C, Rubinsztein DC. Plasma membrane contributes to the formation of pre-autophagosomal structures. Nat Cell Biol 12:747-57. Erratum in: Nat Cell Biol (2010) 12:1021.

[33] Novikoff PM, Novikoff AB, Quintana N, Hauw JJ. Golgi apparatus, GERL, and lysosomes of neurons in rat dorsal root ganglia, studied by thick section and thin section cytochemistry. J Cell Biol (1971) 50:859-86.

[34] Broadwell RD, Cataldo AM. The neuronal endoplasmic reticulum: its cytochemistry and contribution to the endomembrane system. II. Axons and terminals. J Comp Neurol (1984) 230:231-48.

[35] Matthews MR, Raisman G. A light and electron microscopic study of the cellular response to axonal injury in the superior cervical ganglion of the rat. Proc R Soc Lond B Biol Sci (1972) 181:43-79.

[36] Wang QJ, Ding Y, Kohtz DS, Mizushima N, Cristea IM, Rout MP, Chait BT, Zhong Y, Heintz N, Yue Z. Induction of autophagy in axonal dystrophy and degeneration. J Neurosci (2006) 26:8057-68.

[37] Yue Z. Regulation of neuronal autophagy in axon. Implication of autophagy in axonal function and dysfunction/degeneration. Autophagy (2007) 3:139-141.

[38] Liang XH, Jackson S, Seaman M, Brown K, Kempkes B, Hibshoosh H, Levine B. Induction of autophagy and inhibition of tumorigenesis by beclin-1. Nature (1999) 402:672-6.

[39] Selimi F, Lohof AM, Heitz S, Lalouette A, Jarvis CI, Bailly Y, Mariani J. Lurcher GRID2-induced death and depolarization can be dissociated in cerebellar Purkinje cells. Neuron (2003) 37:813-9.

[40] Yue Z, Horton A, Bravin M, DeJager PL, Selimi F, Heintz N. A novel protein complex linking the delta 2 glutamate receptor and autophagy: implications for neurodegeneration in lurcher mice. Neuron (2002) 35:921-33.

[41] Yang DS, Stavrides P, Mohan PS, Kaushik S, Kumar A, Ohno M, Schmidt SD, Wesson DW, Bandyopadhyay U, Jiang Y, Pawlik M, Peterhoff CM, Yang AJ, Wilson DA, St George-Hyslop P, Westaway D, Mathews PM, Levy E, Cuervo AM, Nixon RA. Therapeutic effects of remediating autophagy failure in a mouse model of Alzheimer disease by enhancing lysosomal proteolysis. Autophagy (2011) 7:788-9.

[42] Jarheiss L, Menzies FM, Rubinsztein DC. The itinerary of autophagosomes: from peripheral formation to kiss-and-run fusion with lysosomes. Traffic (2008) 9:574-587.

[43] Halpain J, Dehmelt L. The MAP1 family of microtubule-associated proteins Genome biology (2006) 7:224-230.

[44] Köchl R, Hu XW, Chan EY, Tooze SA. Microtubules facilitate autophagosome formation and fusion of autophagosomes with endosomes. Traffic (2006) 7:129-45.

[45] Fass E, Shvets E, Degani I, Hirschberg K, Elazar Z. Microtubules support production of starvation-induced autophagosomes but not their targeting and fusion with lysosomes. J Biol Chem (2006) 281:36303-16.

[46] Gonzalez-Billault C, Jimenez-Mateos EM, Caceres A, Diaz-Nido J, Wandosell F, Avila J. Microtubule-associated protein 1B function during normal development, regeneration, and pathological conditions in the nervous system. J Neurobiol (2004) 58:48-59.

[47] Nixon RA, Wegiel J, Kumar A, Yu WH, Peterhoff C, Cataldo A, Cuervo AM. Extensive involvement of autophagy in Alzheimer disease: an immuno-electron microscopy study. J Neuropathol Exp Neurol (2005) 64:113-22.

[48] Boland B & Nixon RA. Neuronal macroautophagy: from development to degeneration. Mol Aspects Med (2006) 27:503-19.

[49] Boland B, Kumar A, Lee S, Platt FM, Wegiel J, Yu WH, Nixon RA. Autophagy induction and autophagosome clearance in neurons: relationship to autophagic pathology in Alzheimer's disease. J Neurosci (2008) 28:6926-37.

[50] Lee S, Sato Y, Nixon RA. Primary lysosomal dysfunction causes cargo-specific deficits of axonal transport leading to Alzheimer-like neuritic dystrophy. Autophagy (2011a) 7:1562-1563.

[51] Sigmond T, Fehér J, Baksa A, Pásti G, Pálfia Z, Takács-Vellai K, Kovács J, Vellai T, Kovács AL. Qualitative and quantitative characterization of autophagy in Caenorhabditis elegans by electron microscopy. Methods Enzymol (2008) 451:467-91.

[52] Okazaki N, Yan J, Yuasa S, Ueno T, Kominami E, Masuho Y, Koga H, Muramatsu M. Interaction of the Unc-51-like kinase and microtubule-associated protein light chain 3 related proteins in the brain: possible role of vesicular transport in axonal elongation. Brain Res Mol Brain Res (2000) 85:1-12.

[53] Komatsu M, Wang QJ, Holstein GR, Friedrich VL Jr, Iwata J, Kominami E, Chait BT, Tanaka K, Yue Z. Essential role for autophagy protein Atg7 in the maintenance of axonal homeostasis and the prevention of axonal degeneration. Proc Natl Acad Sci USA (2007a) 104:14489-94.

[54] Nishiyama J, Miura E, Mizushima N, Watanabe M, Yuzaki M. Aberrant membranes and double-membrane structures accumulate in the axons of Atg5-null Purkinje cells before neuronal death. Autophagy (2007) 3:591-6.

[55] Komatsu M, Waguri S, Koike M, Sou YS, Ueno T, Hara T, Mizushima N, Iwata J, Ezaki J, Murata S, Hamazaki J, Nishito Y, Iemura S, Natsume T, Yanagawa T, Uwayama J, Warabi E, Yoshida H, Ishii T, Kobayashi A, Yamamoto M, Yue Z, Uchiyama Y, Kominami E, Tanaka K. Homeostatic levels of p62 control cytoplasmic inclusion body formation in autophagy-deficient mice. Cell (2007b) 131:1149-63.

[56] Pankiv S, Clausen TH, Lamark T, Brech A, Bruun JA, Outzen H, Øvervatn A, Bjørkøy G, Johansen T. p62/SQSTM1 binds directly to Atg8/LC3 to facilitate degradation of ubiquitinated protein aggregates by autophagy. J Biol Chem (2007) 282:24131-45.

[57] Zhou X, Babu JR, da Silva S, Shu Q, Graef IA, Oliver T, Tomoda T, Tani T, Wooten M, Wang F. Unc-51-like kinase 1/2-mediated endocytic processes regulate filopodia extension and branching of sensory axons. Proc Natl Acad Sci USA (2007) 104:5842-7.

[58] Coleman M. Axon degeneration mechanisms: commonality amid diversity. Nat Rev Neurosci (2005) 6:889-98.

[59] Raff MC, Whitmore AV, Finn JT. Axonal self-destruction and neurodegeneration. Science (2002) 296:868-71.

[60] Yang Y, Fukui K, Koike T, Zheng X. Induction of autophagy in neurite degeneration of mouse superior cervical ganglion neurons. Eur J Neurosci (2007) 26:2979-2988.

[61] Shehata M, Matsumura H, Okubo-Suzuki R, Ohkawa N, Inokuchi K. Neuronal stimulation induces autophagy in hippocampal neurons that is involved in AMPA receptor degradation after chemical long-term depression. J Neurosci (2012) 32:10413-22.

[62] Gordon PB, Seglen PO. Prelysosomal convergence of autophagic and endocytic pathways. Biochem Biophys Res Commun (1988) 151:40-7.

[63] Rusten TE, Vaccari T, Lindmo K, Rodahl LM, Nezis IP, Sem-Jacobsen C, Wendler F, Vincent JP, Brech A, Bilder D, Stenmark H. ESCRTs and Fab1 regulate distinct steps of autophagy. Curr Biol (2007) 17:1817-25.

[64] Filimonenko M, Stuffers S, Raiborg C, Yamamoto A, Malerød L, Fisher EM, Isaacs A, Brech A, Stenmark H, Simonsen A. Functional multivesicular bodies are required for autophagic clearance of protein aggregates associated with neurodegenerative disease. J Cell Biol (2007) 179:485-500.

[65] Larsen KE, Fon EA, Hastings TG, Edwards RH, Sulzer D. Methamphetamine-induced degeneration of dopaminergic neurons involves autophagy and upregulation of dopamine synthesis. J Neurosci (2002) 22:8951-60.

[66] Ravikumar B, Duden R, Rubinsztein DC. Aggregate-prone proteins with polyglutamine and polyalanine expansions are degraded by autophagy. Hum Mol Genet (2002) 11:1107-17.

[67] Rubinsztein DC. The roles of intracellular protein-degradation pathways in neurodegeneration. Nature (2006) 443:780-6.

[68] Rubinsztein DC, Gestwicki JE, Murphy LO, Klionsky DJ. Potential therapeutic applications of autophagy. Nat Rev Drug Discov (2007) 6:304-12.

[69] Nezis IP, Simonsen A, Sagona AP, Finley K, Gaumer S, Contamine D, Rusten TE, Stenmark H, Brech A. Ref(2)P, the Drosophila melanogaster homologue of mammalian p62, is required for the formation of protein aggregates in adult brain. J Cell Biol (2008) 180:1065-71.

[70] Collins CA, Wairkar YP, Johnson SL, DiAntonio A. Highwire restrains synaptic growth by attenuating a MAP kinase signal. Neuron (2006) 51:57-69.

[71] Shen W, Ganetzky B. Autophagy promotes synapse development in Drosophila. J Cell Biol (2009) 187:71-9.

[72] Wan HI, DiAntonio A, Fetter RD, Bergstrom K, Strauss R, Goodman CS. Highwire regulates synaptic growth in Drosophila. Neuron (2000) 26:313-29.

[73] Rowland AM, Richmond JE, Olsen JG, Hall DH, Bamber BA. Presynaptic terminals independently regulate synaptic clustering and autophagy of GABAA receptors in Caenorhabditis elegans. J Neurosci (2006) 26:1711-20.

[74] Zhang XD, Wang Y, Wang Y, Zhang X, Han R, Wu JC, Liang ZQ, Gu ZL, Han F, Fukunaga K, Qin ZH. p53 mediates mitochondria dysfunction-triggered autophagy activation and cell death in rat striatum. Autophagy (2009) 5:339-50.

[75] Wairkar YP, Toda H, Mochizuki H, Furukubo-Tokunaga K, Tomoda T, Diantonio A. Unc-51 controls active zone density and protein composition by downregulating ERK signaling. J Neurosci (2009) 29:517-28.

[76] Hernandez D, Torres CA, Setlik W, Cebrián C, Mosharov EV, Tang G, Cheng HC, Kholodilov N, Yarygina O, Burke RE, Gershon M, Sulzer D. Regulation of presynaptic neurotransmission by macroautophagy. Neuron (2012) 74:277–284.

[77] Kamada Y, Yoshino K, Kondo C, Kawamata T, Oshiro N, Yonezawa K, Ohsumi Y. Tor directly controls the Atg1 kinase complex to regulate autophagy. Mol Cell Biol (2010) 30:1049-58.

[78] Huang J & Manning BD. A complex interplay between Akt, TSC2 and the two mTOR complexes. Biochem Soc Trans (2009) 37:217-22.

[79] Long X, Müller F, Avruch J. TOR action in mammalian cells and in Caenorhabditis elegans. Curr Top Microbiol Immunol (2004) 279:115-38.

[80] Richter JD, Klann E. Making synaptic plasticity and memory last: mechanisms of translational regulation. Genes Dev (2009) 23:1-11.

[81] Hu JY, Chen Y, Schacher S. Protein kinase C regulates local synthesis and secretion of a neuropeptide required for activity-dependent long-term synaptic plasticity. J Neurosci (2007) 27:8927-8939.

[82] Weragoda RM, Walters ET. Serotonin induces memory-like, rapamycin-sensitive hyperexcitability in sensory axons of aplysia that contributes to injury responses. J Neurophysiol (2007) 98:1231-9.

[83] Lee S, Sato Y, Nixon RA. Lysosomal proteolysis inhibition selectively disrupts axonal transport of degradative organelles and causes an Alzheimer's-like axonal dystrophy. J Neurosci (2011) 31(21):7817-30.

[84] Bunge MB. Fine structure of nerve fibers and growth cones of isolated sympathetic neurons in culture. J Cell Biol (1973) 56:713-35.

[85] Zanjani SH, Selimi F, Vogel MW, Haeberlé AM, Boeuf J, Mariani J, Bailly YJ. Survival of interneurons and parallel fiber synapses in a cerebellar cortex deprived of Purkinje cells: studies in the double mutant mouse Grid2Lc/+;Bax(-/-). J Comp Neurol (2006) 497:622-35.

[86] Eskelinen EL. Maturation of autophagic vacuoles in mammalian cells. Autophagy (2005) 1:1-10.

[87] Mizushima N, Klionsky DJ.Protein turnover via autophagy: implications for metabolism. Annu Rev Nutr (2007) 27:19-40.

[88] Ramesh Babu J, Lamar Seibenhener M, Peng J, Strom AL, Kemppainen R, Cox N, Zhu H, Wooten MC, Diaz-Meco MT, Moscat J, Wooten MW. Genetic inactivation of p62 leads to accumulation of hyperphosphorylated tau and neurodegeneration. J Neurochem (2008) 106:107-20.

[89] Jiang J, Parameshwaran K, Seibenhener ML, Kang MG, Suppiramaniam V, Huganir RL, Diaz-Meco MT, Wooten MW. AMPA receptor trafficking and synaptic plasticity require SQSTM1/p62. Hippocampus (2009) 19:392-406.

[90] Jobim PF, Pedroso TR, Werenicz A, Christoff RR, Maurmann N, Reolon GK, Schröder N, Roesler R. Impairment of object recognition memory by rapamycin inhibition of mTOR in the amygdala or hippocampus around the time of learning or reactivation. Behav Brain Res (2012) 228:151-8.

[91] Cota D, Proulx K, Smith KA, Kozma SC, Thomas G, Woods SC, Seeley RJ. Hypothalamic mTOR signaling regulates food intake. Science (2006) 312:927-30.

[92] Young JE, La Spada AR. Development of selective nutrient deprivation as a system to study autophagy induction and regulation in neurons. Autophagy (2009) 5:555-7.

[93] Sarkar S, Ravikumar B, Floto RA, Rubinsztein DC. Rapamycin and mTOR-independent autophagy inducers ameliorate toxicity of polyglutamine-expanded huntingtin and related proteinopathies. Cell Death Differ (2009) 16(1):46-56.

[94] Du L, Hickey RW, Bayir H, Watkins SC, Tyurin VA, Guo F, Kochanek PM, Jenkins LW, Ren J, Gibson G, Chu CT, Kagan VE, Clark RS. Starving neurons show sex difference in autophagy. J Biol Chem (2009) 284:2383-96.

[95] Koike M, Shibata M, Tadakoshi M, Gotoh K, Komatsu M, Waguri S, Kawahara N, Kuida K, Nagata S, Kominami E, Tanaka K, Uchiyama Y. Inhibition of autophagy prevents hippocampal pyramidal neuron death after hypoxic-ischemic injury. Am J Pathol (2008) 172:454-69.

[96] Adhami F, Liao G, Morozov YM, Schloemer A, Schmithorst VJ, Lorenz JN, Dunn RS, Vorhees CV, Wills-Karp M, Degen JL, Davis RJ, Mizushima N, Rakic P, Dardzinski BJ, Holland SK, Sharp FR, Kuan CY. Cerebral ischemia-hypoxia induces intravascular coagulation and autophagy. Am J Pathol (2006) 169:566-83.

[97] Borsello T, Croquelois K, Hornung JP, Clarke PG. N-methyl-d-aspartate-triggered neuronal death in organotypic hippocampal cultures is endocytic, autophagic and mediated by the c-Jun N-terminal kinase pathway. Eur J Neurosci (2003) 18:473-485.

[98] Wang Y, han R, Liang ZQ, Wu JC,ZhangXD, Gu ZI, Qin ZH. An autophagic mechanism is involved in apoptotic death of rat striatal neurons induced by the non N-methyl-D-aspartate receptor agonist kainic acid. Autophagy (2008) 4:214-226.

[99] Høyer-Hansen M, Jäättelä M. Connecting endoplasmic reticulum stress to autophagy by unfolded protein response and calcium. Cell Death Differ (2007) 14:1576-82.

[100] Castino R, Lazzeri G, Lenzi P, Bellio N, Follo C, Ferrucci M, Fornai F, Isidoro C. Suppression of autophagy precipitates neuronal cell death following low doses of methamphetamine. J Neurochem (2008) 106:1426-39.

[101] Zhu JH, Horbinsky C, Guo F, Watkins S, Uchiyama Y, Chu CT. Regulation of autophagy by extracellular signal-regulated protein kinases during 1-methyl-4-phenylpyridinium-induced cell death. Am J Pathol (2007) 170:75-86.

[102] Ding Q, Dimayuga E, Martin S, Bruce-Keller AJ, Nukala V, Cuervo AM, Keller JN. Characterization of chronic low-level proteasome inhibition on neural homeostasis. J Neurochem (2003) 86:489-97.

[103] Pandey UB, Nie Z, Batlevi Y, McCray BA, Ritson GP, Nedelsky NB, Schwartz SL, DiProspero NA, Knight MA, Schuldiner O, Padmanabhan R, Hild M, Berry DL, Garza D, Hubbert CC, Yao TP, Baehrecke EH, Taylor JP. HDAC6 rescues neurodegeneration and provides an essential link between autophagy and the UPS. Nature (2007) 447:859-63.

[104] Bedford L, Hay D, Devoy A, Paine S, Powe DG, Seth R, Gray T, Topham I, Fone K, Rezvani N, Mee M, Soane T, Layfield R, Sheppard PW, Ebendal T, Usoskin D, Lowe J, Mayer RJ. Depletion of 26S proteasomes in mouse brain neurons causes neurodegeneration and Lewy-like inclusions resembling human pale bodies. J Neurosci (2008) 28:8189-98.

[105] Koike M, Shibata M, Waguri S, Yoshimura K, Tanida I, Kominami E, Gotow T, Peters C, von Figura K, Mizushima N, Saftig P, Uchiyama Y. Participation of autophagy in storage of lysosomes in neurons from mouse models of neuronal ceroid-lipofuscinoses (Batten disease). Am J Pathol (2005) 167:1713-28.

[106] Liao G, Yao Y, Liu J, Yu Z, Cheung S, Xie A, Liang X, Bi X. Cholesterol accumulation is associated with lysosomal dysfunction and autophagic stress in Npc1 -/- mouse brain. Am J Pathol (2007) 171:962-75.

[107] Pacheco CD, Lieberman AP. Lipid trafficking defects increase Beclin-1 and activate autophagy in Niemann-Pick type C disease. Autophagy (2007) 3:487-9.

[108] Vergarajauregui S, Connelly PS, Daniels MP, Puertollano R. Autophagic dysfunction in mucolipidosis type IV patients. Hum Mol Genet (2008) 17:2723-2737.

[109] Clausen TH, Lamark T, Isakson P, Finley K, Larsen KB, Brech A, Øvervatn A, Stenmark H, Bjørkøy G, Simonsen A, Johansen T. p62/SQSTM1 and ALFY interact to facilitate the formation of p62 bodies/ALIS and their degradation by autophagy. Autophagy (2010) 6:330-44.

[110] Knaevelsrud H, Simonsen A. Fighting disease by selective autophagy of aggregate-prone proteins. FEBS Lett (2010) 584:2635-45.

[111] Webb JL, Ravikumar B, Atkins J, Skepper JN, Rubinsztein DC. Alpha-synuclein is degraded by both autophagy and the proteasome. J Biol Chem (2003) 278:25009-25013.

[112] Fortun J, Dunn WA Jr, Joy S, Li J, Notterpek L. Emerging role for autophagy in the removal of aggresomes in Schwann cells. J Neurosci (2003) 23:10672-80.

[113] Bjørkøy G, Lamark T, Brech A, Outzen H, Perander M, Overvatn A, Stenmark H, Johansen T. p62/SQSTM1 forms protein aggregates degraded by autophagy and has a protective effect on huntingtin-induced cell death. J Cell Biol (2005) 171:603-14.

[114] Kabuta T, Suzuki Y, Wada K. Degradation of amyotrophic lateral sclerosis-linked mutant Cu/Zn-superoxide dismutase proteins by macroautophagy and the proteasome. J Biol Chem (2006) 281:30524-33.

[115] Berger Z, Ravikumar B, Menzies FM, Oroz LG, Underwood BR, Pangalos MN, Schmitt I, Wullner U, Evert BO, O'Kane CJ, Rubinsztein DC. Rapamycin alleviates toxicity of different aggregate-prone proteins. Hum Mol Genet (2006) 15:433-42.

[116] Petersén A, Larsen KE, Behr GG, Romero N, Przedborski S, Brundin P, Sulzer D. Expanded CAG repeats in exon 1 of the Huntington's disease gene stimulate dopamine-mediated striatal neuron autophagy and degeneration. Hum Mol Genet (2001) 10:1243-54.

[117] Rubinsztein DC, DiFiglia M, Heintz N, Nixon RA, Qin ZH, Ravikumar B, Stefanis L, Tolkovsky A. Autophagy and its possible roles in nervous system diseases, damage and repair. Autophagy (2005) 1:11-22.

[118] Ventruti A, Cuervo AM. Autophagy and neurodegeneration. Curr Neurol Neurosci Rep (2007) 7:443-51.

[119] Nixon RA, Yang DS, Lee JH. Neurodegenerative lysosomal disorders: a continuum from development to late age. Autophagy (2008) 4:590-9.

[120] Lee JH, Yu WH, Kumar A, Lee S, Mohan PS, Peterhoff CM, Wolfe DM, Martinez-Vicente M, Massey AC, Sovak G, Uchiyama Y, Westaway D, Cuervo AM, Nixon RA. Lysosomal proteolysis and autophagy require presenilin 1 and are disrupted by Alzheimer-related PS1 mutations. Cell (2010) 141:1146-1158.

[121] Mariño G, Madeo F, Kroemer G. Autophagy for tissue homeostasis and neuroprotection. Curr Opin Cell Biol (2011) 23:198-206.

[122] Bence NF, Sampat RM, Kopito RR. Impairment of the ubiquitin-proteasome system by protein aggregation. Science (2001) 292:1552-5.

[123] Fleming A, Noda T, Yoshimori T, Rubinsztein DC. Chemical modulators of autophagy as biological probes and potential therapeutics. Nat Chem Biol (2011) 7:9-17.

[124] Nixon RA. Autophagy, amyloidogenesis and Alzheimer disease. J Cell Sci (2007) 120:4081-91.

[125] Spilman P, Podlutskaya N, Hart MJ, Debnath J, Gorostiza O, Bredesen D, Richardson A, Strong R, Galvan V. Inhibition of mTOR by rapamycin abolishes cognitive deficits

and reduces amyloid-beta levels in a mouse model of Alzheimer's disease. PLoS One (2010) 5:e9979.

[126] Heiseke A, Aguib Y, Riemer C, Baier M, Schatzl HM. Lithium induces clearance of protease resistant prion protein in prion-infected cells by induction of autophagy. J Neurochem (2009) 109:25-34.

[127] Cortes CJ, Qin K, Cook J, Solanki A, Mastrianni JA. Rapamycin delays disease onset and prevents PrP plaque deposition in a mouse model of Gerstmann-Sträussler-Scheinker disease. J Neurosci (2012) 32:12396-12405.

[128] Massey AC, Kaushik S, Cuervo AM. Lysosomal chat maintains the balance. Autophagy (2006) 2:325-7.

[129] Iwata A, Riley BE, Johnston JA, Kopito RR. HDAC6 and microtubules are recquired for autophagic degradation of aggregated huntingtin. J Biol Chem (2005) 280:40282-40292.

[130] Martinez-Vicente M, Cuervo AM. Autophagy and neurodegeneration: when the cleaning crew goes on strike. Lancet Neurol (2007) 6:352-61.

[131] Shimizu S, Kanaseki T, Mizushima N, Mizuta T, Arakawa-Kobayashi S, Thompson CB, Tsujimoto Y. Role of Bcl-2 family proteins in a non-apoptotic programmed cell death dependent on autophagy genes. Nat Cell Biol (2004) 6:1221-8.

[132] Veneault-Fourrey C, Talbot NJ. Autophagic cell death and its importance for fungal developmental biology and pathogenesis. Autophagy(2007) 3:126-7.

[133] Yu WH, Kumar A, Peterhoff C, Shapiro Kulnane L, Uchiyama Y, Lamb BT, Cuervo AM, Nixon RA. Autophagic vacuoles are enriched in amyloid precursor protein-secretase activities: implications for beta-amyloid peptide over-production and localization in Alzheimer's disease. Int J Biochem Cell Biol (2004) 36:2531-40.

[134] Yu L, Wan F, Dutta S, Welsh S, Liu Z, Freundt E, Baehrecke EH, Lenardo M. Autophagic programmed cell death by selective catalase degradation. Proc Natl Acad Sci USA (2006) 103:4952-7.

[135] Scott RC, Juhász G, Neufeld TP. Direct induction of autophagy by Atg1 inhibits cell growth and induces apoptotic cell death. Curr Biol (2007) 17:1-11.

[136] Wullschleger S, Loewith R, Hall MN. TOR signaling in growth and metabolism. Cell (2006) 124:471-84.

[137] Arico S, Petiot A, Bauvy C, Dubbelhuis PF, Meijer AJ, Codogno P, Ogier-Denis E. The tumor suppressor PTEN positively regulates macroautophagy by inhibiting the phosphatidylinositol 3-kinase/protein kinase B pathway. J Biol Chem (2001) 276:35243-6.

[138] Jin S. Autophagy, mitochondrial quality control, and oncogenesis. Autophagy (2006) 2:80-4.

[139] Martin DN, Baehrecke EH. Caspases function in autophagic programmed cell death in Drosophila. Development (2004) 131:275-84.

[140] Chu CT. Autophagic stress in neuronal injury and disease. J Neuropathol Exp Neurol (2006) 65:423-32.

[141] Alirezaei M, Jelodar G, Niknam P, Ghayemi Z, Nazifi S. Betaine prevents ethanol-induced oxidative stress and reduces total homocysteine in the rat cerebellum. J Physiol Biochem (2011) 67:605-12.

[142] Prusiner SB. Prions. Proc Natl Acad Sci USA(1998) 95:13363-13383.

[143] Weissmann C. The state of prion. Nat Rev Microbiol (2004) 2:861-871.

[144] Aguzzi A, Polymenidou M. Mammalian prion biology: one century of evolving concepts. Cell (2004) 116:313-327.

[145] Collinge J. Molecular neurology of prion disease. J Neurol Neurosurg Psychiatry (2005) 76:906-919.

[146] Prusiner SB. Novel proteinaceous infectious particles cause scrapie. Science (1982) 216:136-144.

[147] Aguzzi A, Haass C.Games played by rogue proteins in prion disorders and Alzheimer's disease. Science (2003) 302:814-818.

[148] Brandner S, Raeber A, Sailer A, Blättler T, Fischer M, Weissmann C, Aguzzi A. Normal host prion protein (PrPC) is required for scrapie spread within the central nervous system. Proc Natl Acad Sci USA (1996) 93:13148-51.

[149] Büeler H, Aguzzi A, Sailer A, Greiner RA, Autenried P, Aguet M, Weissmann C. Mice devoid of PrP are resistant to scrapie. Cell (1993) 73:1339-47.

[150] Radford HE, Mallucci GR. The role of GPI-anchored PrP(C) in mediating the neurotoxic effect of scrapie prions in neurons. Curr Issues Mol Biol (2010) 12:119-128.

[151] Harris DA, True HL. New insights into prion structure and toxicity. Neuron (2006) 50:353-7.

[152] Winklhofer KF, Tatzelt J, Haass C. The two faces of protein misfolding: gain- and loss-of-function in neurodegenerative diseases. EMBO J (2008) 27:336-49.

[153] Brown DR, Qin K, Herms JW, Madlung A, Manson J, Strome R, Fraser PE, Kruck T, von Bohlen A, Schulz-Schaeffer W, Giese A, Westaway D, Kretzschmar H. The cellular prion protein binds copper in vivo. Nature (1997) 390:684-7.

[154] Gauczynski S, Peyrin JM, Haïk S, Leucht C, Hundt C, Rieger R, Krasemann S, Deslys JP, Dormont D, Lasmézas CI, Weiss S. The 37-kDa/67-kDa laminin receptor acts as the cell-surface receptor for the cellular prion protein. Embo J (2001) 20:5863-75.

[155] Schmitt-Ulms G, Legname G, Baldwin MA, Ball HL, Bradon N, Bosque PJ, Crossin KL, Edelman GM, DeArmond SJ, Cohen FE, Prusiner SB. Binding of neural cell adhesion molecules (N-CAMs) to the cellular prion protein. J Mol Biol (2001) 314:1209-25.

[156] Loubet D, Dakowski C, Pietri M, Pradines E, Bernard S, Callebert J, Ardila-Osorio H, Mouillet-Richard S, Launay JM, Kellermann O, Schneider B. Neuritogenesis: the prion protein controls b1 integrin signaling activity. FASEB J (2012) 26:678-90.

[157] Brown DR, Schulz-Schaeffer WJ, Schmidt B, Kretzschmar HA. Prion protein-deficient cells show altered response to oxidative stress due to decreased SOD-1 activity. Exp Neurol (1997) 146:104-12.

[158] Kuwahara C, Takeuchi AM, Nishimura T, Haraguchi K, Kubosaki A, Matsumoto Y, Saeki K, Matsumoto Y, Yokoyama T, Itohara S, Onodera T.Prions prevent neuronal cell-line death. Nature (1999) 400:225-6.

[159] Milhavet O, Lehmann S. Oxidative stress and the prion protein in transmissible spongiform encephalopathies. Brain Res Brain Res Rev (2002) 38:328-39.

[160] Spudich A, Frigg R, Kilic E, Kilic U, Oesch B, Raeber A, Bassetti CL, Hermann DM. Aggravation of ischemic brain injury by prion protein deficiency: role of ERK-1/-2 and STAT-1. Neurobiol Dis (2005) 20:442-449.

[161] Weise J, Sandau R, Schwarting S, Crome O, Wrede A, Schulz-Schaeffer W, Zerr I, Bähr M. Deletion of cellular prion protein results in reduced Akt activation, enhanced postischemic caspase-3 activation, and exacerbation of ischemic brain injury. Stroke (2006) 37:1296-300.

[162] Mitteregger G, Vosko M, Krebs B, Xiang W, Kohlmannsperger V, Nölting S, Hamann GF, Kretzschmar HA. The role of the octarepeat region in neuroprotective function of the cellular prion protein. Brain Pathol (2007) 17:174-83.

[163] Zanata SM, Lopes MH, Mercadante AF, Hajj GN, Chiarini LB, Nomizo R, Freitas AR, Cabral AL, Lee KS, Juliano MA, de Oliveira E, Jachieri SG, Burlingame A, Huang L, Linden R, Brentani RR, Martins VR. Stress-inducible protein 1 is a cell surface ligand for cellular prion that triggers neuroprotection. Embo J (2002) 21:3307-16.

[164] Schneider B, Mutel V, Pietri M, Ermonval M, Mouillet-Richard S, Kellermann O. NADPH oxidase and extracellular regulated kinases 1/2 are targets of prion protein signaling in neuronal and nonneuronal cells. Proc Natl Acad Sci USA (2003) 100:13326-31.

[165] Pradines E, Loubet D, Schneider B, Launay JM, Kellermann O, Mouillet-Richard S. CREB-dependent gene regulation by prion protein: impact on MMP-9 and beta-dystroglycan. Cell Signal (2008) 20:2050-2058.

[166] Pradines E, Loubet D, Mouillet-Richard S, Manivet P, Launay JM, Kellermann O, Schneider B. Cellular prion protein coupling to TACE-dependent TNF-alpha shed-

ding controls neurotransmitter catabolism in neuronal cells. J Neurochem (2009) 110:912-23.

[167] Linden R, Martins VR, Prado MA, Cammarota M, Izquierdo I, Brentani RR. Physiology of the prion protein. Physiol Rev (2008) 88:673-728.

[168] Moreno JA, Radford H, Peretti D, Steinert JR, Verity N, Martin MG, Halliday M, Morgan J, Dinsdale D, Ortori CA, Barrett DA, Tsaytler P, Bertolotti A, Willis AE, Bushell M, Mallucci GR. Sustained translational repression by eIF2α-P mediates prion neurodegeneration. Nature (2012) 485:507-511.

[169] Ashe KH, Aguzzi A. Prions, prionoids and pathogenic proteins in Alzheimer disease. Prion (2013) 7: in press.

[170] Kristiansen M, Deriziotis P, Dimcheff DE, Jackson GS, Ovaa H, Naumann H, Clarke AR, van Leeuwen FW, Menéndez-Benito V, Dantuma NP, Portis JL, Collinge J, Tabrizi SJ. Disease-associated prion protein oligomers inhibit the 26S proteasome. Mol Cell (2007) 26:175-88.

[171] Pietri M, Caprini A, Mouillet-Richard S, Pradines E, Ermonval M, Grassi J, Kellermann O, Schneider Schneider B. Overstimulation of PrPC signaling pathways by prion peptide 106-126 causes oxidative injury of bioaminergic neuronal cells. J Biol Chem (2006) 281:28470-9.

[172] Heiseke A, Aguib Y, Schatzl HM. Autophagy, prion infection and their mutual interactions. Curr Issues Mol Biol (2010) 12:87-98.

[173] Sikorska B, Liberski PP, Giraud P, Kopp N, Brown P. Autophagy is a part of ultrastructural synaptic pathology in Creutzfeldt-Jakob disease: a brain biopsy study. Int J Biochem Cell Biol (2004) 36:2563-73.

[174] Liberski PP, Brown DR, Sikorska B, Caughey B, Brown P. Cell death and autophagy in prion diseases (transmissible spongiform encephalopathies). Folia Neuropathol (2008) 46:1-25.

[175] Boellaard JW, Schlote W, Tateishi J. Neuronal autophagy in experimental Creutzfeldt-Jakob's disease. Acta Neuropathol (1989) 78:410-418.

[176] Boellaard JW, Kao M, Schlote W, Diringer H. Neuronal autophagy in experimental scrapie. Acta Neuropathol (1991) 82:225-228.

[177] Schätzl HM, Laszlo L, Holtzman DM, Tadzelt J, DeArmond SJ, Weiner RI, Mobley WC, Prusiner SB. A hypothalamic neuronal cell line persistenly infected with scrapie prions exhibits apoptosis. J Virol (1997) 71:8821-8831.

[178] Dron M, Dandoy-Dron F, Guillo F, Benboudjema L, Haw J-J, Lebon P, Dormont D, Tovey MG. Characterization of the human analogue of a scrapie-responsive gene. J Biol Chem (1998) 273:18015-18018.

[179] Dandoy-Dron F, Guillo F, Benboudjema L, Deslys J-P, Lasmézas C, Dormont D, Tovey MG, Dron M. Gene expression in scrapie. Cloning of a new scrapie-responsive

gene and the identification of seven other mRNA transcripts. J Biol Chem (1998) 273:7691-7697.

[180] Dandoy-Dron F, Benboudjema L, Guillo F, Jaegly A, Jasmin C, Dormont D, Tovey MG, Dron M. Enhanced levels of scrapie responsive gene mRNA in BSE-infected mouse brain. Brain Res Mol Brain Res (2000) 76:173-179.

[181] Dron M, Bailly Y, Beringue V, Haeberlé A-M, Griffond B, Risold P-Y, Tovet MG, Laude H, Dandoy-Dron F. Scrg1 is induced in TSE and brain injuries, and associated with autophagy. Eur J Neurosci (2005) 22:133-146.

[182] Dron M, Bailly Y, Beringue V, Haeberlé A-M, Griffond B, Risold P-Y, Tovey MG, Laude H, Dandoy-Dron F. SCRG1, a potential marker of autophagy in transmisible spongiform encephalopathies. Autophagy (2006) 2:58-60.

[183] Xu Y, Tian C, Wang SB, Xie WL, Guo Y, Zhang J, Shi Q, Chen C, Dong XP. Activation of the macroautophagic system in scrapie-infected experimental animals and human genetic prion diseases. Autophagy (2012) 8 (in press).

[184] Hung SY, Huang WP, Liou HC, Fu WM. Autophagy protects neurons from Abeta-induced cytotoxicity. Autophagy (2009) 5:502-510.

[185] Doh-Ura K, Iwaki T, Caughey B. Lysosomotropic agents and cysteine protease inhibitors inhibit scrapie-associated prion protein accumulation. J Virol (2000) 74:4894-4897.

[186] Ertmer A, Gilch S, Yun SW, Flechsig E, Klebl B, Stein-Gerlach M, Klein MA, Schätzl HM. The tyrosine kinase inhibitor STI571 induces cellular clearance of PrPsc in prion-infected cells. J Biol Chem (2004) 279:41918-41927.

[187] Oh JM, Shin HY, Park SJ, Kim BH, Choi JK, Choi EK, Carp RI, Kim YS. The involvement of cellular prion protein in the autophagy pathway in neuronal cells. Mol Cell Neurosci (2008) 39:238-47.

[188] Mok SW, Riemer C, Madela K, Hsu DK, Liu FT, Gültner S, Heise I, Baier M. Role of galectin-3 in prion infections of the CNS. Biochem Biophys Res Commun (2007) 359:672-8.

[189] Ertmer A, Huber V, Gilch S, Yoshimori T, Erfle V, Duyster J, Elsässer HP, Schätzl HM. The anticancer drug imatinib induces cellular autophagy. Leukemia (2007) 21:936-942.

[190] Yun SW, Ertmer A, Flechsig E, Gilch S, Riederer P, Gerlach M, Schatzl HM, Klein MA. The tyrosine kinase inhibitor imatinib mesylate delays prion neuroinvasion by inhibiting prion propagation in the periphery. J Neurovirol (2007) 13:328-337.

[191] Sarkar S, Floto RA, Berger Z, Imarisio S, Cordenier A, Pasco M, Cook LJ, Rubinsztein DC. Lithium induces autophagy by inhibiting inositol monophosphatase. J Cell Biol (2005) 170:1101-11.

[192] Sarkar S, Davies JE, Huang Z, Tunnacliffe A, Rubinztein DC. Trehalose, a novel mTOR-independent autophagy enhancer, accelerates the clearance of mutant huntingtin and alpha-synuclein. J Biol Chem (2007) 282:5641-5652.

[193] Aguib Y, Heiseke A, Gilch S, Riemer C, Baier M, Schätzl HM, Ertmer A. Autophagy induction by trehalose counteracts cellular prion infection. Autophagy (2009) 5:361-9.

[194] Sunyach C, Jen A, Deng J, Fitzgerald KT, Frobert Y, Grassi J, McCaffrey MW, Morris R. The mechanism of internalization of glycosylphosphatidylinositol-anchored prion protein. EMBO J (2003) 22:3591-3601.

[195] Marella M, Lehmann S, Grassi J, Chabry J. Filipin prevents pathological prion protein accumulation by reducing endocytosis and inducing cellular PrP release. J Biol Chem (2002) 277:25457-25464.

[196] Parkin ET, Watt NT, Turner AJ, Hooper NM. Dual mechanisms for shedding of the cellular prion protein. J Biol Chem (2004) 279:11170-11178.

[197] Aguib Y, Gilch S, Krammer C, Ertmer A, Groschup MH, Schätzl HM. Neuroendocrine cultured cells counteract persistent prion infection by downregulation of PrPc. Mol Cell Neurosci (2008) 38:98-109.

[198] Heiseke A, Schöbel S, Lichtenthaler SF, Vorberg I, Groschup MH, Kretzschmar H, Schätzl HM, Nunziante M. The novel sorting nexin SNX33 interferes with cellular PrP formation by modulation of PrP shedding. Traffic (2008) 9:1116-1129.

[199] Seo JS, Moon MH, Jeong JK, Seol JW, Lee YJ, Park BH, Park SY. SIRT1, a histone deacetylase, regulates prion protein-induced neuronal cell death. Neurobiol Aging (2012) 33:1110-1120.

[200] Jeong JK, Moon MH, Lee YJ, Seol JW, Park SY. Autophagy induced by the class III histone deacetylase Sirt1 prevents prion peptide neurotoxicity. Neurobiol Aging (2013) 34:146-156.

[201] Fornai F, Longone P, Cafaro L, Kastsiuchenka O, Ferrucci M, Manca ML, Lazzeri G, Spalloni A, Bellio N, Lenzi P, Modugno N, Siciliano G, Isidoro C, Murri L, Ruggieri S, Paparelli A. Lithium delays progression of amyotrophic lateral sclerosis. Proc Natl Acad Sci USA 105:2052-7. Erratum in: Proc Natl Acad Sci USA (2008) 105:16404-7.

[202] Ravikumar B, Vacher C, Berger Z, Davies JE, Luo S, Oroz LG, Scaravilli F, Easton DF, Duden R, O'Kane CJ, Rubinsztein DC. Inhibition of mTOR induces autophagy and reduces toxicity of polyglutamine expansions in fly and mouse models of Huntington disease. Nat Genet (2004) 36:585-95.

[203] Williams A, Jahreiss L, Sarkar S, Saiki S, Menzies FM, Ravikumar B, Rubinsztein DC. Aggregate-prone proteins are cleared from the cytosol by autophagy: therapeutic implications. Curr Top Dev Biol (2006) 76:89-101.

[204] Sakaguchi S, Katamine S, Nishida N, Moriuchi R, Shigematsu K, Sugimoto T, Nakatani A, Kataoka Y, Houtani T, Shirabe S, Okada H, Hasegawa S, Miyamoto T, Noda

T. Loss of cerebellar Purkinje cells in aged mice homozygous for a disrupted PrP gene. Nature (1996) 380:528-531.

[205] Moore RC, Redhead NJ, Selfridge J, Hope J, Manson JC, Melton DW. Double replacement gene targeting for the production of a series of mouse strains with different prion protein gene alterations. Biotechnology (1995) 13:999-1004.

[206] Rossi D, Cozzio A, Flechsig E, Klein MA, Rulicke T, Aguzzi A, Weissmann C. Onset of ataxia and Purkinje cell loss in PrP null mice inversely correlated with Dpl level in brain. EMBO J (2001) 20:694-702.

[207] Yokoyama T, Kimura KM, Ushiki Y, Yamada S, Morooka A, Nakashiba T, Sassa T, Itohara S. In vivo conversion of cellular prion protein to pathogenic isoforms, as monitored by conformation-specific antibodies. J Biol Chem (2001) 276:11265-71.

[208] Flechsig E, Hegyi I, Leimeroth R, Zuniga A, Rossi D, Cozzio A, Schwarz P, Rülicke T, Götz J, Aguzzi A, Weissmann C. Expression of truncated PrP targeted to Purkinje cells of PrP knockout mice causes Purkinje cell death and ataxia. EMBO J (2003) 22:3095-3101.

[209] Anderson L, Rossi D, Linehan J, Brandner S, Weissmann C. Transgene-driven expression of the Doppel protein in Purkinje cells causes Purkinje cell degeneration and motor impairment. Proc Natl Acad Sci USA (2004) 101:3644-3649.

[210] Yamaguchi N, Sakaguchi S, Shigematsu K, Okimura N, Katamine S. Doppel-induced Purkinje cell death is stoichiometrically abrogated by prion protein. Biochem Biophys Res Commun (2004) 319:1247-1252.

[211] Nishida N, Tremblay P, Sugimoto T, Shigematsu K, Shirabe S, Petromilli C, Erpel SP, Nakaoke R, Atarashi R, Houtani T, Torchia M, Sakaguchi S, DeArmond SJ, Prusiner SB, Katamine S. A mouse prion protein transgene rescues mice deficient for the prion protein gene from Purkinje cell degeneration and demyelination. Lab Invest (1999) 79:689-697.

[212] Cui T, Holme A, Sassoon J, Brown DR. Analysis of doppel protein toxicity. Mol Cell Neurosci (2003) 23:144-155.

[213] Atarashi R, Nishida N, Shigematsu K, Goto S, Kondo T, Sakaguchi S, Katamine S. Deletion of N-terminal residues 23-88 from prion protein (PrP) abrogates the potential to rescue PrP-deficient mice from PrP-like protein/doppel-induced neurodegeneration. J Biol Chem (2003) 278:28944-28949.

[214] Shmerling D, Hegyi I, Fischer M, Blattler T, Brandner S, Gotz J, Rulicke T, Flechsig E, Cozzio A, von Mering C, Hangartner C, Aguzzi A, Weissmann C. Expression of amino-terminally truncated PrP in the mouse leading to ataxia and specific cerebellar lesions. Cell (1998) 93:203-214.

[215] Sakudo A, Lee DC, Nakamura I, Taniuchi Y, Saeki K, Matsumoto Y, Itohara S, Ikuta K, Onodera T. Cell-autonomous PrP-Doppel interaction regulates apoptosis in PrP gene-deficient neuronal cells. Biochem Biophys Res Commun (2005) 333:448-454.

[216] Wong BS, Liu T, Paisley D, Li R, Pan T, Chen SG, Perry G, Petersen RB, Smith MA, Melton DW, Gambetti P, Brown DR, Sy MS. Induction of HO-1 and NOS in doppel-expressing mice devoid of PrP: implication for doppel function. Mol Cell Neurosci (2001) 17:768-775.

[217] Diarra-Mehrpour M, Arrabal S, Jalil A, Pinson X, Gaudin C, Pietu G, Pitaval A, Ripoche H, Eloit M, Dormont D, Chouaib S. Prion protein prevents human breast carcinoma cell line from tumor necrosis factor alpha-induced cell death. Cancer Res (2004) 64:719-727.

[218] Paitel E, Sunyach C, Alves da Costa C, Bourdon JC, Vincent B, Checler F. Primary cultured neurons devoid of cellular prion display lower responsiveness to staurosporine through the control of p53 at both transcriptional and post-transcriptional levels. J Biol Chem (2004) 279:612-618.

[219] Solforosi L, Criado JR, McGavern DB, Wirz S, Sanchez-Alavez M, Sugama S, DeGiorgio LA, Volpe BT, Wiseman E, Abalos G, Masliah E, Gilden D, Oldstone MB, Conti B, Williamson RA. Cross-linking cellular prion protein triggers neuronal apoptosis in vivo. Science (2004) 303:1514-1516.

[220] Bounhar Y, Zhang Y, Goodyer CG, LeBlanc A. Prion protein protects human neurons against Bax-mediated apoptosis. J Biol Chem (2001) 276:39145-39149.

[221] Roucou X, Guo Q, Zhang Y, Goodyer CG, LeBlanc A. Cytosolic prion protein is not toxic and protects against Bax-mediated cell death in human primary neurons. J Biol Chem (2003) 278:40877-40881.

[222] Roucou X, Giannopoulos PN, Zhang Y, Jodoin J, Goodyer CG, LeBlanc A. Cellular prion protein inhibits proapoptic Bax conformational change in human neurons and in breast carcinoma MCF-7 cells. Cell Death Differ (2005) 12:783-795.

[223] Didonna A, Sussman J, Benetti F Legname G. The role of Bax and caspase-3 in doppel-induced apoptosis of cerebellar granule cells. Prion (2012) 6:309-316.

[224] Heitz S, Lutz Y, Rodeau J-L, Zanjani H, Gautheron V, Bombarde G, Richard F, Fuchs J-P, Vogel MW, Mariani J, Bailly Y. BAX contributes to Doppel-induced apoptosis of prion-deficient Purkinje cells. Dev Neurobiol (2007) 67:670-686.

[225] Heitz S, Gautheron V, Lutz Y, Rodeau J-L, Zanjani H, Sugihara I, Bombarde G, Richard F, Fuchs J-P, Vogel MW, Mariani J, Bailly Y. BCL-2 counteracts Doppel-induced apoptosis of prion protein-deficient Purkinje cells in the Ngsk $Prnp^{0/0}$ mouse. Dev Neurobiol (2008) 68:332-348.

[226] Yin XM, Oltvai ZN, Korsmeyer SJ. BH1 and BH2 domains of Bcl-2 are required for inhibition of apoptosis and heterodimerisation with Bax. Nature (1994) 369:321-323.

[227] Roucou X, Gains M, LeBlanc A. Neuroprotective functions of prion protein. J Neurosci Res (2004) 75:153-161.

[228] Oltvai ZN, Milliman CL, Korsmeyer SJ. Bcl-2 heterodimerizes in vivo with a conserved homolg, Bax, that accelerates programmed cell death. Cell (1993) 74:609-619.

[229] Gross A, McDonnell JM, Korsmeyer SJ. BCL-2 family members and the mitochondria in apoptosis. Genes Dev (1999) 13:1899-1911.

[230] Cheng EH, Wei MC, Weiler S, Flavell RA, Mak TW, Lindsten T, Korsmeyer SJ. BCL-2, BCL-X(L) sequester BH3 domain-only molecules preventing BAX- and BAK-mediated mitochondrial apoptosis. Mol Cell (2001) 8:705-711.

[231] Li A, Barmada S, Roth K, Harris D. N-terminally deleted forms of the prion protein activate both Bax-dependent and Bax-independent neurotoxic pathways. J Neurosci (2007) 27:852-859.

[232] Chiesa R, Piccardo P, Ghetti B, Harris DA. Neurological illness in transgenic mice expressing a prion protein with an insertional mutation. Neuron (1998) 21:1339-1351.

[233] Lee DC, Sakudo A, Kim CK, Nishimura T Saeki K, Matsumoto Y, Yokoyama T, Chen SG, Itohara S, Onodera T. Fusion of doppel to octapeptide repeat and N-terminal half of hydrophobic region of prion protein confers resistance to serum deprivation. Microbiol Immunol (2006) 50:203-209.

[234] Heitz S, Grant NJ, Bailly Y. Doppel induces autophagic stress in prion protein-deficient Purkinje cells. Autophagy (2009) 5:422-424.

[235] Heitz S, Grant NJ, Leschiera R, Haeberlé A-M, Demais V, Bombarde G, Bailly Y. Autophagy and cell death of Purkinje cells overexpressing Doppel in Ngsk Prnp-deficient mice. Brain Pathol (2010) 20:119-132.

[236] Dole S, Heitz S, Bombarde G, Haeberlé A-M, Demais V, Grant NJ, Bailly Y. New insights into Doppel neurotoxicity using cerebellar organotypic cultures from prion-protein-deficient mice. Prion 2010. Medimond International Proceedings Eds. (2010) pp7-14.

[237] Metzger F, Kapfhammer JP. Protein kinase C: its role in activity-dependent Purkinje cell dendritic development and plasticity. Cerebellum (2003) 2:206-214.

[238] Falsig J, Sonati T, Herrmann US, Saban D, Li B, Arroyo K, Ballmer B, Liberski PP, Aguzzi A. Prion pathogenesis is faithfully reproduced in cerebellar organotypic slice cultures. PLoS Pathog (2012) 8:e1002985.

[239] Barbieri G, Palumbo S, Gabrusiewicz K, Azzalin A, Marchesi N, Spedito A, Biggiogera M, Sbalchiero E, Mazzini G, Miracco C, Pirtoli L, Kaminska B, Comincini S. Silencing of cellular prion protein (PrPc) expression by DNA-antisens oligonucleotides induces autophagy-dependent cell death in glioma cells. Autophagy (2011) 7:840-853.

[240] Brown DR, Wong BS, Hafiz F, Clive C, Haswell SJ, Jones IM. Normal prion protein has an activity like that of superoxide dismutase. Biochem J (1999) 345:1-5.

[241] Brown DR, Besinger A. Prion protein expression and superoxide dismutase activity. Biochem J (1998) 334:423-429.

[242] Oh JM, Choi EK, Carp RI, Kim YS. Oxidative stress impairs autophagic flux in prion protein-deficient hippocampal cells. Autophagy (2012) 8:1448-1461.

[243] Baumann F, Tolnay M, Brabeck C, Pahnke J, Kloz U, Niemann HH, Heikenwalder M, Rulicke T, Burkle A, Aguzzi A. Lethal recessive myelin toxicity of prion protein lacking its central domain. Embo J (2007) 26:538-547.

[244] Baumann F, Pahnke J, Radovanovic I, Rulicke T, Bremer J, Tolnay M, Aguzzi A. Functionally relevant domains of the prion protein identified in vivo. PLoS One (2009) 4:e6707.

[245] Lemaire-Vieille C, Bailly Y, Erlich P, Loeuillet C, Brocard J, Haeberlé A-M, Bombarde G, Rak C, Demais V, Dumestre-Pérard C, Gagnon J, Cesbron J-Y. Ataxia with cerebellar lesions in mice expressing chimeric PrP-Dpl protein. J Neurosci (2013) (in press).

Section 8

Autophagy and Cell Death

Chapter 19

Role of Autophagy in the Ovary Cell Death in Mammals

M.L. Escobar, O.M. Echeverría and G.H. Vázquez-Nin

Additional information is available at the end of the chapter

http://dx.doi.org/10.5772/54777

1. Introduction

The process of cell death is implicated in several other processes, such as tissue homeostasis, embrionary development, and the elimination of unwanted cells. Programmed cell death is classified first according to the morphological characteristics of the cells observed, and then by the molecular machinery involved in the process. To date, programmed cell death is known to involve apoptosis and autophagy, two processes with different morphological and molecular characteristics.

In mammals, germinal cells are contained in the follicles, specialized structures that develop through several phases of maturation. During follicular growth, cell proliferation and cell death are present simultaneously. During ovarian follicular development, the follicles not selected for the ovulation process are physiologically eliminated. Several studies have shown that in mammalian ovaries follicular atresia is governed by granulosa cell apoptosis (Manabe et al., 2004); however, recent evidence from studies of pre-pubertal (Ortíz et al., 2006; Escobar et al., 2008 and 2010) and adult rats (Escobar et al., 2012) shows that autophagy is an alternative route taken by some germinal cells to induce follicular atresia in the ovary. The emerging importance of autophagy in cellular elimination in the mammalian ovary is a very interesting development.

2. Autophagy as a cell death program

Autophagy is an evolutionary process that eliminates damaged cellular proteins and organelles (Ferraro and Cecconi, 2007). It also plays an important role in bioenergetic management during periods of starvation (Othman et al., 2009), and is the major pathway for the degradation and recycling of intracellular contents.

The autophagy process occurs at a basal level in normal cells under certain adverse conditions, such as starvation, low oxygen levels, and growth factor withdrawal, among others. Under these conditions, autophagy functions as a cytoprotective program that helps maintain cellular homeostasis by recycling the cytoplasmic contents. Another function of autophagy is to eliminate damaged organelles so as to maintain correct cellular functions. Thus, all the features of autophagy in cells perform cytoprotective functions.

In eukaryotic cells, autophagy has been characterized according to the way in which it is carried out: microautophagy, chaperone-mediated autophagy, and macroautophagy (Klionsky, 2006; Massey et al., 2005). In microautophagy, the lysosomal surface directly engulfs the cytoplasm that is to be degraded (Figure 1A). In chaperone-mediated autophagy, the material to be degraded crosses the lysosomal membrane directly (Figure 1B), while macroautophagy, commonly referred to simply as autophagy, is characterized by a double-membrane vesicle (Figure 1C) that encloses (sequesters) organelles and portions of the cytosol (reviewed in Yang and Klionsky, 2009).

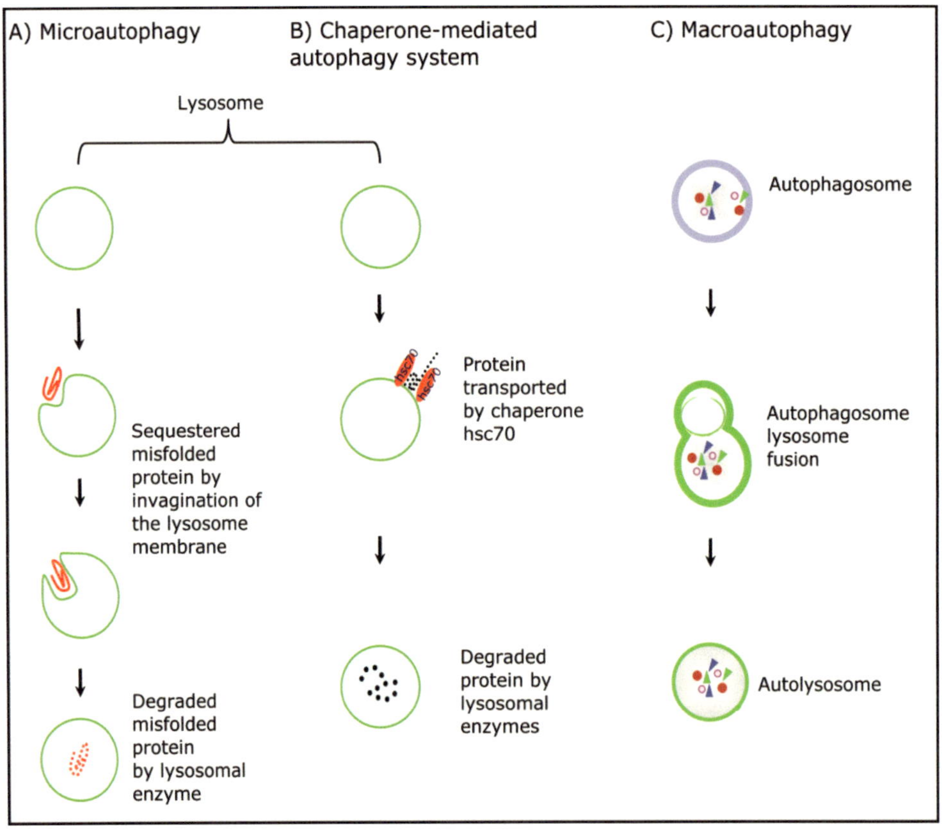

Figure 1. Schematic drawing of autophagic routes. A) Microautophagy: the cytoplasmic contents is directly enclosed by direct invagination from the lysosomal membrane. B) Chaperone-mediated autophagy: the components to be degraded are selectively transported toward the lysosome after interacting with the chaperone hsc70. C) Macroautophagy (commonly called autophagy): the autophagosome containing various cytosolic proteins fuses with lysosomes; subsequently, the contents of the autophagosome is degraded by the lysosomal enzymes.

http://dx.doi.org/10.5772/54777

Morphologically, autophagy is evidenced by the presence of autophagic vesicles, characterized by a double membrane structure. In mammalian cells, autophagy is initiated by the formation or elongation of the isolation membrane, also called a phagophore. Autophagy entails a sequence of events that includes sequestering the cytoplasmic contents in the double membrane vesicle. Once formed, the autophagosomes are conducted toward lysosomes to constitute the autolysosomes in which the sequestered cellular material is degraded. To avoid degradation itself, the lysosomal membrane is enriched by specific membrane proteins called lysosomal associated membrane proteins (Lamp1 and Lamp2) (Fukuda, 1991).

Though autophagy was first identified in mammalian cells, its molecular characteristics were discovered in yeast. Identification of the participation of the autophagy *Atg* genes and the subsequent documentation of their homologues in mammals (Yang and Klionsky, 2009) made it possible to determine the molecular machinery involved in the formation and maturation of autophagosomes. TOR kinase is considered an important element in autophagy. When TOR is inhibited under stress conditions, autophagy is induced upon the activation of this kinase; then Atg13 is quickly dephosphorylated, causing a higher affinity for Atg1 and Atg17 that results in an increase in the activity of the Atg1 protein kinase (Kamada et al., 2000; Kabeya et al., 2005). Atg1 kinase plays a pivotal role in controlling autophagy, and its activity is required for the switch from cytoplasm formation to vacuole targeting vesicles (Cvt) and the emergence of autophagosomes (Scott et al., 1996; Matsuura et al., 1997). In mammals, the microtubule-associated protein 1 light chain 3 (LC3) homolog of the Atg8 yeast is an important protein involved in the autophagy process. LC3 is present in autophagosomes and is synthetized in an inactive form called LC3-I, which is later converted into an active membranous form: LC3-II, the lapidated form, which means that it bonds to phosphatidylethanolamine (Wang et al., 2009; Maiuri et al., 2007). LC3 is lapidated via an ubiquitylation-like system that is targeted to the early autophagosome membrane (Kabeya et al., 2004).

During autophagy induction, LC3 is converted from the LC3-I to the LC3-II form. It has been suggested that the amount of LC3-II correlates to the number of autophagosomes present. Autophagy is involved in stress response, developmental remodeling, organelle homeostasis, and disease pathophysiology, and this process may also be used as a host-cell response against bacteria and viruses (reviewed in Kindergaar, 2004).

Additionally, it has been suggested that the effects of autophagy can be either deleterious or protective, depending on the specific cellular context and the stage of the pathological process (reviewed in Rubinsztein et al., 2005). At present, we know that autophagy functions as a form of programmed cell death, classified as type II. One essential difference between physiological autophagy and autophagic cell death is that the levels of autophagy in dying cells are excessive. The role of autophagy as a process of cell death is interesting because it has been observed under certain experimentally manipulated systems. When the pro-apoptotic proteases are inhibited, autophagic levels increase (Yu et al, 2004). Some neurodegenerative diseases, such as Parkinson's disease, have also been associated with the autophagic cell death process (Anglade et al., 1997), and autophagic cell death has been observed as well in remodeling tissues.

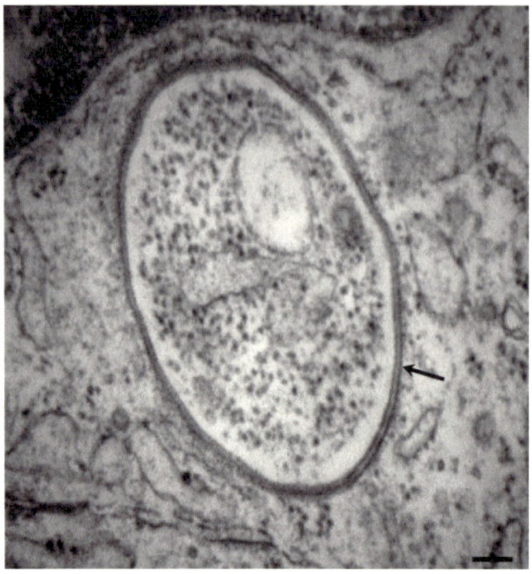

Figure 2. Electron microscopy showing an autophagosome. The arrow points to the two lipid bilayers that surround the cytoplasmatic content. Scale bar 100 nm.

Morphological evidence of autophagy has been found in electron microscopy studies that have shown vesicular structures surrounded by two lipid layers known as autophagosomes (Figure 2). Autophagosomes may contain cytoplasm and cytoplasmic organelles, such as mitochondria and peroxisomes, etc. Autophagy can be evidenced by the immune-microscopic localization of the proteins involved in this process, including LC3/Atg8, or the Lamp1 proteins (Figure 3).

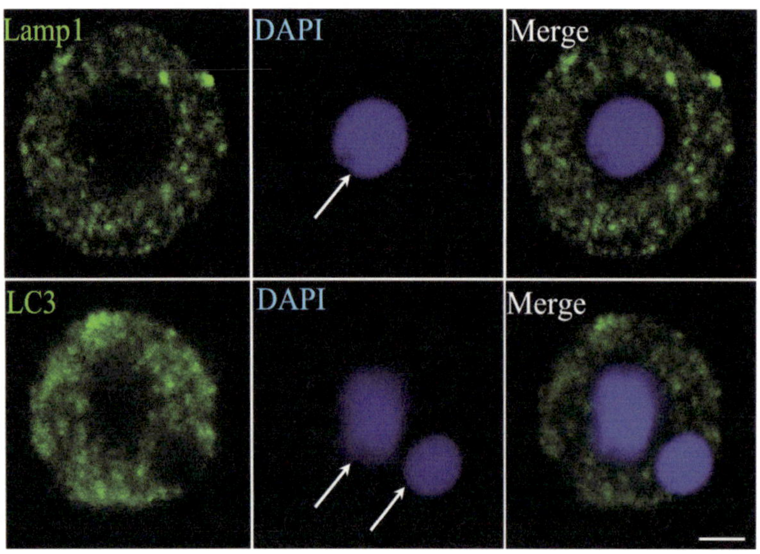

Figure 3. Immunodetection of Lamp1 and LC3. Confocal observations of cellular fractions: DAPI is evident in the nucleus (arrows). The green punctuate fluorescence distinguishes the cytoplasmatic localization of the Lamp1 and LC3 proteins. Scale bar 10 microns.

http://dx.doi.org/10.5772/54777

The role of autophagy as a process of cell death in diverse pathologies, including cancers, has been evaluated widely, but the results from the different studies are somewhat controversial, because at first autophagy functions as a pro-survival strategy, as in the case of tumor cells under certain stimuli; for example, low oxygen or a lack of nutrients (Lefranc et al., 2007). But cancer cells can also use autophagy as a strategy for evading cell death and a means of adapting to an adverse environment. On the other hand, under certain conditions, tumor cells use autophagy as a mechanism of cell death.

3. Follicular atresia

In mammals, the ovary is a paired organ whose principal functions are oocyte production and hormone synthesis. Structurally, the ovary is made up of a medulla and a cortical region where the follicles are generally located. The mammalian ovary is the site of oocyte maturation, which takes place inside a complex structure, the follicle, which is made up of a germinal cell –the oocyte– surrounded by somatic granulosa cells.

Follicles go through several steps before attaining maturation. During this process, various morphological and functional changes occur in the follicle that have led to its development being classified in stages: primordial, primary, secondary, early antral, and antral, according to the number of granulosa layers that surround the oocyte, the size of the follicle, and the presence of the antrum.

Primordial follicles consist of a single flattened cell layer surrounding the oocyte (Figure 4A). In a primary follicle, the granulosa cells around the oocyte acquire a cubical shape in a single-layer cell (Figure 4B). Secondary follicles are characterized by the presence of two or more granulosa cell layers (Figure 5). In this stage, the oocyte increases in size and the granulosa cell layers emerge through intensive proliferation that leads to the formation of the theca interna cell layer (Knight and Glister, 2006). In the secondary follicular phase, the specialized structure associated with the oocyte, called the zona pellucida, is completely discernible. Early-antral and antral follicles (Figure 6) are characterized by the development of a fluid-filled space among the granulosa cells that forms the antral cavity. Granulosa cells are in intercellular contact with neighboring cells via gap junctions (Figure 7), which allow metabolic exchange and the transport of molecules between follicular cells.

During the follicular maturation process, only a few follicles are selected for ovulation, while more than 99% are eliminated via a process denominated follicular atresia (Kaipia and Hsueh, 1977). In ovarian physiology, follicular atresia is a key mechanism for removing the follicles that are not selected for ovulation.

Numerous morphological and biochemical studies have revealed the frequent participation of apoptosis in follicular atresia; indeed, apoptosis came to be considered the cellular route that underlies this process. In caprine ovaries, ultrastructural changes in the granulosa cells show the classic morphological characteristics of apoptosis (Sharma and Bardwaj, 2009). Several pro-apoptotic factors have been identified in granulosa cells, including the FasL-Fas system, TNF-a, and members of the Bcl-2 family of proteins (reviewed in Matsuda et al., 2006). In fact, follicular atresia has been attributed to the alteration of granulosa cells, since studies have demonstrated that these cells synthetize molecules that are essential for follicular maintenance and growth. Furthermore, the death of granulosa cells due to an apoptotic process results in follicular elimination (Matsuda et al., 2012).

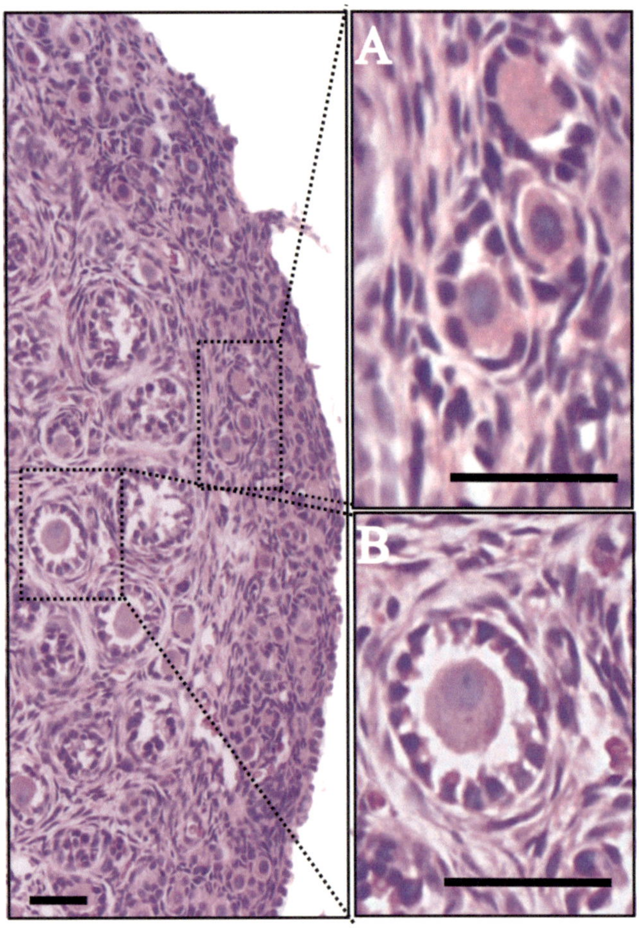

Figure 4. Histological images of a rat ovary. The dotted squares are magnified in the right panel. A) Primordial follicles with flattened pre-granulosa cells. B) Primary follicle with a single layer of cubical granulosa cells. Scale bars 20 microns.

http://dx.doi.org/10.5772/54777

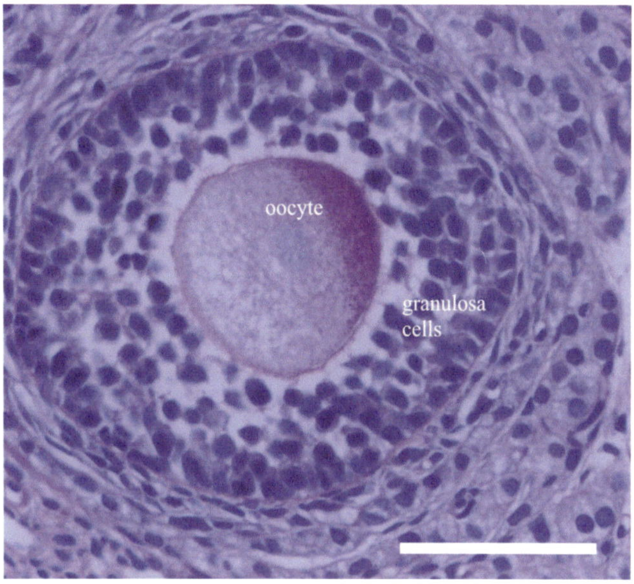

Figure 5. Secondary follicle. The oocyte is surrounded by several layers of granulosa cells. Scale bar 50 microns.

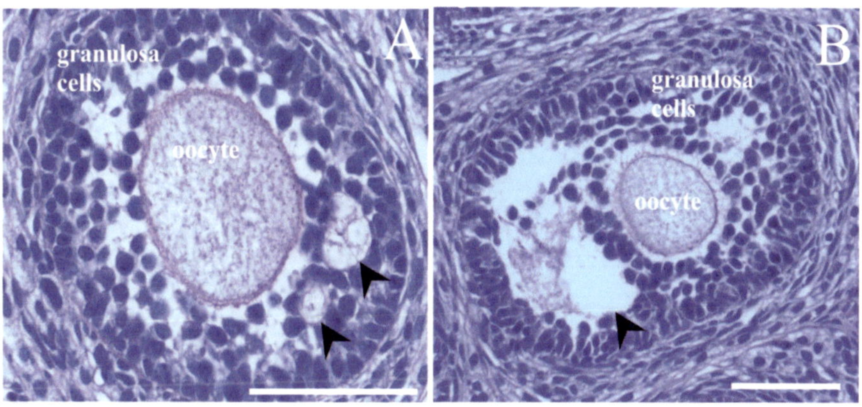

Figure 6. Histological images of a rat ovary. A) Early antral follicle: The arrows show the growing spaces that will form the antral cavity. B) Antral follicle: The arrow points to the antral cavity. Scale bars 50 microns.

4. Granulosa cell death via autophagy

While numerous studies have shown that the process of granulosa cell death is carried out mainly by apoptosis (Feranil et al. 2005; Hurst et al. 2006; Matsuda-Minehata et al. 2006, Lin and Rui, 2010), in some conditions autophagy may be induced in granulosa cells by the process of apoptosis, a process that in rat ovaries is gonadotropin-dependent. These results suggest that both apoptosis and autophagy are gonadotropin-dependent in rat ovaries, and that both processes are involved in regulating granulosa cell death during ovarian follicular development and atresia (Choi et al., 2010). Despite the obvious differences between apoptosis and autophagy, they are now thought to represent points on a continuum of mechanisms of cell death, because the induction of apoptotic cell death is regulated by the process of autophagy. Autophagic cell death is induced by inhibiting the accumulation of autophagosomes in various carcinoma cells, which suggests that the autophagic process may prevent apoptotic death.

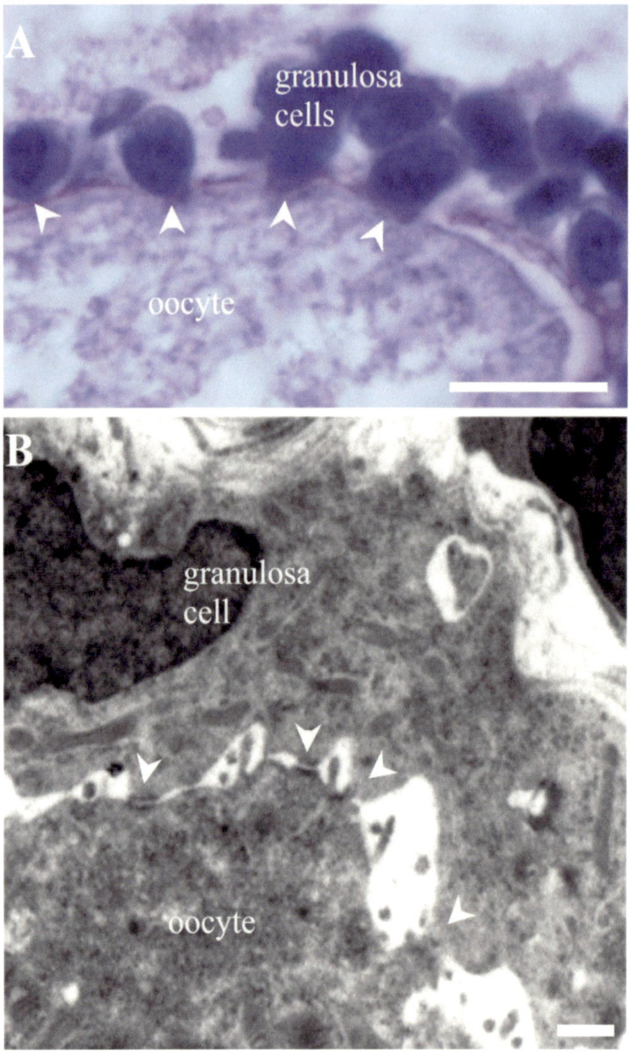

Figure 7. Gap junctions between granulosa cells and the oocyte. A) Optical micrograph showing granulosa cells in strong contact with the oocyte (arrowheads). B) Electron micrograph showing the gap junctions of granulosa cells and the oocyte (arrowheads). Scale Bars: A-50 microns; B-100 nanometers.

In order to investigate the involvement of autophagy in folliculogenesis, and its correlation with apoptosis, isolated rat granulosa cells from immature animals primed with pregnant mare serum gonadotropin were studied. LC3 and autophagic vacuoles were used as markers of autophagy, while cleaved caspase-3 served as the marker of apoptosis. In these conditions, LC3 was expressed by isolated granulosa cells in all developmental stages, and showed a similar expression pattern to the cleaved caspase-3. These results indicate that autophagy is induced in granulosa cells during folliculogenesis in correlation with apoptosis (Choi et al., 2010).

In the human ovary, lectin-like oxidized low-density lipoprotein (LOX) is localized in regressing antral follicles. Treatment with oxLDL (oxidized low-density lipoprotein) causes autophagy in granulosa cells. The process of cell death is characterized by the reorganization of the actin cytoskeleton, abundant vacuoles, autophagosome formation, the absence of apoptotic bodies, and cleaved caspase-3; thus, the reduction of granulosa cells may be mediated by autophagy (Duerrschmidt et al. 2006).

5. Oocyte cell death via autophagy

During the first two trimesters of pregnancy, the number of oocytes in human fetal ovaries increases from approximately 7,200 to 4,933,000 (Mamsen et al., 2011). However, oocyte death begins during the fetal and perinatal stages and continues in newborn, pre-pubertal (Hulas-Stassiak and Gawron, 2011) and adult mammals.

Autophagy is not only a process of cell death; it is also required for cells to survive in conditions of nutrient depletion (Han et al. 2011). Moreover, in murine ovaries it is a cell survival mechanism that maintains the endowment of female gem cells prior to establishing primordial follicle pools (Gawriluk et al. 2011). Several genes have been described as regulators of autophagy; many of them have been conserved from yeast to mammals. In vertebrates, autophagic defects may be lethal if the mutated gene is involved in the early stages of development. However, in different eukaryotes autophagy seems to be crucial during embryogenesis in a way that parallels apoptosis. The earliest autophagic event in mammalian development is observed in fertilized oocytes (Mizushima and Levine, 2010). The identification of *ATG* genes that mediate the initiation and assembly of autophagosomes and their fusion with lysosomes to form autolysosomes brought important advances in our understanding of the various functions of autophagy (Randall-Armant, 2011).

Thus, autophagy seems to be crucial during embryogenesis by acting in tissue remodeling, parallel to apoptosis (Di Bartolomeo et al., 2010). Studies in different organisms indicate that the autophagy pathway in the amoeba *Dictyostelium discoideum* is much more similar to that of mammalian cells than that of *S. cerevisiae*, despite its earlier evolutionary divergence. This indicates that in mammals the autophagic pathway is much older than was previously thought (King, 2012). MicroRNAs are involved in autophagy and are also important regulators of the crosstalk between autophagy and apoptosis (Xu et al. 2012).

ATG genes are also essential for the autophagic pathway in mammalian development (Mizushima and Levine 2010). The oocyte-specific deletion of Atg5, which removes the maternal stores of this protein, produces oocytes that fail to develop past the eight-cell stage, thus demonstrating that autophagy is required during pre-implantation development (Randall Armant 2011). An important increase in the number of autophagosomes takes place immediately after fertilization, which shows the need for autophagosomes after fertilization, in all likelihood to destroy the existing proteins and provide amino-acids for subsequent development (Randall Armant 2011).

In 1-to-28-day-old –i.e., newborn to pre-pubertal– rats, numerous follicles undergo atresia and oocytes are eliminated by processes that include, simultaneously, features of both apoptosis and autophagy. Elements of apoptosis are present in adjacent sections of the same dying oocyte, in the form of active caspase-3 and DNA breaks, as well as large increases of the Lamp1 protein and acid phosphatase, which are present in autophagosomes (Escobar et al., 2008; Escobar et al., 2010). Studies carried out in adult rats have also demonstrated that in all phases of the estrous cycle oocytes die by processes involving features of apoptosis and autophagy simultaneously (Escobar et al., 2012). Morphological changes in atretic oocytes include vacuolization of the cytoplasm, condensation of the mitochondria and segmentation, alterations that are not involved in classic apoptosis (Devine et al. 2000). These analyses were carried out using classic markers of apoptosis, such as the TUNEL reaction that reveals DNA fragmentation, immunolocalization of active caspase-3, and markers of autophagy like a large increase of acid phosphatase, lysosomal hydrolase, and immunodetection of Lamp1, a protein of the lysosomal membrane. These markers are located in the same regions of the oocyte's cytoplasm that present clear vacuoles which correspond to the autophagosomes that became visible using adjacent, semi-thin sections of the same oocyte (Escobar et al., 2008).

In newborn and pre-pubertal spiny mouse oocytes, follicular atresia was studied using markers of apoptosis like the TUNEL reaction, which demonstrate DNA fragmentation and active caspase-3, as well as with markers of autophagy, such as immunodetection of Lamp1. Numerous small clear vacuoles, autophagosomes and Lamp1 staining were found in all follicle types, especially in primordial and primary samples (Figure 8). Active caspase-3 and the TUNEL reaction were detected only in the granulosa cells, showing that both apoptosis and autophagy are involved in follicular atresia, and that these processes are both cell- and developmental-stage specific (Hulas-Stasiak and Gawron, 2011).

Follicular atresia has also been studied in fish ovaries during early and advanced stages of follicular regression. The main events assessed using light microscopy were splits in the zona radiata, yolk degradation and reabsorption, hypertrophy of follicular cells and the accumulation of autophagy vacuoles. Labeling for Bcl-2 and cathepsin-D was pronounced in follicular cells when they were involved in yolk phagocytosis. Immunofluorescence for Beclin-1 was significant in the follicular cells that often surround autophagic vacuoles during the advanced stages of follicular regression. TUNEL-positive reactions and immunostaining for Bax and caspase-3 showed the participation of apoptosis in advanced stages of follicular regression. These observations show that both autophagy and apoptosis are activated in some stages of follicular regression in fish ovaries (Morais et al., 2012). Inhibition of the increase of proliferating cell nuclear antigen (PCNA) markedly reduces the apoptosis of oocytes and down-regulates known pro-apoptotic genes, such as Bax, caspase-3, and TNFα, while up-regulating known anti-apoptotic genes like Bcl-2 (Xu et al. 2011).

Retraction of the prolongations of the granulosa cells that normally contacts the surface of the oocyte is one of the early signs of follicular atresia (Devine et al. 2000). Numerous unpublished observations by the authors of this chapter show that the microvilli of the oocyte are elongated after retraction of the prolongations of the granulosa cells during the process of atresia (Figure 9).

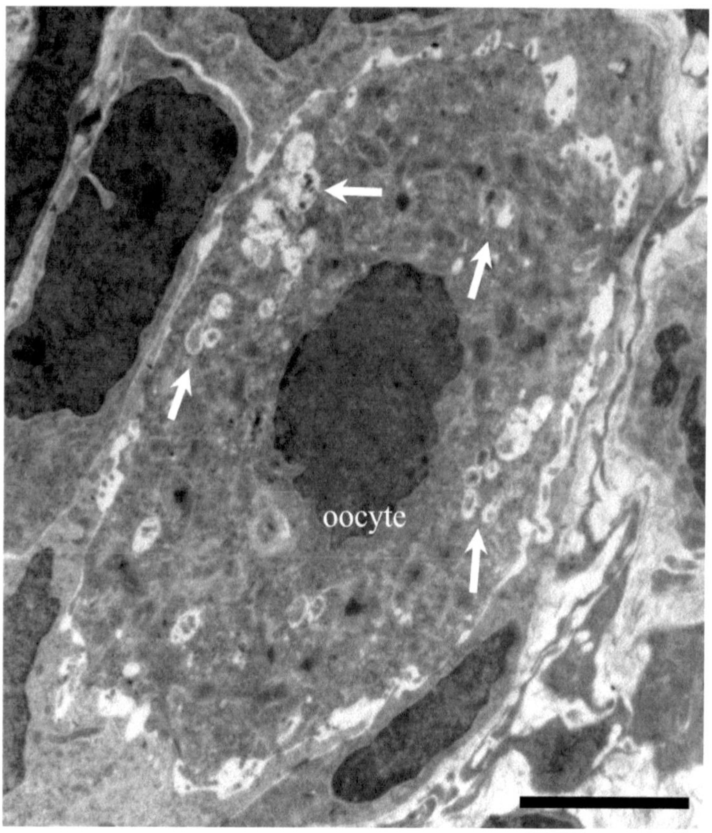

Figure 8. Primordial follicle. The cytoplasm has numerous vacuoles with cytoplasmic contents in different degrees of degradation. Scale bar 2 microns.

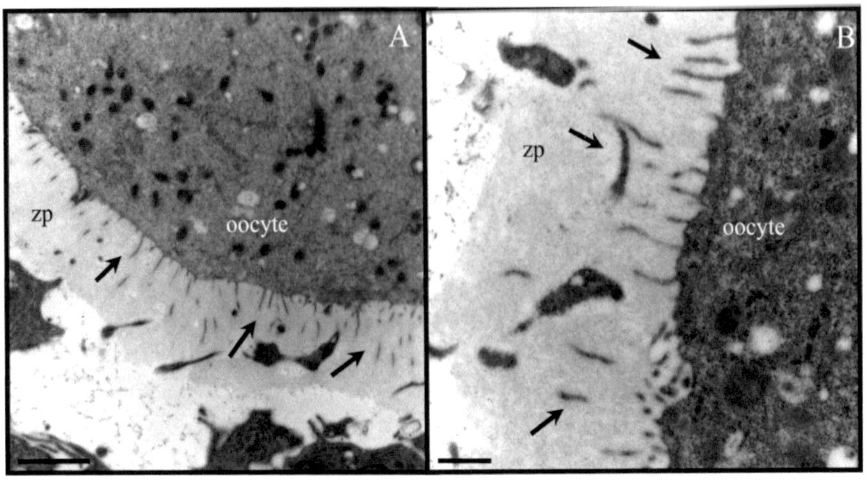

Figure 9. Retraction of the prolongation of the granulosa cells. The arrows point to several microvilli of the oocyte that are elongated after retraction of the prolongations of granulosa cells during atresia. Zp: zona pellucida. Scale bars: A-2 microns; B-500 nanometers.

6. Autophagic cell death in *corpus luteum* regression

The corpus luteum is a transitory ovarian structure formed by cells of the ovulated follicle (figure 10). After an initial proliferation of granulosa cells and closing of the antral cavity, capillaries and theca cells invade the region that was once the granulosa layer of the follicle. During its life span, the corpus luteum undergoes a period of rapid growth that involves hypertrophy, proliferation and differentiation of steroidogenic cells with extensive angiogenesis. After that, it engages in a large production of steroids. Growth factors including insulin-like factor, vascular endothelial growth factor and fibroblast growth factor are important for the development and completion of the dense network of capillaries during the formation of the corpus luteum (Berisha and Schams 2005). There is evidence to suggest that the luteinizing hormone, growth hormones and local regulators such as growth factors, peptides, steroids and prostaglandins are all important regulators of the luteal function. During early corpus luteum development, and up to the mid-luteal stage, oxytocin, prostaglandins and progesterone itself stimulate luteal cell proliferation and functioning, supported by the luteotropic action of several growth factors. High mRNA expression, protein concentration and localization of vascular endothelial growth factor, fibroblast growth factor and members of the family of insulin-like growth factors suggest that they play important roles in the maintenance of the corpus luteum. Progesterone regulates the length of survival of the corpus luteum (Berisha and Shams 2005). In addition, progesterone increases Bcl-2 expression in different stages of the estrous cycle. Treatment of luteal cells with progesterone and prostaglandin PGE2 for 24 hours decreased active caspase-3, while aminoglutethimide, spermine and staurosporine increased caspase-3 activity in luteal cells. These results suggest that progesterone concentrates

http://dx.doi.org/10.5772/54777

in luteal cells to protect against apoptosis, while disruption of steroidogenesis and the reduced ability of luteal cells to produce progesterone can induce cell death (Liszewska et al. 2005).

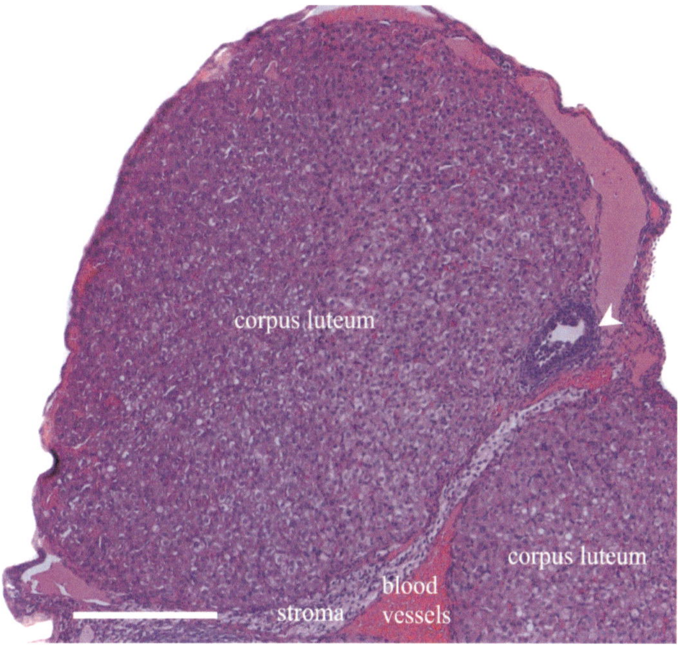

Figure 10. Corpus luteum from a rat ovary. The arrow points to a secondary follicle. Scale bar: 50 microns.

In non-fertile cycles, uterine release of prostaglandin (PG)F(2a) initiates a cascade of events that result in a rapid loss of steroidogenesis and destruction of the luteal tissue (Pate et al. 2012). Periodically, the corpus luteum regresses (luteolysis) and numerous luteal cells undergo cell death processes, mainly through apoptosis and autophagy.

Studies on the role of autophagy in corpus luteum regression have shown an increase of the protein microtubule-associated protein light chain 3 (LC3), a marker of autophagy. Apoptosis was evaluated by measuring cleaved caspase-3 expression (Choi et al. 2011). LC3 expression increases slightly from the early to the mid-luteal stage in steroidogenic cells. The expression levels of the membrane form of LC3 (LC3-II) also increase during luteal stage progression. In the same period, the expression of cleaved caspase-3 also increases. LC3-II expression rises, as do the levels of active caspase-3 in luteal cells cultured with prostaglandin F(2α), which is known to induce corpus luteum regression. These facts suggest that autophagy of luteal cells is directly involved in corpus luteum regression, and correlates with an increase of apoptosis (Choi et al. 2011). When autophagosome degradation by fusion with lysosomes was inhibited using bafilomycine A1 (Baf A1) increased apoptotic cell death. Moreover, inhibition of autophagosome formation using 3-methyladenine decreased apoptosis and cell death, suggesting that the accumulation of autophagosomes induces luteal cell apoptosis. The accumulation of autophagosomes increased apoptotic luteal cell death via an increase in the Bax/Bcl-2 ratio and subsequent caspase activation. Therefore, autophagy plays an important role in regulating apoptotic luteal cell death by controlling the Bax-to-Bcl-2 ratio and the subsequent activation of caspases. These experimental results indicate that autophagy is involved in rat luteal cell death through apoptosis, and that it is most prominent during corpus luteum regression (Choi et al. 2011).

Luteal cell regression during the normal postpartum involution of the corpora lutea is characterized by a large increase in the number of lysosomes and the appearance of numerous double-walled autophagic vacuoles, which become evident under electron microscope cytochemistry (Paavola, 1978).

Compelling evidence indicates that both apoptotic and autophagic cell death programs are involved in corpus luteum regression in primates. Beclin-1, an autophagy-related protein, is involved in the relation between apoptosis and autophagy through interaction with the anti-apoptotic protein Bcl-2. In ovarian follicles, Beclin 1 has been found in the theca layer, but granulosa cells are negative. After ovulation, Beclin-1 is present in theca-lutein and granulosa-lutein cells. The expression of Beclin 1 is related to the functional and structural status of the corpus luteum, as it is a factor in cell survival and plays important roles in the life span of the human corpus luteum (Gaytán et al. 2008).

An endocrine type, voltage-activated sodium channel was identified in the human ovary and human luteinized cells. Whole-cell patch-clamp studies showed that the voltage-activated sodium channels in granulosa cells are functional and tetrodotoxin-sensitive. Luteotrophic hormone was found to decrease the peak amplitude of the sodium current within seconds. Treatment with hGC (human chorionic gonadotropin) for 24-48 hours suppressed not only the mRNA levels in voltage-activated sodium channels, but also the mean sodium peak currents and resting potentials. Tetrodotoxin preserves a highly differentiated cellular phenotype, whereas veratridine not only increases the number of secondary lysosomes but also leads to a reduced progesterone production. In luteinized granulosa cells in culture, abundant secondary lysosomes were evident in the regressing corpus luteum, suggesting a functional link between the voltage-activated sodium channel activity and autophagic cellular regression *in vivo* (Bulling et al., 2000).

Taken together, these data show that several factors are involved in corpus luteum regression. One type of factors includes the process of eliminating the different types of cells that form the corpus luteum, while other types of factors are those involved in destroying the structure of this transitory organ. The normal programmed cell death processes –apoptosis and autophagy– are involved in cell elimination in the corpus luteum. Most authors have found that the most frequent process of cell death is apoptosis; however, very detailed studies demonstrate that both processes are often present simultaneously, as in the case of cell elimination in other organs.

7. Conclusions

Recent years have seen interest grow in the different routes of cell death. Today, two types of programmed cell death are known: apoptosis and autophagy. Cell death in follicle structures is a continuous event during the life of female organisms. Several studies have demonstrated the active participation of apoptosis in this process, but recent biochemical and morphological evidence has revealed the participation of autophagic cell death in oocyte elimination during this physiological process. In granulosa cell death and corpus luteum regression, experimental evidence has shown that autophagy is an active route in the process of cellular elimination. Future studies should test for different stimuli and molecular mechanisms involved in the autophagic cell death process in follicular atresia in vertebrates.

Acknowledgements

The authors would like to thank the grants CONACYT 180526 and PAPIIT IN212912. They also thank Paul C. Kersey Johnson for reviewing the English word usage and grammar.

Author details

M.L. Escobar, O.M. Echeverría and G.H. Vázquez-Nin

Departamento de Biología Celular. Facultad de Ciencias. Universidad Nacional Autónoma de México, Mexico

References

[1] Anglade, P, Vyas, S, Javoy-agid, F, Herrero, M. T, Michel, P. P, Marquez, J, Mouatt-prigent, A, Ruberg, M, Hirsch, E. C, & Agid, Y. (1997). Apoptosis and autophagy in nigral neurons of patients with Parkinson's disease. *Histol Histopathol 12*, 25-31.

[2] Berisha, B, & Schams, D. (2005). Ovarian function in ruminants. *Domest Anim Endocrinol* , 29, 307-317.

[3] Bulling, A, Berg, F. D, Berg, U, Duffy, D. M, Ojeda, S. R, Gratzi, M, & Mayerhofer, A. (2000). Identification of an ovarian voltage-activated Na+ channel type: hints to involvement in luteolysis. *Mol Endocrinol* , 7, 1064-1074.

[4] Choi, J, Jo, M, Lee, E, & Choi, D. (2011). The role of autophagy in corpus luteum regression in the rat. *BiolReproduct 85*, 465-472.

[5] Choi, J. Y, Jo, M, Lee, E. Y, & Choi, D. (2011). Induction of apoptotic cell death via accumulation of autophagosomes in rat granulosa cells. *Fertility and Sterility* , 95, 1482-1485.

[6] Devine, P. J, Payne, C. M, Mccusney, M. K, & Hoyer, P. B. (2000). Ultrastructural evaluation of oocytes during atresia in rat ovarian follicles. *Biol Reprod* , 63, 1245-1252.

[7] Di Bartolomeo SNazio F, Cecconi F. ((2010). The role of autophagy during development in higher eukaryotes. *Traffic* , 10, 1280-1289.

[8] Duerrschmidt, N, Zabirnyk, O, Nowicki, M, Hmeidan, F. A, Blumenauer, V, Burlak, J, & Spanel-borowski, K. (2006). Lectin-like oxidized low-density lipoprotein receptor-1-mediated autophagy in human granulosa cells as an alternative of programmed cell death. *Endocrinology* , 147, 3851-3860.

[9] Escobar, M. L, Echeverría, O. M, Ortiz, R, & Vázquez-nin, G. H. (2008). Combined apoptosis and autophagy, the process that eliminates the oocyte of atretic follicles in immature rats. *Apoptosis* , 13, 1253-1266.

[10] Escobar, M. L, Echeverría, O. M, Sánchez-sánchez, L, Méndez, C, & Pedernera, E. Vázquez-Nin

[11] Escobar Sánchez MLEcheverría Martínez OM, Vázquez-Nin GH. ((2012). Immunohistochemical and ultrastructural visualization of different routes of oocyte elimination in adult rats. *Eur J Histochemistry* , 56, 102-110.

[12] Feranil, J, Isobe, N, & Nakao, T. (2005). Apoptosis in the antral follicles of swamp buffalo and cattle ovary: TUNEL and caspase-3 histochemistry. *Reprod Domest Anim* , 40, 111-116.

[13] Ferraro, E, & Cecconi, F. (2007). Autophagic and apoptotic response to stress signals in mammalian cells. *Arch Biochem Biophys* , 462, 210-219.

[14] Fukuda, M. (1991). Lysosomal membrane glycoproteins. Structure, biosynthesis and intracellular trafficking. *J Biol Chem* , 266, 21327-21330.

[15] Gawriluk, T. R, Hale, A. N, Flews, J. A, Dillon, C. P, Green, D. R, & Rucker, E. B. (2011). Autophagy is a cell survival program for female germ cells in the murine ovary. *Reproduction* , 141, 759-765.

[16] Gaytán, M, Morales, C, Sánchez-criado, J. E, & Gaytán, F. (2008). Immunolocalization of beclin 1, a bcl-2-binding, autophagy-related protein, in human ovary: possible relation to life span of corpus luteum. *Cell Tissue Res* , 331, 509-517.

[17] GH. (2010). Analysis of different cell death processes of prepubertal rat oocytes in vitro. *Apoptosis* 15(4):511-526.

[18] Han, Y. K, Ha, T. K, Lee, S. J, Lee, J. S, & Lee, G. M. (2011). Autophagy and apoptosis of recombinant Chinese hamster ovary cells during fed-batch culture; effect of nutrient supplementation. *Biotechnology Bioeng* , 108, 2182-2192.

[19] Hulas-stasiak, M, & Gawron, A. (2011). Follicular atresia in the prepubertal spiny mouse (*Acomys cahirrinus*) ovary. *Apoptosis* , 10, 967-975.

[20] Hurst, P. R, Mora, J. M, & Fenwick, M. A. (2006). Caspase-3 TUNEL and ultrastructural studies of small follicles in adult human ovarian biopsies. *Hum Reprod* , 21, 1974-1980.

[21] Kabeya, Y, Mizushima, N, Yamamoto, A, Oshitani-okamoto, S, Ohsumi, Y, & Yoshimori, T. and GATE16 localize to autophagosomal membrane depending on form-II formation. *J Cell Sci* , 117, 2805-2812.

[22] Kaipia, A, & Hsueh, A. J. (1997). Regulation of ovarian follicle atresia. *Annu Rev Physiol 59*, 349-363.

[23] Kamada, Y, et al. (2000). Tor-mediated induction of autophagy via an Apg1 protein kinase complex. *J Cell Biol 150*, 1507-1513.

[24] King, J. S. (2012). Autophagy across the eukaryote: Is S. cerevisiae the odd one out? *Autophagy* , 7, 1159-1162.

[25] Kirkegaard, K, Taylor, M. P, & Jackson, W. T. (2004). Cellular autophagy: surrender, avoidance and subversion by microorganisms. *Nat Rev Microbiol* , 2, 301-314.

[26] Klionsky, D. J. (2005). The molecular machinery of autophagy: unanswered questions. *J Cell Sci* , 118, 7-18.

[27] Knight, P. G, & Glister, C. (2006). TGF-β superfamily and ovarian follicle development. *Reproduction* , 132, 191-206.

[28] Lefranc, F, Facchini, V, & Kiss, R. (2007). Proautophagic drugs: A novel means to combat apoptosis-resistant cancers, with a special emphasis on glioblastomas. *Oncologist* , 12, 1395-1403.

[29] Lin, P, & Rui, R. (2010). Effects of follicular size and FSH on granulosa cell apoptosis and atresia in porcine antral follicles. *Mol Reprod Dev 77*(8), 670-678.

[30] Liszewska, E, Rekawiecki, R, & Kotwica, L. (2005). Effect of progesterone on the expression of bax and bcl-2 on caspase activity in bovine luteal cells. *Prostaglandins Other Lipid Mediat* 78(1-4):67-81.

[31] Maiuri, M. C, Zalckvar, E, Kimchi, A, & Kroemer, G. (2007). Self-eating and self-killing: crosstalk between autophagy and apoptosis. *Nat Rev Mol Cell Biol* , 8, 741-752.

[32] Mamsen, L. S, Lutterodt, M. C, Andersen, E. W, Byskov, A. G, & Andersen, C. Y. (2011). Germ cell numbers in human embryonic and fetal gonads during the first two trimesters of pregnancy: analysis of six published studies. *Hum Reprod* , 8, 2140-2145.

[33] Manabe, N, Goto, Y, Matsuda-minehata, F, Inoue, N, Maeda, A, Sugimoto, M, Sakamaki, K, & Miyano, T. (2004). Regulation mechanism of selective atresia in porcine follicles: Regulation of granulosa cell apoptosis during atresia. *J Reprod Dev* , 50, 493-514.

[34] Massey, A. C, Zhang, C, & Cuervo, A. M. (2006). Chaperone-mediated autophagy in aging and disease. *Curr Top Dev Biol* , 73, 205-235.

[35] Matsuda, F, Inoue, N, Manabe, N, & Ohkura, S. (2012). Follicular growth and atresia in mammalian ovaries: regulation by survival and death of granulosa cells. *J Reprod Dev* , 58(1), 44-50.

[36] Matsuda-minehata, F, & Inoue, N. Goto Yasufumi, Manabe N. ((2006). The regulation of ovarian granulosa cell death by pro- and anti-apoptotic molecules. *J Reprod Dev* 52(6), 695-705.

[37] Matsuura, A, Tsukada, M, Wada, Y, & Ohsumi, Y. a novel protein kinase required for the autophagic process in Saccharomyces cerevisiae. Gene , 192, 245-250.

[38] Mcgee, E. A, & Hsueh, A. J. (2000). Initial and cyclic recruitment of ovarian follicles. *Endocr Rev* , 21(2), 200-214.

[39] Mizushima, N, & Levine, B. (2010). Autophagy in mammalian development and differentiation. *Nat Cell Biol* , 12, 823-830.

[40] Morais, R. D, Thomé, R. G, Lemos, F. S, Bazzoli, N, & Rizzo, E. (2012). Autophagy and apoptosis interplay during follicular atresia in fish ovary: a morphological and immunocytochemical study. *Cell Tissue Res* , 347, 467-478.

[41] Ortiz, R, Echeverría, O. M, Salgado, R, Escobar, M. L, & Vázquez-nin, G. H. (2006). Fine structural analysis of the processes of cell death of oocytes in atretic follicles in new born and prepubertal rats. *Apoptosis* , 11, 25-37.

[42] Othman, E. Q, Kaur, G, Mutee, A. F, Muhammad, T. S, & Tan, M. L. (2009). Immunohistochemical expression of MAP1LC3A and MAP1LC3B protein in breast carcinoma tissues. *J Clin Lab Anal* 23(4), 249-258.

[43] Paavola, L. G. (1978). The corpus luteum of guinea pig. III. Cytochemical studies on the Golgi complex and GERL during normal postpartum regression of luteal cells, emphasizing the origin off lysosomes and autophagic vacuoles. *J Cell Biol* , 79, 59-73.

[44] Pate, J, Johnson-larson, C, & Ottobre, J. (2012). Life and death in the corpus luteum. *Reprod Domest Anim* , 47, 297-303.

[45] Randall Armant D(2011). Autophagy's expanding role in development: implantation is next. *Endocrinology* , 152, 11739-11741.

[46] Rubinsztein, D. C. DiFiglia M, Heintz N, Nixon RA, Qin ZH, Ravikumar B, Stefanis L, Tolkovsky A. ((2005). Autophagy and its possible roles in nervous system diseases, damage and repair. *Autophagy* , 1, 11-22.

[47] Scott, S. V, Hefner-gravink, A, Morano, K. A, Noda, T, Ohsumi, Y, & Klionsky, D. J. (1996). Cytoplasm-tovacuole targeting and autophagy employ the same machinery to deliver proteins to the yeast vacuole. *Proc Natl Acad Sci USA* , 93, 12304-12308.

[48] Wang, A. L, & Boulton, M. E. Dunn WA Jr, Rao HV, Cai J, Lukas TJ, Neufeld AH. ((2009). Using LC3 to monitor autophagy flux in the retinal pigment epithelium. *Autophagy 5*, 1190-1193.

[49] Xu, B, Hua, J, Zhang, Y, Jiang, X, Zhang, H, Ma, T, Zheng, W, Sun, R, Shen, W, Sha, J, Cooke, H. J, & Shi, Q. (2011). Proliferating cell nuclear antigen (PCNA) regulates primordial follicle assembly by promoting apoptosis of oocytes in fetal and neonatal mouse ovaries. *Plos ONE* 6(1):e16046.

[50] Xu, J, Wang, Y, Tan, X, & Jing, H. (2012). Micro RNAs in autophagy and their emerging roles in crosstalk with apoptosis. *Autophagy* , 8(6), 873-882.

[51] Yang, Z. Klionsky DJ: An overview of the molecular mechanism of autophagy. *Curr Top Microbiol Immunol* , 335, 1-32.

[52] Yu, L, Alva, A, Su, H, Dutt, P, Freundt, E, Welsh, S, Baehrecke, E. H, & Lenardo, M. J. (2004). Regulation of an ATG7-beclin 1 program of autophagic cell death by caspase-8. *Science* , 304, 1500-1502.

Chapter 20

Autophagy in Development and Remodelling of Mammary Gland

Malgorzata Gajewska, Katarzyna Zielniok and Tomasz Motyl

Additional information is available at the end of the chapter

http://dx.doi.org/10.5772/54558

1. Introduction

Mammary gland is a unique organ, which undergoes the majority of its development in the postnatal life of mammals, especially during puberty and pregnancy. It is a complex tissue comprised of many cell types. The actual glandular part is formed by two epithelial cell subtypes: outer myoepithelia and inner luminal epithelia, which form a complicated net of ducts and lobules involved in milk synthesis and secretion during lactation. The glandular epithelium is embedded in stroma composed of mesenchymal cells, such as fibroblasts, adipocytes, immune cells, and extracellular matrix (ECM). The general features of mammary gland development is universal for animals and humans, however some differences in growth rate and hormonal control of the process can be distinguished between species.

During embryogenesis emergence of epithelial buds from ectoderm into mammary mesenchyme initiates formation of a rudimentary system of ducts, which continue their moderate elongation after birth, simultaneously with the increases in the body weight. In rodents prepubertal mammary gland consist of long, infrequently branching ducts terminated by highly proliferative structures, called terminal end buds (TEBs). TEBs contain two distinct cell types: cap cells organized as a single layer at the leading edge of these structures, staying in direct contact with the thin layer of basal lamina, and body cells, which form the multicellular bulk of the TEB [1] (Figure 1). In ruminants mammary ductal network develops as compact, highly branched structure within loose connective tissue called the terminal ductal units (TDUs), consisting of solid cords of epithelial cells that penetrate into mammary stroma [2]. An accelerated, hormone-dependent expansion of the glandular epithelium occurs at puberty, however the final stages of functional development do not take place until gestation. At the time of pregnancy mammary growth becomes exponential, and driven by pregnancy hor-

INTECH
open science | open minds

mones, giving rise not only to more extensive ductal branching, but also to the development of alveolar structures required for milk production. Mammary alveoli are build by functionally differentiated secretory epithelial cells, showing the ability to synthesize and secrete milk components during lactation. These characteristic structures with hollow cavity are formed in the last stage of mammary gland morphogenesis, termed alveologenesis (Figure 1). Weaning terminates the lactation period causing programmed cell death of a substantial part of the secretory epithelium, which leads to mammary gland involution. The cycle of proliferation, differentiation, and regression can be repeated many times in female's life. That is why mammary gland has become a very good and convenient model for studying processes involved in development and differentiation.

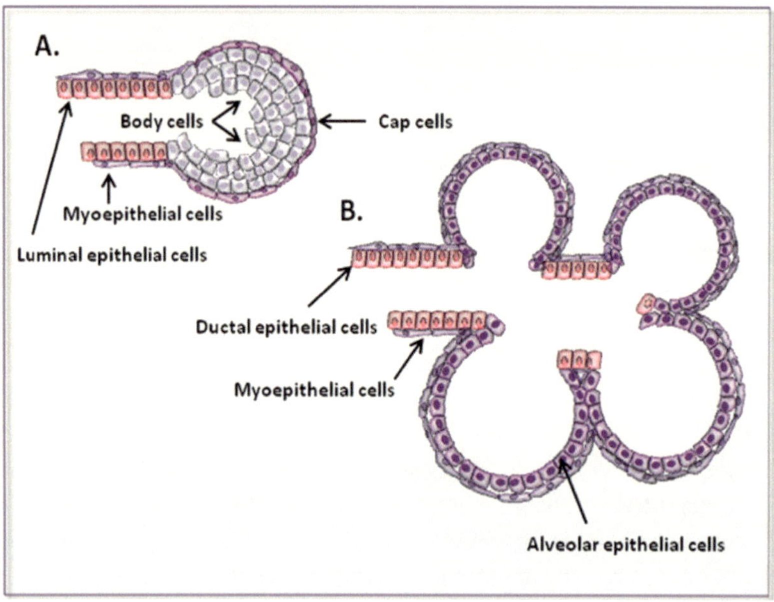

Figure 1. Schematic representation of the structure of mammary epithelium on different stages of differentiation. (A) Pre-pubertal and pubertal mammary gland consists of ducts terminated by highly proliferative terminal end buds (TEBs) that comprise cap cells in direct contact with the basal lamina, and body cells forming multicellular bulk of the TEB. (B) During pregnancy more extensive ductal branching and formation of alveolar structures required for milk production takes place. In functionally active mammary gland alveoli are built by a single layer of milk secreting luminal epithelial cells surrounded by myoepithelial cells, and basement membrane. Myoepithelial cells contractions release the milk to the ducts, and further to the nipple, whereas basement membrane provides cell contact with extracellular environment.

The lactation cycle includes periods of intensive proliferation of mammary epithelial cells (MECs), their functional differentiation during lactogenesis, and tissue involution caused by death of the secretory cells. Nowadays it is well established, that apoptosis plays a crucial role in all stages of mammary gland development. It is involved in lumen formation during ductal and alveolar morphogenesis, in replacement of cells during lactation, when MECs show high secretory activity, and in the involution of mammary gland. However, proper growth, development and remodelling require also well controlled balance between protein synthesis and organelle biogenesis versus protein degradation and organelle turnover in the cells. Since autophagy is the major cellular pathway for degradation of long-lived proteins and control of the cytoplasmic organelles, this process is particularly important during development and under certain stress conditions.

2. Role of autophagy in mammary gland development

2.1. Role of autophagy in mammogenesis and its relation with apoptosis

During all stages of mammary gland development lumen formation is essential for building functional network of ducts, terminated by milk-producing alveoli at the time of lactation. Lumen formation follows the process of branching morphogenesis, and it is said to be based mainly on the clearance of cells via apoptosis of an inner cell population within newly branched epithelial cords or newly formed acini, creating a cavity [3]. When TEBs invade the mammary stroma they are built by multilayers of epithelial cells, however, the primary ducts behind them posses only a single outer layer of myoepithelial cells and an inner layer of luminal cells surrounding an empty hollow lumen. It was noted that the body cells of TEBs show high rates of apoptosis indicating that this type of programmed cell death contributes to lumen formation [4]. The hypothesis was additionally confirmed with the use of transgenic mice overexpressing an antipoptotic protein Bcl-2 in the mammary gland, because the mammary glands of these rodents showed delayed lumen formation [4].

2.1.1. Autophagy, apoptosis and lumen formation

More information on the mechanisms regulating the lumenization process were obtained in the studies using a three dimensional (3D) cell culture model. This *in vitro* culture system was developed on the basis of the first observations showing, that epithelial cells are able to maintain their tissue structure when grown on ECM components which more closely mimic the *in vivo* microenvironment then the rigid plastic surfaces used in the classic monolayer cultures [5, 6]. It has been shown that mammary epithelial cells (MECs) cultured on laminin-rich reconstituted basement membrane (rBM) are able to recapitulate numerous features of mammary epithelium *in vivo,* including the formation of acini-like spheroids, with a hollow lumen, apicobasal polarization of cells, basal deposition of basement membrane components (collagen IV, laminin I), and the ability to produce milk proteins [5, 7, 8]. Upon seeding within the rBM, normal MECs first undergo a few cycles of proliferation forming small organoids. Next, the structures develop an axis of apicobasal polarity, illustrated among others by basal localization of integrin receptors which are in direct contact with ECM, and lateral (e.g. E-cadherin) or apico-lateral (e.g. ZO-1) localization of junctional complexes [7, 8]. The spherical structures subsequently become unresponsive to proliferative signals, and a bona fide lumen is formed by cavitation, involving the removal of centrally localized cells by death processes [9, 10, 1] (Figure 2). Lack of cell contact with ECM is regarded to be the direct cause of apoptosis initiation in the MECs placed in the centre of developing spheroid. This type of apoptotic death program is termed anoikis [11, 12]. However, it has been shown that overexpression of antiapoptotic proteins Bcl-2 or Bcl-XL in acini formed by human mammary epithelial cell line

MCF-10A only delayed lumen clearance for a few days, although apoptosis was inhibited, which pointed at a possibility that other processes may contribute to lumen formation [9]. In the same study electron microscopy analysis revealed the presence of numerous autophagic vacuoles in the central cells of developing acinar structures. Since autophagy was observed also in spheroids overexpressing Bcl-2 it was concluded that this process proceeds independently of apoptosis. Thus, it was initially proposed that autophagy may also promote lumen formation by initiating type 2 of programmed cell death [9, 13]. However, this hypothesis had some flaws, as it ignored the fact, that the cells lacking contact with ECM may initially induce autophagy as a cytoprotective mechanism against this stressful condition.

2.1.2. Role of autophagy in cells lacking contact with ECM

Integrin receptors are responsible for sustaining cell-matrix interactions, and mediating signal transduction from ECM into the cells [14]. β1-integrin has been shown critical for the alveolar morphogenesis of a glandular epithelium and for maintenance of its differentiated function [15]. Inhibition of β1-integrin in human MECs using blocking antibodies resulted in induction of autophagy. On the contrary, when laminin-rich basement membrane was added to the cells cultured in suspension autophagy was not induced, which points at a direct relationship between autophagy induction and the loss of cell contact with ECM [16]. Furthermore, it was demonstrated that depletion of some of the major Atg genes responsible for autophagy induction and autophagosome formation, namely: Atg5, Atg6 (beclin1) and Atg7, using siRNA technique resulted in reduced autophagy and enhanced apoptosis in suspended cells. A reduced clonogenic viability upon reattachment of MECs was also observed, indicating the prosurvival function of autophagy in cells lacking the direct contact with ECM during acini formation [16]. Interestingly, these studies have shown that inhibition of autophagy either pharmacologically (using 3-MA) or by knocking down Atg5 or Atg7 genes failed to elicit long-term luminal filling even when combined with inhibition of apoptosis. Similar results were obtained by Karantza-Wadsworth and co-workers [17], who worked on immortalized mouse mammary epithelial cells lacking one allel of beclin1 (beclin1$^{+/-}$). These cells when grown in 3D culture formed acini, which exhibited accelerated lumen formation compared to wild type controls that had both allels of beclin1 gene (beclin1 $^{+/+}$). The authors concluded that defective autophagy may sensitize MECs to metabolic stress, leading to accelerated lumen formation. Central acinar cells of Bcl-2-expressing beclin1$^{+/-}$ spheroids exhibited signs of necrotic cells death, suggesting that necrosis may be the default cell death mechanism upon apoptosis and autophagy inactivation [17]. However, their study also indicated, that defective autophagy compromises the ability of cells to adapt to metabolic stress, which may lead to insufficient ATP generation, accumulation of damaged mitochondria with excessive reactive oxygen species (ROS), and this in turn may cause accumulation of DNA damage, resulting in genome instability and increased risk of cancer progression.

The role of autophagy as the first line survival mechanism of cells centrally localized in the acinar structures was also proven by Sobolewska et al. [18] in the studies on bovine MECs. Bovine BME-UV1 mammary epithelial cells cultured on rBM behave in a similar manner to other described MECs, forming acinar structures composed of an outer-layer of polarized cells and a hollow lumen in the centre of the spheroids within 16 days of 3D culture. Autophagy was observed on the basis of the punctuated pattern of GFP-LC3 protein in the centre of developing acini by the end of the first week of 3D culture. The induction of autophagy preceded apoptosis, as the expression of apoptosis executor enzyme - cleaved caspase-3 was detected starting from the 9th day of cell culture. Thus, autophagy was observed in the acinar structures when a clear distinction of two populations of cells within the structures could be determined – the outer polarized layer of cells with direct contact with rBM, and the centrally localized cells lacking this contact. Subsequent intensive apoptosis eliminated the inner cells forming hollow lumen of the acini [18]. The importance of autophagy and the time of its activation during formation of spherical structures was further confirmed in the experiments on bovine MECs cultured in 3D system in the presence of 3-MA – the inhibitor of early autophagosome formation. 3MA caused formation of small, underdeveloped organoids, and the cells forming these structures showed signs of apoptosis (cleaved caspase-3 activity) before the process of polarization was completed [18]. However, others have shown that the addition of 3-MA to the 3D culture in the later time points, when minimal luminal apoptosis was already observed, caused only increased luminal cell death, not influencing the shape of the acini [16]. Thus, not only localization of autophagic cells, but also the time of autophagy induction, determines the proper development of mammary alveoli

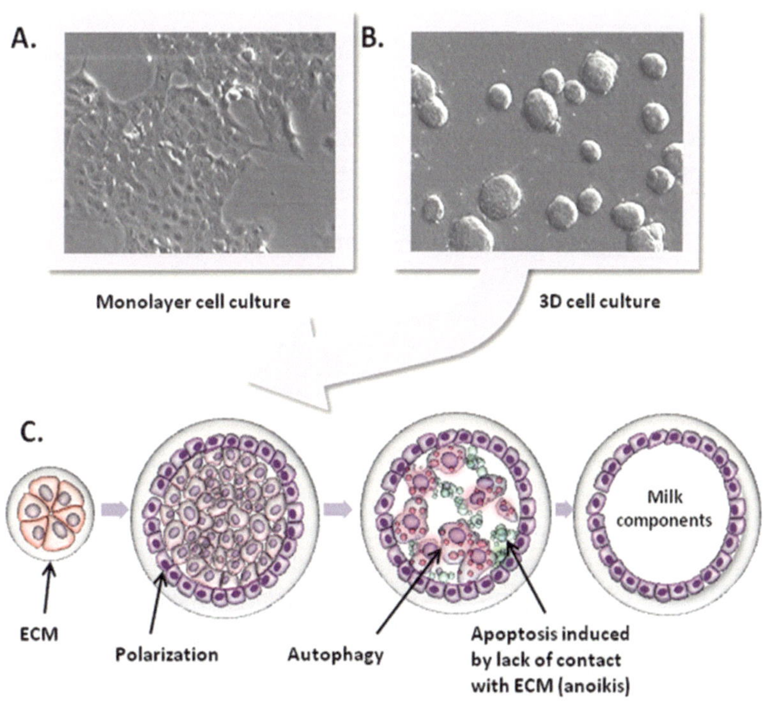

Figure 2. Formation of alveoli-like structures by mammary epithelial cells (MECs). (A) Image of bovine BME-UV1 MECs forming *in vitro* a monolayer on plastic surface. (B) Image of BME-UV1 cells forming 3D spherical structures *in vitro* when cultured on extracellular matrix (ECM) components. (C) Schematic representation of processes that take place

after seeding MECs within ECM: at the beginning of 3D culture cells undergo a few cycles of proliferation, forming small organoids. Next the outer layer of cells in direct contact with ECM develops an axis of apicobasal polarity, while the centrally localized cells lacking the necessary signals from the matrix undergo metabolic changes. At first the inner cells induce autophagy as a survival mechanism, but the sustained stress conditions subsequently lead to lumen formation by apoptotic cell death. The developed alveoli-like structures are able to secrete milk components into the luminal space.

2.1.3. Summary

In the process of mammary alveoli formation the outer layer of epithelial cells, which is in direct contact with ECM undergoes proper apicobasal polarization, and in the later stages develops specific secretory abilities. During alveologenesis autophagy is induced in the cells localized in the centre of the developing alveoli, as a results of the lack of contact of those inner cells with ECM. Autophagy is activated as a survival mechanism under the stress conditions connected with insufficient nutrient and energy supplies, and its main role is cells protection from potential damage of mitochondria and genome instability. The sustained stress conditions in the centre of the alveolar structures lead to apoptosis induction, and elimination of the inner cells by programmed cell death, which results in formation of hollow lumens of the alveoli (Figure 2).

2.2. Extracellular and intracellular factors regulating autophagy in mammary epithelial cells during mammogenesis

2.2.1. Role of endoplasmic reticulum kinase — PERK in the induction of autophagy during alveoli formation

Lack of cell contact with ECM in the centre of developing acinar structures leads to a rapid decrease in glucose intake, which correlates with a drop in ATP levels, and progressive accumulation of reactive oxygen species [19]. In this context the subsequent induction of autophagy supports the hypothesis about the primary adaptive and survival function of this process in the inner population of MECs. Studies on the potential mechanisms taking part in the induction of autophagy during acini formation pointed at the role of endoplasmic reticulum kinase – PERK in this process. PERK kinase is known to attenuate the initiation of translation by phosphorylating eIF2α (eukaryotic initiation factor 2α), when an accumulation of misfolded proteins in endoplasmic reticulum (ER) lumen occurs. It has been shown that upon loss of adhesion MECs activate the canonical PERK-eIF2 α signalling pathway, which serves as an important transcriptional regulator of multiple autophagic genes (ATGs), such as: Atg5, beclin1, Atg8/LC3, involved in autophagosomes formation [20]. PERK not only takes part in the induction of autophagy, but also contributes to the maintenance of ATP production and stimulation of a ROS detoxification response. All together these mechanism protect cells, until the adhesion can be restored, however if the stressful conditions persist the cells finally undergo apoptotic or autophagic death. An evidence for this hypothesis was obtained in the experiment on human MCF10A mammary epithelial cells cultured on rBM. Avivar-Valderas and co-workers [20] observed that enforced PERK activation during the late stages of acinar structures development allowed the centrally localized cells to persistently occupy the luminal space. At the same time an increase in the number of basal cells was noted, suggesting that some of the surviving MECs reattached to ECM in the outer/basal layer of acinar cells. Complementary observations were made *ex vivo* on murine mammary glands isolated at the lactation period. Immunohistochemical analysis of the expression of activated/phosphorylated PERK (p-PERK) and autophagic marker: LC3 in mammary tissue from lactating mice revealed that PERK was highly activated in the cells found in the luminal space of the mammary alveoli, as well as in the luminal epithelium, whereas LC3 was detected only in the detached cells. On the other hand, the expression of pro-apoptotic protein BimEL was weakly detected in the cells found in the luminal space of the mammary tissue, however and increased staining was observed in the epithelium of female mice with conditional deletion of PERK gene. Simultaneously, autophagy was decreased in the tissue samples from the genetically modified animals, suggesting that activation of PERK promotes autophagy and inhibits induction of apoptosis enabling a sustained survival of mammary epithelial cells during lactation [20].

2.2.2. *Regulation of autophagy by signalling pathway mediated by mTOR kinase*

Another kinase that may be involved in the induction of autophagy in the inner cells of the developing alveoli is AMP-activated protein kinase (AMPK). This enzyme is activated through the upstream kinase LKB1 when the cellular energy levels are reduced due to intracellular metabolic stress, leading to an increased AMP to ATP ratio. Activation of AMPK in turn leads to phosphorylation of the tuberous sclerosis complex (TSC1/2 complex), causing inhibition of mTOR. mTOR (mammalian target of rapamycin) is a conserved Ser/Thr protein kinase that regulates cell growth, cell cycle progression, protein synthesis and nutrient import [21]. In the nutrient and energy rich conditions it is also considered an inhibitor of autophagy [22]. Thus, the reduction of energy levels leads to mTOR inhibition via the LKB1 and AMPK kinases activation, and stimulation of autophagy in the cells [23]. Since AMPK activity was shown to be significantly increased in the MECs lacking contact with ECM it is highly probable that this pathway is also involved in autophagy regulation during alveolar lumen formation.

The activity status of mTOR constitutes an important switch in cell metabolism and fate. As mentioned above, when the energy and nutrient supply is sufficient the signalling pathway activating mTOR and its downstream targets is involved in translation regulation, mRNA turnover, protein stability, actin cytoskeletal organization, cell cycle progression and inhibition of autophagy [21]. This kinase is a target of a macrolide antibiotic called rapamycin, which specifically inhibits mTOR. Rapamycin was used in the studies on the mechanisms regulating the development of acinar structures formed by bovine mammary epithelial cells cultured on rBM [18]. Addition of the drug from the first day of 3D culture resulted in formation of small, underdeveloped spheroids, because rapamycin blocked cell proliferation. At the same time autophagy was induced in all cells forming the acini, as judged by high expression of the active form of LC3-II protein. The induction of autophagy prevented cells from immediate cell death, since the levels of the apotosis executor enzyme – cleaved caspase 3 were reduced in the rapamycin treated acinar structures. Results of this experiment further confirmed the protective role of autophagy in MECs, but also showed the importance of the proper timing of autophagy induction, which should not precede the period of intensive growth of the developing acini and the proper polarization of the outer layer of epithelial cells.

2.2.3. Mitogenic function of IGF-I and its effect on the rate of differentiation of mammary alveolar structures

During the time of mammary gland development several growth factors synthesized locally in the stromal or mesenchymal compartment of the gland, such as: IGF-I (insulin-like growth factor-I), EGF (epidermal growth factor), or HGF (hepatocytes growth factor), induce cell proliferation and survival, leading to expansion of the glandular epithelium. IGF-I plays a pivotal role in mammary tissue homeostasis, stimulating cell proliferation and differentiation during mammogenesis and lactogenesis. This growth factor exerts both endocrine and local actions. It is produced in the liver in response to pituitary growth hormone (GH), but is also synthesized and secreted by the cells of the mammary gland. Signals from IGF-I are transmitted into the cell via type I IGF receptor (IGFIR) located on the surface membranes of epithelial cells. When IGF-I binds to its receptor, IGFIR associates with the p85α subunit of phosphatidylinoinositol-3-kinase (PI3K) and activates another downstream target – Akt kinase [24]. Active Akt (phosphorylated at Ser^{473}) initiates other downstream signalling components involved in initiation of proliferation or activation of survival mechanisms. Moreover, the effect of IGF-I on cell growth and metabolism, mediated by Akt involves also activation of mTOR signalling pathway. Therefore, during normal mammary gland development the mitogenic signals from IGF-I must be under control of other locally produced growth factors and systemic hormones in order to maintain the proper homeostasis in the mammary gland. In fact, studies have shown that mammary epithelial cells grown on ECM components in the presence of IGF-I formed large spheroids lacking a hollow lumen in the centre [18]. The MECs showed prolonged proliferative activity, and decreased apoptosis measured on the basis of cleaved caspase-3 expression. At the same time an increased autophagy was observed in the centrally localized cells. The intensive autophagy of these inner cells, however, might have been induced by stressful conditions evoked by the lack of contact with ECM components, and decreased availability of nutrients inside of the large organoids, rather than directly by IGF-I. In fact, another study with the use of human mammary epithelial cell line MCF-10A over-expressing IGFIR, showed that the cells formed large, misshapen acinar structures with filled lumens and disrupted apico-basal polarisation in the presence of IGF-I [25]. The investigators observed that the MECs over-expressing IGFIR showed increased proliferation and decreased apoptosis, which was connected with increased activity of Akt, as well as mTOR. The phenotype of large misshapen spheroids could also be obtained, when MCF10A cells expressed a conditionally active variant of Akt [26]. Sustained Akt activation caused enhanced proliferation, and increased cell size, along with variability in size and shape of the cells forming the large spheroids. However, when rapamycin was added to the 3D culture the morphological disruption was prevented, indicating that mTOR function is required for the biological effect of Akt action during acinar development, and that the activity of mTOR also needs to be tightly regulated [26]. Although the described studies did not examine the role of autophagy in these processes one can expect that the effect of rapamycin addition not only resulted in inhibition of the proliferative signals induced by conditionally active Akt, but also induced autophagy, which could participate in lumen clearance.

2.2.4. 17β-estradiol and progesterone control mammogenesis by stimulation of mammary epithelial cells proliferation, regulation of gene expression, and induction of autophagy

The proper development of mammary gland is possible thanks to interactions of many cellular signalling pathways induced by intramammary factors, as well as endocrine hormones. Sex steroids belong to the important regulators of normal mammogenesis. Throughout puberty and gestation 17β-estradiol (E2) and progesterone (P4) induce proliferation of mammary epithelium and act as survival factors. Biological responses induced by both hormones are mediated by their receptors, which are located inside the cells, and translocate from the cytoplasm to the nucleus upon activation. There are two types of estrogen receptors (ERα, and ERβ), however ERα is shown to play the major function in the mammary gland. The expression of ERα was found both in the epithelial and stromal compartment of the mammary gland in many species, although in humans and heifers only mammary epithelium express ER. Progesterone can also act through two types of specific receptors: PR-A and PR-B, and the ratio in expression of both isoforms in the mammary gland is critical for the normal response to P4 [27]. PR is expressed only by mammary epithelial cells, and not all MECs show its expression, thus a paracrine interaction occurs between the PR-positive and PR-negative cells, which activates the proliferative and survival signals.

E2 is considered to be responsible for ductal morphogenesis, while P4 is critical for lobulo-alveolar development and transition from ductal to lobulo-alveolar morphology [27]. However, E2 also plays an indirect role in the alveologenesis by stimulating the PR expression in mammary epithelial cells [28]. Additionally, recent studies using the 3D culture system have shown that sex steroids may be involved in autophagy induction during mammogenesis [18]. When bovine MECs were cultured on rBM in the presence of E2 or P4 the cells formed proper acinar structures. During the development of these acini an intensified induction of autophagy was observed in the centre of the structures, as judged by the higher fluorescence intensity of the condensed pattern of GFP-LC3 autophagy marker. Additionally, apoptosis was also elevated in these cells, which led to a faster formation of the hollow lumen inside the spherical structures. Moreover, in case of 17β-estradiol it was shown that this hormone not only accelerated formation of the membrane-bound form of LC3 (LC3-II), but also increased the level of the LC3-I protein [18]. It is well established that both sex steroids exert an influence on target cells through a genomic pathway after binding to their receptors that translocate to the nucleus. Inside the nucleus the activated receptors associate with co-activators, or co-repressors, and finally regulate gene transcription by binding to target genes on the specific sites of the promoter regions, called response elements (ERE – estrogen response element, PRE – progesterone response element) [29, 30]. The observed increase in the total amount of LC3 protein indicates that E2 could enhance the expression of *LC3* gene. More recently steroids, especially estrogen (E2), have been found to exert rapid, non-genomic effects via membrane-bound receptors (mER), causing stimulation of cytoplasmic signalling pathways, such as: MAPK, and PI3K/Akt [31, 30]. So far the non-genomic molecular mechanism of steroids has not been investigated in regard to their possible influence on autophagy induction. However, John and co-workers [32] reported that beclin1 is able to bind with ERα, and the interactions between these proteins may modulate their action. Thus, sex steroid play a major role in the control of mammary gland development not only by acting as prosurvival factors, and stimulating epithelial cells proliferation, but also by regulating autophagy during alveoli formation. Furthermore, it is possible that E2 and P4 may regulate the action of other intra-mammary factors in the mammary epithelium by interactions of the signalling pathways induced by these endocrine and local factors (Figure 3).

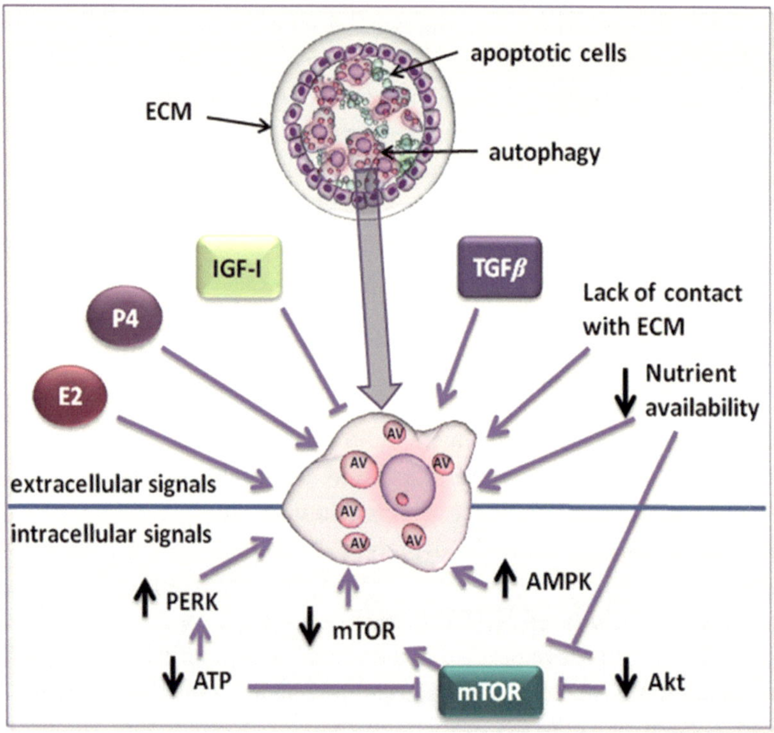

Figure 3. Extracellular and intracellular factors regulating autophagy induction in mammary epithelial cells during the process of lumen formation in mammary alveoli.

2.2.5. *Summary*

Induction of autophagy in the centrally localized cells of developing acini is regulated by several intracellular pathways and extracellular factors. The stress caused by insufficient nutrient and energy supply in the centre of the alveolar structures activates PERK kinase, which is an important transcriptional regulator, controlling the expression of autophagic (Atg) genes. Increased intracellular metabolic stress inside the developing alveoli also activates AMPK kinase, which inhibits the mTOR mediated signalling pathway leading to autophagy induction. During mammogenesis, the balance between proliferation and cell death processes is also controlled by locally secreted growth factors (i.a. IGF-I), and endocrine hormones (i.a.

http://dx.doi.org/10.5772/54558

sex steroids). IGF-I is an important survival factor, involved in stimulation of the enhanced growth of mammary epithelium during mammogenesis, whereas E2 and P4 play a major role in the control of mammary gland development by stimulating epithelial cells proliferation, as well as regulating autophagy induction during alveoli formation (Figure 3).

2.3. Role of autophagy in mammary gland involution

Mammary gland shows full functional activity during lactation, when the lobules contain fully developed alveoli formed by differentiated mammary epithelial cells secreting milk components into the luminal space. The surrounding myoepithelial cells contract, releasing the milk further to the ductal network, which delivers the milk to the nipple. The process of milk synthesis is under control of galactopoetic hormones (prolactin – PRL, growth hormone – GH), which stimulate the expression of milk proteins, survival of the glandular epithelium and contractions of the alveoli (oxitocin). After the period of functional activity, when females stop feeding their offspring the mammary gland regresses and returns to the state of development similar to the one prior pregnancy. This stage of remodelling is termed involution. Involution can be gradually initiated in the mammary gland, starting from the peak of lactation, because the young are progressively weaned. In case of dairy animals it starts with the natural, progressive decline in the milk yield. Alternatively, the mammary gland involution can be induced by litter removal (forced weaning), or in diary animals by termination of milking.

The withdrawal of suckling or cessation of milking results in the interruption of the release of galactopoetic hormones, which leads to milk stasis, and a rapid decline in milk synthesis caused by downregulation of genes involved in this process. In rodents, which have been extensively used as a model for studying the progress of mammary gland involution, forced weaning very quickly (within 24h after pup removal) leads to the first signs of apoptotic cell death of the epithelium, as some of the MECs are shed into the lumens of the alveoli. This stage, however, is still reversible, and the renewal of suckling preserves the structure of the secretory tissue. When the involution progresses several other processes take place, leading to the regression of the glandular epithelium.

At the time when the first apoptotic cells can be observed in the luminal space, MECs which remain within the alveoli begin reabsorbing the residual milk. Additionally, it has been shown that these cells undergo a change in their phenotype from secretory to phagocytic, which enables them to actively reabsorb also the apoptotic cells from the lumens by a process resembling efferocytosis [33]. During efferocytosis the cell membrane of phagocytic cell engulfs the apoptotic cell, forming a large fluid-filled vesicle, called efferosome or phagosome, which contain the dead cell. The efferosome subsequently fuses with lysosome, causing degradation of the engulfed material. The change in the phenotype of MECs requires changes in the expression of many genes, and thus, is thought to be transcriptionally-mediated. It has been shown, that more than 20 traditional markers of lysosomal activity are upregulated within 24h of forced weaning, and LC3 was detected among these upregulated proteins [33]. Since it is one of the key proteins involved in autophagosomes formation and fusion of lysosomes with autophagic vacuoles, these results indicate that autophagy is induced during the early stages of involution. Although there is no additional information on the role of autophagy during the initial phase of involution, there are evidence showing participation of this process when the regression of the mammary gland progresses.

In the second phase of involution proteolysis of the extracellular matrix and further apoptosis of the secretory epithelium takes place, causing the alveoli to collapse. There is an increase in expression of the protease genes, such as plasminogen activators (serine proteases), that induce the formation of active plasmin from plasminogen. Subsequently, plasmin activates matrix metalloproteinases (MMPs), which are responsible for the proteolytic degradation of basement membrane and ECM of the mammary gland. Removal of ECM induces apoptosis of the epithelial cells, that failed to respond to the first phase of apoptotic signals. The large number of apoptotic cells and debris are removed by phagocytosis performed by professional and non-professional phagocytes (macrophages, and epithelial cells, respectively) [34]. Finally, in the last stage of involution the regrowth of stromal adipose tissue is observed, filling the space of the regressed epithelium. The described course of mammary gland regression concerns the situation when lactation is separated from gestation by a dry period, during which the mammary gland remains in a quiescent state. It was extensively studied in rodents, and is often considered to reflect the general changes during the remodelling of the mammary gland in mammalian species. However, it is well documented that the mammary gland involution in ruminants differs in a significant manner. In cows and goats there is a characteristic overlap between the periods of lactation and next pregnancy, which means that these animals are typically pregnant when the involution is induced by termination of milking. Thus, the high levels of pregnancy hormones stimulating the development of new secretory tissue oppose the stimuli for mammary involution initiated by the milk stasis.

2.3.1. Role of autophagy in the regenerative involution of ruminant mammary gland

The nonlactating period before parturition in diary animals is termed the dry period. During this time the morphological changes in the mammary gland of ruminants (especially cows) are less pronounced than those occurring in the involuting glands of mice or rats. They reflect the change in the secretory state of the mammary epithelium, rather than the characteristic features of the tissue regression. The alveolar structure remains mostly intact during bovine mammary gland involution, even after several weeks of the dry period, although about 30 days after milking cessation the luminal area in the mammary tissue decrease and epithelial cells exhibit few secretory vacuoles [35]. Some of the bovine MECs undergo apoptotic cell death during involution, however, this subpopulation is significantly smaller than in rodents. It is considered that the nonlactating period in dairy cows serves to enhance the replacement of the senescent mammary epithelial cells prior to the next lactation, and thus, the processes taking place during that time are described as regenerative involution. Interestingly, studies have shown that bovine mammary tissue during the dry period shows signs of autophagy. It is manifested by increased expression of beclin1 and a high number of cells with typical morphological features of autophagy (autophagosomes and autophagolysosomes) [36]. Furthermore, *in vitro* studies on BME-UV1 bovine mammary epithelial cells revealed that cells partially devoid of nutritional factors and bioactive compounds induce formation of the autophagosome membrane-bound LC3-II form [37]. These experiments aimed at reflecting the conditions observed in the bovine mammary gland during dry period, when enhanced competition of intensively developing fetus and mother organism for nutritional and bioactive compounds creates a state of temporary malnutrition of MECs. When the concentration of fetal bovine serum (FBS) was significantly reduced in the culture medium of BME-UV1 cells (from standard 10% to 0.5%), the activity of mTOR kinase was significantly decreased, which corresponded with the induction of autophagy. Moreover, autophagy induced by FBS-withdrawal was inhibited by an addition of IGF-I, or EGF. Both growth factors play a prosurvival role in MECs during mammary development, whereas at the dry period their activity is decreased, similarly to the decreased levels of lactogenic hormones. Simultaneously the levels of sex steroids are elevated due to the pregnancy overlapping the period of glandular involution. The *in vitro* studies demonstrated, that in the presence of E2 or P4 bovine MECs cultured in FBS-deficient conditions showed higher levels of autophagy, which suggests, that these hormones additionally stimulate the induction of this process [37]. Thus, autophagy may be induced in bovine mammary epithelial cells as an additional survival process, which participates in preservation of the glandular morphology during involution.

Additionally, it was shown that autophagy can also be stimulated by transforming growth factor – beta 1 (TGF-β1), a cytokine classified as local growth inhibitor and apoptosis inducer in many cell types, including MECs. TGF-β1 expression was shown to be high during puberty and involution, low during gestation, and undetectable at the time of lactation. This growth factor can regulate cellular processes by a specific signalling pathway, which is induced upon binding of TGF-β1 to its membrane receptors (TβRI and TβRII). The receptors then form heterocomplexes and activate downstream components - the Smad proteins [38]. Smads transmit the signal to the nucleus, where they play a role of transcription factors and bind to DNA on the promoter region regulating the transcription of specific genes. TGF-β1 was shown to regulate the expression patterns of cyclins involved in the cell cycle progression, cell adhesion elements, such as integrins, and IGF binding proteins (IGFBP-3,4 and 5), which regulate the activity of IGF within the mammary gland [27, 39]. For example, IGFBP5 prevents binding of IGF-I to its receptor and inhibits the prosurvival signals. TGF-β1 is also able to induce apoptosis in mammary epithelial cells through the mitochondrial pathway involving: activation and translocation of the proapoptotic protein Bax to mitochondrial membranes, release of cytochrome c, and activation of the executor enzyme caspase-3 [40]. The experiments on bovine MECs revealed, that this cytokine also increased the level of LC3 and beclin1 proteins, indicating the direct role of TGF-β1 in autophagy induction. Moreover, it was found that the high expression of TGF-β1 receptors in the involuting bovine mammary tissue correlated with increased levels of beclin1 and downregulation of growth hormone receptor (GH-R) and IGF-I receptor (IGF-IRα) in this tissue [36].

The induction of autophagy by TGF-β1 was also observed during mammary acini formation in the studies with the use of 3D culture system [18]. These results correspond with other findings, showing that TGF-β1 is responsible for regulation of growth and pattering of the mammary ductal tree during mammogenesis, and can partially act by modulation of the effect of IGF-I on the developing tissue [41, 39].

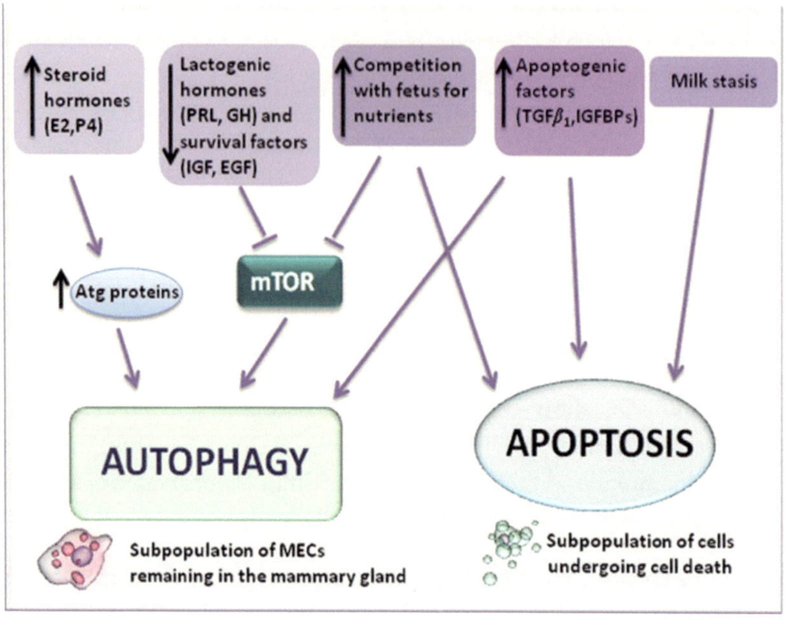

Figure 4. Endocrine hormones, auto/paracrine factors and intramammary conditions inducing autophagy and apoptosis in bovine mammary epithelial cells during regenerative involution.

2.3.2. Summary

In the period of mammary gland involution, during which regression of the secretory epithelium takes place returning the gland to the quiescent state, autophagy seems to be involved in the efferocytosis of the apoptotic epithelial cells. In case of the regenerative involution of bovine mammary gland autophagy is induced as a survival mechanism participating in the preservation of the glandular epithelium prior to next lactation. When milking is terminated a small population of mammary epithelial cells undergo apoptotic cell death, due to milk stasis and increased levels of proapoptotic factors, such as: TGF-β1. The remaining cells down-regulate milk secreting pathways and await parturition by inducing autophagy, as a mechanism which stabilizes intracellular supplies of energy and amino acids at the time of enhanced competition of intensively developing fetus and mother organism for nutritional and bioactive compounds. Additionally, local factors, such as TGF-β1, and pregnancy hormones (17β-estradiol and progesterone) stimulate autophagy during the dry period, suggesting a possible role of these factors in the control of the balance between apoptosis and survival of the epithelial cells in the involuting bovine mammary gland (Figure 4).

3. Autophagy and breast cancer

Most of breast malignancies arise in the terminal duct lobular units (TDLUs). In general, carcinomas are characterized by the loss of epithelial polarity and tissue organization. Cancer cells, which remain within the basement membrane of the mammary ductal-lobular system are classified as benign in situ carcinomas, whereas when neoplatic cells invade into the adjacent stroma the tumour becomes malignant [42]. Early premalignant breast cancer lesions, such as hyperplastic lesions with atypia and carcinoma *in situ* are characterized by a complete or partially filled lumen [1]. Moreover, *in vitro* experiments with the use of 3D culture system have shown, that human breast tumour cell lines are not able to form acinar structures with the centrally localized hollow lumen, and polarisation of cells surrounding this lumen. Instead they develop into nonpolarized clusters with limited differentiation [43, 44]. Both apoptosis and autophagy have been shown to be involved in the process of lumen clearance, however, autophagy is thought to be induced first as a survival mechanism in the central acinar cells, which are under increased metabolic stress connected with the lack of contact with ECM, hypoxia and decreased nutrient and energy supplies. Karantza-Wadsworth and co-workers [17] have shown that monoallelic deletion of beclin1 (beclin1$^{+/-}$) that leads to defective autophagy, causes increased DNA damage, and genome instability in cells. When defective autophagy is synergized with defects in apoptosis machinery, mammary tumorigenesis can be promoted.

Perturbations in autophagy has been implicated in the pathogenesis of diverse disease states, including cancer. The monoallelic deletion of beclin1 is observed in 50% of breast tumours [45]. Human breast carcinoma cell lines, as well as tumour tissue have decreased beclin1 levels, while mammary tissue from beclin1$^{+/-}$ mice shows hyperproliferative, preneoplastic changes [46]. Moreover, beclin1$^{+/-}$ immortalized mouse mammary epithelial cells, which exhibit compromised autophagy under metabolic stress, cause accelerated tumorigenesis after allogeneic transplantation into nude mice [17]. On the contrary, beclin1 ectopic expression in MCF-7 breast cancer cells, which are tetraploid but have only three beclin1 copies, led to a slower proliferation of these cells *in vitro*, as well as *in vivo* in the xenograft tumours. These findings indicate that beclin1 is a haploinsufficient tumour suppressor [47]. Furthermore, when deficiency in autophagy synergizes with defective apoptosis the response to the environmental stress is impaired and tumorigenecity is increased, promoting tumour growth.

On the other hand, autophagy as a known survival mechanism preserves cell viability during periods of nutrient limitation and hypoxia, which suggests that it can sustain cellular metabolism within the tumour. Metabolic stress is a common feature of solid tumours, resulting from inadequate vascularisation, and causing nutrient, growth factors, and oxygen deprivation [48]. It has been shown that solid tumours formed by cells with defective apoptosis are able to survive the metabolic stress by inducing autophagy [49]. Thus, when tumour cells have intact autophagy it may be induced as an adaptive response to anticancer agents, in which case autophgay may act as a treatment resistance mechanism prolonging tumour cell survival. It is especially important in apoptosis-defective cancers, which rely on autophagy under stressful conditions. In this case inhibition of autophagy should inhibit cancer cells' survival and enhance the efficacy of anticancer treatment [45]. Attempts to use autophagy inhibitors to sensitize cancer cells to treatment have been recently reported. For example, knockdown of autophagic genes in MCF-7 and T-47D breast cancer cells, combined with tamoxifen or 4-hydroxy-tamoxifen treatment, resulted in decreased viability of these cells [50, 51]. Hydroxychloroquine, which is a lysosomotropic agent causing increase in intralysosomal pH, and impairing autophagic protein degradation, has been used in clinical trials to modulate autophagy in metastatic breast cancer [45]. On the contrary, autophagy-deficient cancer cells with intact apoptotic machinery are shown to be particularly sensitive to therapeutically agents inducing metabolic stress (e.g.anti-angiogenic drugs), as these drugs cause apoptosis of the tumour cells. However, in a situation, when tumour cells show defective autophagy and apoptosis the approach to treatment should be different. As mentioned previously, simultaneous deficiency in autophagy and apoptosis makes the tumour cells susceptible to metabolic stress and DNA damage, leading to genome instability. Thus, use of metabolic or replication stress-inducing agents may cause further DNA damage in these cells, resulting in enhanced tumorigenic potential and development of drug resistance [17].

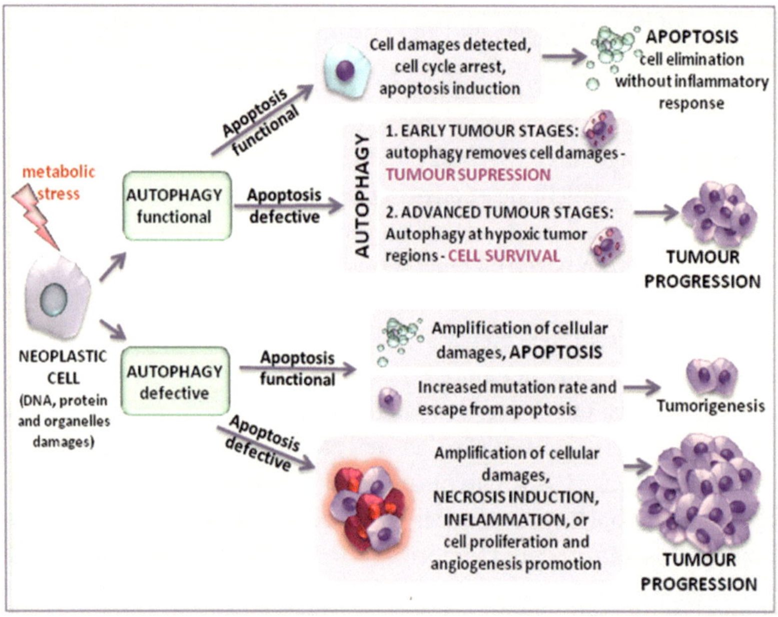

Figure 5. Different scenarios of response to metabolic stress in mammary tumorigenesis, depending on the functional status of autophagy and apoptosis in the tumour cells.

Finally, recent studies have pointed at a possible, yet still unexplored role of autophagy during invasion and metastasis. The studies with the use of 3D culture system revealed, that autophagy is induced in cell lacking the direct contact with ECM. The ability to survive in the absence of normal ECM is critical for metastasis, since cancer cells in the bloodstream or secondary tissue sites are either deprived of matrix or exposed to foreign matrix components [52]. The metastatic, secondary tumours are often resistant to therapy. Disseminated tumour cells, prior to development into the secondary tumours can remain in a dormant state for many years. *In vitro* and *in vivo* studies on mice have shown, that inhibition of β1-integrin, one of the subunits of integrin receptor responsible for cell contact with ECM, prevents proliferation of tumour cells, but not their viability, leading to induction of dormant state [53, 54]. Other studies have

http://dx.doi.org/10.5772/54558

shown that cell detachment from ECM induces autophagy [16, 18], and blocking β1-integrin function in attached human MECs is also sufficient for autophagy induction [16]. Therefore it is possible that detachment-induced autophagy in disseminated tumour cells may promote survival of these cells in the dormant state.

All presented results show the complexity of the role of autophagy in cancer development and progression. The possible role of activated, as well as defective autophagy in tumorigenesis is summarized in Figure 5. Research work continues to determine the molecular pathways regulating autophagy in tumour cells on different stages of tumour development. Results of the future studies may be beneficial for proper modulation of autophagy during cancer treatment and prevention.

Nomenclature

AMPK – AMP-activated protein kinase

E2 - 17β-estradiol

ECM – extracellular matrix

eIF2α – eukaryotic translation initiation factor 2 alpha

ER – endoplasmic reticulum

GFP-LC3 – LC3 (Atg8) protein fused with green fluorescence protein (a marker of autophagy)

GH – growth hormone

IGFBPs - insulin growth factor binding proteins

IGF-I – insulin-like growth factor-I

MECs – mammary epithelial cells

mTOR – mammalian target of rapamycin

P4 – progesterone

PERK – endoplasmic reticulum kinase

rBM – reconstituted basement membrane

TDLU - terminal duct lobular unit

TDU – terminal ductal unit

TEB – terminal end bud

TGF- β1 – transforming growth factor beta1

TSC1/2 – tuberous sclerosis complex (taking part in inhibition of mTOR kinase pathway)

Author details

Malgorzata Gajewska*, Katarzyna Zielniok and Tomasz Motyl

*Address all correspondence to: malgorzata_gajewska@sggw.pl

Department of Physiological Sciences, Faculty of Veterinary Medicine, Warsaw University of Life Sciences, Warsaw, Poland

References

[1] Reginato, M. J, & Muthuswamy, S. K. Illuminating the center: mechanisms regulating lumen formation and maintenance in mammary morphogenesis. Journal of Mammary Gland Biology and Neoplasia (2006). , 11, 205-211.

[2] Capuco, A. V, & Ellis, S. Bovine mammary progenitor cells: current concepts and future directions. Journal of Mammary Gland Biology and Neoplasia (2005).

[3] Mailleux, A. A, Overholtzer, M, & Brugge, J. S. Lumen formation during mammary epithelial morphogenesis: insights from in vitro and in vivo models. Cell Cycle. (2008).

[4] Humphreys, R. C, Krajewska, M, Krnacik, S, Jaeger, R, Weiher, H, Krajewski, S, Reed, J. C, & Rosen, J. M. Apoptosis in the terminal endbud of the murine mammary gland: a mechanism of ductal morphogenesis. Development (1996). , 122(12), 4013-4022.

[5] Barcellos-hoff, M. H, Aggeler, J, Ram, T. G, & Bissell, M. J. Functional differentiation and alveolar morphogenesis of primary mammary cultures on reconstituted basement membrane. Development (1989). , 105, 223-235.

[6] Li, M. L, Aggeler, J, Farson, D. A, Hatier, C, Hassell, J, & Bissell, M. J. Influence of a reconstituted basement membrane and its components on casein gene expression and secretion in mouse mammary epithelial cells. Proceedings of the National Academy of Sciences U S A. (1987).

[7] Debnath, J, Muthuswamy, S. K, & Brugge, J. S. Morphogenesis and oncogenesis of MCF-10A mammary epithelial acini grown in three-dimensional basement membrane cultures. Methods (2003). , 30, 256-268.

[8] Kozlowski, M, Wilczak, J, Motyl, T, & Gajewska, M. Role of extracellular matrix and prolactin in functional differentiation of bovine BME-UV1 mammary epithelial cells. Polish Journal of Veterinary Sciences (2011).

[9] Debnath, J, Mills, K. R, Collins, N. L, Reginato, M. J, Muthuswamy, S. K, & Brugge, J. S. The role of apoptosis in creating and maintaining luminal space within normal and oncogene-expressing mammary acini. Cell (2002). , 111, 29-40.

[10] Shaw KRMWrobel CN, Brugge JS. Use of three-dimensional basement membrane cultures to model oncogene-induced changes in mammary epithelia. Journal of Mammary Gland Biology and Neoplasia (2004). , 9, 297-310.

[11] Frisch, S. M, & Francis, H. Disruption of epithelial cell-matrix interactions induces apoptosis. Journal of Cell Biology (1994).

[12] Gilmore, A. P. Anoikis. Cell Death and Differentiation. (2005). Suppl , 2, 1473-1477.

[13] Mills, K. R, Reginato, M, Debnath, J, Queenan, B, & Brugge, J. S. Tumor necrosis factor related apoptosis-inducing ligand (TRAIL) is required for induction of autophagy during lumen formation in vitro. Proceedings of the National Academy of Sciences U S A. (2004). , 101, 3438-3443.

[14] Hynes, R. O. Integrins: bidirectional, allosteric signaling machines. Cell. (2002). , 110, 673-687.

[15] Naylor, M. J, Li, N, Cheung, J, Lowe, E. T, Lambert, E, Marlow, R, Wang, P, Schatzmann, F, Wintermantel, T, Schüetz, G, Clarke, A. R, Mueller, U, Hynes, N. E, & Streuli, C. H. Ablation of beta1 integrin in mammary epithelium reveals a key role for integrin in glandular morphogenesis and differentiation. Journal of Cell Biology (2005).

[16] Fung ChLock R, Gao S, Salas E, Debnath J. Induction of autophagy during Extracellular Matrix detachment promotes cell survive. Molecular Biology of the Cell (2008). , 19, 797-806.

[17] Karantza-wadsworth, V, Patel, S, Kravchuk, O, Chen, G, Mathew, R, Jin, S, & White, E. Autophagy mitigates metabolic stress and genome damage in mammary tumorigenesis. Genes and Development (2007). , 21, 1621-1635.

[18] Sobolewska, A, Motyl, T, & Gajewska, M. Role and regulation of autophagy in the development of acinar structures formed by bovine BME-UV1 mammary epithelial cells. European Journal of Cell Biology (2011). , 90, 854-864.

[19] Schafer, Z. T, Grassian, A. R, Song, L, Jiang, Z, Gerhart-hines, Z, Irie, H. Y, Gao, S, Puigserver, P, & Brugge, J. S. Antioxidant and oncogene rescue of metabolic defects caused by loss of matrix attachment. Nature. (2009). , 461(7260), 109-113.

[20] Avivar-valderas, A, Salas, E, Bobrovnikova-marjon, E, Diehl, J. A, Nagi, C, Debnath, J, & Aguirre-ghiso, J. A. PERK integrates autophagy and oxidative stress responses to promote survival during extracellular matrix detachment. Molecular and Cellular Biology (2011).

[21] Wullschleger, S, Loewith, R, & Hall, M. N. TOR signaling in growth and metabolism. Cell. (2006).

[22] Pattingre, S, Espert, L, Biard-piechaczyk, M, & Codogno, P. Regulation of macroautophagy by mTOR and Beclin 1 complexes. Biochimie. (2008).

[23] Liang, J, Shao, S. H, Xu, Z. X, Hennessy, B, Ding, Z, Larrea, M, Kondo, S, Dumont, D. J, Gutterman, J. U, Walker, C. L, Slingerland, J. M, & Mills, G. B. The energy sensing LKB1-AMPK pathway regulates kip1) phosphorylation mediating the decision to enter autophagy or apoptosis. Nature Cell Biology (2007). , 27.

[24] Fleming, J. M, Desury, G, Polanco, T. A, & Cohick, W. S. Insulin growth factor-I and epidermal growth factor receptors recruit distinct upstream signaling molecules to enhance AKT activation in mammary epithelial cells. Endocrinology. (2006).

[25] Yanochko, G. M, & Eckhart, W. (2006). Type I insulin-like growth factor receptor over-expression induces proliferation and anti-apoptotic signaling in a three-dimensional culture model of breast epithelial cells. Breast cancer Research 8, R18.

[26] Liu, H, Radisky, D. C, Wang, F, & Bissell, M. J. Polarity and proliferation are controlled by distinct signaling pathways downstream of PI3-kinase in breast epithelial tumor cells. Journal of Cell Biology (2004). , 164, 603-612.

[27] Lamote, I, Meyer, E, Massart-leen, A. M, & Burnvenich, C. Sex steroids and growth factors in the regulation of mammary gland proliferation, differentiation and involution. Steroids (2004). , 69, 145-159.

[28] Atwood, C. S, Hovey, R. C, Glover, J. P, Chepko, G, Ginsburg, E, Robison, W. G, & Vonderhaar, B. K. Progesterone induces side-branching of the ductal epithelium in the mammary glands of peripubertal mice. Journal of Endocrinology (2000).

[29] Nilsson, S, Mäkelä, S, Treuter, E, Tujague, M, Thomsen, J, Andersson, G, Enmark, E, Pettersson, K, Warner, M, & Gustafsson, J. A. Mechanisms of estrogen action. Physiological Reviews (2001). , 81, 1535-1565.

[30] Fox, E. M, Andrade, J, & Shupnik, M. A. Novel actions of estrogen to promote proliferation: integration of cytoplasmic and nuclear pathways. Steroids (2009). , 74, 622-627.

[31] Levin, E. R. Bidirectional signaling between the estrogen receptor and the epidermal growth factor receptor. Molecular Endocrinology (2003). , 17, 309-317.

[32] John, S, Nayvelt, I, Hsu, H. C, Yang, P, Liu, W, Das, G. M, Thomas, T, & Thomas, T. J. Regulation of estrogenic effects by beclin1 in breast cancer cells. Cancer Research (2008). , 68, 7855-7863.

[33] Monks, J, & Henson, P. M. Differentiation of the mammary epithelial cell during involution: implications for breast cancer. Journal of Mammary Gland Biology and Neoplasia (2009).

[34] Monks, J, Geske, F. J, Lehman, L, & Fadok, V. A. Do inflammatory cells participate in mammary gland involution? Journal of Mammary Gland Biology and Neoplasia (2002).

[35] Capuco, A. V, & Akers, R. M. Mammary involution in diary Animals. Journal of Mammary Gland and Neoplasia. (1999). , 4(2), 137-144.

[36] Zarzynska, J, Gajkowska, B, Wojewódzka, E, Dymnicki, E, & Motyl, T. Apoptosis and autophagy in involuting bovine mammary gland is accompanied by up-regulation of TGF-β1 and suppression of somatotropic pathway. Polish Journal of Veteterinary Sciences (2007). , 10, 1-9.

[37] Sobolewska, A, Gajewska, M, Zarzynska, J, Gajkowska, B, & Motyl, T. IGF-I, EGF and sex steroids regulate autophagy in bovine mammary epithelial cells via the mTOR pathway. European Journal of Cell Biology (2009). , 88, 117-130.

[38] Mulder, K. M. Role of Ras Mapks in TGF-beta signaling. Cytokine and Growth Factor Reviews (2000). , 11, 23-35.

[39] Gajewska, M, & Motyl, T. IGF-binding proteins mediate TGF-βinduced apoptosis in bovine mammary epithelial BME-UV1 cells. Comperative Biochemistry and Physiology Part C (2004). , 1.

[40] Kolek, O, Gajkowska, B, Godlewski, M. M, & Motyl, T. Colocalization of apoptosis-regulating proteins in mouse mammary epithelial HC11 cells exposed to TGF-β1. European Journal of Cell Biology (2003). , 82, 303-312.

[41] Daniel, C. W, Robinson, S, & Silberstein, G. B. The transforming growth factors beta in development and functional differentiation of the mouse mammary gland. Advances in Experimental Medicine and Biology (2001). , 501, 61-70.

[42] Weigelt, B, & Bissell, M. J. Unraveling the microenvironmental influences on the normal mammary gland and breast cancer. Seminars in Cancer Biology (2008).

[43] Petersen, O. W, Rønnov-jessen, L, Howlett, A. R, & Bissell, M. J. Interaction with basement membrane serves to rapidly distinguish growth and differentiation pattern of normal and malignant human breast epithelial cells. Proceedings of the National Academy of Sciences U S A. (1992).

[44] Debnath, J, & Brugge, J. S. Modelling glandular epithelial cancers in three-dimensional cultures. Nature Reviews Cancer (2005).

[45] Chen, N, & Karantza-wadsworth, V. Role and regulation of autophagy in cancer. Biochimica et Biophysica Acta. (2009).

[46] Qu, X, Yu, J, Bhagat, G, Furuya, N, Hibshoosh, H, Troxel, A, Rosen, J, Eskelinen, E. L, Mizushima, N, Ohsumi, Y, Cattoretti, G, & Levine, B. Promotion of tumorigenesis by heterozygous disruption of the beclin 1 autophagy gene. Journal of Clinical Investigation (2003).

[47] Yue, Z, Jin, S, Yang, C, Levine, A. J, & Heintz, N. Beclin 1, an autophagy gene essential for early embryonic development, is a haploinsufficient tumor suppressor. Proceedings of the National Academy of Sciences U S A. (2003).

[48] Jin, S, & White, E. Role of autophagy in cancer: management of metabolic stress. Autophagy. (2007).

[49] Degenhardt, K, Mathew, R, Beaudoin, B, Bray, K, Anderson, D, Chen, G, Mukherjee, C, Shi, Y, Gélinas, C, Fan, Y, Nelson, D. A, Jin, S, & White, E. Autophagy promotes tumor cell survival and restricts necrosis, inflammation, and tumorigenesis. Cancer Cell. (2006).

[50] Qadir, M. A, Kwok, B, Dragowska, W. H, To, K. H, Le, D, Bally, M. B, & Gorski, S. M. Macroautophagy inhibition sensitizes tamoxifen-resistant breast cancer cells and enhances mitochondrial depolarization. Breast Cancer Research and Treatment (2008).

[51] Samaddar, J. S, Gaddy, V. T, Duplantier, J, Thandavan, S. P, Shah, M, Smith, M. J, Browning, D, Rawson, J, Smith, S. B, Barrett, J. T, & Schoenlein, P. V. A role for macroautophagy in protection against hydroxytamoxifen-induced cell death and the development of antiestrogen resistance. Molecular Cancer Therapeutics (2008). , 4.

[52] Debnath, J. Detachment-induced autophagy during anoikis and lumen formation in epithelial acini. Autophagy (2008). , 4, 352-335.

[53] White, D. E, Kurpios, N. A, Zuo, D, Hassell, J. A, Blaess, S, Mueller, U, & Muller, W. J. Targeted disruption of betaintegrin in a transgenic mouse model of human breast cancer reveals an essential role in mammary tumor induction. Cancer Cell. (2004). , 1.

[54] Lock, R, & Debnath, J. Extracellular matrix regulation of autophagy. Current Opinion in Cell Biology (2008). , 20, 583-588.

Chapter 21

Integrin and Adhesion Regulation of Autophagy and Mitophagy

Eric A. Nollet and Cindy K. Miranti

Additional information is available at the end of the chapter

http://dx.doi.org/10.5772/55398

1. Introduction

Cell differentiation is a dynamic process that generates a functionally distinct cell from its progenitor. For example, human erythrocytes lack most organelles - including a nucleus - while erythrocyte precursors have a complete set of organelles. Autophagy plays a critical role in organelle elimination in differentiating erythrocytes. On the other hand, most other differentiated cells in the body do not lose their organelles, and would not seem to heavily depend on autophagy during differentiation. However, evidence indicates that these cells require regulated autophagy during the differentiation process. For example, during keratinocyte differentiation a basal cell detaches from the basement membrane and is pushed to the upper strata, and as the cell ascends its intracellular contents are replaced with copious amounts of keratin. Drastic changes such these require cells to eliminate or turn over a large amount of biomass. Differentiation also causes a shift in signaling and survival pathways. An example of this is the prostate gland where the luminal cells require PI3K for survival while the undifferentiated basal cells do not; they depend on MAPK signaling for their survival. As we will discuss below, several different cell types use and require the autophagic pathway to properly differentiate and survive. While links between autophagy and differentiation are rapidly being identified, the mechanisms that trigger autophagic processes during differentiation are poorly understood. Genetics has served as a powerful tool for identifying the components of the autophagy machinery, but how they are integrated with cellular cues that trigger differentiation need further characterization. We will discuss the recently identified link between autophagy and cell adhesion, and its role in cellular differentiation and survival.

2. Role of autophagy during differentiation

2.1. Erythrocytes

Erythrocytes are especially unique cells; they lack a nucleus, internal membrane-bound organelles, and ribosomes, and they are packed with the oxygen transporter hemoglobin. The cell arises from reticulocytes, a nucleated progenitor capable of generating the needed surplus hemoglobin and the erythrocyte itself [1,2]. Generating such a simple cell requires processes that eliminate organelles not essential for fully differentiated erythrocytes. A series of publications from 2008 to present not only elucidated the involvement of autophagy during reticulocyte differentiation, but also discovered that the mitochondria are specifically targeted for autophagy by a protein called Nix [3-7].

Nix and its related family member, Bnip3, are unique mitochondrial-localized BH3-only proteins. Although they can induce apoptosis when over expressed like most of the BH3-only proteins, Nix and Bnip3 function to stimulate autophagy and mitophagy as well. The mitophagic function of Nix was discovered in differentiating reticulocytes. Researchers knew that Nix expression dramatically increases in the terminal stage of reticulocyte differentiation, but the purpose for this remained unknown for several years [8]. In a relatively simple experiment, researchers harvested erythrocytes from $Nix^{-/-}$ mice and using mitotrophic dyes quantified mitochondria-containing erythrocytes. A significant population of the $Nix^{-/-}$ erythrocytes still contained mitochondria. The absence of Nix did not affect LC3 levels, and autophagosomes were still present. However, the autophagosomes in $Nix^{-/-}$ erythrocytes contained significantly less mitochondria than the wild type controls [4].

Erythrocytes from $Ulk1^{-/-}$ mice retained mitochondria and ribosomes, indicating the necessity of the general autophagy program in reticulocyte differentiation. Inhibiting autophagy with the class III-PI3K inhibitor 3-methyladenine (3-MA) in wild type reticulocytes prevented mitochondrial clearance as well. The peripheral blood in $Ulk^{-/-}$ mice contained an increased number of reticulocytes indicating a reduced capacity to be converted to erythrocytes [3]. Transplantation of fetal liver cells from $Atg7^{-/-}$ mice into irradiated wild type mice resulted in overall fewer erythrocytes, and these erythrocytes contained mitochondria.[7]. Thus, both general macroautophagy, and organelle-specific autophagy is required for erythrocyte differentiation. The above studies focused on mitochondria, but whether specific autophagic targeting of other organelles, such as ribosomes and ER, is mediated by Nix or other organelle-specific factors during reticulocyte differentiation remains unanswered. Bnip3, like Nix, localizes to the ER, and researchers recently demonstrated that Bnip3 targets the ER for autophagy [9]. This raises the possibility that Nix may also target the ER for autophagy. Targeted autophagic degradation of ribosomes, termed ribophagy, occurs in S. cerevisiae. However, whether selective ribophagy occurs in higher eukaryotes remains unknown [10]. These questions would be interesting to answer in the reticulocyte differentiation model.

http://dx.doi.org/10.5772/55398

2.2. Lymphocytes

In the same Atg7$^{-/-}$ transplantation model, there was a four-fold reduction in the number of white blood cells and nine-fold reduction of lymphoid cells [7]. This observation was further validated in hematopoietic-specific stem, fetal, and adult cell Atg7 knockout mouse models [9, 10, 11]. Mitochondrial accumulation was observed in both CD8^{+} and CD4^{+} T-cells, and these cells succumbed to apoptosis in vitro more readily than the wild type controls [11]. T-cell specific knockouts of Atg5 and Atg7 also displayed increased amounts of mitochondria and were more susceptible to apoptosis [12,13]. Considering that the overall population of mature T lymphocytes decreased and sensitivity to apoptosis increased in the autophagy-deficient T lymphocyte models, it seems that the lymphopenia occurs due to excess cell death. Puo et al tested this by stimulating T-cell proliferation in wild type and Atg5$^{-/-}$ T-cells and quantifying the amount of daughter cells produced by using the stable dye, 5,6-carboxyfluorescein diacetate succinimidyl ester (CFSE). CFSE diffuses into both cells during mitosis, causing a decrease in individual cell fluorescent intensity with every division. The analysis demonstrated that stimulated Atg5$^{-/-}$ T-cells did not produce daughter cells after three days and the authors concluded that Atg5 is necessary for T-cell proliferation. However, CFSE cannot differentiate between a dead or permeabilized cell and a non-proliferating cell, and over the three day period there was a moderate increase in apoptosis [14,15]. In this model, it appears that autophagy is required during T-cell hematopoiesis, at least in part, to prevent death of the newly differentiated cells. It would be interesting to investigate this in an inducible model, to test whether the dependence on autophagy is transient, i.e. only during early differentiation stages, or required to maintain the differentiated population.

2.3. Adipocytes

Evidence also supports a role for autophagy in adipocyte differentiation. Treatment of mouse embryonic fibroblasts (MEFs) with a cocktail of differentiation-inducing factors over the course of two weeks causes accumulation of lipid analogous to mature adipocytes. In the absence of Atg5 the MEFs fail to accumulate lipid droplets or display the molecular or structural phenotype of differentiated adipocytes. Furthermore, Atg5$^{-/-}$ mouse pups have far fewer adipocytes [16]. Knocking down Atg7 in 3T3-L1 preadipocytes attenuated the accumulation of triglycerides and several markers of mature adipocytes as well. This was further validated in vivo in an adipose-specific Atg7 knock out mouse; the total fat in the knock out mouse weighed less than half that of the wild type mice. Interestingly, the adipose-specific Atg7$^{-/-}$ mouse had increased amounts of brown adipose tissue (BAT) and less white adipose tissue (WAT) relative to the wild type mouse, and both the lipid droplets and adipocytes were smaller and more densely packed than normal adipocytes [17,18]. The brown appearance of BAT is due to high quantities of densely packed mitochondria. BAT can uncouple the mitochondrial ATPase and continue oxidizing nutrients to generate thermal energy in place of the chemical energy provided by ATP. Zhang et al. validated these results, but went further and confirmed an increased presence of mitochondria [18]. Thus, while autophagy deficient adipocytes fail to express differentiation markers and lack lipid droplet morphology, some of the effects seen in autophagy-deficient adipocytes may result from the accumulation of mitochondria. Indeed, increased lipid catabolism in the form of β-oxidation occurred in the $Atg7^{-/-}$ adipocytes [17]. These findings points to a potentially functionally significant role for mitophagy during adipocyte differentiation. WAT primarily stores fatty acids for use under starvation conditions. Perhaps mitophagy is a necessary step for preventing wasteful energy expenditure through lipid catabolism in adipocytes. However, loss of the pro-mitophagic protein Bnip3 causes hepatocytes to accumulate mitochondria like adipocytes, but at the same time actually increase anabolic processes such as fatty acid synthesis [19]. It would be interesting to identify the cause of these differences.

2.4. Epithelial cells

Autophagy plays a role in different aspects of epithelial differentiation. One aspect resembles that seen in erythrocytes. Keratinocyte basal cells begin differentiation by detaching from the basement membrane and moving apically toward their fate of cornification and exfoliation. In the granulation stage, keratinocytes lose their organelles and nucleus [20]. Increased proteasome activity accompanies granulation and accounts for some of the degradative action, but entire organelles may require the autophagy pathway. Granulating keratinocytes are packed with lysosome bodies, and examination of expression patterns during differentiation revealed increased expression of the pro-autophagic proteins Beclin and Sirt1 [21,22]. Whether these events are more specifically controlled by organelle-specific autophagy mechanisms as is observed in erythrocytes has yet to be determined.

A question that remains is whether there is a role for autophagy in earlier epithelial differentiation stages, such as during the first emergence of suprabasal cells in striated epithelium. In one study, 3-MA-treated keratinocytes failed to express the differentiation marker involucrin when stimulated to differentiate in low glucose and high calcium [22]. However, this study should be interpreted with caution, because the reduction in glucose used to induce differentiation is a metabolic stressor. Because the metabolic state of cells regulates rates of autophagy, the increase in autophagy seen here may be an independent coincidence caused by the metabolic stress used to induce differentiation. Additionally, the sole use of 3-MA to inhibit autophagy brings up the question of selectivity. 3-MA also inhibits class I PI3K [23], an important regulator of keratinocyte differentiation [24,25].

The most striking link between autophagy and early epithelial differentiation comes from studies in the mammary gland. Autophagy promotes the survival of differentiated breast epithelial cells in 3D models of breast acinar formation [26]. Like keratinocytes, the breast epithelial cells lose adhesion to the ECM and push away from the basement membrane as differentiation occurs. Autophagy is increased in the differentiating cells in the center of the acini. However, inhibiting autophagy with 3-MA during differentiation increases luminal cell caspase-3 cleavage and death rather than blocking differentiation per se [27]. This suggests autophagy controls the rate of lumen cell loss, but the reason for needing this intermediate step is not clear.

Another example of autophagy in epithelial differentiation is in development of the embryoid body (EB). Embryoid bodies develop spontaneously from cultured embryonic stem cells, forming a spherical body of cells that can differentiate into various cell types [28]. The inner

http://dx.doi.org/10.5772/55398

cells of the EB display increased amounts of autophagic vacuoles similar to the breast epithelia 3D acini. However, unlike what happens in the breast acini, autophagy inhibition resulted in the accumulation of dead cell bodies in the EB lumen, suggesting that in this case increased autophagy is responsible for clearing out the dead cell debris accumulating at the center of the EB spheroid. In EB, cavitation occurs in response to increased hypoxia. Qi et al concluded that the hypoxic environment increased apoptosis inducing factor (AIF), which in turn increased ROS production and HIF2α signaling, and subsequently up-regulated pro-autophagic Bnip3. Knocking down any of the aforementioned proteins delayed cavitation of the EB. However, unlike the 3D breast acini model, inhibiting autophagy delayed caspase cleavage [29].

The examples in breast and embryoid bodies indicate that one of the roles of autophagy is to control when and how cell death occurs during differentiation, whereas autophagy is required for organelle clearing and survival in granulocytic keratinocytes and erythrocytes [21]. Thus, depending on the situation or cell state, autophagy can play distinct roles in differentiation.

A key question that remains in many of these models is why autophagy is important during differentiation. Mitochondrial clearance is likely important for the functional aspect of erythrocytes; if erythrocytes still contained mitochondria and carried out oxidative phosphorylation they would probably deplete much of their own oxygen in circulation before reaching tissues needing oxygen. Failure to eliminate mitochondria during adipocyte differentiation results in increased catabolism and a poor ability to store fat. However, in many systems autophagy seems to allow the cells to simply survive differentiation. Perhaps the metabolic stress of differentiating takes a toll on the integrity of the mitochondria and other intracellular components and autophagy increases just to keep up with the increased demand for maintenance. On the other hand, autophagy may have specific tumor suppressive effects during differentiation. Loss or reduction in autophagy appears to be required for tumor initiation and it is well established that tumorigenesis is linked with aberrant differentiation. One possibility is that autophagy actually promotes terminal differentiation, creating an antagonist to tumorigenesis. The sacrifice the tumor cells make is to create a greater dependence on anti-apoptotic survival pathways to compensate for the lack of the survival benefit of autophagy. However, as cancer progresses, some aspects of the autophagy pathway re-emerge as a mechanism to escape death under stress, but this must be balanced by preventing differentiation and reduced dependence on mitochondria for energy.

3. Adhesion and regulation of autophagy

In the case of multilayered epithelia, such as skin, bladder, prostate, and breast, differentiation occurs as the cells detach from the basement membrane and ascend apically. Such major changes require extensive reprogramming of signaling networks and gene expression, and cells must eliminate or inhibit the former cellular programming machinery. Failure to efficiently modify the programming can cause cellular stress and/or oncogenesis. Perhaps an immediate increase in autophagy aids in temporal separation of the undifferentiated and differentiated cell signaling and programming pathways. Coincident with this process is the need to maintain cell survival. This brings to question what signaling molecules in differentiation regulate autophagy.

3.1. Integrins and detachment-induced autophagy

Integrins, a family of heterodimeric transmembrane proteins, mediate cell adhesion to the extracellular matrix (ECM). Integrins regulate a variety of functions at the cellular and molecular level [30]. In stratified epithelial cell differentiation the integrins holding basal cells to the basement membrane become internalized and eliminated as the cell stratifies. Integrins form focal points with complexes of signaling molecules to maintain the connection between the matrix and the contracting cytoskeleton [31]. Cell survival is a key function of integrins. Loss of cell adhesion due to disengagement of integrins from their ECM ligand simultaneously activates extrinsic and intrinsic apoptotic cell death termed anoikis [32]. This signaling pathway prevents exfoliated or damaged cells from surviving detachment and adhering at an improper location.

Detachment and anoikis also aids in lumen clearing in secretory epithelium. As epithelial cells become contact inhibited, they force some cells off the basement membrane [33]. In undifferentiated epithelium this loss of adhesion would normally trigger anoikis and apoptosis. However, in differentiating epithelium, autophagy is triggered which delays the rapid onset of anoikis. This was demonstrated in both 3D acini and in suspended cells. Autophagy was induced in a subpopulation of human mammary epithelial MCF10A and canine kidney epithelial cells when they were placed in suspension, which allowed them to survive anoikis. Inhibiting autophagy by knocking down the autophagy proteins ATG5, ATG6, and ATG7 in suspended MCF-10A cells increased cleaved caspase-3 positive cells and decreased replating efficiency [27]. Further studies demonstrated that the decision to undergo autophagy, as opposed to anoikis, is controlled in part by up-regulation PERK to suppress ROS, through activation of the ER stress pathway, and resulting in increased expression of ATG6 and ATG8 [34]. When squamous cell carcinomas (SCCs) were isolated from FAK-/- mice (which presumably have altered integrin-based adhesion), or when SCCs isolated from wild type mice were placed in suspension, the rate of autophagy was significantly increased. This resulted in the targeting of Src to autophagosomes through c-Cbl in an E3-ligase-independent mechanism [35]. The major conclusion drawn from these studies is that loss of matrix adhesion induces autophagy.

3.2. Attachment-induced autophagy

This latter conclusion is partially contradicted by several studies demonstrating that adhesion to matrix is required to promote autophagy. In a study on human primary basal prostate epithelial cells (PrEC) researchers found that blocking integrin interactions with the ECM inhibited autophagy induced by starvation. In culture, PrECs secrete and adhere to a laminin 5, via integrins $\alpha3\beta1$ and $\alpha6\beta4$. These integrins were necessary for maintaining cell survival through Src and ligand-independent EGFR activation. Blocking integrin $\alpha3$, $\alpha6$, or $\beta4$ with antibodies inhibited both survival and autophagy in these cells. Particularly interesting is that inhibiting autophagy with 3-MA or blocking integrin $\alpha3$ antibodies induced caspase cleavage, but inhibiting EGFR alone did not. Since a class I-specific inhibitor could not induce apoptosis in PrECs, it is unlikely that off target inhibition of class I PI3K by 3-MA resulted in caspase cleavage. In addition, cell adhesion or inhibition of EGFR had no effect on Akt phosphorylation, which suggests that integrin-mediated maintenance of autophagy does not occur through the PI3K/Akt pathway in PrECs [36]. This study demonstrated that integrin adhesion mediates autophagy. While the exact downstream signaling pathway from integrins to autophagy remains unclear, survival was dependent on Erk signaling which is known to promote autophagy. A subsequent study demonstrated that in prostate cancer cells the androgen receptor promoted survival via integrin $\alpha6\beta1$ activation of IKK/NF-κB signaling, independent of the PI3K/Akt pathway [37]. A similar dependence on the IKK/NF-κB pathway for survival is mediated by integrin $\alpha6\beta4$ in polarized cells of 3D breast acinar structures and over activation of this pathway is sufficient to overcome suspension-induced anoikis and apoptosis [38-40]. Given that activation of the IKK/NF-κB pathway can induce autophagy [41], it is reasonable to suspect that integrins may regulate autophagy in part through the IKK/NF-κB pathway (Figure 1A). Further insight into how integrins regulate NF-κB and autophagy is needed.

Stimulation of smooth muscle cells with the integrin ligand osteopontin, stimulated autophagy related-genes, autophagosome formation, and ultimately cell death [42]. This effect was mediated by p38-MAPK through integrin and CD44 (Figure 1C). In a Drosophila genetic screen, paxillin, a downstream target of integrins, was found to associate with Atg1 and to be required for starvation-induced autophagy in MEFs [43]. However, in the fly, the ability of paxillin to facilitate autophagy was not linked with integrins. Whether this is true in the mammalian model has not been determined. Thus, several studies implicate integrin-based adhesion in both suppressing as well as stimulating autophagy. Since all these studies were conducted in different model systems and under different culture conditions, it is difficult to draw a definitive conclusion. Therefore, when studying different biological events it is necessary to consider several possible mechanisms and signaling pathways that could influence autophagy.

Some consideration should be given to a potential alternative mechanism to explain detachment-induced autophagy. Although it was quite thoroughly demonstrated that the cells increased the rate of autophagy after losing ECM contact, in both the EB and acini models the cells are undergoing a differentiation program and remaining in contact with neighboring cells. In the MCF-10A breast model where cells were put into suspension and 25% survived anoikis, these cells existed in large aggregates [27]. In light of the fact that increased cell density also increases autophagy in vitro, one could hypothesize that the proteins mediating cell-cell contact may be signaling to increase autophagy in the detached cells [23].

3.3. Cell-cell adhesion-induced autophagy

A relatively new process called entosis, whereby cell-cell adhesion induces the entry of one cell into a neighboring cell, was found to trigger an autophagic response [44] by mechanisms that resemble pathogen engulfment by immune cells. The ability of cells to entose depends on adherens junctions in the absence of integrin adhesion [45]. The major protein that mediates

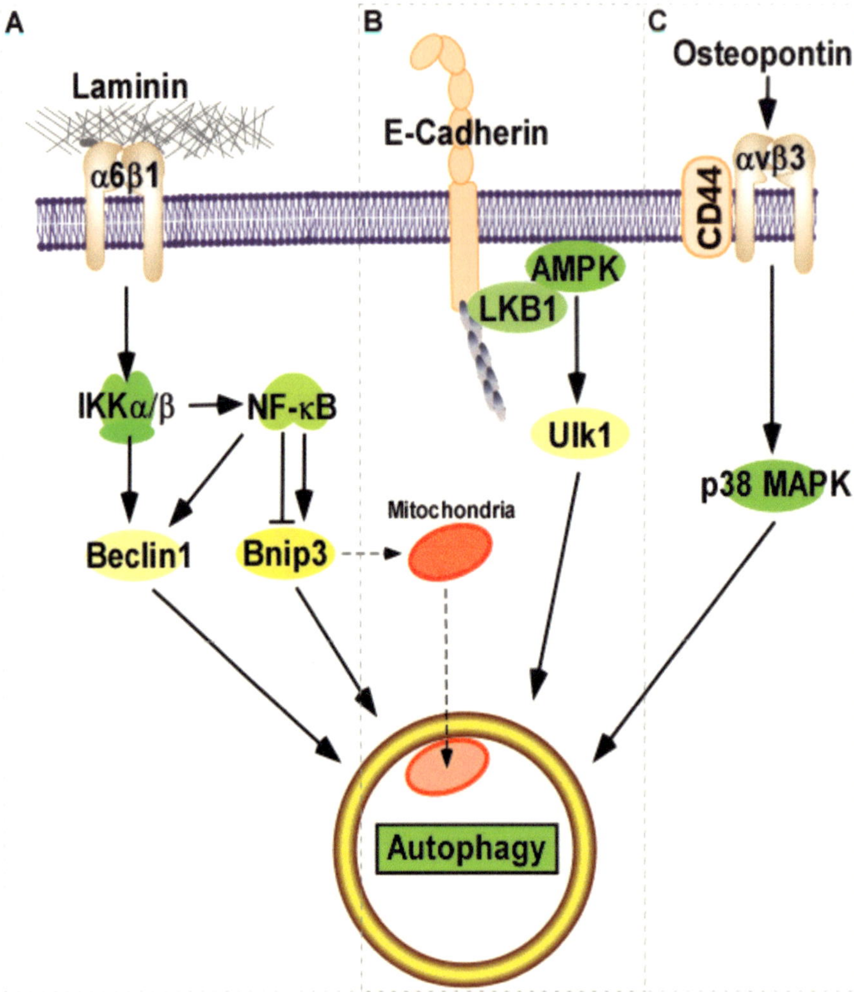

Figure 1. Cell Adhesion Pathways that Promote Autophagy. (A) Integrin α6β1 activates NF-κB through IKK signaling. Together, IKK and NF-κB can up-regulate and activate Beclin, and NF-κB in turn regulates pro-mitophagic Bnip3 expression. (B) AMPK localizes with LKB1 at adherens junctions where LKB1 activates AMPK signaling, which can activate autophagy through Ulk1. (C) Osteopontin activates p38 MAPK-induced autophagy through CD44 and αvβ3.

epithelial cell-cell contact through adherens junctions is E-cadherin. This vital protein holds the epithelium together and promotes the survival of cells [46-48]. Adherens junctions mediate the connection between cells by essentially bridging the actin cytoskeleton of two different cells. A plethora of signaling events are controlled by E-cadherin. In the case of autophagy, one point of interest is regulation of AMPK.

In response to low ATP levels, AMPK directly phosphorylates and activates Ulk1 in parallel to deactivating the autophagy suppressor mTOR [49]. Membrane bound E-cadherin localizes with the AMPK activator LKB1 (Figure 1B), and by doing so increases AMPK activation by bringing it into close proximity with LKB1 [50]. In one study AMPK was required for the drug

http://dx.doi.org/10.5772/55398

Atorva to induce an endoplasmic reticulum stress response, autophagy, and phosphorylation of eIF2α [51]. Interestingly, in the MCF-10A detachment-induced autophagy model, suspended cells had increased levels of activated PERK, a well-documented cytoprotective ER stress-response protein that induces autophagy, and increased phosphorylation of the downstream PERK target eIF2α in conjunction with increased autophagic flux [34]. Considering the overlapping pathways and the correlation of detachment-induced autophagy with cell-cell contact this could be a promising area to investigate.

3.4. Genetic evidence for integrin adhesion and autophagy

Deficiency of the laminin α2 chain (LAMA2) in laminin 2 causes a form of congenital muscular dystrophy known as MDC1A. MDC1A patients present with severe muscle weakness, peripheral neuropathy, and joint contractures [52]. The LAMA2-deficient mouse model (dy^{3K}/dy^{3K}) develops muscle weakness two weeks after birth and dies by 5 weeks with severe muscular dystrophy [53,54]. The dy^{3K}/dy^{3K} mice feature increases in myocyte apoptosis, degeneration/regeneration cycles, variable fiber sizes, and connective tissue hyperproliferation. dy^{3K}/dy^{3K} mice exhibit significantly increased expression of the autophagy related genes Bnip3, Bnip3l, p62, LC3B, GABARAP1, Vps34, Atg4b, Cathepsin L, Lamp2a, and Beclin. Administration of 3-MA to the dy^{3K}/dy^{3K} mice increased lifespans, increased average muscle fiber diameter, decreased the presence of caspase-3 in myocytes, and partially restored muscular morphology. In addition, immunofluorescent staining of muscle biopsies from MDC1A patients showed an increased amount of LC3B in the myocytes. Thus, patients with MDC1A may essentially suffer from excessive muscular autophagy due to a lack of interaction with the matrix protein laminin 2 (Figure 2A). This example supports a role for matrix detachment in inducing autophagy.

On the other hand, one of the LAMA2 receptors, dystroglycan which binds laminin extracellularly and dystrophin intracellularly, may promote adhesion-dependent autophagy. One function of the dystroglycan/dystrophin complex is to bind F-actin to anchor the cytoskeleton in place. Like the dy^{3K}/dy^{3K} mice, the dystrophin mutant (mdx) mice display a muscular dystrophy phenotype. However, in this case, the mdx mice accumulate damaged mitochondria, and inducing autophagy through AMPK activation in these mice ameliorates the disease phenotype, indicating there is a defect in muscle autophagy in mdx mice. This further suggests that adhesion to laminin through the dystroglycan/dystrophin connection is required to maintain autophagy in myocytes [55,56]. It is worth noting that mutations in integrin α7 are associated with muscular dystrophy as well. The integrin dimer $\alpha 7\beta 1$ is another receptor that binds laminin-α2 and resides on myocytes [57]. However, whether $\alpha 7\beta 1$ interactions affect autophagy in these diseases remains unknown. Further evidence supporting the concept that cell adhesion mediates autophagy comes from the study of Ulrich's Congenital Muscular Dystrophy (UCMD) and Bethlem Myopathy. Both diseases involve chronic weakening and degradation of skeletal muscle due to the spontaneous death of myocytes, which ultimately leads to death. Genetic studies discerned that mutations in the gene coding for collagen VI correlate with the condition [58]. Furthermore, genetically modified mice lacking the collagen VI gene (Col6-/-) develop myopathies similar to UCMD and Bethlem Myopathy. The $Col6^{-/-}$

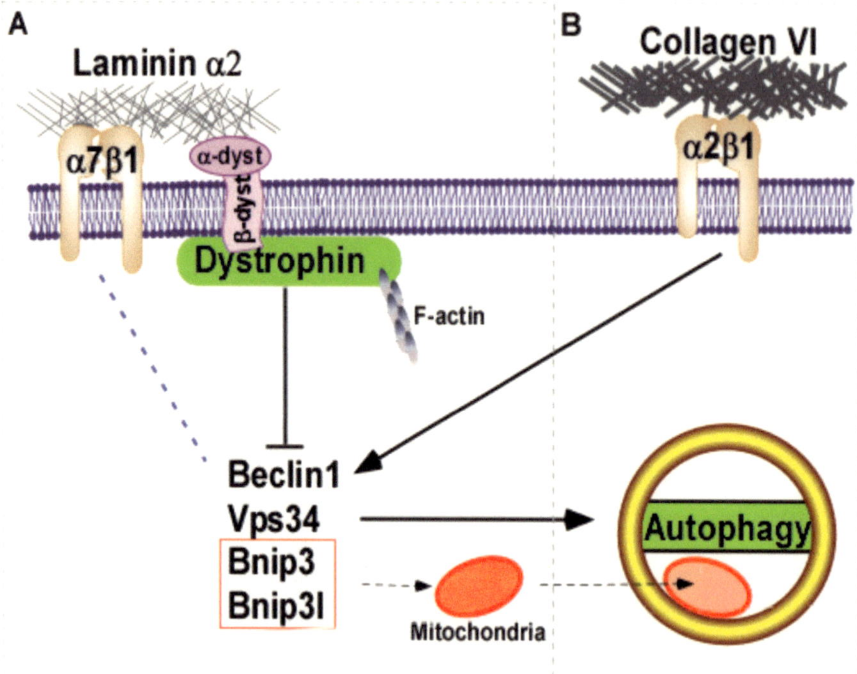

Figure 2. Matrix Adhesion Pathways and Autophagy in Muscle Cells. (A) Laminin α2 is bound by the α- and β-dystroglycan complex, which in turn binds dystrophin, a protein that anchors the complex onto F-actin. Loss of Laminin α2 caused increase expression of Beclin, Bnip3, Bnip3l, Vps34, and several other autophagy related proteins. Both Bnip3 and Bnip3l can induce mitophagy as well. Integrin α7β1 binds laminin α2 as well, but a connection with autophagy has not been established. (B) Loss of Collagen VI increased pro autophagic and pro mitophagic genes and caused decreased autophagy along with decreased expression of Beclin and Bnip3. The collagen receptor integrin α2β1 may be responsible for this.

mice have dysmorphic and dysfunctional myofibers characterized by increased apoptosis and reduced muscle strength compared to wild type mice [59]. Electron microscopy of $Col6^{-/-}$ myocytes revealed damaged and misshapen mitochondria and biochemical assays demonstrated that the mitochondria are dysfunctional. Inhibiting mitochondrial permeablization with Cyclosporine A in $Col6^{-/-}$ mice and in patients with homologous myopathies ameliorated the diseases effects [60], and genetic ablation of the cyclosporine A target, Cyclophilin D, also mitigated the symptoms in $Col6^{-/-}$ mice [61]

Later research found that $Col6^{-/-}$ mice muscle tissue expressed lower levels of the pro-autophagic proteins Beclin and Bnip3 and displayed a significant decrease in autophagy (Figure 2B). Forced expression of Beclin and Bnip3 partially reversed the disease phenotype in $Col6^{-/-}$ myocytes, and inducing autophagy by fasting the $Col6^{-/-}$ mice reestablished autophagy and decreased apoptosis in the myocytes. Furthermore, fasting or treating the $Col6^{-/-}$ mice with Rapamycin, to block the autophagy inhibitor mTor, decreased the presence of dysfunctional mitochondria [62]. These latter studies on Col6-related myopathies indicate that myocyte cell adhesion to collagen maintains autophagic flux and cell survival, while unregulated detachment inhibits autophagy.

http://dx.doi.org/10.5772/55398

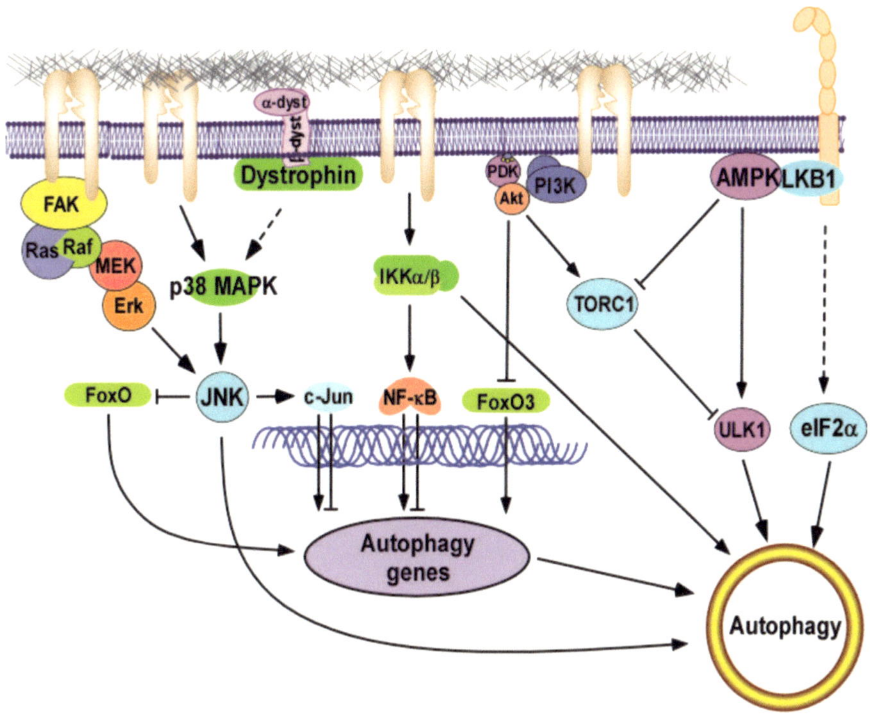

Figure 3. Mechanisms for Cell Adhesion-Induced Autophagy. The intracellular domain of integrins binds to FAK, which activates the Ras/Raf/MEK/Erk signaling cascade. Integrins and dystrophin both can activate p38 MAPK. Both Erk and p38 can phosphorylate JNK and induce autophagy. JNK in turn can either inhibit autophagy through suppression of FoxO or promote autophagy through activation of c-Jun, which can positively and negatively regulate autophagy. Integrins can also activate the IKKα/β–NF-κB pathway. IKKα/β induces autophagy both dependently and independently of NF-κB; however, NF-κB can also inhibit autophagy in some cases. Integrins also activate the PI3K/Akt/TORC1, which inhibit autophagy by phosphorylating and inhibiting FoxO3 and Ulk1. LKB1 binds to E-cadherin and activates AMPK. AMPK both activates Ulk1 and inhibits TORC1, causing an increase in autophagy. Some evidence indicates that cell-cell adhesion may also activate the autophagy-inducer eIF2α.

3.5. Cell adhesion signaling pathways that regulate autophagy

Adhesion proteins interact with a host of proteins and activate several well characterized pathways that affect autophagy. Recall that the dystrophin glycoprotein complex promotes adhesion dependent signaling in myocytes, and the importance of this was highlighted in the Mdx and dy^{3K}/dy^{3K} mice (Figure 2A). When the dystrophin glycoprotein complex is bound to laminin, an intracellular component of the complex called syntrophin binds and activates the Rac1 signaling complex, leading to activation of JNK, a protein that can both positively and negatively regulate autophagy [63] (Figure 3). JNK can induce autophagy by phosphorylating Bcl-2 and thus preventing Bcl-2 from inhibiting Beclin [64]. JNK also regulates several transcription factors that in turn mediate expression or repression of autophagy related genes. For example, genetic ablation of all three JNK isoforms results in constitutive activation of the transcription factor FoxO, causing an increase in autophagy and the effector proteins Bnip3 and Beclin. JNK can also activate the transcription factor c-Jun, a protein which when overexpressed can inhibit autophagy under starvation conditions [65]. On the other hand, another study showed that JNK was required for ceramide induced autophagy and expression of Beclin [66]. Furthermore, Jnk1 can phosphorylate Bcl-2 to release Beclin and promote autophagy [67]. Integrins may also regulate autophagy via p38 MAPK. Osteopontin binds integrins and CD44 and induces autophagy through activation of p38 MAPK [42]. Although some evidence indicates that p38 inhibits autophagy, p38 can activate JNK [68], which could provide a potential mechanism for both positive and negative regulation of autophagy by integrins. Finally, integrins stimulate Erk1/2 signaling as well, which activates autophagy through numerous pathways including activation of JNK [68].

Signaling through IKKα/β–NF-κB is another node for promoting autophagy (Figure 3). IKKα/β regulates the expression of autophagy-related genes by activating NF-κB, which in turn has been shown to both positively and negatively regulate autophagy [41,69,70]. IKKα/β can also activate autophagy independently of NF-κB [70]. Cells expressing constitutively active subunits of the IKK complex had increased autophagy and increased levels of phosphorylated active AMPK and JNK [71]. Considering that blocking integrin adhesion in PrECs subsequently attenuated autophagy [36], and that integrin mediated adhesion to laminin increased IKKα/β - NF-κB signaling enough to significantly increase survival in a prostate cancer line [46], signaling through IKKα/β - NF-κB may be a mechanism by which integrin-mediated-cell-adhesion promotes autophagy. It would be interesting to determine whether this mechanism occurs and whether it is dependent or independent of NF-κB.

PI3K/Akt signaling potently suppresses autophagy [72,73] (Figure 3). Akt activates the TORC1 complex, which in turn phosphorylates and inactivates the autophagy-initiating kinase Ulk1 [73]. Akt itself phosphorylates and inactivates the FoxO3a transcription factor. This prevents FoxO3a from inducing the expression of numerous autophagy- and Mitophagy-related genes [74]. Integrins can activate PI3K/Akt signaling [30], making it possible that integrins could suppress autophagy through this pathway as well. Perhaps this explains the cases in which detachment from the basement membrane actually promoted autophagy. However, it was shown that detachment of breast epithelial cells both increased autophagy and induced phosphorylation of eIF2α, a target of the ER kinase PERK and a protein capable of inducing autophagy [34].

AMPK activates autophagy by inhibiting TORC1 and activating Ulk1 via direct phosphorylation [49]. LKB1 localizes with engaged E-cadherin and activates AMPK signaling when ATP levels are low (Figure 3), providing a potential mechanism for cell-cell-interaction-mediated autophagy that might be relevant in detachment induced autophagy mechanisms in which cell-cell contact is maintained [50].

4. Conclusion

In many of the studies discussed above and in other diseases and models the effects of autophagy are very mitochondria-centric. For example, the $Col6^{-/-}$ mice phenotype was rescued by inhibiting mitochondrial depolarization, re-expressing Beclin or the pro-mitophagic protein Bnip3, or by starvation-induced autophagy [61,62]. Although not discussed in this chapter, it is interesting to note that Parkinson's disease is associated with mutations in several genes that target damaged mitochondria for mitophagy, and increasing evidence implicates mitophagy as having a role in retarding aging in general [75,76]. Considering that mitochondria can induce apoptosis and regulate the levels of ROS, one could imagine mitophagy as being a mechanism of survival during detachment or a method of maintaining cell integrity during adhesion.

Cell interaction with ECM and other cells of the same organism is imperative for - and essentially defines - multicellular organisms. Aside from being the scaffolding that organizes cells into functional macrostructures, the ECM extensively regulates cellular behavior [30]. The importance of these interactions is well demonstrated in the mounting evidence indicating that varied interactions with the ECM can actually differentially dictate the fate of stem cell differentiation [77]. Understanding the effects of the ECM on cell biology not only increases the understanding of normal organismal development and biology, but also diseases such as cancer or the various forms of muscular dystrophy [78,79]. With the current rise in understanding of autophagy in diseases and development, understanding how extracellular interactions and differentiation affect autophagy is an important avenue to explore in the future.

Nomenclature

3-MA - 3-methyladenine

AIF - apoptosis inducing factor

BAT - brown adipose tissue

Col6 - collagen VI gene

CFSE - 5,6-carboxyfluorescein diacetate succinimidyl ester

EB - embryoid body

ECM - extracellular matrix

LAMA2 - laminin α2 chain

Mdx - dystrophin gene

MEF - mouse embryo fibroblast

PrEC - prostate epithelial cell

SSC- squamous cell carcinomas

UCMD - Ulrich's Congenital Muscular Dystrophy

WAT- white adipose tissues

Acknowledgements

We wish to thank the Van Andel Research Institute (EAN, CKM) and the Van Andel Institute Graduate School (EAN) for their generous support. This work was also supported in part by an NIH/NCI grant CA154835 (CKM).

Author details

Eric A. Nollet[1,2] and Cindy K. Miranti[1*]

*Address all correspondence to: cindy.miranti@vai.org

1 Laboratory of Integrin Signaling and Tumorigensis, Van Andel Research Institute, Grand Rapids, Michigan, USA

2 Van Andel Institute Graduate School, Grand Rapids, Michigan, USA

References

[1] Gronowicz, G, Swift, H, & Steck, T. L. Maturation of the reticulocyte in vitro. J Cell Sci. (1984). http://www.ncbi.nlm.nih.gov/pubmed/6097593, 71, 177-97.

[2] Koury, M. J, Koury, S. T, Kopsombut, P, & Bondurant, M. C. In vitro maturation of nascent reticulocytes to erythrocytes. Blood. (2005). http://www.ncbi.nlm.nih.gov/pubmed/15528310, 105(5), 2168-74.

[3] Kundu, M, Lindsten, T, Yang, C. Y, Wu, J, Zhao, F, Zhang, J, et al. Ulk1 plays a critical role in the autophagic clearance of mitochondria and ribosomes during reticulocyte maturation. Blood. (2008). http://www.ncbi.nlm.nih.gov/pubmed/18539900, 112(4), 1493-502.

[4] Schweers, R. L, Zhang, J, Randall, M. S, Loyd, M. R, Li, W, Dorsey, F. C, et al. NIX is required for programmed mitochondrial clearance during reticulocyte maturation. Proc Natl Acad Sci U S A. (2007). http://www.ncbi.nlm.nih.gov/pubmed/18048346, 104(49), 19500-5.

[5] Ney, P. A. Normal and disordered reticulocyte maturation. Curr Opin Hematol. (2011). http://www.ncbi.nlm.nih.gov/pubmed/21423015, 18(3), 152-7.

[6] Zhang, J, Loyd, M. R, Randall, M. S, Waddell, M. B, Kriwacki, R. W, & Ney, P. A. A short linear motif in BNIP3L (NIX) mediates mitochondrial clearance in reticulocytes. Autophagy. (2012). http://www.ncbi.nlm.nih.gov/pubmed/22906961, 8(9), 1325-32.

[7] Zhang, J, Randall, M. S, Loyd, M. R, Dorsey, F. C, Kundu, M, Cleveland, J. L, et al. Mitochondrial clearance is regulated by Atg7-dependent and-independent mechanisms during reticulocyte maturation. Blood. (2009). http://www.ncbi.nlm.nih.gov/pubmed/19417210, 114(1), 157-64.

[8] Aerbajinai, W, Giattina, M, Lee, Y. T, Raffeld, M, & Miller, J. L. The proapoptotic factor Nix is coexpressed with Bcl-xL during terminal erythroid differentiation. Blood. (2003). http://www.ncbi.nlm.nih.gov/pubmed/12663450, 102(2), 712-7.

[9] Hanna, R. A, Quinsay, M. N, Orogo, A. M, Giang, K, Rikka, S, & Gustafsson, A. B. Microtubule-associated protein 1 light chain 3 (LC3) interacts with Bnip3 protein to selectively remove endoplasmic reticulum and mitochondria via autophagy. J Biol Chem. (2012). http://www.ncbi.nlm.nih.gov/pubmed/22505714, 287(23), 19094-104.

[10] Cebollero, E, Reggiori, F, & Kraft, C. Reticulophagy and ribophagy: regulated degradation of protein production factories. Int J Cell Biol. (2012). http://www.ncbi.nlm.nih.gov/pubmed/22481944

[11] Mortensen, M, Ferguson, D. J, Edelmann, M, Kessler, B, Morten, K. J, Komatsu, M, et al. Loss of autophagy in erythroid cells leads to defective removal of mitochondria and severe anemia in vivo. Proc Natl Acad Sci U S A. (2010). http://www.ncbi.nlm.nih.gov/pubmed/20080761, 107(2), 832-7.

[12] Pua, H. H, Guo, J, Komatsu, M, & He, Y. W. Autophagy is essential for mitochondrial clearance in mature T lymphocytes. J Immunol. (2009). http://www.ncbi.nlm.nih.gov/pubmed/19299702, 182(7), 4046-55.

[13] Stephenson, L. M, Miller, B. C, Ng, A, Eisenberg, J, Zhao, Z, Cadwell, K, et al. Identification of Atg5-dependent transcriptional changes and increases in mitochondrial mass in Atg5-deficient T lymphocytes. Autophagy. (2009). http://www.ncbi.nlm.nih.gov/pubmed/19276668, 5(5), 625-35.

[14] Dumitriu, I. E, Mohr, W, Kolowos, W, Kern, P, Kalden, J. R, & Herrmann, M. 5,6-carboxyfluorescein diacetate succinimidyl ester-labeled apoptotic and necrotic as well as detergent-treated cells can be traced in composite cell samples. Anal Biochem. (2001). http://www.ncbi.nlm.nih.gov/pubmed/11730350, 299(2), 247-52.

[15] Pua, H. H, Dzhagalov, I, Chuck, M, Mizushima, N, & He, Y. W. A critical role for the autophagy gene Atg5 in T cell survival and proliferation. The Journal of experimental medicine. (2007). http://www.ncbi.nlm.nih.gov/pubmed/17190837, 204(1), 25-31.

[16] Baerga, R, Zhang, Y, Chen, P. H, Goldman, S, & Jin, S. Targeted deletion of autophagy-related 5 (atg5) impairs adipogenesis in a cellular model and in mice. Autophagy. (2009). http://www.ncbi.nlm.nih.gov/pubmed/19844159, 5(8), 1118-30.

[17] Singh, R, Xiang, Y, Wang, Y, Baikati, K, Cuervo, A. M, Luu, Y. K, et al. Autophagy regulates adipose mass and differentiation in mice. The Journal of clinical investigation. (2009). http://www.ncbi.nlm.nih.gov/pubmed/19855132, 119(11), 3329-39.

[18] Zhang, Y, Goldman, S, Baerga, R, Zhao, Y, Komatsu, M, & Jin, S. Adipose-specific deletion of autophagy-related gene 7 (atg7) in mice reveals a role in adipogenesis. Proc Natl Acad Sci U S A. (2009). http://www.ncbi.nlm.nih.gov/pubmed/19910529, 106(47), 19860-5.

[19] Glick, D, Zhang, W, Beaton, M, Marsboom, G, Gruber, M, Simon, M. C, et al. BNip3 regulates mitochondrial function and lipid metabolism in the liver. Mol Cell Biol. (2012). http://www.ncbi.nlm.nih.gov/pubmed/22547685, 32(13), 2570-84.

[20] Lippens, S, Denecker, G, Ovaere, P, Vandenabeele, P, & Declercq, W. Death penalty for keratinocytes: apoptosis versus cornification. Cell Death Differ. (2005). Suppl http://www.ncbi.nlm.nih.gov/pubmed/16247497, 2, 1497-508.

[21] Gosselin, K, Deruy, E, Martien, S, Vercamer, C, Bouali, F, Dujardin, T, et al. Senescent keratinocytes die by autophagic programmed cell death. Am J Pathol. (2009). http://www.ncbi.nlm.nih.gov/pubmed/19147823, 174(2), 423-35.

[22] Aymard, E, Barruche, V, Naves, T, Bordes, S, Closs, B, Verdier, M, et al. Autophagy in human keratinocytes: an early step of the differentiation? Exp Dermatol. (2011). http://www.ncbi.nlm.nih.gov/pubmed/21166723, 20(3), 263-8.

[23] Klionsky, D. J, Abdalla, F. C, Abeliovich, H, Abraham, R. T, Acevedo-arozena, A, Adeli, K, et al. Guidelines for the use and interpretation of assays for monitoring autophagy. Autophagy. (2012). http://www.ncbi.nlm.nih.gov/pubmed/22966490, 8(4), 445-544.

[24] Xie, Z, & Bikle, D. D. The recruitment of phosphatidylinositol 3-kinase to the E-cadherin-catenin complex at the plasma membrane is required for calcium-induced phospholipase C-gamma1 activation and human keratinocyte differentiation. J Biol Chem. (2007). http://www.ncbi.nlm.nih.gov/pubmed/17242406, 282(12), 8695-703.

[25] Calautti, E, Li, J, Saoncella, S, Brissette, J. L, & Goetinck, P. F. Phosphoinositide 3-kinase signaling to Akt promotes keratinocyte differentiation versus death. J Biol Chem. (2005). http://www.ncbi.nlm.nih.gov/pubmed/16036919, 280(38), 32856-65.

[26] Kenny, P. A, Lee, G. Y, Myers, C. A, Neve, R. M, Semeiks, J. R, Spellman, P. T, et al. The morphologies of breast cancer cell lines in three-dimensional assays correlate with their profiles of gene expression. Molecular oncology. (2007). http://www.ncbi.nlm.nih.gov/pubmed/18516279, 1(1), 84-96.

[27] Fung, C, Lock, R, Gao, S, Salas, E, & Debnath, J. Induction of autophagy during extracellular matrix detachment promotes cell survival. Mol Biol Cell. (2008). http://www.ncbi.nlm.nih.gov/pubmed/18094039, 19(3), 797-806.

[28] Desbaillets, I, Ziegler, U, Groscurth, P, & Gassmann, M. Embryoid bodies: An in vitro model of mouse embryogenesis. Exp Physiol. (2000). http://www.ncbi.nlm.nih.gov/pubmed/11187960, 85(6), 645-51.

[29] Qi, Y, Tian, X, Liu, J, Han, Y, Graham, A. M, Simon, M. C, et al. Bnip3 and AIF cooperate to induce apoptosis and cavitation during epithelial morphogenesis. J Cell Biol. (2012). http://www.ncbi.nlm.nih.gov/pubmed/22753893, 198(1), 103-14.

[30] Miranti, C. K, & Brugge, J. S. Sensing the environment: a historical perspective on integrin signal transduction. Nature cell biology. (2002). Ehttp://www.ncbi.nlm.nih.gov/pubmed/11944041, 83-90.

[31] Sieg, D. J, Hauck, C. R, & Schlaepfer, D. D. Required role of focal adhesion kinase (FAK) for integrin-stimulated cell migration. J Cell Sci. (1999). Pt 16):2677-91. http://www.ncbi.nlm.nih.gov/pubmed/10413676

[32] Frisch, S. M, & Francis, H. Disruption of epithelial cell-matrix interactions induces apoptosis. J Cell Biol. (1994). http://www.ncbi.nlm.nih.gov/pubmed/8106557, 124(4), 619-26.

[33] Mizushima, N, & Levine, B. Autophagy in mammalian development and differentiation. Nature cell biology. (2010). http://www.ncbi.nlm.nih.gov/pubmed/20811354, 12(9), 823-30.

[34] Avivar-valderas, A, Salas, E, Bobrovnikova-marjon, E, Diehl, J. A, Nagi, C, Debnath, J, et al. PERK integrates autophagy and oxidative stress responses to promote survival during extracellular matrix detachment. Mol Cell Biol. (2011). http://www.ncbi.nlm.nih.gov/pubmed/21709020, 31(17), 3616-29.

[35] Sandilands, E, Serrels, B, Mcewan, D. G, Morton, J. P, Macagno, J. P, Mcleod, K, et al. Autophagic targeting of Src promotes cancer cell survival following reduced FAK signalling. Nature cell biology. [10.1038/ncb2386]. (2012). http://www.ncbi.nlm.nih.gov/pubmed/22138575, 14(1), 51-60.

[36] Edick, M. J, Tesfay, L, Lamb, L. E, Knudsen, B. S, & Miranti, C. K. Inhibition of integrin-mediated crosstalk with epidermal growth factor receptor/Erk or Src signaling pathways in autophagic prostate epithelial cells induces caspase-independent death. Mol Biol Cell. (2007). http://www.ncbi.nlm.nih.gov/pubmed/17475774, 18(7), 2481-90.

[37] Lamb, L. E, Zarif, J. C, & Miranti, C. K. The androgen receptor induces integrin a6b1 to promote prostate tumor cell survival via NF-kB and Bcl-xL Independently of PI3K signaling. Cancer Res. (2011). http://www.ncbi.nlm.nih.gov/pubmed/21310825, 71(7), 2739-49.

[38] Friedland, J. C, Lakins, J. N, Kazanietz, M. G, Chernoff, J, Boettiger, D, & Weaver, V. M. a. a6b4 integrin activates Rac-dependent kinase 1 to drive NF-kB-dependent resistance to apoptosis in 3D mammary acini. J Cell Sci. (2007). Pt 20):3700-12. http://www.ncbi.nlm.nih.gov/pubmed/17911169, 21.

[39] Weaver, V. M, Lelievre, S, Lakins, J. N, Chrenek, M. A, Jones, J. C, Giancotti, F, et al. b4 integrin-dependent formation of polarized three-dimensional architecture confers

resistance to apoptosis in normal and malignant mammary epithelium. Cancer Cell. (2002). http://www.ncbi.nlm.nih.gov/pubmed/12242153, 2(3), 205-16.

[40] Zahir, N, Lakins, J. N, Russell, A, Ming, W, Chatterjee, C, Rozenberg, G. I, et al. Autocrine laminin-5 ligates a6b4 integrin and activates RAC and NF-kB to mediate anchorage-independent survival of mammary tumors. J Cell Biol. (2003). http://www.ncbi.nlm.nih.gov/pubmed/14691145, 163(6), 1397-407.

[41] Copetti, T, Bertoli, C, Dalla, E, Demarchi, F, & Schneider, C. p. RelA modulates BECN1 transcription and autophagy. Mol Cell Biol. (2009). http://www.ncbi.nlm.nih.gov/pubmed/19289499, 29(10), 2594-608.

[42] Zheng, Y. H, Tian, C, Meng, Y, Qin, Y. W, Du, Y. H, Du, J, et al. Osteopontin stimulates autophagy via integrin/CD44 and MAPK signaling pathways in vascular smooth muscle cells. J Cell Physiol. (2012). http://www.ncbi.nlm.nih.gov/pubmed/21374592, 38.

[43] Chen, G. C, Lee, J. Y, Tang, H. W, Debnath, J, Thomas, S. M, & Settleman, J. Genetic interactions between Drosophila melanogaster Atg1 and paxillin reveal a role for paxillin in autophagosome formation. Autophagy. (2008). http://www.ncbi.nlm.nih.gov/pubmed/17952025, 4(1), 37-45.

[44] Florey, O, Kim, S. E, Sandoval, C. P, Haynes, C. M, & Overholtzer, M. Autophagy machinery mediates macroendocytic processing and entotic cell death by targeting single membranes. Nature cell biology. (2011). http://www.ncbi.nlm.nih.gov/pubmed/22002674, 13(11), 1335-43.

[45] Overholtzer, M, Mailleux, A. A, Mouneimne, G, Normand, G, Schnitt, S. J, King, R. W, et al. A nonapoptotic cell death process, entosis, that occurs by cell-in-cell invasion. Cell. (2007). http://www.ncbi.nlm.nih.gov/pubmed/18045538, 131(5), 966-79.

[46] Lamb, L. E, & Knudsen, B. S. Miranti CK. E-cadherin-mediated survival of androgen-receptor-expressing secretory prostate epithelial cells derived from a stratified in vitro differentiation model. J Cell Sci. (2010). Pt 2):266-76. http://www.ncbi.nlm.nih.gov/pubmed/20048343

[47] Boussadia, O, Kutsch, S, Hierholzer, A, & Delmas, V. Kemler R. E-cadherin is a survival factor for the lactating mouse mammary gland. Mech Dev. (2002). http://www.ncbi.nlm.nih.gov/pubmed/12049767

[48] Li, L, Bennett, S. A, & Wang, L. Role of E-cadherin and other cell adhesion molecules in survival and differentiation of human pluripotent stem cells. Cell Adh Migr. (2012). http://www.ncbi.nlm.nih.gov/pubmed/22647941, 6(1), 59-70.

[49] Kim, J, Kundu, M, Viollet, B, & Guan, K. L. AMPK and mTOR regulate autophagy through direct phosphorylation of Ulk1. Nature cell biology. (2011). http://www.ncbi.nlm.nih.gov/pubmed/21258367, 13(2), 132-41.

[50] Sebbagh, M, Santoni, M. J, Hall, B, Borg, J. P, & Schwartz, M. A. Regulation of LKB1/STRAD localization and function by E-cadherin. Current biology : CB. (2009). http://www.ncbi.nlm.nih.gov/pubmed/19110428, 19(1), 37-42.

[51] Yang, P. M, Liu, Y. L, Lin, Y. C, Shun, C. T, Wu, M. S, & Chen, C. C. Inhibition of autophagy enhances anticancer effects of atorvastatin in digestive malignancies. Cancer Res. (2010). http://www.ncbi.nlm.nih.gov/pubmed/20876807, 70(19), 7699-709.

[52] Allamand, V, & Guicheney, P. Merosin-deficient congenital muscular dystrophy, autosomal recessive (MDC1A, MIM#156225, LAMA2 gene coding for alpha2 chain of laminin). Eur J Hum Genet. (2002). http://www.ncbi.nlm.nih.gov/pubmed/11938437, 10(2), 91-4.

[53] Miyagoe, Y, Hanaoka, K, Nonaka, I, Hayasaka, M, Nabeshima, Y, Arahata, K, et al. Laminin a2 chain-null mutant mice by targeted disruption of the Lama2 gene: a new model of merosin (laminin 2)-deficient congenital muscular dystrophy. FEBS Lett. (1997). http://www.ncbi.nlm.nih.gov/pubmed/9326364, 415(1), 33-9.

[54] Carmignac, V, Svensson, M, Korner, Z, Elowsson, L, Matsumura, C, Gawlik, K. I, et al. Autophagy is increased in laminin a2 chain-deficient muscle and its inhibition improves muscle morphology in a mouse model of MDC1A. Hum Mol Genet. (2011). http://www.ncbi.nlm.nih.gov/pubmed/21920942, 20(24), 4891-902.

[55] Chamberlain, J. S, Metzger, J, Reyes, M, Townsend, D, & Faulkner, J. A. Dystrophin-deficient mdx mice display a reduced life span and are susceptible to spontaneous rhabdomyosarcoma. FASEB J. (2007). http://www.ncbi.nlm.nih.gov/pubmed/17360850, 21(9), 2195-204.

[56] Pauly, M, Daussin, F, Burelle, Y, Li, T, Godin, R, Fauconnier, J, et al. AMPK activation stimulates autophagy and ameliorates muscular dystrophy in the mdx mouse diaphragm. Am J Pathol. (2012). http://www.ncbi.nlm.nih.gov/pubmed/22683340, 181(2), 583-92.

[57] Doe, J. A, Wuebbles, R. D, Allred, E. T, Rooney, J. E, Elorza, M, & Burkin, D. J. Transgenic overexpression of the a7 integrin reduces muscle pathology and improves viability in the dy(W) mouse model of merosin-deficient congenital muscular dystrophy type 1A. J Cell Sci. (2011). Pt 13):2287-97. http://www.ncbi.nlm.nih.gov/pubmed/21652631

[58] Sparks, S, Quijano-roy, S, Harper, A, Rutkowski, A, Gordon, E, Hoffman, E. P, et al. Congenital Muscular Dystrophy Overview. (1993). http://www.ncbi.nlm.nih.gov/pubmed/20301468

[59] Irwin, W. A, Bergamin, N, Sabatelli, P, Reggiani, C, Megighian, A, Merlini, L, et al. Mitochondrial dysfunction and apoptosis in myopathic mice with collagen VI deficiency. Nature genetics. (2003). http://www.ncbi.nlm.nih.gov/pubmed/14625552, 35(4), 367-71.

[60] Merlini, L, Angelin, A, Tiepolo, T, Braghetta, P, Sabatelli, P, Zamparelli, A, et al. Cyclosporin A corrects mitochondrial dysfunction and muscle apoptosis in patients with collagen VI myopathies. Proc Natl Acad Sci U S A. (2008). http://www.ncbi.nlm.nih.gov/pubmed/18362356, 105(13), 5225-9.

[61] Palma, E, Tiepolo, T, Angelin, A, Sabatelli, P, Maraldi, N. M, Basso, E, et al. Genetic ablation of cyclophilin D rescues mitochondrial defects and prevents muscle apoptosis in collagen VI myopathic mice. Hum Mol Genet. (2009). http://www.ncbi.nlm.nih.gov/pubmed/19293339, 18(11), 2024-31.

[62] Grumati, P, Coletto, L, Sandri, M, & Bonaldo, P. Autophagy induction rescues muscular dystrophy. Autophagy. (2011). http://www.ncbi.nlm.nih.gov/pubmed/21543891, 7(4), 426-8.

[63] Oak, S. A, Zhou, Y. W, & Jarrett, H. W. Skeletal muscle signaling pathway through the dystrophin glycoprotein complex and Rac1. J Biol Chem. (2003). http://www.ncbi.nlm.nih.gov/pubmed/12885773, 278(41), 39287-95.

[64] Wei, Y, Pattingre, S, Sinha, S, Bassik, M, & Levine, B. JNK1-mediated phosphorylation of Bcl-2 regulates starvation-induced autophagy. Molecular cell. (2008). http://www.ncbi.nlm.nih.gov/pubmed/18570871, 30(6), 678-88.

[65] Yogev, O, Goldberg, R, Anzi, S, & Shaulian, E. Jun proteins are starvation-regulated inhibitors of autophagy. Cancer Res. (2010). http://www.ncbi.nlm.nih.gov/pubmed/20197466, 70(6), 2318-27.

[66] Li, D. D, Wang, L. L, Deng, R, Tang, J, Shen, Y, Guo, J. F, et al. The pivotal role of c-Jun NH2-terminal kinase-mediated Beclin 1 expression during anticancer agents-induced autophagy in cancer cells. Oncogene. (2009). http://www.ncbi.nlm.nih.gov/pubmed/19060920, 28(6), 886-98.

[67] Wei, Y, Sinha, S. C, & Levine, B. Dual Role of JNK1-mediated phosphorylation of Bcl-2 in autophagy and apoptosis regulation. Autophagy. (2008). http://www.ncbi.nlm.nih.gov/pubmed/18570871, 4(7), 949-51.

[68] Riol-blanco, L, Sanchez-sanchez, N, Torres, A, Tejedor, A, Narumiya, S, Corbi, A. L, et al. The chemokine receptor CCR7 activates in dendritic cells two signaling modules that independently regulate chemotaxis and migratory speed. J Immunol. (2005). http://www.ncbi.nlm.nih.gov/pubmed/15778365, 174(7), 4070-80.

[69] Djavaheri-mergny, M, Amelotti, M, Mathieu, J, Besancon, F, Bauvy, C, & Codogno, P. Regulation of autophagy by NF-kB transcription factor and reactives oxygen species. Autophagy. (2007). http://www.ncbi.nlm.nih.gov/pubmed/17471012, 3(4), 390-2.

[70] Salminen, A, Hyttinen, J. M, Kauppinen, A, & Kaarniranta, K. Context-Dependent Regulation of Autophagy by IKK-NF-kB Signaling: Impact on the Aging Process. Int J Cell Biol. (2012). http://www.ncbi.nlm.nih.gov/pubmed/22899934

[71] Criollo, A, Senovilla, L, Authier, H, Maiuri, M. C, Morselli, E, Vitale, I, et al. The IKK complex contributes to the induction of autophagy. The EMBO journal. (2010). http://www.ncbi.nlm.nih.gov/pubmed/19959994, 29(3), 619-31.

[72] Wu, Y. T, Tan, H. L, Huang, Q, Ong, C. N, & Shen, H. M. Activation of the PI3K-Akt-mTOR signaling pathway promotes necrotic cell death via suppression of autophagy. Autophagy. (2009). http://www.ncbi.nlm.nih.gov/pubmed/19556857, 5(6), 824-34.

[73] Chen, Y, & Klionsky, D. J. The regulation of autophagy- unanswered questions. J Cell Sci. (2011). Pt 2):161-70. http://www.ncbi.nlm.nih.gov/pubmed/21187343

[74] Mammucari, C, Schiaffino, S, & Sandri, M. Downstream of Akt: FoxO3 and mTOR in the regulation of autophagy in skeletal muscle. Autophagy. (2008). http://www.ncbi.nlm.nih.gov/pubmed/18367868, 4(4), 524-6.

[75] Jin, S. M, & Youle, R. J. PINK1- and Parkin-mediated mitophagy at a glance. J Cell Sci. (2012). Pt 4):795-9. http://www.ncbi.nlm.nih.gov/pubmed/22448035

[76] Seo, A. Y, Joseph, A. M, Dutta, D, Hwang, J. C, Aris, J. P, & Leeuwenburgh, C. New insights into the role of mitochondria in aging: mitochondrial dynamics and more. J Cell Sci. (2010). Pt 15):2533-42. http://www.ncbi.nlm.nih.gov/pubmed/20940129

[77] Guilak, F, Cohen, D. M, Estes, B. T, Gimble, J. M, Liedtke, W, & Chen, C. S. Control of stem cell fate by physical interactions with the extracellular matrix. Cell Stem Cell. (2009). http://www.ncbi.nlm.nih.gov/pubmed/19570510, 5(1), 17-26.

[78] Lu, P, Weaver, V. M, & Werb, Z. The extracellular matrix: a dynamic niche in cancer progression. J Cell Biol. (2012). http://www.ncbi.nlm.nih.gov/pubmed/22351925, 196(4), 395-406.

[79] Davies, K. E, & Nowak, K. J. Molecular mechanisms of muscular dystrophies: old and new players. Nat Rev Mol Cell Biol. (2006). http://www.ncbi.nlm.nih.gov/pubmed/16971897, 7(10), 762-73.

Chapter 22

Time Flies: Autophagy During Ageing in *Drosophila*

Sebastian Wolfgang Schultz, Andreas Brech and
Ioannis P. Nezis

Additional information is available at the end of the chapter

http://dx.doi.org/10.5772/55396

1. Introduction

1.1. Ageing

The process of ageing compromises the age-associated decrease in fertility, gradual loss of function, and increased vulnerability to disease, which progressively diminishes the capability of an organism to survive [1-3]. Unsurprisingly, in the past years it has been of great interest to understand which factors influence this inevitable and complex process. As a result a wide array of molecular and cellular damages has been identified and shown to accumulate during ageing. The lifelong accumulation of such damages will eventually result in frailty and disease [4]. The variety of identified age-dependent damages has given rise to different theories for molecular ageing mechanisms. These mechanisms include decreased cellular capacity to deal with DNA damage, and decline in cellular division capacity, which is linked to the progressive shortening of telomeres upon each cell cycle. Also an increased accumulation of damaged mitochondria and the involved increase in reactive oxygen species (ROS) production and decline in ATP synthesis has been shown to occur over time (reviewed in [5]). One of the phenotypic hallmarks of aged cells is the intracellular accumulation of damaged proteins and therefore protein turnover/protein degradation has attracted attention over the last years [2].

At the same time, forward genetics have allowed to investigate single gene alterations and their influence on lifespan of whole organisms. Even though the ageing process is without doubt influenced by stochastic and environmental factors, single gene mutations were shown to extend lifespan in worms, flies, and mice, suggesting the existence of a central process of ageing [6, 7]. Many of the genetic manipulations that alter longevity affect metabolism, nutrient sensing and stress response pathways. As all these pathways are connected to autophagy (an important player also in protein turnover), the question about the role of autophagy in ageing has come more and more to the fore. In this chapter we will focus on how research conducted in the excellent genetic model system *Drosophila melanogaster* has contributed to understand more about the interplay of autophagy and ageing.

2. Autophagy

Autophagy, which literally means "self-eating" (coined by Nobel Laureate Christian de Duve in 1963), allows cells to digest cytosolic components via lysosomal degradation. Autophagy and the Ubiquitin Proteasome System (UPS) constitute together the main cellular pathways for protein and organelle turnover [8, 9]. Today, three different classes of autophagy are distinguished: microautophagy, chaperone-mediated autophagy (CMA), and macroautophagy.

During microautophagy, which is mainly studied in yeast (containing vacuoles instead of lysosomes), cytoplasmic material is delivered to the vacuolar lumen by direct invagination of the vacuolar boundary membrane and budding of autophagic bodies into the vacuolar lumen [10]. The molecular mechanisms underlying microautophagy in eukaryotic cells are largely unknown. However, Cuervo and colleagues described a microautophagy-like process (named endosomal microautophagy, e-MI) in mammalian cells, whereby soluble cytosolic proteins are selectively taken up by late endosomes/multivesicular bodies (MVBs). The cargo selection in e-MI depends on the chaperone Hsc70 and electrostatic interactions with the endosomal membrane [11]. Hsc70 is also involved in chaperone-mediated autophagy (CMA), in which cytosolic cargo is selectively recognized, bound by the lysosome-associated membrane type protein 2A (LAMP-2A) and finally taken up by the lysosome, thereby allowing for direct lysosomal degradation of cytosolic proteins. The requirement of protein unfolding and the binding of LAMP2-A is characteristic for CMA and thereby distinguishes CMA from e-MI [11, 12]. So far, CMA has not been investigated in *Drosophila melanogaster*. The third common type of autophagy, macroautophagy (henceforth referred to as autophagy), is highly conserved from yeast to mammalian cells [8]. Autophagy allows for cytosolic bulk degradation of long-lived macromolecules and organelles. Morphologically this process was already described in the 1960s but it was not before several decades later when genetic screens in *Saccharomyces cerevisiae* identified multiple genes involved in autophagy and thereby allowed to investigate the molecular mechanisms in further detail [13, 14]. Genetic screens in *S. cerevisiae* have since then led to the identification of numerous autophagy-related (ATG) genes and many homologs have been identified and characterized in higher eukaryotes [15]. In general, autophagy can be divided into three steps: 1) induction/nucleation; 2) expansion; and 3) maturation [16].

The formation of a cytosolic double membrane structure called the phagophore (also called isolation membrane) is an important step of autophagy initiation. It is subject of discussion about the origin of this initial autophagic membrane. Independent experiments identified ER, Golgi, or the outer membrane of mitochondria to contribute to the phagophore double membrane [17, 18]. Cytosolic components are enwrapped during the growth of the phagophore. Closure of the phagophore completes this engulfment and gives rise to a new structure called the autophagosome. These newly formed autophagosomes will further mature and subsequently fuse with lysosomes where the captured cytosolic constituents will be degraded. Autophagy can achieve several purposes; it scavenges the cytosol from macromolecules and organelles but also provides a way to supply the cells with amino acids and if necessary with energy once the recycled amino acids are converted into intermediates of the tricarboxylic acid cycle (TCA) [15, 18-20]. It is therefore of little surprise that the autophagic machinery, which under normal conditions is running on low basal levels, can be set in motion by several intra- and extracellular stress factors, such as starvation, ER-stress, hypoxia and pathogen invasion [15]. Besides non-selective cytosolic bulk-degradation, autophagy is also implicated in selective turnover in yeast, a pathway known as the cytoplasm-to-vacuole targeting (CVT) pathway [21]. In analogy, cargo selective degradation of aggregated proteins (aggrephagy [22]), mitochondria (mitophagy [23]), ribosomes (ribophagy [24]), peroxisomes (pexophagy [25]), endoplasmic reticulum (reticulophagy [26]) and many more have been reported for mammalian systems [27]. The role of selective autophagy in ageing will be further addressed in a separate section of this chapter.

Several protein complexes are involved along the path from initiation to completion of autophagy. Induction of autophagy in *Drosophila* requires the Ser/Thr kinase Atg1 that forms a complex with Atg13. Phosphorylation of Atg13 by Atg1 directs phagophore initiation through a complex containing the class III PI(3)-kinase Vps34, the Ser/Thr kinase Vps15 and Atg6 (Beclin1 in mammals). The activation of this complex leads to localized generation of phosphatidylinositol-3-phosphate (PI3P), a critical step in autophagy. In mammalian systems, this core complex has several known interaction partners, e.g. Atg14, Ambra1, UVRAG, or Rubicon, that are all involved in autophagy. Several of these mammalian genes have orthologues in *Drosophila*, however their involvement in autophagy remains to be shown (reviewed in [28]). UVRAG has recently been found to be important in the regulation of Notch levels in the context of organ rotation during development. This role of UVRAG is coupled to endocytic degradation of Notch and, in this context, not to autophagy [29]. Autophagosome formation requires the ubiquitin-like proteins Atg12 and Atg8 and their respective ubiquitin-like conjugation systems [30]. Atg8 is processed by the cysteine protease Atg4 and covalently linked to phosphatidylethanolamine (PE) through the action of the E1 activating enzyme Atg7, the E2 activating enzyme Atg3 and the E3 like Atg12-Atg5-Atg16 complex, which is found at the phagophore membrane. The E3 like Atg12-Atg5-Atg16 complex itself requires also Atg7 and the E2 activating enzyme Atg10 for its assembly (reviewed in [31]). Once Atg8 is activated and lipid-conjugated it is localized to both sides of the phagophore and Atg4 later only removes the portion residing at the cytosolic side prior to autophagosome-lysosome/endosome fusion. It has also been reported that Atg8 can modulate the size of autophagosomes by influencing membrane curvature. For all these reasons, activation of Atg8/LC3 is widely used to monitor autophagy [15, 32]. The process from autophagy initiation until autolysosome formation is schematically illustrated in figure 1.

It is believed that stepwise fusion of autophagosomes with different endosomal populations account for maturation and culminates in the fusion with lysosomes, the organelle responsible for degradation [33]. Such stepwise fusion is supported by the findings that impairment of ESCRT machinery results in reduced autolysosome formation, measured as decrease in lysotracker staining and accumulation of Atg8 positive punctate respectively [34, 35]. Similar accumulation of autophagosomes can be seen in flies with mutant *Drosophila* deep orange (dor) and dvps16A [36, 37]. Both proteins are known to play important roles in endocytic trafficking.

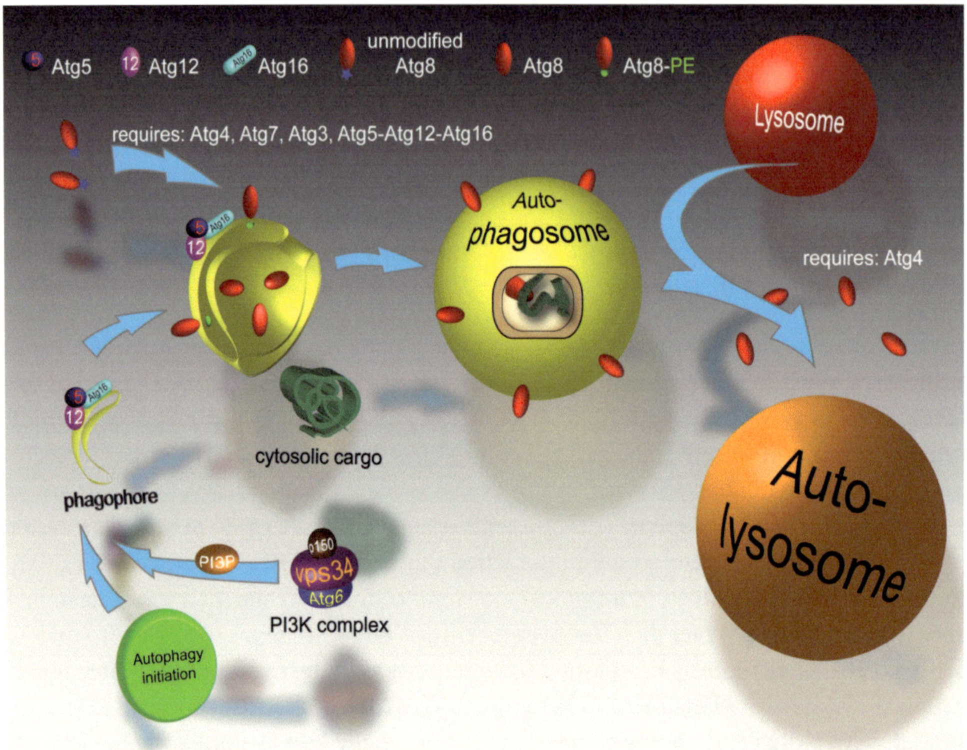

Figure 1. Schematic illustration of macroautophagy. Upon autophagy initiation the PI3K complex generates PI3P, which is then provided at high local concentrations at the initial step of phagophore membrane formation. The ubiquitin like proteins Atg12 and Atg8 with their respective conjugation system are recruited and activated once the phagophore is formed. Membrane expansion leads to phagophore maturation, which is finalized by vesicle closure and thereby autophagosome formation. This vesicle can fuse with different endocytic compartments or directly with lysosomes, forming autolysosomes. There, phagophore-sequestered cytosolic cargo is degraded and macromolecules can be recycled back to the cytosol. For further details see section 2 and references therein.

3. Role of autophagy in development, homeostasis and ageing

In general, the role of autophagy is predominantly described as cytoprotective. Intensive research over the last decade has increased our understanding of multiple cellular events that involve autophagy, e.g. dealing with low nutrient levels, development and morphogenesis, response to oxidative stress, turnover of protein aggregates and damaged organelles, immune response and lately also cell signaling (figure 2). Altogether the picture has emerged that the role and regulation of autophagy is extremely dependent on the cellular context [20, 38, 39].

http://dx.doi.org/10.5772/55396

With this review we therefore want to highlight what is known from research conducted in *Drosophila*, a model organism, which allows for elegant genetic manipulations of the cellular setup in a multi cellular organism.

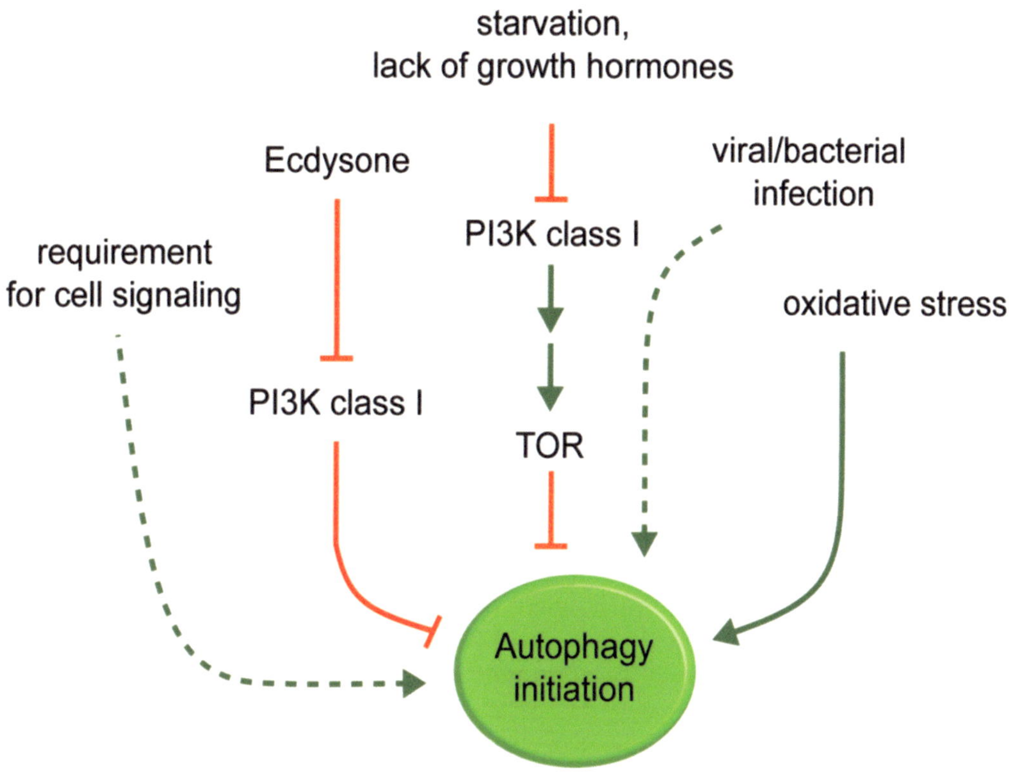

Figure 2. Autophagy can be initiated by multiple ways. Autophagy is involved in a variety of different cellular events (e.g. development, survival under conditions of low nutrient levels, oxidative stress response, immune response, and cell signaling), which requires several ways to initiate the core autophagy machinery (dashed lines: the exact pathway is still uncertain, however autophagy is shown to be upregulated as downstream effect). For further details see section 3 and references therein.

In *Drosophila*, autophagy plays a pivotal role during development and is crucial for a wide range of developmental processes. Cell growth depends on nutrients provided by autophagy as seen in the fat body. On the other hand autophagy has been reported to be necessary for targeted cell death and removal of tissue, e.g. during oogenesis and development of gut salivary glands.

More than ten years ago, both the class I phosphatidylinositol 3-kinase (PI3K) and the serine/threonine kinase Target of Rapamycin (TOR) have been shown to control a signaling network that is important for development (reviewed in [40]). The growth of cells and tissues does require energy and building blocks. Hormones, such as insulin have been identified as important signals in order to meet these requirements by e.g. upregulation of protein synthesis. Already in 2003, Tom Neufeld speculated about the role of catabolic processes, such as autophagy, to be important in development. This idea was supported by previous findings that established a connection between reduced basal autophagic protein turnover and cellular growth as well as that Apg6p, the yeast homologue of the tumor suppressor gene *Beclin 1*, is required for autophagy in yeast (reviewed in [40]). Furthermore, it was already shown that insulin, as well as class I PI3Ks can, besides their effect on protein synthesis, inhibit autophagic protein turnover, providing a plausible molecular link between autophagy and cell growth [41, 42]. Therewith the stage was set for two important findings published in 2004, revealing the regulation of programmed autophagy in the fat body and the importance for functional autophagy in cell growth [43, 44]. The levels of the hormone 20-hydroxyecdysone (ecdysone) rise during the development of *Drosophila*, leading to inactivation of the class I PI3K and subsequent autophagy activation [43]. This initiation of autophagy is necessary in order to supply the developing *Drosophila* larva/pupa with nutrients and to maintain survival and growth. The protective role of autophagy in this context is dominant over its otherwise known role in growth suppression [44, 45]. Noteworthy, dor (Deep orange), the *Drosophila* homolog to Vps18, can influence autophagy in the fat body in two separate ways. Dor is necessary for secretion of ecdysone from the salivary glands, thereby influencing the levels of this hormone. However, dor is also important in the fusion of autophagosomes with lysosomes, thereby directly controlling autophagy [36]. Autophagy in the fat body is dependent on the PI3K Vps34 [34]. Vps34 was initially identified to be involved in vacuolar protein sorting (Vps) in yeast [46]. Flies lacking Vps34 or its regulatory subunit, the protein kinase Vps15 (also referred to as p150), are hampered in their ability to initiate autophagy upon starvation in the fat body and die during development [34, 47]. Interestingly, the absence of Atg7 does not lead to lethality in the developing fly. Atg7 deficient flies have severe defects in autophagy but nevertheless are viable. However, such flies are short lived, show signs of accelerated ageing in the form of ubiquitin-positive aggregates in degenerating neurons and have very low resistance to nutrient deprivation and oxidative stress. This underscores the necessity of functional autophagy for cellular homeostasis and stress survival in the adult fly [48].

A very different aspect of autophagy during development has been revealed in the context of programmed cell death. Autophagy is upregulated during the reorganisation of the salivary gland and gut [49, 50]. Inhibition of autophagy in salivary glands by activating the class I PI3K pathway reduces salivary gland cell degradation. In contrast, induction of autophagy in salivary gland cells results in premature cell death and it was shown that this cell death is dependent on both caspases and autophagy [49]. Similar events can be seen in the midgut. Even though caspases are highly expressed, the canonical apoptotis pathway is not required for midgut removal. Inhibition of autophagy on the other hand, impairs midgut degradation and simultaneously decreases caspase activity [50]. Additional ways how cell death and autophagy are connected are pointed out by the findings that autophagy can selectively degrade survival factors and thereby initiate cell death. During late oogenesis, autophagy is necessary to degrade the apoptosis inhibitor dBruce in nurse cells. Nurse cells lack the, under normal conditions typical, fragmentation of DNA and caspase-3 activity in the absence of autophagy [51]. A similar principle for cell death control is suggested by the finding that the valosin-containing protein (vcp), a ubiquitin-selective AAA chaperone, is required for degradation of the apoptosis inhibitor DIAP1 during regulated degeneration of dendrites of class IV dendritic arborisation neurons [52]. It was already shown before that vcp is necessary for autophagy [53]. Altogether, this implies a role for autophagy in activating apoptosis by selective degradation of apoptosis inhibitors. It will be interesting to see if such a mechanism is limited to the programmed reorganization events during development or if this is a strategy employed even in other cellular contexts. If this is a general mechanism to initiate cell death, autophagic degradation of apoptosis inhibitors might become an interesting strategy for developing drugs aimed for cancer treatment.

The role of autophagy in *Drosophila* is not limited to development but instead autophagy is also important for various aspects during lifetime of eclosed flies. Any organism needs to be able to cope with oxidative stress, which itself is tightly linked to ageing [5]. In *Drosophila*, Jun N-terminal kinase (JNK) can protect the gut from oxidative toxicity due to feeding on paraquat, a well-established oxidative stress inducer. In addition, genetic upregulation of the JNK pathway extends lifespan of flies in a Foxo dependent manner [54, 55]. This cell protective effect of JNK is mediated by the transcriptional activation of autophagy. JNK cannot protect flies from oxidative toxicity when Atg1 or Atg6 activity is reduced [56].

A different putative way for autophagy to protect cells from oxidative stress is given by its involvement in the selective degradation of damaged mitochondria, termed mitophagy. So far, mitophagy has not been directly shown to occur in *Drosophila*, nevertheless several findings indicate that mitophagy also happens in flies. Studies in *Drosophila* have suggested that the E3 ubiquitin ligase Parkin normally facilitates mitochondrial fission and/or inhibits fusion [57]. In addition the PTEN-induced putative kinase protein 1 (PINK1) has been shown to genetically interact with Parkin in flies, and results from experiments in *Drosophila* S2 cells revealed that PINK1 is required for the recruitment of Parkin to damaged mitochondria leading to their degradation [58]. Interestingly, the finding that the level of the protrusion factor mitofusin (mfn) increased in the absence of PINK1 or Parkin, suggests that mfn might be ubiquitinated by Parkin, which can serve as putative label and targeting signal for degradation of damaged mitochondria [58]. In yeast and mammals it has been shown that ubiquitination of mitochondrial proteins by Parkin results in autophagic degradation of mitochondria (reviewed in [23]). This role of Parkin in the removal of damaged mitochondria might also explain the muscle degeneration, mitochondrial pathology and reduced lifespan in *parkin* mutant flies [59]. Lately, it has been reported that mitochondrial protein misfolding in *Drosophila* leads to degradation of mitochondria and that accumulation of an unfolded protein in the mitochondria phenocopies flies with mutations in PINK1 and Parkin. The requirement of Ref(2)P (refractory to Sigma P, the *Drosophila* homolog of p62) for this mitochondrial turnover resembles mitophagy as described in mammalian systems [60]. However, it remains to be proven that the turnover of damaged mitochondria in flies really is conducted by autophagy, hence that mitophagy also occurs in *Drosophila*.

Without doubt autophagy is crucial for cellular homeostasis and it is therefore of no surprise that autophagy is also induced upon viral or bacterial infections as both lead to changes in the intracellular environment. Flies with impaired autophagy are hampered in their immune defence. Even though this role of autophagy is much more studied in mammalian system, there are 4 different reports that highlight an involvement of autophagy in the *Drosophila* immune response. When autophagy was impaired by the expression of RNAi against Atg5, Atg7, or Atg12, *Drosophila* displays a decreased resistance to injected *Escherichia coli*, which manifests in higher titers of *E. coli* and reduced survival rates. Interestingly, knockdown of any of these three Atg genes did not shorten lifespan of uninjected flies [61]. The latter finding is not in line with findings from Atg7 deficient flies, which show a significant shortening of life span [48]. Even though the conditional knockdown of Atg7 did lead to a decrease in lysotracker staining, a sign for reduced autophagy, it cannot be excluded that some remaining Atg7 activity is enough in order to allow for basal autophagy and thereby not altering lifespan. It can be expected that such basal autophagy is more severely affected in flies completely missing the gene for Atg7. On the other hand there is also the possibility that $Atg7^{-/-}$ flies already accumulate cell damage during development that might allow them to hatch normally but still will give them a severe survival disadvantage right from the start.

Autophagy does not only protect against bacterial but also against viral infection as shown in the case of the mammalian viral pathogen vesicular stomatitis virus (VSV) [62]. Autophagy protects flies against VSV by decreasing viral replication. Repression of autophagy has the contrary effect, increased viral replication and pathogenesis. The authors of this study were able to pinpoint the PI3K/Akt pathway to be responsible for autophagy regulation upon VSV infection [62]. Flies infected with *Mycobacterium marinum* are dependent on autophagy in order for mycobacteria drug treatment to be successful. *Drosophila* lacking the gene for Atg7 had a reduced survival rate upon *Mycobacterium marinum* infection and this phenotype could not be rescued with the help of antimycobacterial treatment [63].

An additional involvement of autophagy in immunity was found in the cortical remodelling of hemocytes (*Drosophila* blood cells). Integrin-mediated hemocyte spreading and Rho1-induced cell protrusions require continuous autophagy. As a consequence, flies with impaired autophagy in their hemocytes show severe defects in recruiting hemocytes to epidermal wounds. Furthermore, this study identified Ref(2)P to be crucial for functional autophagy, which suggests selective autophagy (see below) to be involved in this process [64]. The requirement for selective autophagic turnover of single proteins to maintain cellular homeostasis has been implicated in several different cellular contexts. E.g. activated rhodopsin is degraded via the endosomal pathway and mutations in rhodopsin leading to hampered endocytic turnover results in retinal degeneration [65]. Autophagy has also been connected to the turnover of activated rhodopsin and mutations in Atg7 or Atg8, or genes necessary for proper autophagosome formation, result in light-dependent retinal degeneration [66].

Another example for the necessity of functional selective autophagic degradation of proteins for proper homeostasis is given in muscle tissue maintenance. There, chaperone-assisted selective autophagy is necessary to remove contraction-induced damaged filamin from Z-discs in order to prevent Z disk disintegration and progressive muscle weakness in flies [67].

Autophagy also serves several functions in neuron plasticity and homeostasis. An interesting finding was that synapse development is controlled by autophagy via the E3 ubiquitin ligase highwire. Highwire inhibits neuromuscular junction growth and is itself a substrate for selective autophagic turnover, indicating that autophagy activity might lead to synaptic overgrowth [68]. Tian et al. identified Rae1 to bind to highwire and thereby protecting highwire from autophagic degradation [69]. The link between autophagy and synaptic growth at the neuromuscular junction is further strengthened by the observation that ROS can act as signaling molecules and mediate synaptic growth. At the same time, high ROS levels activate the JNK pathway, a previously reported activator of autophagy. As impairment of autophagy results in decreased synaptic size in a *Drosophila* model, whereas activation of autophagy has the opposite effect, one can speculate that ROS mediated synaptic growth is mediated by activation of JNK and subsequent autophagy (reviewed in [70]).

Off course, autophagy also plays a major role in simply keeping the cells "clean" by enabling the cells to turn over the cytosol. This recycling effect is especially pronounced in post-mitotic cells. Flies that are mutant for Atg8a are severely hampered in their efficiency to eliminate cellular material, which can be observed as an increase in ubiquitinated proteins and the increased presence of electron dense protein aggregates in young fly brains when investigated with transmission electron microscopy. Moreover, such Atg8a mutant flies display a drastic decrease of lifespan [71]. In addition, a very recent report by Fouillet et al., reveals an autophagy-mediated decrease of apoptosis in neurons upon mild ER-stress and further underscore a cytoprotective role of autophagy, that potentially can prolong survival of the whole organism [72]. Taken together, these data outline the versatile role of autophagy in homeostasis and normal survival of flies.

4. Autophagy and neurodegeneration

Ageing is a major risk factor for the development of neurodegenerative diseases and over the last decade autophagy has been implicated in many neurodegenerative diseases, such as Huntington's disease (HD), Parkinson's disease (PD), amyotrophic lateral sclerosis (ALS), or Alzheimer's disease (AD). Many neurodegenerative diseases share the common phenotype of accumulations of protein aggregates [73]. Before reviewing the role of autophagy in *Drosophila* models for neurodegeneration we first want to give a short overview of key findings on the link between autophagy and neurodegeneration as known from mammalian systems and patient data.

Both, HD and PD are connected to elevated autophagy. In case of HD, autophagy can only be triggered by a mutant form of huntingtin that is prone to aggregate but not by wildtype huntingtin. Cytosolic aggregates of α-synuclein, the protein involved in PD, can be degraded by macroautophagy and CMA [74-77]. In ALS loss of motor neurons deprives patients of voluntary controlled muscle movements. The disease is associated with ubiquitinated, p62 positive protein inclusions of TDP-43 (TAR DNA binding protein 43) or SOD1 (superoxide dismutase 1) or rare mutations in a subunit of the ESCRT complex [78, 79]. A defective ESCRT complex in its turn has been shown to result in autophagosome accumulation [80], but also point mutations of the p150 subunit of dynactin resulting in defects in the transport machinery along microtubules have been implicated in ALS. Transport along microtubules is necessary for autophagosome-lysosome fusion and therefore crucial for functional autophagy [81, 82]. Extensive alterations in macroautophagy can also be found in patients with AD. An immuno-electron microscopy study on neocortical biopsies from AD patients identified autophagosomes, multivesicular bodies, multilamellar bodies, and cathepsin-containing autophagolysosomes as the predominant organelles that occupied most of the cytosol of dystrophic neurites. Autophagy was detected in cell bodies with neurofibrillary pathology and associated with a relative depletion of mitochondria and other organelles. The authors of this study speculated that the accumulation of immature autophagic vacuoles results from impaired transport to and fusion with lysosomes thereby hampering the protective effects of autophagy [83]. Disruption of lysosomal proteolysis in primary mouse cortical neurons by inhibiting cathepsins, or by supressing lysosomal acidification, impairs transport of autolysosomes, endosomes and lysosomes, and leads to accumulation of these structures within dystrophic axonal swellings. Such a phenotype can also be seen in numerous mouse models of AD. The phenotype is not caused by general disruption of the axonal transport machinery, as mitochondria and cathepsin-lacking organelles were not influenced in their movements. Axonal dystrophy is reversed once lysosomal function is restored [84].

In the past, several independent groups have established *Drosophila* models for neurodegenerative diseases and/or investigated the role of aggregating proteins implied in neurodegenerative diseases in flies. Remarkably, already in 1982 Stark and Carlson characterized the degenerative phenotypes evoked by a mutant form of the rdgB (retinal-degeneration-B) protein in the fly compound eye and found amongst others lysosome-like bodies and vacuoles suggesting involvement of autophagy [85]. The compound eye of flies displays a highly structured order and degenerative properties of protein aggregates can easily be monitored as impairments of this structure. Expression of mutant huntingtin containing a polyQ-expansion of 120 glutamine leads to degeneration of the eye. However, treatment with rapamycin, an activator of autophagy, reduces this phenotype [86]. Treatment with new small-molecule enhancers (SMER) of the cytosolic effects of rapamycin, which were shown to induce autophagy in mammalian cells, also protected flies from polyQ huntingtin induced neurodegeneration [87]. Instead of treating flies with rapamycin in order to inhibit TOR by pharmacological means, Wang et al. highlighted the importance of TOR in neurodegeneration by genetical manipulations. Hyperactivation of TOR, achieved by expression of the TOR kinase activator Ras homologue enriched in brain protein (Rheb) or introduction of mutations in the TOR inhibitor dTsc1 increased age- and light-dependent photoreceptor loss [88]. The authors of this study were able to exclude TORs effects on growth to be responsible for this photoreceptor degeneration but instead pointed out autophagy as the downstream signaling of TOR mediating photoreceptor cell death. Activation of autophagy by overexpressing Atg1 protected not only cells from age- and light dependent photoreceptor degeneration, but also photoreceptor cells which either produced 120 polyQ-huntingtin or lacked a functional *Drosophila phospholipase C* gene *norpA* respectively. Both latter manipulations are commonly used to model neurodegeneration [88]

Macroautophagy in flies can also be upregulated by Rab5 over-expression and this approach also mitigates polyQ-huntingtin mediated degeneration in the eye [89]. However, there is a fine line between beneficial and detrimental consequences of autophagy activation in the context of neurodegeneration as shown in a dentatorubralpallidoluysian atrophy (DRPLA) fly model [90]. This model is built upon the expression of atrophin with a polyQ expansion and is characterized by lysosomal dysfunction and blocked autophagosome-lysosome fusion, hence reduced autophagic flux [90]. Even though introduction of a mutant form of Atg1 intensified the neurodegenerative phenotype, upregulation of autophagy in this system had no rescuing effect but, in some case, even had the opposite outcome and increased neurodegeneration [90]. In other words, autophagy plays an important role in scavenging polyQ atrophin from the cytosol, but is only of beneficial nature as long as autophagy can proceed all the way to lysosomal degradation. Reaching a rate-limiting step in the autophagy cycle can have negative effects on the outcome of autophagy initiation. Work from the same lab also identified a mechanism how polyQ atrophin itself impairs autophagic flux. PolyQ-atrophin inhibits the tumor suppressor fat, which under normal conditions protects from neurodegeneration through the Hippo kinase cascade and subsequent increases autophagy [91]. How Hippo exactly activates autophagy is not completely understood yet [92]. Data obtained from studies in the salivary gland suggests that the phosphorylation of Warts (wts), a substrate of the Hippo kinase, acts upstream of TOR and thereby regulates autophagy [93]. It also has been reported that the Hippo pathway can directly interact with LC3 (the mammalian homolog of Atg8) and thereby initiate autophagy [92].

An additional, interesting link between polyQ sequence derived neurodegeneration and macroautophagy is given by puromycin-sensitive aminopeptidase (PSA). PSA is the only cytosolic enzyme capable of digesting polyQ sequences and it is therefore not surprising that there is inverse correlation between PSA expression and severity of neurodegeneration, e.g. over-expression of PSA has protective effects in cells expressing polyQ expanded ataxin-3, mutant α-synuclein and mutant superoxide dismutase (SOD) [94]. It comes as a surprise though that this beneficial role of PSA is mediated by its activation of macroautophagy rather than its role in degrading polyQ aggregates and thereby making them available for proteasomal degradation, although the putative involvement of the proteasome in cell protection in this process remains to be further understood [94].

A different way to induce neurodegeneration is to inhibit proteasomal function. Interestingly, proteasome impairment can be compensated for by autophagy, a rescue that depends on the histone deacetylase 6 (HDAC6) [95]. A protective role of autophagy in context of neurodegeneration was also demonstrated in a genetic screen conducted in *Drosophila* with pathogenic Ataxin-3-induced neurodegeneration. Knockdown of Atg5 in these flies reverts the polyQ containing Ataxin-3 mediated toxicity. Testing the effects of identified neurodegeneration-suppressors on autophagy revealed that these factors had different impact on autophagy. The authors of this study proposed a model in which some neurodegeneration-suppressors induce autophagy, thereby contributing to protein clearance whereas others mitigate autophagy in order to counteract autophagic cell death [96]. The role of autophagy in removal of protein aggregates in neurodegenerative diseases was further confirmed by the finding that depletion of subunits of the ESCRT complex in flies intensifies the toxic effects exerted by polyQ-expanded huntingtin [97]. Depletion of ESCRT subunits has autophagy inhibition as consequence, which manifests in accumulation of protein aggregates containing ubiquitinated proteins, p62 and Alfy [98].

The Alzheimer's disease related peptide $A\beta_{1-42}$ also induces neurodegeneration, mediated by age-dependent autophagy-lysosomal injury in a *Drosophila* model of AD [99]. The age dependence was shown to be of high importance as brain ageing is accompanied by an increasingly defective autophagy-lysosomal system and accumulation of dysfunctional autophagosomes and autolysosomes. As a consequence intracellular membranes and organelles are damaged. The expression of $A\beta_{1-42}$ resulted in similar changes already in young *Drosophila* and this raised the question if chronic deterioration of the autophagy-lysosomal system by $A\beta_{1-42}$ simply accelerates brain ageing [100]. This concept is supported the finding that expression of autophagy genes decreases with age, and disruption of the autophagy pathway reduces lifespan of flies [71].

5. Autophagy and its role in lifetime extension

The rate of ageing is reciprocally linked to lifespan and therefore are interventions that extend longevity of an organism the most direct indication that ageing is slowed down [101]. One well established, and long known intervention that extends lifespan is dietary restriction (DR), the limitation of food intake below the *ad libidum* level without malnutrition. DR has successfully been proven to extend lifespan in every organism tested, including yeast, worms, flies and rodents. In addition, DR not only extends lifespan, even the occurrence of age-associated pathologies, e.g. cardiovascular disease, multiple kinds of cancer, neurodegeneration, are drastically reduced or at least postponed in animal models [102]. The possibility to perform forward genetics in different model organisms has boosted the general understanding of underlying molecular mechanisms how DR, and other life extending interventions, can execute their effects. Studies in *Caenorhabditis elegans* by Cynthia Kenyon and co-workers have already almost two decades ago showed how mutations in the single gene daf-2 (the insulin receptor homologue in *C. elegans*) can increase survival by more than two-fold and that such extended survival is dependent on a second gene, namely daf-16 (a forkhead transcription factor) [103, 104]. Since then the role of nutrient-sensing pathways in ageing has been addressed by many independent groups, which has helped to identify numerous proteins that are crucial in lifespan determination. Amongst other pathways, both the insulin/insulin-like growth factor (IGF) and the Target of Rapamycin (TOR) network have been shown to be important modulators of longevity (reviewed in [101, 105, 106]). The fact that both these networks also are involved in the regulation of autophagy emphasizes a putative role of autophagy in lifetime extension and has been addressed in *Drosophila* by several groups.

Simonsen and co-workers showed that downregulation of autophagy genes in *Drosophila* neural tissue is part of the normal ageing process. This is accompanied by accumulation of insoluble ubiquitinated proteins (IUPs). Impairment of autophagy due to mutations in Atg8a aggravates the occurrence of IUPs at earlier time points and lowers survival rates [71]. As lipid-conjugation of Atg8 is essential for nucleation and phagophore elongation it can be speculated that Atg8 is a limiting factor in autophagic turnover. The over-expression of Atg8 in the central nervous system of *Drosophila* indeed extends average and maximum life span by approx. 50% [71]. Flies not only live longer upon Atg8a over-expression, but also showed a higher tolerance to oxidative stress and lower occurrence of IUPs [71]. Interestingly, the longevity promoting effect of Atg8a over-expression cannot be seen when over-expression is initiated during development but decreases over time as seen in flies where Atg8a expression was driven by the early pan-neural driver line *Elav-Gal4* [71]

The question if IUPs are cause or a consequence of the ageing process remains to be answered though. Albeit, the age-dependent accumulation of ubiquitinated proteins that are positive for Ref(2)P, a protein necessary for cargo recognition in selective autophagy, can be employed as conserved marker of neuronal ageing and progressive autophagic defects [107].

Also Atg7 was recently reported to extend life span when over-expressed in neuronal tissues of flies [108]. The life-extending effect of Atg7 is not as pronounced when compared to Atg8. This might be due to different capabilities in inducing autophagic turnover, or non-autophagy related side effects of either Atg7 or Atg8a.

Proteostasis is not only important in neuronal tissues but also in muscles of flies. With increasing age polyubiquitinated proteins accumulate that co-localise with Ref(2)P in muscles and the cumulative appearance of such aggregates has been demonstrated to impair muscle fitness [109]. The build-up of such aggregates can be reverted in muscles by the constitutive activation of the transcription factor FOXO and its target 4E-BP (eukaryotic translation initiation factor 4E binding protein). Interestingly, the activation of FOXO/4E-BP signaling in muscles is sufficient to extend lifespan of the whole organism [109]. Furthermore it has been shown that the FOXO/4E-BP dependent delay in protein aggregate accumulation in muscles depends on functional autophagy, suggesting promotion of basal autophagy upon FOXO/4E-BP signaling [109]. The autophagy dependent beneficial effect of FOXO is well in line with earlier findings that revealed FOXO to be capable to upregulate autophagy [110]. In addition, the translational repressor 4E-BP is known to be upregulated upon DR and to mediate enhanced mitochondrial function and life span extension in *Drosophila* [111]. As already mentioned earlier, autophagy has a known role in the selective turnover of damaged mitochondria in yeast and mammals, and it is therefore tempting to speculate that autophagy can promote longevity by improving mitochondrial function in a FOXO/4E-BP dependent manner, however this remains to be proven.

Ageing in *Drosophila* can also be manipulated by pharmacological means. Feeding the TOR inhibitory drug rapamycin, a well-described drug for human use, significantly increases lifespan and resistance to starvation as well as the oxidative stress inducer paraquat [112]. Rapamycin fails to extend the lifespan of flies with downregulated Atg5 suggesting that autophagy has to be active in order for rapamycin to slow down ageing [112]. The finding that inhibition of TOR increases lifespan in *Drosophila* is well in line with earlier studies demonstrating that mutant, inactive TOR or over-expression of the TOR inhibitors dTsc1 or dTsc2 extend longevity [113, 114]. However, the specific role of autophagy was not addressed in those two studies.

Keeping *Drosophila* on food supplemented with the polyamine spermidine promotes increased longevity and this effect has been shown to be autophagy dependent, since depletion of Atg7 abrogates this anti-ageing effect [115].

Taken together, all these data indicate an anti-ageing effect of autophagy, however caution is advised in trying to merely upregulate autophagy pharmacologically in order to counter-act ageing. Autophagy is essential for the recycling of cellular content, which can serve two general purposes: autophagy can unburden cells from hazards by removal of those and autophagy can provide cells with new building blocks for cellular survival. During the lifetime of an organism, autophagy will most certainly switch forth and back between those roles. In order to completely understand the complex role of autophagy in ageing it is therefore important to understand the regulation and cellular outcome of autophagy in a tissue and time dependent manner.

6. Selective autophagy and ageing

In the following section we want to shed some light on the current knowledge about the selective removal of cellular contents by autophagy in *Drosophila melanogaster*. Above, we have already discussed some examples of selective autophagy in normal ageing and homeostasis. We therefore will focus more on the mechanistic insights of selective autophagy and what is known so far about the role of selective autophagy explicitly in ageing of *Drosophila*.

Selective autophagy in the form of CVT has been known in yeast for a long time and has gained major attention in mammalian systems over the last years. Selectivity requires crucial, additional steps to the above described autophagy process: cargo has to be recognized by specific receptors and must be delivered to the autophagic machinery.

Ubiquitin has emerged as a molecule to tag proteins that are determined for degradation [116]. Conjugation of ubiquitin depends on a complex reaction cascade that requires activation of ubiquitin (by E1 enzymes), conjugation (E2 ubiquitin conjugating enzyme), and ligation of ubiquitin with a target substrate (E3 ubiquitin ligase). As a result, ubiquitin is covalently bound via an isopeptide bond between the C-terminal glycine of ubiquitin and the ε-amino group of a lysine residue on the substrate protein. Substrate specificity is given by the E3 ubiquitin ligase that specifically recognizes a protein substrate and brings it to the E2 ubiquitin conjugating enzyme. A wide spectrum of E1, E2, and E3 enzymes provide cells with selectivity for this signaling machinery [117]. Ubiquitin itself contains seven lysine residues enabling ubiquitin to self-attach, thereby forming a polyubiquitin tag. The best-characterised linkages occur via K48, targeting the substrate for proteasomal degradation, and via K63, which is preferred by ubiquitin-binding autophagy receptors. Furthermore, K63 ubiquitination has been reported to be a potent enhancer of inclusion formation and leads to substrate degradation via the autophagy/lysosome degradation pathway [116, 118-120]. Also more atypical sites for polyubiquitination, such as K6 or K29, have been reported but the exact role of these ubiquitin chains is still poorly understood [121].

Taken together, ubiquitin conjugation offers several possibilities to flag proteins and organelles in different ways by variation of chain length and various sites for ubiquitin self-attachment and thereby act as a signal for distinct subsequent cellular processing. Molecular links between ubiquitinated proteins and autophagy were identified in form of the cargo receptors sequestosome marker SQSTM1/p62 and NBR1 (neighbour of BRCA1 gene) [122]. The conserved functional homologue for p62/NBR1 in *Drosophila* is Ref(2)P. Ref(2)P is a 599 amino acid long protein with an N-terminal Phox Bem1p (PB1) domain, followed by a ZZ-type Zinc finger domain and a C-terminal UBA (ubiquitin-associated) domain [123]. The PB1 domain allows for self- and hetero-oligomerisation, while the UBA domain enables Ref(2)P to recognize and directly interact with ubiquitin. Both domains are necessary for formation of protein aggregates normally found in brains of adult *Drosophila* [124]. Flies mutant for Atg8 display an increased amount of deposited protein aggregates in the brain, however such aggregates are absent in double-mutant Atg8/Ref(2)P flies [124]. This suggests that Ref(2)P is a selective cargo receptor for selective autophagy in *Drosophila*, similar as its homologue p62 in mammals. This is supported by the presence of a putative LIR (LC3 interacting) domain in Ref(2)P as identified by bioinformatics analysis [122]. The LIR domain is known to be essential for p62 to interact with LC3, but it remains to be elucidated if Ref(2)P really interacts with Atg8 via its putative LIR domain. Independent of the absence of final proof of direct interaction between Atg8 and Ref(2)P, protein aggregations containing Ref(2)P serve as excellent markers for neuronal ageing and autophagic defects in *Drosophila* [107].

Filimonenko et al. were able to identify the mammalian phosphatidylinositol-3-phosphate (PI3P) binding protein Alfy (PI3P-binding Autophagy-linked FYVE domain protein) to be actively involved in autophagic degradation of polyglutamine (polyQ) expanded, aggregated proteins [125]. Albeit harbouring a FYVE domain Alfy is usually not found on endosomes but instead resides in the nucleus decorating the nuclear membrane. The presence of ubiquitinated, aggregated proteins in the cytosol leads to relocalization of Alfy to these aggregates [126]. Alfy can directly interact with p62 and Atg5 [125, 127]. *In vitro*, Alfy is necessary to recruit Atg5 to polyQ protein aggregates. In addition, Alfy scaffolds the Atg5-Atg12-Atg16L complex to p62- and ubiquitin-positive polyQ inclusions [125]. The Atg5-Atg12-Atg16L complex on the other hand is important for LC3 lipidation [128]. Taken together, all these interactions allow for LC3 lipidation in close spatial proximity to ubiquitinated, aggregated proteins and explain the absence of other cytosolic components in aggregate filled autophagosomes [125]. Primary neurons expressing polyQ Htt (Huntingtin) have fewer polyQ inclusions upon ectopic Alfy expression. These results were confirmed *in vivo* with a *Drosophila* model where polyQ production provokes a phenotype that is due to toxicity. The outcome of polyQ-mediated toxicity was much milder once bchs (blue cheese, the *Drosophila* homologue of Alfy) was co-expressed [125]. Reduced levels of bchs in mutant flies had opposite effects and led to shortened live span and extensive neurodegeneration [129]. It remains to be elucidated if Alfy/bchs directly recognizes ubiquitinated aggregates or if this interaction is mediated by p62/Ref(2)P [22].

Accumulation of damaged mitochondria and increased production of ROS are generally believed to account for age associated pathologies [5]. The efficiency of selective removal of damaged mitochondria, mitophagy, might therefore play a major role in the outcome of the ageing process. Although several lines of evidence suggest the existence of mitophagy in *Drosophila* (see section 3) the molecular details in flies still have to be further unravelled.

7. Summary and outlook

The cytoprotective role of autophagy has been shown in many different cellular contexts and induction of autophagy by either pharmacological or genetical means has life extending effects. However, research conducted in *Drosophila* has also identified situations during development when autophagy is necessary for controlled tissue removal and cell death initiation. These two rather contrary roles, cytoprotection versus cell death initiation, highlight the complexity of the autophagy pathway and also underscore the importance to understand the molecular mechanisms by which autophagy exerts its role. As autophagy most likely is regulated in a tissue and time dependent manner it is of great interest to pinpoint those time points and tissues in which autophagy has the biggest impact on the general ageing processes.

Ageing is not only influenced by one single pathway but in contrary is a multifaceted process. Age is a major risk factor for a variety of diseases, e.g. neurodegenerative diseases, metabolic syndrome, cancer and more. In the past, extensive research has been undertaken to model neurodegenerative diseases in the fruitfly and has helped to push our understanding, not the least concerning the involvement of autophagy, to new levels. Today, *Drosophila* is getting growing attention as cancer model and it will be exciting to follow future research in order to get new insights from *Drosophila melanogaster* about the complex role of autophagy in cancer. By putting several different pieces of puzzle together, *Drosophila* already has helped us to get a clearer picture about the role of autophagy in various aspects of ageing and for sure the fruitfly will continue to help the research community to reveal more of this complex picture in the future.

Author details

Sebastian Wolfgang Schultz[1,2], Andreas Brech[1,2] and Ioannis P. Nezis[3]

*Address all correspondence to: I.Nezis@warwick.ac.uk

1 Department of Biochemistry, Institute for Cancer Research, Oslo University Hospital, The Norwegian Radium Hospital, Oslo, Norway

2 Centre for Cancer Biomedicine, Faculty of Medicine, University of Oslo, Oslo, Norway

3 School of Life Sciences, University of Warwick, Coventry, United Kingdom

References

[1] Kirkwood, T. B, & Austad, S. N. Why do we age? Nature, (2000). , 233-238.

[2] Vellai, T. Autophagy genes and ageing. Cell death and differentiation, (2009). , 94-102.

[3] Lionaki, E, Markaki, M, & Tavernarakis, N. Autophagy and ageing: Insights from invertebrate model organisms. Ageing research reviews, (2012).

[4] Kirkwood, T. B. A systematic look at an old problem. Nature, (2008). , 644-647.

[5] Kirkwood, T. B. Understanding the odd science of aging. Cell, (2005). , 437-447.

[6] Guarente, L, & Kenyon, C. Genetic pathways that regulate ageing in model organisms. Nature, (2000). , 255-262.

[7] Hekimi, S, & Guarente, L. Genetics and the specificity of the aging process. Science, (2003). , 1351-1354.

[8] Reggiori, F, & Klionsky, D. J. Autophagy in the eukaryotic cell. Eukaryotic cell, (2002). , 11-21.

[9] De Duve, C. The lysosome. Scientific American, (1963). , 64-72.

[10] Uttenweiler, A, et al. The vacuolar transporter chaperone (VTC) complex is required for microautophagy. Molecular biology of the cell, (2007). , 166-175.

[11] Sahu, R, et al. Microautophagy of cytosolic proteins by late endosomes. Developmental cell, (2011). , 131-139.

[12] Cuervo, A. M. Chaperone-mediated autophagy: selectivity pays off. Trends in endocrinology and metabolism: TEM, (2010). , 142-150.

[13] Tsukada, M, & Ohsumi, Y. Isolation and characterization of autophagy-defective mutants of Saccharomyces cerevisiae. FEBS letters, (1993). , 169-174.

[14] Thumm, M, et al. Isolation of autophagocytosis mutants of Saccharomyces cerevisiae. FEBS letters, (1994). , 275-280.

[15] He, C, & Klionsky, D. J. Regulation mechanisms and signaling pathways of autophagy. Annual review of genetics, (2009). , 67-93.

[16] Yamamoto, A, & Simonsen, A. The elimination of accumulated and aggregated proteins: A role for aggrephagy in neurodegeneration. Neurobiol Dis, (2010).

[17] Mcewan, D. G, & Dikic, I. Not all autophagy membranes are created equal. Cell, (2010). , 564-566.

[18] Tooze, S. A, & Yoshimori, T. The origin of the autophagosomal membrane. Nature cell biology, (2010). , 831-835.

[19] Seglen, P. O, Gordon, P. B, & Holen, I. Non-selective autophagy. Seminars in cell biology, (1990). , 441-448.

[20] Mizushima, N, & Komatsu, M. Autophagy: renovation of cells and tissues. Cell, (2011). , 728-741.

[21] Yorimitsu, T, & Klionsky, D. J. Autophagy: molecular machinery for self-eating. Cell death and differentiation, (2005). Suppl 2: , 1542-1552.

[22] Knaevelsrud, H, & Simonsen, A. Fighting disease by selective autophagy of aggregate-prone proteins. FEBS letters, (2010). , 2635-2645.

[23] Youle, R. J, & Narendra, D. P. Mechanisms of mitophagy. Nature reviews. Molecular cell biology, (2011). , 9-14.

[24] Kraft, C, et al. Mature ribosomes are selectively degraded upon starvation by an autophagy pathway requiring the Ubp3ubiquitin protease. Nature cell biology, (2008). p. 602-10., Bre5p.

[25] Dunn, W. A, et al. Pexophagy: the selective autophagy of peroxisomes. Autophagy, (2005). , 75-83.

[26] Bernales, S, Mcdonald, K. L, & Walter, P. Autophagy counterbalances endoplasmic reticulum expansion during the unfolded protein response. PLoS biology, (2006). , e423.

[27] Klionsky, D. J, et al. How shall I eat thee? Autophagy, (2007). , 413-416.

[28] Chang, Y. Y, & Neufeld, T. P. Autophagy takes flight in Drosophila. FEBS letters, (2010). , 1342-1349.

[29] Lee, G, et al. UVRAG is required for organ rotation by regulating Notch endocytosis in Drosophila. Developmental biology, (2011). , 588-597.

[30] Ohsumi, Y. Molecular dissection of autophagy: two ubiquitin-like systems. Nature reviews. Molecular cell biology, (2001). , 211-216.

[31] Mcphee, C. K, & Baehrecke, E. H. Autophagy in Drosophila melanogaster. Biochimica et biophysica acta, (2009). , 1452-1460.

[32] Simonsen, A, & Tooze, S. A. Coordination of membrane events during autophagy by multiple class III PI3-kinase complexes. The Journal of cell biology, (2009). , 773-782.

[33] Razi, M, Chan, E. Y, & Tooze, S. A. Early endosomes and endosomal coatomer are required for autophagy. The Journal of cell biology, (2009). , 305-321.

[34] Juhasz, G, et al. The class III PI(3)K Vps34 promotes autophagy and endocytosis but not TOR signaling in Drosophila. The Journal of cell biology, (2008). , 655-666.

[35] Rusten, T. E, et al. ESCRTs and Fab1 regulate distinct steps of autophagy. Current biology : CB, (2007). , 1817-1825.

[36] Lindmo, K, et al. A dual function for Deep orange in programmed autophagy in the Drosophila melanogaster fat body. Experimental cell research, (2006). , 2018-2027.

[37] Pulipparacharuvil, S, et al. Drosophila Vps16A is required for trafficking to lysosomes and biogenesis of pigment granules. Journal of cell science, (2005). Pt 16): , 3663-3673.

[38] Backues, S. K, & Klionsky, D. J. Autophagy gets in on the regulatory act. Journal of molecular cell biology, (2011). , 76-77.

[39] Rubinsztein, D. C, Marino, G, & Kroemer, G. Autophagy and aging. Cell, (2011). , 682-695.

[40] Neufeld, T. P. Body building: regulation of shape and size by PI3K/TOR signaling during development. Mechanisms of development, (2003). , 1283-1296.

[41] Petiot, A, et al. Distinct classes of phosphatidylinositol 3'-kinases are involved in signaling pathways that control macroautophagy in HT-29 cells. The Journal of biological chemistry, (2000). , 992-998.

[42] Pfeifer, U. Inhibition by insulin of the formation of autophagic vacuoles in rat liver. A morphometric approach to the kinetics of intracellular degradation by autophagy. The Journal of cell biology, (1978). , 152-167.

[43] Rusten, T. E, et al. Programmed autophagy in the Drosophila fat body is induced by ecdysone through regulation of the PI3K pathway. Developmental cell, (2004). , 179-192.

[44] Scott, R. C, Schuldiner, O, & Neufeld, T. P. Role and regulation of starvation-induced autophagy in the Drosophila fat body. Developmental cell, (2004). , 167-178.

[45] Lee, S. B, et al. ATG1, an autophagy regulator, inhibits cell growth by negatively regulating S6 kinase. EMBO reports, (2007). , 360-365.

[46] Banta, L. M, et al. Organelle assembly in yeast: characterization of yeast mutants defective in vacuolar biogenesis and protein sorting. The Journal of cell biology, (1988). , 1369-1383.

[47] Lindmo, K, et al. The PI 3-kinase regulator Vps15 is required for autophagic clearance of protein aggregates. Autophagy, (2008). , 500-506.

[48] Juhasz, G, et al. Atg7-dependent autophagy promotes neuronal health, stress tolerance, and longevity but is dispensable for metamorphosis in Drosophila. Genes & development, (2007). , 3061-3066.

[49] Berry, D. L, & Baehrecke, E. H. Growth arrest and autophagy are required for salivary gland cell degradation in Drosophila. Cell, (2007). , 1137-1148.

[50] Denton, D, et al. Autophagy, not apoptosis, is essential for midgut cell death in Drosophila. Current biology : CB, (2009). , 1741-1746.

[51] Nezis, I. P, et al. Autophagic degradation of dBruce controls DNA fragmentation in nurse cells during late Drosophila melanogaster oogenesis. The Journal of cell biology, (2010)., 523-531.

[52] Rumpf, S, et al. Neuronal remodeling and apoptosis require VCP-dependent degradation of the apoptosis inhibitor DIAP1. Development, (2011)., 1153-1160.

[53] Ju, J. S, et al. Valosin-containing protein (VCP) is required for autophagy and is disrupted in VCP disease. The Journal of cell biology, (2009)., 875-888.

[54] Wang, M. C, Bohmann, D, & Jasper, H. JNK signaling confers tolerance to oxidative stress and extends lifespan in Drosophila. Developmental cell, (2003)., 811-816.

[55] Wang, M. C, Bohmann, D, & Jasper, H. JNK extends life span and limits growth by antagonizing cellular and organism-wide responses to insulin signaling. Cell, (2005)., 115-125.

[56] Wu, H, Wang, M. C, & Bohmann, D. JNK protects Drosophila from oxidative stress by trancriptionally activating autophagy. Mechanisms of development, (2009). , 624-637.

[57] Deng, H, et al. The Parkinson's disease genes pink1 and parkin promote mitochondrial fission and/or inhibit fusion in Drosophila. Proceedings of the National Academy of Sciences of the United States of America, (2008)., 14503-14508.

[58] Ziviani, E, Tao, R. N, & Whitworth, A. J. Drosophila parkin requires PINK1 for mitochondrial translocation and ubiquitinates mitofusin. Proceedings of the National Academy of Sciences of the United States of America, (2010)., 5018-5023.

[59] Greene, J. C, et al. Mitochondrial pathology and apoptotic muscle degeneration in Drosophila parkin mutants. Proceedings of the National Academy of Sciences of the United States of America, (2003)., 4078-4083.

[60] Pimenta de Castro I., et al., Genetic analysis of mitochondrial protein misfolding in Drosophila melanogaster. Cell death and differentiation, (2012)., 1308-1316.

[61] Ren, C, Finkel, S. E, & Tower, J. Conditional inhibition of autophagy genes in adult Drosophila impairs immunity without compromising longevity. Experimental gerontology, (2009)., 228-235.

[62] Shelly, S, et al. Autophagy is an essential component of Drosophila immunity against vesicular stomatitis virus. Immunity, (2009)., 588-598.

[63] Kim, J. J, et al. Host cell autophagy activated by antibiotics is required for their effective antimycobacterial drug action. Cell host & microbe, (2012)., 457-468.

[64] Kadandale, P, et al. Conserved role for autophagy in Rho1-mediated cortical remodeling and blood cell recruitment. Proceedings of the National Academy of Sciences of the United States of America, (2010)., 10502-10507.

[65] Chinchore, Y, Mitra, A, & Dolph, P. J. Accumulation of rhodopsin in late endosomes triggers photoreceptor cell degeneration. PLoS genetics, (2009). , e1000377.

[66] Midorikawa, R, et al. Autophagy-dependent rhodopsin degradation prevents retinal degeneration in Drosophila. The Journal of neuroscience : the official journal of the Society for Neuroscience, (2010). , 10703-10719.

[67] Arndt, V, et al. Chaperone-assisted selective autophagy is essential for muscle maintenance. Current biology : CB, (2010). , 143-148.

[68] Shen, W, & Ganetzky, B. Autophagy promotes synapse development in Drosophila. The Journal of cell biology, (2009). , 71-79.

[69] Tian, X, et al. Drosophila Rae1 controls the abundance of the ubiquitin ligase Highwire in post-mitotic neurons. Nature neuroscience, (2011). , 1267-1275.

[70] Milton, V. J, & Sweeney, S. T. Oxidative stress in synapse development and function. Developmental neurobiology, (2012). , 100-110.

[71] Simonsen, A, et al. Promoting basal levels of autophagy in the nervous system enhances longevity and oxidant resistance in adult Drosophila. Autophagy, (2008). , 176-184.

[72] Fouillet, A, et al. ER stress inhibits neuronal death by promoting autophagy. Autophagy, (2012). , 915-926.

[73] Knaevelsrud, H, & Simonsen, A. Fighting disease by selective autophagy of aggregate-prone proteins. FEBS Lett, (2010). , 2635-2645.

[74] Nagata, E, et al. Autophagosome-like vacuole formation in Huntington's disease lymphoblasts. Neuroreport, (2004). , 1325-1328.

[75] Ravikumar, B, et al. Inhibition of mTOR induces autophagy and reduces toxicity of polyglutamine expansions in fly and mouse models of Huntington disease. Nat Genet, (2004). , 585-595.

[76] Webb, J. L, et al. Alpha-Synuclein is degraded by both autophagy and the proteasome. J Biol Chem, (2003). , 25009-25013.

[77] Cuervo, A. M, et al. Impaired degradation of mutant alpha-synuclein by chaperone-mediated autophagy. Science, (2004). , 1292-1295.

[78] Parkinson, N, et al. ALS phenotypes with mutations in CHMP2B (charged multivesicular body protein 2B). Neurology, (2006). , 1074-1077.

[79] Talbot, K, & Ansorge, O. Recent advances in the genetics of amyotrophic lateral sclerosis and frontotemporal dementia: common pathways in neurodegenerative disease. Human molecular genetics, (2006). Spec (2), R182-R187.

[80] Rusten, T. E, & Stenmark, H. How do ESCRT proteins control autophagy? Journal of cell science, (2009). Pt 13): , 2179-2183.

[81] Munch, C, et al. Point mutations of the subunit of dynactin (DCTN1) gene in ALS. Neurology, (2004). p. 724-6., 150.

[82] Rubinsztein, D. C. The roles of intracellular protein-degradation pathways in neurodegeneration. Nature, (2006). , 780-786.

[83] Nixon, R. A, et al. Extensive involvement of autophagy in Alzheimer disease: an immuno-electron microscopy study. Journal of neuropathology and experimental neurology, (2005). , 113-122.

[84] Lee, S, Sato, Y, & Nixon, R. A. Lysosomal proteolysis inhibition selectively disrupts axonal transport of degradative organelles and causes an Alzheimer's-like axonal dystrophy. The Journal of neuroscience : the official journal of the Society for Neuroscience, (2011). , 7817-7830.

[85] Stark, W. S, & Carlson, S. D. Ultrastructural pathology of the compound eye and optic neuropiles of the retinal degeneration mutant (w rdg BKS222) Drosophila melanogaster. Cell and tissue research, (1982). , 11-22.

[86] Ravikumar, B, et al. Inhibition of mTOR induces autophagy and reduces toxicity of polyglutamine expansions in fly and mouse models of Huntington disease. Nature genetics, (2004). , 585-595.

[87] Sarkar, S, et al. Small molecules enhance autophagy and reduce toxicity in Huntington's disease models. Nature chemical biology, (2007). , 331-338.

[88] Wang, T, Lao, U, & Edgar, B. A. TOR-mediated autophagy regulates cell death in Drosophila neurodegenerative disease. The Journal of cell biology, (2009). , 703-711.

[89] Ravikumar, B, et al. Rab5 modulates aggregation and toxicity of mutant huntingtin through macroautophagy in cell and fly models of Huntington disease. Journal of cell science, (2008). Pt 10): , 1649-1660.

[90] Nisoli, I, et al. Neurodegeneration by polyglutamine Atrophin is not rescued by induction of autophagy. Cell death and differentiation, (2010). , 1577-1587.

[91] Napoletano, F, et al. Polyglutamine Atrophin provokes neurodegeneration in Drosophila by repressing fat. The EMBO journal, (2011). , 945-958.

[92] Calamita, P, & Fanto, M. Slimming down fat makes neuropathic hippo: the Fat/Hippo tumor suppressor pathway protects adult neurons through regulation of autophagy. Autophagy, (2011). , 907-909.

[93] Dutta, S, & Baehrecke, E. H. Warts is required for PI3K-regulated growth arrest, autophagy, and autophagic cell death in Drosophila. Current biology : CB, (2008). , 1466-1475.

[94] Menzies, F. M, et al. Puromycin-sensitive aminopeptidase protects against aggregation-prone proteins via autophagy. Human molecular genetics, (2010). , 4573-4586.

[95] Pandey, U. B, et al. HDAC6 rescues neurodegeneration and provides an essential link between autophagy and the UPS. Nature, (2007). , 859-863.

[96] Bilen, J, & Bonini, N. M. Genome-wide screen for modifiers of ataxin-3 neurodegeneration in Drosophila. PLoS genetics, (2007). , 1950-1964.

[97] Rusten, T. E, et al. ESCRTing autophagic clearance of aggregating proteins. Autophagy, (2007).

[98] Filimonenko, M, et al. Functional multivesicular bodies are required for autophagic clearance of protein aggregates associated with neurodegenerative disease. The Journal of cell biology, (2007). , 485-500.

[99] Ling, D, et al. Abeta42-induced neurodegeneration via an age-dependent autophagic-lysosomal injury in Drosophila. PloS one, (2009). , e4201.

[100] Ling, D, & Salvaterra, P. M. Brain aging and Abeta$_{1-42}$ neurotoxicity converge via deterioration in autophagy-lysosomal system: a conditional Drosophila model linking Alzheimer's neurodegeneration with aging. Acta neuropathologica, (2011). , 183-191.

[101] Partridge, L, et al. Ageing in Drosophila: the role of the insulin/Igf and TOR signaling network. Experimental gerontology, (2011). , 376-381.

[102] Bishop, N. A, & Guarente, L. Genetic links between diet and lifespan: shared mechanisms from yeast to humans. Nature reviews. Genetics, (2007). , 835-844.

[103] Kenyon, C, et al. A C. elegans mutant that lives twice as long as wild type. Nature, (1993). , 461-464.

[104] Lin, K, et al. daf-16: An HNF-3/forkhead family member that can function to double the life-span of Caenorhabditis elegans. Science, (1997). , 1319-1322.

[105] Kapahi, P, et al. With TOR, less is more: a key role for the conserved nutrient-sensing TOR pathway in aging. Cell metabolism, (2010). , 453-465.

[106] Kenyon, C. J. The genetics of ageing. Nature, (2010). , 504-512.

[107] Bartlett, B. J, et al. Ref(2)P and ubiquitinated proteins are conserved markers of neuronal aging, aggregate formation and progressive autophagic defects. Autophagy, (2011). p. 572-83., 62.

[108] Chen, S. F, et al. Autophagy-related gene 7 is downstream of heat shock protein 27 in the regulation of eye morphology, polyglutamine toxicity, and lifespan in Drosophila. Journal of biomedical science, (2012). , 52.

[109] Demontis, F, & Perrimon, N. FOXO/4E-BP signaling in Drosophila muscles regulates organism-wide proteostasis during aging. Cell, (2010). , 813-825.

[110] Juhasz, G, et al. Gene expression profiling identifies FKBP39 as an inhibitor of autophagy in larval Drosophila fat body. Cell death and differentiation, (2007). , 1181-1190.

[111] Zid, B. M, et al. E-BP extends lifespan upon dietary restriction by enhancing mitochondrial activity in Drosophila. Cell, (2009). , 149-160.

[112] Bjedov, I, et al. Mechanisms of life span extension by rapamycin in the fruit fly Drosophila melanogaster. Cell metabolism, (2010). , 35-46.

[113] Kapahi, P, et al. Regulation of lifespan in Drosophila by modulation of genes in the TOR signaling pathway. Current biology : CB, (2004). , 885-890.

[114] Luong, N, et al. Activated FOXO-mediated insulin resistance is blocked by reduction of TOR activity. Cell metabolism, (2006). , 133-142.

[115] Eisenberg, T, et al. Induction of autophagy by spermidine promotes longevity. Nature cell biology, (2009). , 1305-1314.

[116] Kirkin, V, et al. A role for ubiquitin in selective autophagy. Molecular cell, (2009). , 259-269.

[117] Hershko, A, & Ciechanover, A. The ubiquitin system. Annual review of biochemistry, (1998). , 425-479.

[118] Tan, J. M, et al. Lysine 63-linked ubiquitination promotes the formation and autophagic clearance of protein inclusions associated with neurodegenerative diseases. Human molecular genetics, (2008). , 431-439.

[119] Welchman, R. L, Gordon, C, & Mayer, R. J. Ubiquitin and ubiquitin-like proteins as multifunctional signals. Nature reviews. Molecular cell biology, (2005). , 599-609.

[120] Wooten, M. W, et al. Essential role of sequestosome 1/in regulating accumulation of Lys63-ubiquitinated proteins. The Journal of biological chemistry, (2008). p. 6783-9., 62.

[121] Ikeda, F, & Dikic, I. Atypical ubiquitin chains: new molecular signals.'Protein Modifications: Beyond the Usual Suspects' review series. EMBO reports, (2008). , 536-542.

[122] Johansen, T, & Lamark, T. Selective autophagy mediated by autophagic adapter proteins. Autophagy, (2011). , 279-296.

[123] Nezis, I. P. Selective autophagy in Drosophila. International journal of cell biology, (2012). , 146767.

[124] Nezis, I. P, et al. Ref(2)P, the Drosophila melanogaster homologue of mammalian is required for the formation of protein aggregates in adult brain. The Journal of cell biology, (2008). p. 1065-71., 62.

[125] Filimonenko, M, et al. The selective macroautophagic degradation of aggregated proteins requires the PI3P-binding protein Alfy. Molecular cell, (2010). , 265-279.

[126] Simonsen, A, et al. Alfy, a novel FYVE-domain-containing protein associated with protein granules and autophagic membranes. Journal of cell science, (2004). Pt 18): , 4239-4251.

[127] Clausen, T. H, et al. SQSTM1 and ALFY interact to facilitate the formation of p62 bodies/ALIS and their degradation by autophagy. Autophagy, (2010). p. 330-44., 62.

[128] Fujita, N, et al. The Atg16L complex specifies the site of LC3 lipidation for membrane biogenesis in autophagy. Molecular biology of the cell, (2008). , 2092-2100.

[129] Finley, K. D, et al. blue cheese mutations define a novel, conserved gene involved in progressive neural degeneration. The Journal of neuroscience : the official journal of the Society for Neuroscience, (2003). , 1254-1264.